ESO ASTROPHYSICS SYMPOSIA
European Southern Observatory

Series Editor: Bruno Leibundgut

Springer
Berlin
Heidelberg
New York
Hong Kong
London
Milan
Paris
Tokyo

Physics and Astronomy ONLINE LIBRARY

http://www.springer.de/phys/

W. Hillebrandt B. Leibundgut (Eds.)

From Twilight
to Highlight:
The Physics
of Supernovae

Proceedings of the ESO/MPA/MPE Workshop
Held at Garching, Germany, 29-31 July 2002

 Springer

Volume Editors

Wolfgang Hillebrandt
Max-Planck-Institut für Astrophysik
Postfach 1317
85741 Garching, Germany

Bruno Leibundgut
European Southern Observatory
Karl-Schwarzschild-Strasse 2
85748 Garching, Germany

Series Editor

Bruno Leibundgut
European Southern Observatory
Karl-Schwarzschild-Strasse 2
85748 Garching, Germany

Cataloging-in-Publication Data applied for

A catalog record for this book is available from the Library of Congress.

Bibliographic information published by Die Deutsche Bibliothek
Die Deutsche Bibliothek lists this publication in the Deutsche Nationalbibliografie;
detailed bibliographic data is available in the Internet at http://dnb.ddb.de

ISBN 3-540-00483-1 Springer-Verlag Berlin Heidelberg New York

Springer-Verlag Berlin Heidelberg New York
a member of BertelsmannSpringer Science+Business Media GmbH

http://www.springer.de

© Springer-Verlag Berlin Heidelberg 2003

Typesetting: Camera-ready by the authors/editors
Cover design: Erich Kirchner, Heidelberg

Printed on acid-free paper   55/3141/du - 5 4 3 2 1 0

Preface

Supernovae are by far the most energetic events observed in our cosmic neighbourhood. They are responsible for the formation of most of the chemical elements, they determine to a large extent the energy budget of the interstellar matter and because of their extreme luminosity they can be used, in principle, to measure our distance to the most remote galaxies.

Understanding the physics of supernova explosions has always been a challenge, despite the fact that the most likely sources of energy needed to drive such powerful stellar outbursts, i.e. gravitational binding and/or thermonuclear burning, have been known for decades. But also, over several decades, the problem of supernova physics was not so much a lack of models but their reliability. One of the problems is the fact that for most supernova models hydrodynamic instabilities play a key role and numerical 'experiments' to prove or disprove certain ideas have to be 2- or 3-dimensional, and they have become feasible only recently. Other problems include the complexity of radiation transport which links the models to the still insufficient observational database.

With increased attention to supernovae for their use in cosmology and their possible connection to Gamma-Ray Bursts (GRB) these stellar explosions have entered into the limelight of the astronomical stage over the past few years. However, our understanding of the underlying physics advances slowly and its progress has to be evaluated regularly. It was the goal of this meeting to focus on the basics of the explosions and our current knowledge of these cosmic highlights.

The programme covered all aspects of supernova research, but excluded 'associated' fields, like cosmology or the vast array of GRB observations. The focus on the known events themselves led to a lively programme which is reflected in the contributions to this book. All known aspects of supernova explosions were investigated. The topics covered range from the theoretical ideas of stellar evolution without metals, core collapses and jet explosions, the observational facts of a lack of suitable progenitor systems and a manifest variety of thermonuclear supernovae, traces of the enriched end products in the most metal-poor stars to search programmes for finding more supernovae.

The warm summer days at the end of July were a good match to the hot debates and discussions, which took place during the breaks and the evening beer sessions.

As usual, a conference like this lives from the support it receives. The organizational logistics were expertly handled by Cornelia Rickl, Maria Depner and

Gabriele Kratschmann. They turned the programme into a successful workshop where the scientists could concentrate on the research rather than battle with reservations or transportation worries. The manuscript from the authors were assembled by Pamela Bristow into the book you hold in your hands. We are extremely grateful to these people who made the workshop possible.

Garching,   *Wolfgang Hillebrandt*
November 2002   *Bruno Leibundgut*

Contents

Part III   Progenitors of Thermonuclear Supernovae

Part IV   Models of Thermonuclear Supernova Explosions

Part VI   Supernovae and Their Environment

List of Participants

Miguel Aloy
MPI für Astrophysik
maa@mpa-garching.mpg.de

Athem Alsabti
London Observatory
aalsabti@ulo.ucl.ac.uk

Giuseppe Altavilla
Osservatorio Astronomico di Padova
altavilla@pd.astro.it

Bernd Aschenbach
MPI für extraterrestrische Physik
bra@mpe.mpg.de

Thomas Behrens
MPI für Astrophysik
tbehrens@mpa-garching.mpg.de

Stefano Benetti
Osservatorio Astronomico di Padova
benetti@pd.astro.it

Sergey Blinnikov
ITEP Moscow
blinn@sai.msu.su

Hans Böhringer
MPI für extraterrestrische Physik
hxb@mpe.mpg.de

Christophe Bonnaud
Centre de Recherche Astronomique
de Lyon
cbonnaud@obs.univ-lyon1.fr

Sebastien Bongard
IPNL Villeurbanne
bongard@in2p3.fr

Eduardo Bravo
Universitat Politècnica de Catalunya
eduardo.bravo@upc.es

Adam Burrows
University of Arizona
burrows@as.arizona.edu

Ramon Canal
University of Barcelona
ramon@mizar.am.ub.es

Enrico Cappellaro
Osservatorio Astronomico
di Capodimonte
cappellaro@na.astro.it

Roger A. Chevalier
University of Virginia
rac5x@ur.astro.virginia.edu

Melvyn Davies
University of Leicester
mbd@star.le.ac.uk

Roland Diehl
MPI für extraterrestrische Physik
rod@mpe.mpg.de

Luc Dessart
University of Utrecht
l.dessart@phys.uu.nl

Harald Dimmelmeier
MPI für Astrophysik
harrydee@mpa-garching.mpg.de

Douglas Duncan
University of Chicago
duncan@oddjob.uchicago.edu

Abouazza Elmhamdi
SISSA Trieste
elmhamdi@sissa.it

Alex Filippenko
University of California, Berkeley
alex@astro.berkeley.edu

Claes Fransson
Stockholm Observatory
claes@astro.su.se

Domingo Garcia–Senz
Universitat Politècnica de Catalunya
domingo.garcia@upc.es

Christopher Gerardy
Dartmouth College
gerardy@dartmouth.edu

Lisa Germany
ESO Chile
lgermany@eso.org

Per Grönigsson
Stockholm Observatory
per@astro.su.se

Mario Hamuy
Carnegie Observatories
hamuy@ociw.edu

Dieter Hartmann
Clemson University
HDIETER@clemson.edu

Ulrich Heber
Universität Erlangen–Nürnberg
Heber@sternwarte.uni-erlangen.de

Volker Heesen
MPI für Astrophysik
vheesen@mpa-garching.mpg.de

Alexander Heger
University of Chicago
1@2sn.org

Wolfgang Hillebrandt
MPI für Astrophysik
wfh@mpa-garching.mpg.de

Rob Hoffman
Lawrence Livermore Natl. Laboratory
rdhoffman@llnl.gov

Natalia Ivanova
Northwestern University
nata@northwestern.edu

Anatoli Iyudin
MPI für extraterrestrische Physik
ani@mpe.mpg.de

Hans–Thomas Janka
MPI für Astrophysik
thj@mpa-garching.mpg.de

Paul C. Joss
Massachusetts Institute of Technology
joss@space.mit.edu

Christian Karl
Dr. Remeis Sternwarte Bamberg
karl@sternwarte.uni-erlangen.de

Robert Kirshner
Harvard Center for Astrophysics
kirshner@cfa.harvard.edu

Chiaki Kobayashi
MPI für Astrophysik
chiaki@mpa-garching.mpg.de

Rubina Kotak
Imperial College London
rubina@astro.lu.se

Volodymyr Kryvdyk
Kiev National University
kryvdyk@univ.kiev.ua

Wolfgang Kundt
IAEF, University of Bonn
wkundt@astro.uni-bonn.de

Eric Lentz
University of Georgia
lentz@physast.uga.edu

Bruno Leibundgut
ESO Garching
bleibund@eso.org

Pierre Lesaffre
Institute of Astronomy
lesaffre@ast.cam.ac.uk

Marco Limongi
Osservatorio Astronomico di Roma
marco@mporzio.astro.it

Thorsten Lisker
Dr. Remeis–Sternwarte Bamberg
tlisker@arcor.de

Peter Lundqvist
Stockholm Observatory
peter@astro.su.se

Andrew MacFadyen
California Institute of Technology
andrew@tapir.caltech.edu

Keiichi Maeda
University of Tokyo
maeda@astron.s.u-tokyo.ac.jp

Jon M. Marcaide
University of Valencia
J.M.Marcaide@uv.es

Seppo Mattila
Imperial College London
s.mattila@ic.ac.uk

Paolo Mazzali
Osservatorio Astronomico di Trieste
mazzali@ts.astro.it

Peter Meikle
Imperial College London
p.meikle@ic.ac.uk

O.E. Bronson Messer
University of Tennessee
bmesser@utk.edu

Anthony Mezzacappa
Oak Ridge National Laboratory
mezzacappaa@ornl.gov

Ana Mourao
CENTRA-IST, Lisbon
ana.mourao@ist.utl.pt

Ewald Müller
MPI für Astrophysik
ewald@mpa-garching.mpg.de

Dmitrij K. Nadyozhin
ITEP Moscow
dkn@sai.msu.su

Ralf Napiwotzki
Dr. Remeis Sternwarte Bamberg
napiwotzki
@sternwarte.uni-erlangen.de

Jens Niemeyer
MPI für Astrophysik
jcn@mpa-garching.mpg.de

Ken'ichi Nomoto
University of Tokyo
nomoto@astron.s.u-tokyo.ac.jp

Tanja Nymark
Stockholm Observatory
tanja@astro.su.se

Ulrich Ott
MPI für Chemie, Mainz
ott@mpch-mainz.mpg.de

Reynald Pain
LPNHE, University of Paris
reynald.pain@in2p3.fr

Rodrigo Pascoal
CENTRA-IST, Lisbon
rod@supernova.ist.utl.pt

Andrea Pastorello
University of Padova
pastorello@pd.astro.it

Ferdinando Patat
ESO Garching
fpatat@eso.org

Eva-Maria Pauli
Dr. Remeis Sternwarte Bamberg
pauli@sternwarte.uni-erlangen.de

Emmanuel Pecontal
Centre de Recherche Astronomique
de Lyon
pecontal@obs.univ-lyon1.fr

Christoph Pfrommer
MPI für Astrophysik
pfrommer@mpa-garching.mpg.de

Mark Phillips
Carnegie Institution
mmp@lco.cl

Giuliano Pignata
Osservatorio Astronomico di Padova
pignata@pd.astro.it

Philipp Podsiadlowski
Oxford University
podsi@astro.ox.ac.uk

Francesca Primas
ESO Garching
fprimas@eso.org

Markus Rampp
MPI für Astrophysik
mjr@mpa-garching.mpg.de

Alvio Renzini
ESO Garching
arenzini@eso.org

Marco Riello
Osservatorio Astronomico di Padova
riello@pd.astro.it

Friedrich Roepke
MPI für Astrophysik
fritz@mpa-garching.mpg.de

Evert Rol
University of Amsterdam
evert@science.uva.nl

Pilar Ruiz-Lapuente
University of Barcelona
pilar@mizar.am.ub.es

Katsuhiko Sato
University of Tokyo
sato@phys.s.u-tokyo.ac.jp

Daniel Sauer
MPI für Astrophysik
dsauer@mpa-garching.mpg.de

Leonard Scheck
MPI für Astrophysik
scheck@mpa-garching.mpg.de

Brian Schmidt
Mount Stromlo Observatory
brian@mso.anu.edu.au

Wolfram Schmidt
MPI für Astrophysik
wolfram@mpa-garching.mpg.de

Peter Schneider
IAEF, University of Bonn
peter@astro.uni-bonn.de

Stephen J. Smarrt
Institute of Astronomy
sjs@ast.cam.ac.uk

Jesper Sollerman
Stockholm Observatory
jesper@astro.su.se

Elena Sorokina
Sternberg Astronomical Institute
sorokina@sai.msu.su

Jason Spyromilio
ESO Garching
jspyromi@eso.org

Sumner Starrfield
Arizona State University
starrfield@asu.edu

Matthias Stehle
MPI für Astrophysik
mstehle@mpa-garching.mpg.de

Nicholas Suntzeff
Cerro Tololo Inter-American
Observatory
nsuntzeff@noao.edu

Firoza K. Sutaria
The Open University
f.k.sutaria@open.ac.uk

Friedrich–Karl Thielemann
University of Basel
fkt@quasar.physik.unibas.ch

Claudia Travaglio
MPI für Astrophysik
claudia@mpa-garching.mpg.de

Massimo Turatto
Osservatorio Astronomico di Padova
turatto@pd.astro.it

Victor Utrobin
ITEP Moscow
utrobin@vitep1.itep.ru

Vincenzo Vitale
MPI für Physik
vitale@mppmu.mpg.de

Stanford E. Woosley
University of California, Santa Cruz
woosley@ucolick.org

Luca Zampieri
Osservatorio Astronomico di Padova
zampieri@pd.astro.it

Vyacheslav Zavlin
MPI für extraterrestrische Physik
zavlin@mpe.mpg.de

Part I

Pre-Supernova Evolution of Massive Stars

Massive Star Evolution Through the Ages

Alexander Heger[1], S.E. Woosley[2], C.L. Fryer[3], and Norbert Langer[4]

[1] Department of Astronomy and Astrophysics, Enrico Fermi Institute,
    The University of Chicago, 5640 S. Ellis Ave, Chicago, IL 60637, USA
[2] Department of Astronomy and Astrophysics, University of California,
    Santa Cruz, CA 95064, USA
[3] Theoretical Astrophysics, MS B288, Los Alamos National Laboratories,
    Los Alamos, NM 87545, USA
[4] Astronomical Institute, P.O. Box 80000, NL-3508 TA Utrecht, The Netherlands

Abstract. We review the current basic picture of the evolution of massive stars and how their evolution and structure changes as a function of initial mass. We give an overview of the fate of modern (Pop I) and primordial (Pop III) stars with emphasis on massive and very massive stars. For single stars we show how the type of explosions, the type of remnant and their frequencies changes for different initial metallicities.

1 Massive Star Evolution

As massive stars we denote those that are born with initial masses of more than about 8 $M_\odot$, the minimum mass for a single star to explode as supernova. Once the star has formed, its center generally evolves to increasing central density and temperature. This overall contraction is interrupted by phases of nuclear fusion – hydrogen to helium, helium to carbon and oxygen, then carbon, neon, oxygen and silicon burning, until finally iron is produced and the core collapses. Each fuel burns first in the center, then in a shell. In Table 1 we summarize the burning stages and their durations for a 20 $M_\odot$ star and in Fig. 1 we show the evolution of the interior structure of a 22 $M_\odot$ star. The time scale for helium burning is about ten times shorter than that of hydrogen burning, mostly because of the lower energy release per unit mass. However, the time scale of the burning stages beyond central helium-burning is radically reduced by thermal neutrino losses that carry away energy *in situ*, instead of requiring that it be transported to the stellar surface. These losses increase with temperature, roughly $\propto T^9$. (See [35] for a more extended review.) When the star has built up a large enough iron core, exceeding its Chandrasekhar mass, it collapses to form a neutron star or a black hole. A supernova explosion may result [11], or even, in rare cases, a gamma-ray burst [23,24].

2 Modern Massive Stars' Fates

In Fig. 2 we show the fate of modern stars, formed from a composition comparable to that of our sun (Pop I), as a function of initial mass. Below $\sim 8 M_\odot$

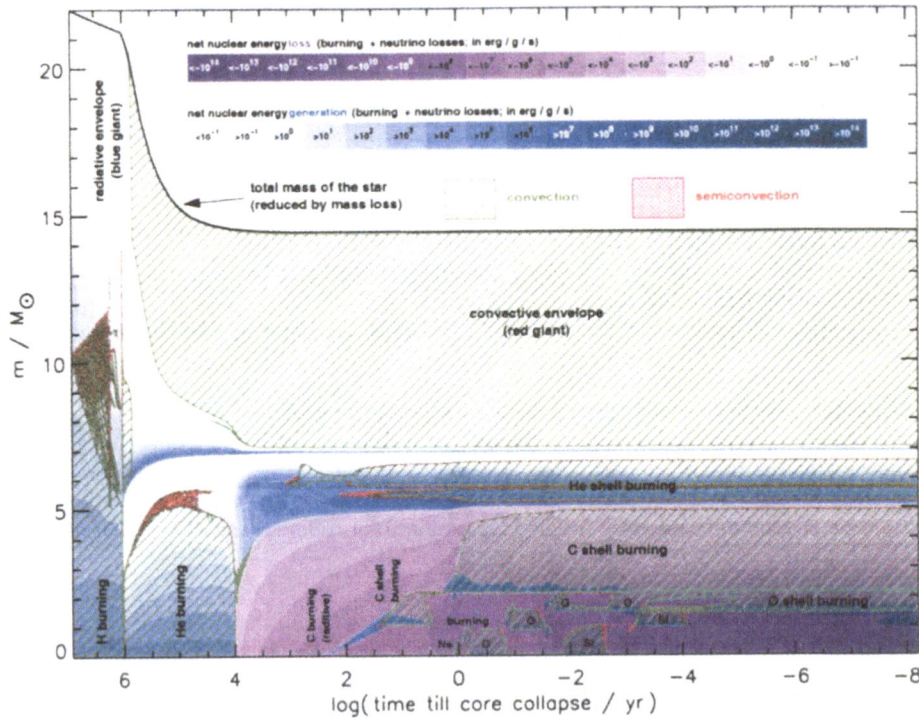

Fig. 1. Interior structure of a $22\,M_\odot$ star of solar composition as a function of time (logarithm of time till core collapse) and enclosed mass. *Green hatching* and *red cross hatching* indicate convective and semiconvective regions. *Blue shading* indicates energy generation and *pink shading* energy loss. Both take into account the sum of nuclear and neutrino loss contributions. The *thick black line* at the top indicates the total mass of the star being reduced by mass loss due to stellar winds. Note that the mass loss rate actually increases at late times of the stellar evolution. The decreasing slope of the total mass of the star in the figure is due to the logarithmic scale chosen for the time axis

initial mass only white dwarfs are formed, neon-oxygen white dwarfs immediately below this limit, and carbon-oxygen white dwarfs at even lower masses. The envelope of such stars is lost during their AGB stage.

The regime of initial masses above $\sim 8\,M_\odot$ we refer to as *massive stars*. Except for the pair-instability supernovae discussed below They leave behind compact remnants. Up to an initial mass of $\sim 20 - 25\,M_\odot$ [10,11] typically neutron stars result. The layers above the neutron star are ejected, consisting of the ashes of the preceding hydrostatic stellar burning phases and explosive burning products (*green cross hatching* indicates partial helium burning, i.e., mostly C and O; *solid green* indicates pure metals, i.e., products of complete helium burning and later burning phases).

At higher masses a large fraction of the core can fall back onto the neutron star after an initial explosion has first driven the matter outward. If the resulting remnant mass exceeds the maximum mass for a neutron star, it collapses to a black hole. In Fig. 2 we indicate this region by "fallback" and "black hole".

The evolution of modern massive stars is significantly affected by mass loss due to stellar winds. The mass loss [27] can become so strong that the final mass of the star may actually decrease as its initial mass increases. Indeed, [35] find a maximum in the final mass around $\sim 20\,M_\odot$. In Fig. 2) the *blue curve* shows the final mass at the time of core collapse for massive stars. Around $\lesssim 35\,M_\odot$ the mass loss becomes so strong that entire hydrogen envelope is lost prior to explosion of the star: in Fig. 2 the curve for the final mass of the star intersects the curve for helium core mass at core collapse. The exact mass where this happens depends on, e.g., the uncertainty of the mass loss rates and can vary with initial stellar rotation [31,30,26,20,15,22,21].

Once the massive stars have lost their hydrogen envelope, they become Wolf-Rayet (WR) stars. The mass loss of these objects is known to be large, but also quite uncertain. Recently, the introduction of "clumping" into modeling of Wolf-Rayet star spectra lead to a reduction of the derived mass loss rates by a factor 2-3 [13]. In Fig. 2 we show for both cases remnant mass and stellar mass at core collapse: the *dotted lines* are for the case of the previously assumed high mass loss rates and the *dashed lines* for the case of the lowered Wolf-Rayet mass loss rates. In the former case, there may be a regime of initial masses in which the core mass at the time of explosion is low enough to form neutron stars, while it is absent in the other case. The final masses may be higher than indicated in the figure if the star is covered by a hydrogen envelope for a significant fraction of central helium burning [6].

Note that the Wolf-Rayet stars can already release products of central helium burning by stellar winds before explosion (*hatching* above the *dashed curve* indicating the final mass of the star) – or even if they do not explode. Such objects are denoted as WC and WO stars. Here we show the metal production only for the case of the low WR mass loss rate.

Table 1. Nuclear burning stages in massive stars. We give typical temperatures and time scales for a $20\,M_\odot$ star (Pop I; similar in Pop III) and a $200\,M_\odot$ star (Pop III)

Burning stages		$20\,M_\odot$ star		$200\,M_\odot$ star	
Fuel	Main product	T (10^9 K)	duration (yr)	T (10^9 K)	duration (yr)
H	He	0.037	8.1×10^6	0.14	2.2×10^6
He	O, C	0.19	1.2×10^6	0.24	2.5×10^5
C	Ne, Mg	0.87	9.8×10^2	1.1[†]	4.5
Ne	O, Mg	1.6	0.60	2.4[†]	1.1×10^{-6}
O	Si, S	2.0	1.3	3.5[†]	3.5×10^{-8}
Si	Fe	3.3	0.031	4.3[‡]	2.7×10^{-7}

[†]central radiative implosive burning   [‡]incomplete silicon burning at bounce

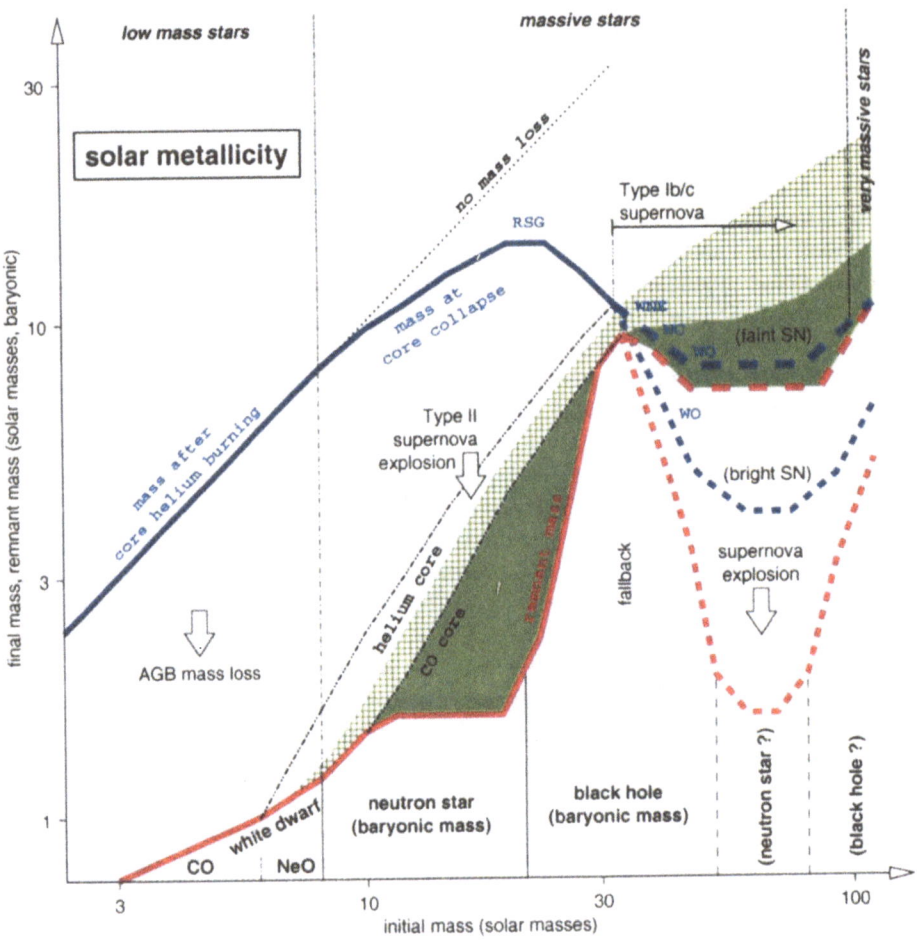

Fig. 2. Stellar mass at time of final explosion/remnant formation (*blue line*), remnant mass (*red line*) and metals released (*green fill* and *hatching*) as a function of initial mass of the star for modern stars (Pop I)

3   Primordial Stars

The lack of initial metals in stars that form from the composition as made in the big bang results in a strongly reduced mass loss, down to the point of negligibility [19]. For simplicity, we can therefore assume that Pop III stars keep almost all of their initial mass through the end of central helium burning (Fig. 3; see also [25]). Recent studies indicate that the first generation of stars may have been rather massive [4,5,1], or may have had at least a very massive component [28] compared to modern stellar populations. But even those stars might keep most of their initial mass till the end of their evolution [2,14].

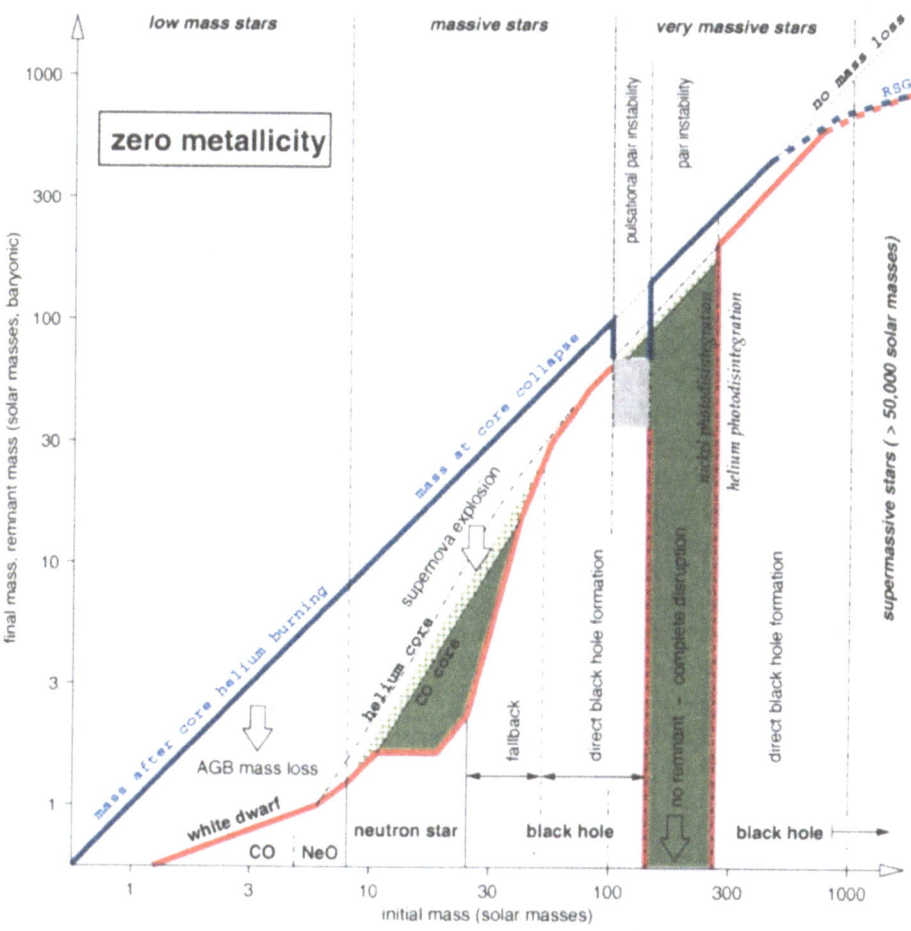

Fig. 3. Stellar mass at time of final explosion/remnant formation (*blue line*), remnant mass (*red line*) and metals released (*green fill and hatching*) as a function of initial mass of the star for primordial (metal-free, Pop III) stars

For masses below $\lesssim 35\,M_\odot$ the fate of the stars is similar to that of modern stars (Fig. 2) except that the mass at core collapse continues increasing. Low-mass stars form white dwarfs, massive stars first form neutron stars, then black holes by fall back. However, as the helium core mass at core collapse increases, eventually a successful supernova shock cannot be launched due to the strong infall of the increasingly larger oxygen and silicon core masses. A black hole is formed directly and no supernova explosion should occur.

Above an initial mass of $\sim 100\,M_\odot$, the stars encounter the electron-positron pair creation instability after central carbon burning [3,34]. This leads to rapid burning of oxygen and silicon (Table 1). Only when enough energy is released, is the infall stopped and reverts into an explosion. Below an initial mass of $\sim 140\,M_\odot$ the explosion energy is not big enough to disrupt the entire star. The outer layers, the hydrogen envelope and maybe part of the helium core, however, are ejected. The rest of the star falls back and may encounter subsequent pulses of this kind within one to several 10,000 yrs until it finally collapses and directly forms a black hole. These stars will not have a hydrogen envelope at the time of collapse, but the ejecta may still be close by. The energy of these eruptions can be as high as 3×10^{51} erg.

For initial masses from $\sim 140 - 260\,M_\odot$ the explosion energy of the first pulse is already sufficient to entirely disrupt the star. In this case no remnant remains and all the metals are ejected. The explosion energies may reach $\sim 10^{53}$ erg and more than $50\,M_\odot$ of radioactive ^{56}Ni may be ejected – both figures are close to 100× that of Type Ia supernovae. These events are rather bright and could be observable out to the edge of the universe in the infrared [16]. Further observational signatures should be their long time scale, already in their rest frame, but additionally boosted by their high redshift, and a Lyman-alpha cut-off corresponding to their redshift, as they should explode when most of the universe is not yet re-ionized.

Above initial masses of $\sim 260\,M_\odot$ photo-disintegration of alpha-particles (which themselves are already the result of photo-disintegration of iron group elements which were made in silicon burning) reduces the pressure enough that the collapse of the star is not turned around but directly continues into a black hole [34,12]. Probably no metals are ejected in this case. Above several $100\,M_\odot$ even primordial stars may evolve into red supergiants, becoming pulsationally unstable and lose mass [2]. Since we have no good estimate of the associate mass loss, we draw the line for the pre-collapse and remnant mass as dashed lines in this regime.

4   Supernova and Remnant Populations

In Figs. 4 and 5 we try to give a rough sketch of the remnant and supernova types as a function of initial mass and metallicity. A very detailed description will be given in [17]. Due to the uncertainty of some mass loss rates and their scaling with metallicity, in particular those of red supergiants and Wolf-Rayet (WR) stars, we do not give an absolute scale on the *ordinate* of these figures, but assume that we can at least outline the rough sequence of populations with increasing metallicity.

Only at sufficiently low metallicities do very massive stars keep all their mass till collapse. As metallicity increases, due to increasing mass loss, first the very massive stars that directly collapse to black holes disappear. Next the pair-instability supernova produce less and less powerful explosions until they disappear. Above the metallicity limit for pair instability all very massive single stars should leave compact remnants. There may be a regime in which pair-instability supernovae can occur in bare helium stars if the hydrogen envelope has been carried away by stellar wind, but the helium core has not been shrunk

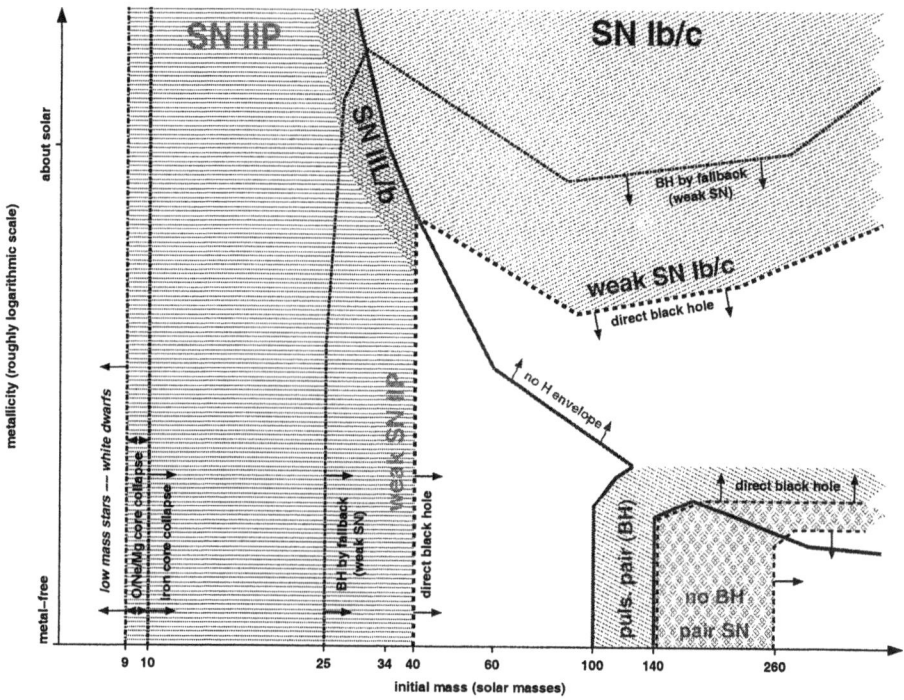

Fig. 4. Supernova types of single stars as a function of stellar initial mass and metallicity. *White regions* show where no supernova explosion is expected. *Horizontal hatching* indicates Type IIL supernovae, *diagonal hatching* Type IIb supernovae, *diagonal hatching in the upper right* Type Ib/c supernovae, and *diagonal hatching* and *cross hatching in the lower right* indicate pulsational and non-pulsational pair-instability supernovae. The *solid curve* is the dividing line between stars that have a hydrogen envelope (below the line) and those that do not

enough by WR winds. At higher metallicities also the pulsational pair-instability supernovae will disappear. They should always collapse as hydrogen-free stars ("window" in the *solid curve* indication the presence/loss of the hydrogen envelope) and their core collapse produces a black hole without launching a supernova explosion. Depending on whether stellar winds have uncovered the helium core before collapse due high enough metallicity or just the pair instability pulsations ejected the hydrogen envelope, the hydrogen may still be in the vicinity of the star or be blown far away.

If mass loss does not uncover the helium core, above $\sim 40\,M_\odot$ black holes are formed directly. In this case we do not expect a supernova explosion display. For stars massive enough, the hydrogen envelope is likely removed by stellar winds, shrinking the stellar core. Since the core collapse and explosion are mostly determined by the core masses, this alters the fate of those stars. This means that above a certain metallicity (and as function of their initial mass) even stars above the direct black hole limit for hydrogen-covered stars, can avoid directly

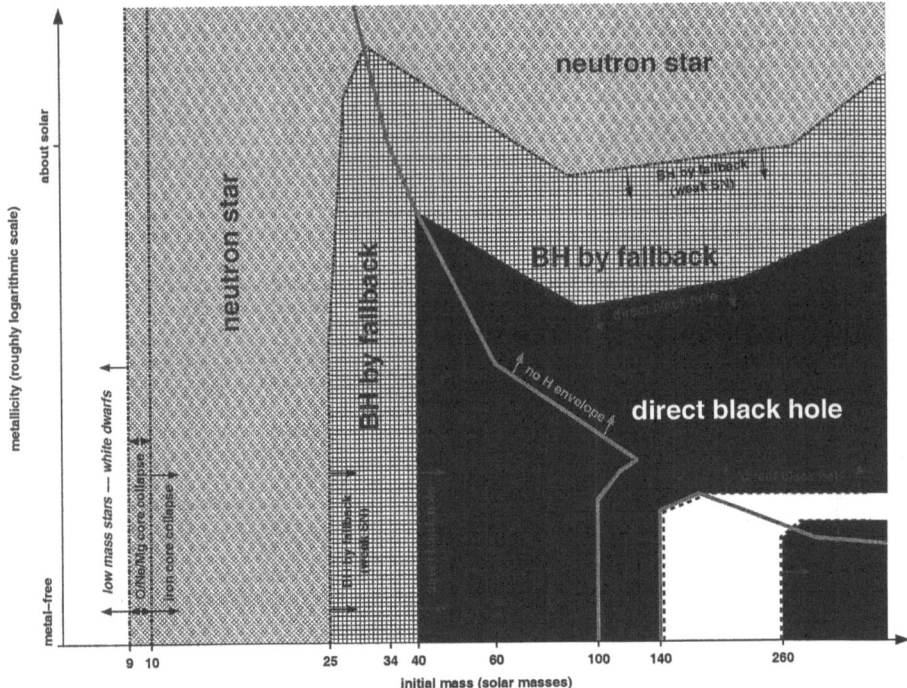

Fig. 5. Stellar remnants of single stars as a function of stellar initial mass and metallicity. *Diagonal cross hatching* indicates where neutron stars are made, *horizontal and vertical cross hatching* indicates regimes where black holes are formed by fall back and *solid black* shows where black holes form directly without launching a supernova shock from a proto-neutron star. In the *white region* at low metallicity and high mass no remnant is left (pair instability supernovae) and in the *white region* on the left hand side white dwarfs are made. We use the same curves and dividing lines as in Fig. 5

forming black holes and may result in hydrogen-free supernova explosions of Type Ib or Ic [8,9] (Fig. 4).

At even higher metallicity, also the limit for black hole formation due to fallback can be avoided in hydrogen-free stars. That is, for high enough metallicity, all stars may result in neutron stars (Fig. 5). Those supernovae, resulting from smaller cores, are potentially brighter than those from the more massive hydrogen-free stars [7] (except pair-instability supernovae).

In stars that keep the hydrogen envelope, down to $\sim 20 - 25\,M_\odot$ the fallback may be big enough to form black holes [10], but yet a Type II supernova [8] may occur. If the fallback is excessive, the supernova may be dim. Below this mass limit neutron stars are made. Stars that have lost most of their hydrogen envelope but about $\lesssim 1\,M_\odot$ at the time of core collapse may result in Type IIL/b supernovae, while those that have left more hydrogen will make Type IIP supernovae (see also [8,9]). Naturally, in Fig. 4 the Type IIL/b supernovae thus occur in a small strip just below the limit for complete loss of the hydrogen envelope as a function of mass and metallicity. Since this curve goes above the limit for successful core collapse supernova explosions at low metallicity, we anticipate that there will be a lower metallicity limit for Type IIL/b supernovae from single stars.

5 Summary and Discussion

We outlined, in a rough scheme, the fate of modern and primordial massive single stars as a function of their initial mass and metallicity. The most important difference caused by the variation of compositions is the strong mass loss in modern stars. Because of this, they may lose their hydrogen envelope above a certain initial mass. When this happens, the exposed helium star may cause even stronger mass loss. Only at very low metallicity can very massive stars retain enough mass to become pair-instability supernovae. On the other hand, certain types of supernovae that require the loss of most or all of the hydrogen envelope, occur in single stars only above a certain level of metal enrichment.

The lines between the different regimes in Figs. 2–5 are only a rough guidance. Due to the interaction of different burning shells (Fig. 1) the core masses (silicon core, iron core, etc.) do not increase monotonously as a function of initial mass or metallicity [35,17]. Therefore one cannot expect a unique transition from one regime to the neighboring, but a rather ragged one, with detached "islands" of one regime within the other.

In contrast to single stars, interacting ("close") binary stars can lead to the loss of the hydrogen envelope even in metal-free stars. Therefore, unfortunately, hydrogen-free stars and supernovae can be present at any metallicity and thus an observational determination of the metallicity limit for the loss of the hydrogen envelope from low metallicity stellar populations is rendered difficult. However, the ratio of supernova types may change when single stars start contributing to Type IIL/b, and Ib/c supernovae. Depending on the initial separation and mass ratio of the binary, the mass exchange can happen at different evolution stages (most prominent: during hydrogen burning, during the transition from hydrogen burning to helium burning, and during or at the end of helium burning). The earlier the mass transfer happens, the smaller is the final mass of the star.

Finally, the evidence for an association of gamma-ray bursts (GRBs) with massive star populations seems to increase [18]. Provided a star that collapses to a black hole has sufficient angular momentum, it may launch a jet [33], blowing up the star (e.g., [24]; "hypernova", see also [29]) and possibly producing a GRB if the jet escapes the star [33,23,37,36]. If any single star can ever have sufficient angular momentum at core collapse, in Fig. 5 such events can occur only in the "black hole by fallback" and "direct black hole" regimes. And only above the line for loss of the hydrogen envelope could GRBs be made.

Acknowledgements. This research has been supported by the NSF (AST 02-06111), the DOE ASCI Program (B347885) and the SciDAC Program of the DOE (DE-FC02-01ER41176). AH is supported in part by the Department of Energy under grant B341495 to the Center for Astrophysical Thermonuclear Flashes at the University of Chicago and acknowledges travel support from MPA to participate in this meeting.

References

1. T. Abel, G.L. Bryan, M.L. Norman: Science **295**, 93 (2002)
2. I. Baraffe, A. Heger, S.E. Woosley: ApJ **550**, 890 (2001)
3. J.R. Bond, W.D. Arnett, B.J. Carr: ApJ **280**, 825 (1984)
4. V. Bromm, A. Ferrara, P.S. Coppi, R.B. Larson: MNRAS **328**, 969 (2001)
5. V. Bromm, P.S. Coppi, R.B. Larson: ApJ **564**, 23 (2002)
6. G.E. Brown, A. Heger, N. Langer, C.-H. Lee, S. Wellstein, H.A. Bethe: New Astronomy **6**, Issue 7, p. 457 (2001)
7. L.M. Ensman, S.E. Woosley: ApJ **333**, 754 (1988)
8. A.V. Filippenko: ARA&A **35**, 309 (1997)
9. A.V. Filippenko: in these proceedings (2002)
10. Fryer, C. L. 1999, ApJ, 522, 413
11. C.L. Fryer, V. Kalogera: ApJ **554**, 548 (2001)
12. C.L. Fryer, S.E. Woosley, A. Heger: ApJ **550**, 372 (2001)
13. W.-R. Hamann, L. Koesterke: A&A **335**, 1003 (1998)
14. A. Heger, S.E. Woosley: ApJ, **567**, 532 (2002)
15. A. Heger, N. Langer, S.E. Woosley: ApJ **528**, 368 (2000)
16. A. Heger, S. E. Woosley, I. Baraffe, T. Abel: "Evolution and Explosion of Very Massive Primordial Stars", in. proc. MPA/ESO/MPE/USM Joint Astronomy Conference 'Lighthouses of the Universe: The Most Luminous Celestial Objects and their use for Cosmology', (Springer: Heidelberg), in press (2002)
17. A. Heger, C.L. Fryer, S.E. Woosley, N. Langer, D.H. Hartmann: in prep. (2002)
18. R.M. Kippen, et al.: ApJ **506**, L27 (1998)
19. R.P. Kudritzki: ApJ **577**, 389 (2002)
20. N. Langer: In "The Eddington Limit in Rotating Massive Stars" eds. A. Nota, H.J.G.L.M. Lamers, ASP Conference Series **120**, 83 (1997)
21. A. Maeder, G. Meynet: A&A **373**, 555 (2001)
22. A. Maeder, G. Meynet: A&A **361**, 159 (2000)
23. A.I. MacFadyen, S.E. Woosley: ApJ **524**, 262 (1999)
24. A.I. MacFadyen, S.E. Woosley, A. Heger: ApJ **550**, 410 (2001)
25. P. Marigo, L. Girardi, C. Chiosi, P.R. Wood: A&A **371**, 152 (2001)
26. G. Meynet, A. Maeder, G. Schaller, D. Schaerer, C. Charbonnel: A&AS **103**, 97 (1994)
27. H. Nieuwenhuijzen, C. de Jager: A&A **231**, 134 (1990)
28. F. Nakamura, M. Umemura: ApJ **548**, 19 (2000)
29. K. Nomoto, K. Maeda, H. Umeda, T. Ohkubo, J. Deng, P. Mazzali: in 'A Massive Star Odyssey, from Main Sequence to Supernova', Proc. IAU Symposium 212, (San Francisco: ASP) eds. K.A. van der Hucht, A. Herrero, C. Esteban, in press (2002)
30. D. Schaerer, G. Meynet, A. Maeder, G. Schaller: A&AS **98**, 523 (1993)
31. G. Schaller, D. Schaerer, G. Meynet, A. Maeder: A&AS **96**, 269 (1992)
32. S. Wellsten, N. Langer: A&A **350**, 148 (1999)
33. S.E. Woosley: ApJ **405**, 273 (1993)
34. S.E. Woosley: in 'Nucleosynthesis and Chemical Evolution', ed. by B. Hauck, A. Maeder, G. Meynet (Switzerland: Geneva Obs.), p. 1 (1986)
35. S.E. Woosley, A. Heger, T.A. Weaver: Rev. Mod. Phys., in press (2002)
36. W. Zhang, S.E. Woosley: in 'proceedings of 3D Stellar Evolution Workshop', Livermore, July, 2002; astro-ph/0209482 (2002)
37. W. Zhang, S.E. Woosley, A.I. MacFadyen: ApJ, subm.; astro-ph/0207436 (2002)

The Progenitor of SN 1987A: A Progress Report

Philipp Podsiadlowski[1] and Natalia Ivanova[2]

[1] University of Oxford, Department of Astrophysics, Oxford OX1 3RH, UK
[2] Northwestern University, Evanston, IL 60209, USA

Abstract. In the last ten years it has becomes increasingly clear that the progenitor of SN 1987A must have experienced a dramatic event some 20,000 yr before the explosion, most likely the merger with a companion star. We here present recent detailed models of such a merger event, using stellar and 2-d hydrodynamical simulations of the various phases in the evolution of the progenitor. We show that in the final merging phase, a stream emanating from the embedded companion star can penetrate deep into the helium core of the progenitor; this leads to the dredge-up of helium and some unusual nucleosynthesis. This can explain most of the major anomalies of the progenitor, its blue colour, the main chemical anomalies in the envelope and the inner ring of the triple-ring nebula. Our simulations suggest that the core of the progenitor should have been rapidly rotating at the time of the explosion which may have affected the explosion and may be responsible for some of the observed asymmetries in the ejecta.

1 Introduction

Supernova 1987A (SN 1987A) in the Large Magellanic Cloud (LMC) was the first naked-eye supernova since Kepler's supernova in 1604 and was undoubtedly one of the most spectacular astronomical events of the decade. It was also a very unusual, even anomalous event: the progenitor was a blue supergiant instead of a red supergiant as had been expected, the envelope showed a number of unexpected chemical anomalies (in particular, a helium enhancement by a factor ~ 2; e.g. [20]), and the supernova is surrounded by a complex triple-ring nebula that was ejected some 20,000 yr before the explosion (see [12] for discussion).

For several years after the supernova, there was a heated debate within the supernova community whether these anomalies required a binary progenitor or whether they could be accommodated within a single-star scenario after some modifications of the stellar input physics, e.g. the convection parameters. In particular, it seemed possible that a rapid-rotation model might be able to explain some of the anomalies [12]. However, subsequently it became clear that with the best stellar input physics and the recent increase of opacities [18], single massive stars with LMC metallicity never end their evolution as blue supergiants for any *plausible* adjustment of the convection parameters [22]. Moreover, it was shown [22] that rapid rotation increases the core mass of the progenitor in such a way to strongly favour a red- over a blue-supergiant progenitor.

Irrespective of the question whether a single massive star in the LMC could ever produce a blue-supergiant progenitor, what settled the argument in favour of a binary scenario in many people's minds was the triple-ring nebula

<table>
</table>

massive, wide binary

dynamical mass-transfer and complete merging
after core helium burning:
formation of a single, rapidly rotating supergiant

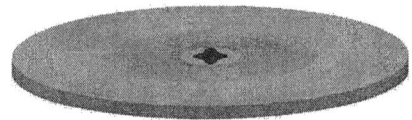

rotationally forced disk outflow in
red-blue transition

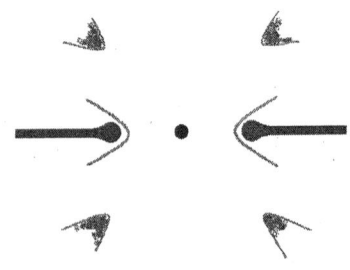

blue-supergiant wind sweeps up previous structures:
formation of the triple-ring nebula

Fig. 1. The merger scenario for SN 1987A

with its axi-symmetric but very non-spherical structure [21,1]: simple angular-momentum considerations imply that even an initially rapidly rotating massive star could not have a significant amount of rotation in its red-supergiant phase and explain the observed degree of non-sphericity. Hence the main question is no longer whether it was a binary but what type of binary. Of the various binary models proposed [12], the favoured model in recent years has been a merger model where two massive stars merged ∼ 20,000 yr before the supernova, linking the various anomalies to a single dramatic event. A merger model was first

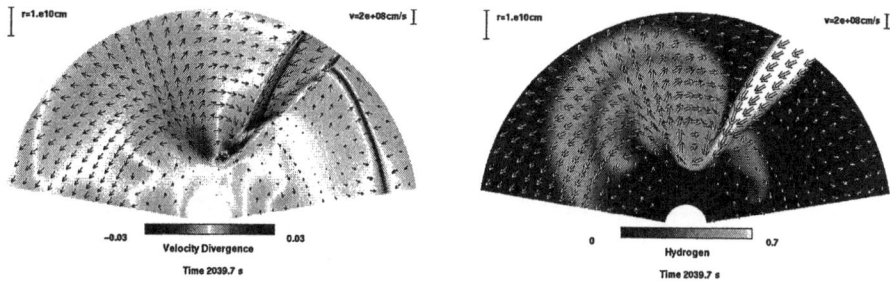

Fig. 2. Example of the steady state situation in a 2-d simulation of the stream–core interaction (left panel: velocity divergence; right panel: hydrogen abundance)

proposed to explain the asymmetry of the supernova ejecta inferred from the observed polarization [2] and was later developed further to explain the other anomalies of the supernova [5,16]. The basic merger model has not changed much in recent years [12,13] and is schematically illustrated in Fig. 1. The system initially consists of a binary with a primary of $15-20\,M_\odot$ and a much less massive companion $(2-10\,M_\odot)$ in a fairly wide orbit (with an orbital period of $\sim 10\,\mathrm{yr}$). The primary fills its Roche lobe on the asymptotic-giant branch (i.e. after helium core burning) and experiences dynamical mass transfer, leading to the complete merger of the two stars and the formation of a rapidly rotating single star. The merger causes the dredge up of part of the helium core, explaining the chemical anomalies and the final blue transition (also see [19]). Since all the orbital angular momentum of the binary is deposited in the envelope of the merger product, which is much more angular momentum than a dynamically stable blue-supergiant can accommodate, the excess angular momentum has to be lost from the system, most likely in the form of an equatorially enforced outflow in the red-blue transition [4], naturally explaining the central ring around the supernova.

While the basic picture has remained the same, it involves several poorly understood phases of stellar evolution (e.g. the merger phase, the formation of the two outer rings). This is where there has been significant process in recent years [6–9] which we briefly describe in this contribution.

2   The Spiral-In and Merger Phases

In the spiral-in phase, the secondary and the helium core of the primary form a binary inside a common envelope, the original envelope of the primary. Friction between this immersed binary and the envelope causes the orbit to shrink. The orbital energy that is released in the process causes an expansion of the envelope and could in principle cause its ejection. Here we are interested in the case where the envelope is not ejected. Then there is a phase of self-regulated spiral-in where all the frictional energy deposited in the envelope can be transported to the surface where it is being radiated away [11,14]. For a massive primary with a mass of $\sim 20\,M_\odot$ filling its Roche lobe as a red supergiant after helium core

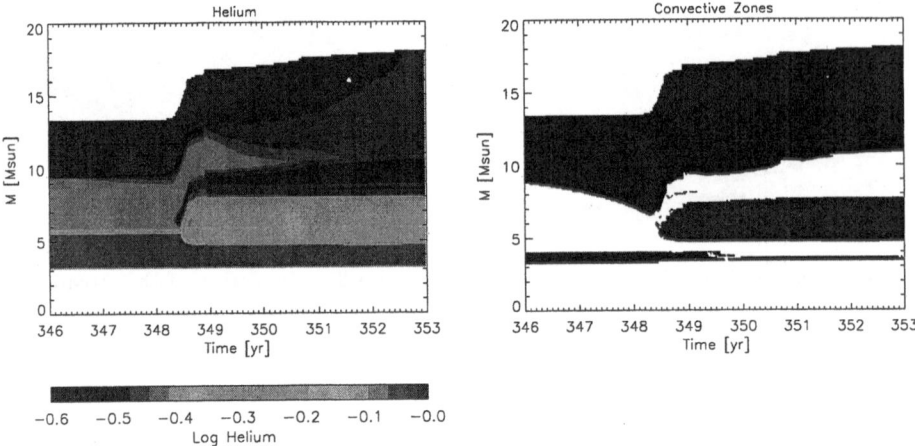

Fig. 3. The evolution of helium abundance (left) and convection zones (right) in a merger simulation applicable to the progenitor of SN 1987A. (The final surface helium abundance by mass in this example was 0.36)

burning, the characteristic time-scale for this phase is several 100 yr. However at some point, the spiraling-in secondary itself will fill its critical tidal lobe and start to transfer mass to the core of the giant. This situation is in many respects similar to the case of normal mass transfer by Roche-lobe overflow except for several fundamental differences. Mass transfer occurs inside a low-density, opaque envelope and is driven by the frictional angular momentum loss with the envelope at a rate of $0.01 - 10\,M_\odot\,\mathrm{yr}^{-1}$, and the secondary may not be in synchronous rotation with the orbit. As a result of this interaction, the secondary is gradually being dissolved inside the envelope. Even though the time-scale for this dissolution is short compared to the evolutionary time-scale of a giant, it is still much longer than the dynamical time-scale of the surrounding envelope. We therefore refer to this process as a 'slow merger' of the components.

In order to model the spiral-in and merger phases, we split the overall complex problem into a number of simpler problems which can be treated with standard numerical methods. The spiral-in phase was modelled with a modified stellar-evolution code [6,8] which models the effects of the spiralling-in binary on the envelope. This determines the density near the binary and sets the time-scale for the final merger. To model the interaction of the stream from the dissolving secondary with the core of the primary, we performed a systematic series of 2-d (and some 3-d) hydrodynamical simulations with the Munich PROMETHEUS code [3]. These calculations allow us to determine the depth of penetration of the stream, the entropy generated in the stream–core interaction and the post-impact conditions, as well as the role of rotation and nuclear burning [9] (see Fig. 2). From the simulations we obtain the region where dredge-up occurs and energy is deposited, which we then use in our modified binary evolution code to

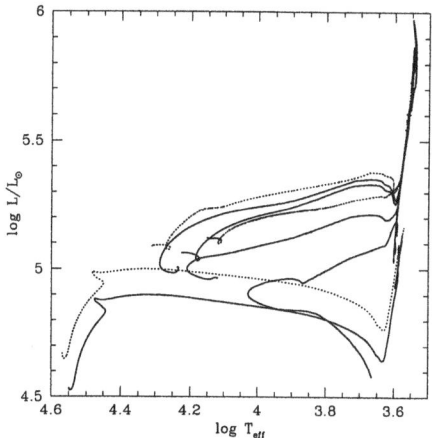

Fig. 4. Evolutionary tracks of stars with initial masses of 18 and 22 $M_\odot$ merging with a secondary of 5 $M_\odot$ after helium core burning ('moderate mergers' [8])

follow the evolution of the merger product (see Fig. 3). We generally can distinguish between three merger types (see [8]). The case applicable to SN 1987A is an intermediate case (a 'moderate merger') where the stream penetrates into the core without reaching the helium-burning layer. This causes significant dredge-up of helium (up to surface helium abundances of ~ 0.4) and of CNO-processed material. Depending on the core mass to total mass ratio, the merger product will evolve to become a blue supergiant or stay as a red supergiant. Figure 4 shows a selection of evolutionary tracks for the merger products, where some of these have surface helium and CNO abundances consistent with the observationally inferred ones (see [12,13] for references). However, unlike the more dramatic 'explosive merger' case [8], we only find a moderate amount of s-processing in these calculations, not sufficient to explain the observed overabundance of barium [10].

3   The Present Status

The detailed modelling of the merging phase presented in this contribution has confirmed that a merger model can satisfactorily explain the main anomalies of the progenitor of SN 1987A: in particular, the blue color of the immediate supernova progenitor, the timing of the final red-blue transition ($\sim 21{,}000\,\mathrm{yr}$ before the explosion), the main chemical anomalies (the He enhancement and the CNO abundances) and the inner ring, which is naturally produced as an equatorial outflow in the red-blue transition [4]. The helium abundance in our self-consistent models can be as high as 0.41 (by mass), which is at the lower end of the abundance range estimate determined from the inner ring [20], but could almost certainly be higher if we had included rotationally induced mixing before and/or after the merger. At present our models cannot explain the barium overabundance in the progenitor [10], but we have not fully explored all the uncertainties in the mixing processes and the treatment of convection.

While there have been a number of suggestions for the origin of the outer rings in the framework of a merger scenario, linking them to interacting winds before, during or after the merging phase, at present none of them seem fully convincing. On the other hand, the chemical composition of the outer rings and the small displacement of the center of symmetry contain important clues to the details and the timing of the ejection of this material which, once properly exploited, will shed further light on the events that occurred $\sim 20,000$ yr ago.

Finally, we note that our stream–core simulations show that the core should have been spun up significantly in the merging phase and that we would expect a rapidly rotating core immediately after the merger. This may have important implication for the supernova explosion itself and may help to explain the asymmetries of the ejecta and perhaps even the mystery spot.

References

1. C. Burrows et al.: ApJ **452**, 680 (1995)
2. R.A. Chevalier, N. Soker: ApJ **341**, 867 (1989)
3. B.A. Fryxell, E. Müller, W.D. Arnett: MPA Report **449** (1989)
4. A. Heger, N. Langer: A&A **334**, 210 (1998)
5. W. Hillebrandt, F. Meyer: A&A **219**, L3 (1989)
6. N. Ivanova: D.Phil. Thesis (Oxford) (2002)
7. N. Ivanova, Ph. Podsiadlowski: in preparation
8. N. Ivanova, Ph. Podsiadlowski: these proceedings
9. N. Ivanova, Ph. Podsiadlowski, H. Spruit: MNRAS **334**, 819 (2002)
10. P.A. Mazzali, N.N. Chugai: A&A **303**, 118 (1995)
11. F. Meyer, E. Meyer-Hofmeister: A&A **78**, 167 (1979)
12. Ph. Podsiadlowski: PASP **104**, 717 (1992)
13. Ph. Podsiadlowski: In *SN 1987A: Ten Years Later*, ed. M.M. Phillips, N.B. Suntzeff, in press
14. Ph. Podsiadlowski: In *Evolution of Binary and Multiple Stars Systems*, ed. Ph. Podsiadlowski et al. ASP Conf. Ser. **Vol. 229**, p. 239
15. Ph. Podsiadlowski, P.C. Joss, J.J.L. Hsu: ApJ **391**, 246 (1992)
16. Ph. Podsiadlowski, P.C. Joss, S. Rappaport: A&A **227**, L9 (1990)
17. Ph. Podsiadlowski, S. Rappaport, E. Pfahl, E.: ApJ **565**, 1107 (2002)
18. F.J. Rogers, C.A. Iglesias: ApJS **79**, 507 (1992)
19. H. Saio, K. Nomoto, M. Kato: ApJ **331**, 388 (1988)
20. G. Sonneborn et al.: ApJ **477**, 848 (1997)
21. E.J. Wampler et al.: ApJ **362**, L13 (1990)
22. S.E. Woosley, A. Heger, T.A. Weaver, N. Langer: In *SN 1987A: Ten Years Later*, ed. M.M. Phillips, N.B. Suntzeff, in press

The Slow Merger of Massive Stars

Natalia Ivanova[1] and Philipp Podsiadlowski[2]

[1] Northwestern University, Evanston IL 60209, USA
[2] University of Oxford, Department of Astrophysics, Oxford OX1 3RH, UK

Abstract. We study the complete merger of two massive stars inside a common envelope and the subsequent evolution of the merger product, a rapidly rotating massive supergiant. Three qualitatively different types of mergers have been identified and investigated in detail, and the post-merger evolution has been followed to the immediate presupernova stage. The "quiet merger" case does not lead to significant changes in composition, and the star remains a red supergiant. In the case of a "moderate merger", the star may become a blue supergiant and end its evolution as a blue supergiant, depending on the core to total mass ratio (as may be appropriate for the progenitor of SN 1987A). In the case of the most effective "explosive merger", the merger product stays a red giant. In the last two cases, the He abundance in the envelope is increased drastically, but significant s-processing is mainly expected in the "explosive merger" case.

1 Introduction

It is evident that the internal structure of the progenitor of a core-collapse supernova (SN) is one of the dominant factors that determines the characteristics of the supernova explosion, such as the light-curve and the abundances produced in the supernova (see e.g. [7]). It has also been shown that the distribution of the angular velocity can produce a strong asymmetry in the nucleosynthesis during the SN explosion and in its ejecta [3]. This makes it necessary to follow the detailed evolution of the abundances and the rotation profile at all stages of the evolution of a massive star before it explodes as a core-collapse SN. Observationally, it is well established that ∼ 40 % of all massive stars are members of binary systems with orbital periods shorter than 1 year and that at least 25 % of these will start to interact by Roche-lobe overflow (RLOF) during the advanced stages of the primary's evolution [4,10]. This implies that a significant fraction of all core-collapse supernova progenitors will have been affected by a previous binary interaction, where one of the most important interactions is the spiral-in of the two binary components inside a common envelope (CE) [10]. The final result of the spiral-in depends on how much of the released orbital energy has been converted into driving the expansion of the envelope relative to its binding energy. Here we study the situation where the deposited energy is not sufficient to eject the common envelope. This leads to the complete merger of the secondary with the core of the primary, forming a rapidly rotating single star in the process. Since the timescale of the merger is much longer than the dynamical timescale of the CE, it cannot be treated with a purely hydrodynamical code.

Mass transfer in the merging phase changes not only the chemical composition profile and the angular velocity distribution, but can also cause the erosion of part of the core, changing the core/envelope ratio. The evolutionary path of the merger product may also differ from that of a normal single star, e.g. by making a late blue loop more probable. Since the resulting supernovae may also differ from those with single-star progenitors, these merged objects are likely to be responsible for some of the variety observed among Type II supernovae.

In this paper we present the results of the modelling of the complete slow merger of a massive binary within a common envelope. In Section 2 we briefly describe the assumptions that were used to model mergers, and in Section 3 we present some of the main results.

2   Method and Initial Models

We used a standard Henyey-type stellar evolution code [8], updated recently [11]. The nuclear reactions rates were taken from Thielemann's library REACLIB [13] and updated as in [2]. In the code OPAL opacities [12] are used, supplemented with contributions from atomic, molecular and grain absorption in the low temperature regime [1].

To model the merger we implemented a number of modifications to the single stellar evolution code. These modifications were made mainly to treat the presence of the secondary inside the primary's envelope, including the mass transfer from the secondary to the core and the associated nucleosynthesis and mixing. A more detailed description of the modifications in the code can be found in [5]. We determine how deep the hydrogen-rich stream penetrates into the core of the primary using the prescription developed in [6].

We considered binaries consisting of a $18 - 22\,M_\odot$ primary and a $1 - 5\,M_\odot$ secondary. At the start of the spiral-in, the primary had already completed core helium burning. The chemical composition was taken as typical for young stars in the LMC ($X = 0.71$ and $Z = 0.01$). We adopted the Schwarzschild criterion for convection and took a mixing-length parameter $\alpha = 2$ and a convective-overshooting parameter equal to $25\,\%$ of a pressure scale height. These parameters were chosen since they are most appropriate for merger models of the progenitor of SN 1987A (see [9]).

3   Results

The qualitative behaviour of the merger and the temporal evolution of the structure of the primary within the secondary's orbit depend on the interaction of the hydrogen-rich material with the surrounding ambient matter. In particular, it depends on how deep and how fast the hydrogen-rich material penetrates into the primary's core and its placement with respect to the hot and/or convective zones within the secondary's orbit. According to our prescription for determining the stream penetration depth, the most effective penetration should occur

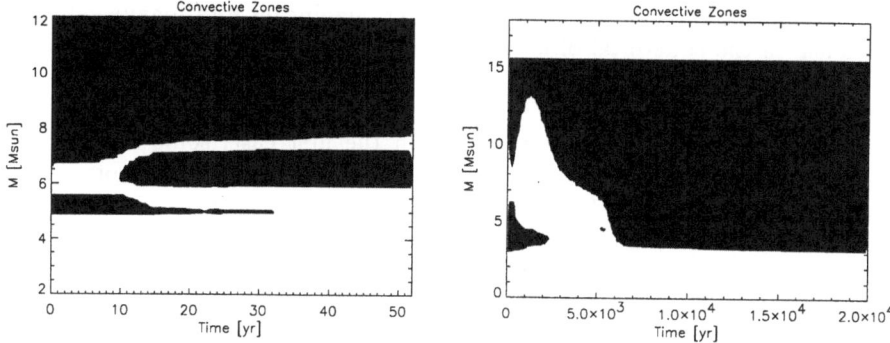

Fig. 1. The moderate merger: the evolution of convective zones (by mass) during the merger (left panel) and after the merger until core carbon ignition (right panel)

near the end of the merger phase, when the mass-loss rate is high and the exposed material from the secondary has low entropy. However, at this time the structure of the primary has changed, and a dense hydrogen-enriched region may already have been built up around the helium core, preventing the stream from penetrating deeper. All these effects combined create a very non-linear picture of the primary's response to mass transfer. In general, as a result of the merger, the primary's core expands, and the central temperature of the core drops (the degree of core cooling depends on the core expansion, i.e. the merger efficiency). This increases the total evolutionary time before core carbon ignition.

Based on our systematic study, we can distinguish three qualitatively different types of merger, where the classification of the mergers can be well explained by considering the temporal evolution of the convective zones during the merger.

- **The quiet merger.** All of the He affected by the penetrating stream is mixed with the outer envelope during the merger by convection, but since most of the He shell is not disturbed, there is only a moderate change in the surface abundances (He at the surface increases only by $4 - 8\%$). The merger product remains a red supergiant, although the progenitor might be significantly spun up compared to a single supergiant evolved in isolation.
This type of merger may happen in systems where the primary is close to carbon ignition at the start of the merger. This implies high pressure and temperature gradients and a correspondingly high entropy dissipation coefficient [6]. In addition, the secondary has to have a mass larger than $\sim 2\,M_\odot$ (due to the larger entropy).

- **The moderate merger.** Here the He core expands significantly, but an extensive He-rich shell remains. During the penetration, the stream creates a hydrogen-rich zone around the core. This zone becomes convective at some point and suppresses the bottom helium convective zone (see Fig. 1). During the merger, the primary core expands more drastically than in the case of the quiet merger. The merger product appears as a red supergiant, rotating rapidly and contracting immediately after the merger. Significant rotationally enhanced mass loss in the equatorial direction during the contraction phase is expected. Depending on the core/envelope ratio, the merger product may then perform a blue loop (if the hydrogen shell source becomes temporarily dominant) or continues its evolution as a red supergiant. In the first case, some dredge-up of helium takes place during the merger, and it can be expected that rotationally induced mixing will cause further significant enhancement of the surface helium abundance. In the second case, a delayed dredge-up phase (a few thousand years after the merger, see Fig. 1) takes place, resulting in a large overabundance of He in the envelope $(20 - 80\%)$.

At the start of the merger, the primary can be either already close to core carbon ignition and has a companion of $\sim 1\,M_\odot$, or the primary is at the start of He shell burning and has a companion of $2\,M_\odot$ or larger. A moderate merger is most appropriate for the progenitor of SN 1987A.

• **The explosive merger.** During the stream-core interaction, the hydrogen-rich material accumulates between two initially convective zones and slowly creates an intermediate zone, which at some point connects to the He burning shell. This leads to a dramatic nuclear flash, resulting in a drastic expansion of the He shell and complete mixing. The duration of this He shell explosion is about 0.25 year. In the case of a strong He shell explosion, even the carbon core may be disturbed. In all cases, there is an immediate significant increase of helium in the envelope, often accompanied by an increase in the carbon abundance. The stripped-off naked carbon core connects with the hydrogen-rich convective envelope and provides the site for efficient s-processing. The merger product continues its evolution as a red supergiant.

This type of the merger can take place in binaries with low-mass secondaries, not very close to core carbon ignition. The anomalous carbon star V Hydrae may provide an example for this merger channel.

References

1. D.R. Alexander, J.W. Ferguson: ApJ **437**, 879 (1994)
2. R.C. Cannon: MNRAS **263**, 817 (1993)
3. C. L. Fryer, A. Heger: ApJ **541**, 1033 (2000)
4. C.D. Garmany, P.S. Conti, P. Massey: ApJ **242**, 1063 (1980)
5. N. Ivanova, Ph. Podsiadlowski: PASP **281**, 191 (2002)
6. N. Ivanova, Ph. Podsiadlowski, H. Spruit: MNRAS **334**, 819 (2002)
7. K. Kifonidis, T. Plewa, H.-Th. Janka, E. Müller, E.: ApJ **531**, L123 (2000)
8. R. Kippenhahn, A. Weigert A., E. Hofmeister, E.: Methods in Computational Physics, **7**, ed. B. Alder et al. (New York: Academic, 1967), 129
9. Ph. Podsiadlowski: PASP **104**, 717 (1992)
10. Ph. Podsiadlowski, P.C. Joss, J.J.L. Hsu: ApJ **391**, 246 (1992)
11. Ph. Podsiadlowski, S. Rappaport, E. Pfahl, E.: ApJ **565**, 1107 (2002)
12. F.J. Rogers, C.A. Iglesias: ApJS **79**, 507 (1992)
13. F.-K. Thielemann, J.W. Truran, M. Arnould: In: *Advances in nuclear astrophysics* (Frontieres, Gif-sur-Yvette, France, 1986), p. 525

Observable Effects of Shocks in Compact and Extended Presupernovae

Sergey Blinnikov[1], Nikolai Chugai[2], Peter Lundqvist[3], Dmitrij Nadyozhin[1], Stan Woosley[4], and Elena Sorokina[5]

[1] ITEP, 117218, Moscow, Russia
[2] Institute of Astronomy, RAS, 109017 Moscow, Russia
[3] Stockholm Observatory, AlbaNova, SE-106 91 Stockholm, Sweden
[4] UCO/Lick Observatory, University of California, Santa Cruz, CA 95064, USA
[5] Sternberg Astronomical Institute, 119992 Moscow, Russia

Abstract. We simulate shock propagation in a wide range of core-collapsing presupernovae: from compact WR stars exploding as SNe Ib/c through very extended envelopes of the narrow-line SNe IIn. We find that the same physical phenomenon of radiating shocks can produce outbursts of X-ray radiation (with photon energy $3kT \sim 1$ keV) lasting only a second in SNe Ib/c, as well as a very high flux of visual light, lasting for months, in SNe IIn.

1 Introduction

Shock waves created at supernova explosions are observed mostly at the stage of supernova remnants due to their X-ray emission. For supernovae themselves shocks are not observed directly: normally the stage of shock break-out through the presupernova surface layers produces a short-lived transient of hard radiation. Yet those transients may have important observational consequences. Moreover, in some supernovae the shock propagation in the surrounding CSM may be decisive in producing their light on time-scales from days to years.

A shock propagating down the profile of decreasing density should accelerate [1]. The acceleration ends when the leakage of hard photons from the shock front into outer space becomes efficient enough. We simulate shock propagation in a variety of core-collapsing presupernova models using a code with hydrodynamics coupled to multi-energy-group time-dependent radiation transfer. In our previous work on extended type II SNe [2], with radii of a few hundred $R_\odot$, we have found the peak effective temperature at the shock break-out, $T_{\text{eff}} \sim 1.5 \times 10^5$K. For SN 1987A, with its presupernova radius of only $\sim 50R_\odot$, we have got $T_{\text{eff}} \sim 6 \times 10^5$K [3]. Due to effects of scattering the maximum color temperature is a factor of 2-3 higher.

In compact presupernovae, like SNe Ib/c, the shock can become relativistic [4] and is able to produce a burst of X-ray and even γ-ray radiation [5–7].

Supernovae of type Ib/c are also interesting for theory due to problems with their light curve and spectral modeling. Their understanding may serve as diagnostics of mass loss from massive stars at the latest phases of evolution.

2   Compact Presupernovae: SNe Ib/c

Numerical modeling of shock breakout in SNe Ib/c was done previously using some simplifying approximations [8]. Our method, realized in the code STELLA, allows us to get more reliable predictions for the outburst. Improvements in the theory done in the current work are: multi-energy-group time-dependent radiation transfer, and taking into account of (some) relativistic effects.

A representative presupernova in our runs was a WR star built by the code KEPLER [9] (model 7A). Late light curves and spectra for this model were studied in [10].

The Model 7A has mass 3.199 $M_\odot$ (including the mass of the collapsed core). Its radius prior to explosion is not strictly fixed because the outer mesh zones in KEPLER output actually model the strong stellar wind and are not in hydrostatic equilibrium. So, we fixed the radius by hand and we got four models with radii from 0.76 up to 2 $R_\odot$ which *are* in hydrostatic equilibrium. Our results are summarized in the Table.

Explosions with KEPLER [10] gave a maximum temperature of photons $T \sim 5 \times 10^5$ K at shock breakout. We have much finer mesh zoning at the edge of the star (down to $1 \times 10^{-12} M_\odot$) and better physics and we find much higher values of T. The left plot in Fig. 1 shows the difference between effective and color temperatures (labels 'e' and 'c', respectively, in the Table). One should note that the peak values of luminosity and temperature given in the table and on the left plot of Fig. 1 do not contain the light travel time correction. Its effect on the luminosity, i.e. smearing the peak on the time-scale of $\sim R/c$ is shown on the right plot of Fig. 1. Models, labeled as 3.5N were computed by one of us (D.K.N.) in 1992 with the equilibrium radiation diffusion hydrocode SNV.

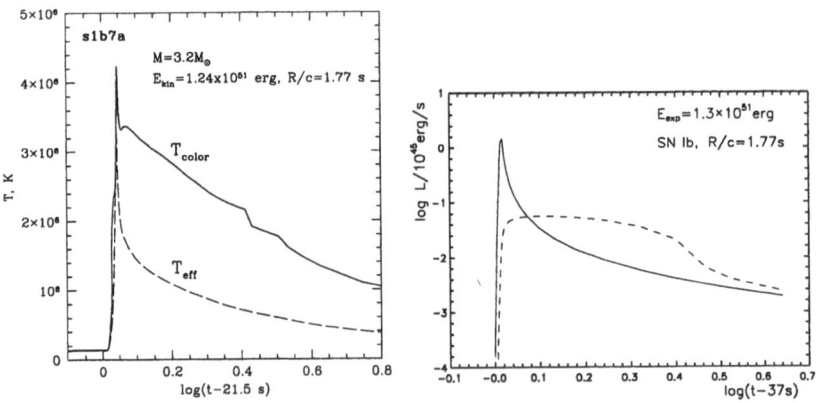

Fig. 1. Effective and color temperatures of emerging radiation (left). Shock breakout luminosity found by the hydrocode SNV (right). Dashed lines demonstrate the effect of averaging the light curve due the light travel time correction

Table 1. Parameters of shock breakouts

M^{a}	R_0^{a}	E_{kin}^{b}	L_p^{c}	T_p^{c}	Δt^{d}
$M_\odot$	$R_\odot$	foe	erg/s	10^6 K	s
3.2	0.76	1.24	4.2×10^{44}	4.2c	0.021
3.2	1.00	1.32	5.8×10^{44}	4.3c	0.026
3.2	1.23	1.30	6.8×10^{44}	4.3c	0.043
3.2	2.	1.39	9.4×10^{44}	4.3c	0.12
3.2	2.	4.36	3.6×10^{45}	5.3c	0.028
3.2	2.	8.86	4.8×10^{45}	7.2c	0.020
3.5N	0.76	1.30	1.4×10^{45}	5.1e	0.028
3.5N	1.23	1.30	8.1×10^{44}	3.5e	0.067

[a] presupernova mass and radius, respectively, in solar units.
[b] kinetic energy at infinity in 10^{51} ergs.
[c] peak luminosity and temperature.
[d] the width of the light curves at 1 stellar magnitude below L_p.

3 Shocks in CSM in Type IIn Supernovae

The narrow-line Type II supernovae (SNe IIn) are embedded in massive circumstellar shells (wind) extending from tens of thousands solar radii (SN 1998S) to $\sim 10^{17}$ cm (SN 1988Z). One of the brightest SNIIn, SN 1994W in NGC 4041, displays a spectrum dominated for $\sim$ 200 days by low-ionization P-Cygni lines with widths of order 10^3 km/s, and at late times by narrower H_α in pure emission. The light curve shows a plateau for $\sim$ 120 days, after which the luminosity drops by $\sim$ 3.5 magnitudes in V in only 12 days. The existence of such a drop, caused by circumstellar shells, was emphasized in paper [12]. The interpretation of spectra and light curve of SN 1994W leads to a coherent model in which the supernova interacts with a massive circumstellar shell ejected roughly 2 years prior to the explosion [11].

Main features of SN 1994W are well reproduced by the radiative shock propagating at relatively low speed for several months in the circumstellar shell. This shock forms its unique light curve with a plateau quite different from a classical SNII-P. (In the latter UBV light curves diverge at the end of the V plateau, while here all colors converge.) The sudden drop of V-flux by $\sim$ 3.5 magnitudes in 12 days is also explained quite naturally. The results presented in Fig. 2 are computed for the supernova model with ejected mass $7M_\odot$ and huge radius of presupernova $R_0 = 2 \times 10^4 R_\odot$ surrounded by a shell with the density $\rho_{wind} = 12/r$ in CGS units. The CS envelope has an outer cut-off radius $R = 6.6 \times 10^4 R_\odot$. The ejection of this dense extended CS envelope might be

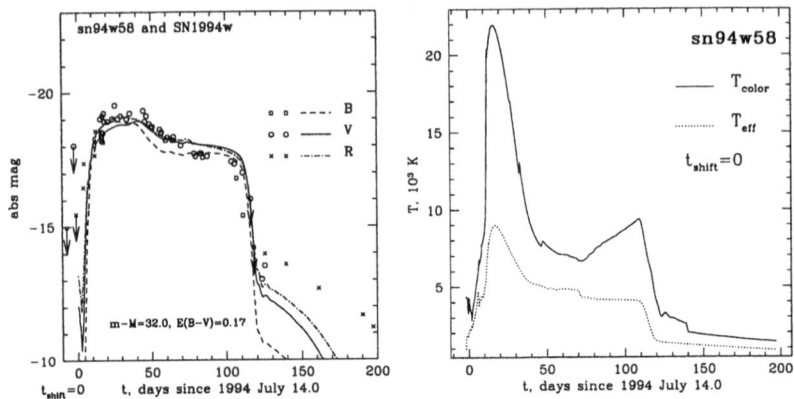

Fig. 2. BVR light curves and T_{ph} for the run sn94w58

related to weak explosions occurring several years prior to the collapse of the core [13] (see also [14]).

We are grateful to W. Hillebrandt and to the MPA staff for permanent and generous support. Support from NSF AST-97 31569 and NASA - NAG5-8128, Russian RBRF 00-02-17230 and RBRF 02-02-16500, the Wenner-Gren Science Foundation and the Royal Swedish Academy is acknowledged.

References

1. G.M. Gandel'man, D.A. Frank-Kamenetskij: Doklady AN SSSR **107**, 811 (1956)
   A. Sakurai: Commun. Pure Appl. Math. **13**, 353 (1960)
2. S.I. Blinnikov, R. Eastman et al.: ApJ, **496**, 454 (1998)
3. S. Blinnikov, P. Lundqvist et al.: ApJ, **532**, 1132 (2000)
4. M.H. Johnson, C.F. McKee: Phys.Rev.D **3**, 858 (1971)
5. S.A. Colgate: Can.J.Phys. **46**, 476 (1969)
6. G.S. Bisnovatyi-Kogan, V.S. Imshennik et al.: Ap.Space Sci. **35**, 23 (1975)
7. S.E. Woosley: Astron.Ap.Suppl., **97**, 205 (1993).
8. L.M. Ensman: Type Ib supernovae and a new radiation hydrodynamics code, PhD Thesis, California Univ., Santa Cruz (1991)
9. S.E. Woosley, N. Langer, T.A. Weaver ApJ, **448**, 315 (1995)
10. S.E. Woosley, R.G. Eastman: Thermonuclear Supernovae, eds. P. Ruiz-Lapuente, R. Canal, J. Isern (1997) p.821
11. N.N. Chugai, S.I. Blinnikov, P. Lundqvist, R.J. Cumming, A.V. Filippenko, A.J. Barth, D.C. Leonard, T. Matheson: Early circumstellar interaction and light curve of supernova 1994W (2002) in preparation.
12. E.K. Grasberg, D.K. Nadezhin: Astron.Zh **64**, 1199; Soviet Astronomy **31**, 629 (1987)
13. Woosley, S. E. 1986. Nucleosynthesis and Stellar Evolution. Saas-Fee Advanced Course 16: Edited by B.Hauck, A. Maeder and G. Meynet. Publisher: Geneva Observatory, CH-1290 Sauverny, Switzerland, 1986, p.1
14. E.K. Grasberg, D.K. Nadezhin: Astron.Zh **68**, 85; Soviet Astronomy **35**, 42 (1991)

Constraining the Masses of the Progenitors of Core-Collapse Supernovae

Stephen J. Smartt

Institute of Astronomy, University of Cambridge, Madingley Road,
Cambridge CB3 0HA, UK; sjs@ast.cam.ac.uk

Abstract. The exact evolution of massive stars up to the point of explosion is still quite uncertain. As massive stars are the most luminous objects in nearby galaxies (<20 Mpc), they can be individually resolved in Hubble Space Telescope and good quality ground-based images. There are now a wealth of multi-colour images available in the HST archive in particular and initiatives in Virtual Observatory science make the searching of multi-telescope, multi-instrument archives very efficient. Four SNe are presented here which have good quality images available before explosion. Although no progenitor was detected in any of the pre-explosion images, strong limits on the luminosities and masses of the stars can be placed. Prospects for continuing this in the future are discussed and how it will directly constrain masses of the progenitors of core-collapse SNe. A continuing programme with HST in Cycles 10 and 11 has been approved which will allow detailed follow up of all nearby core-collapse SNe which have deep, high-resolution archive images available.

1 Which Stars Produce the Core-Collapse SN Types?

The core collapse of massive stars at the end of their nuclear burning lifetimes produce the SNe of Types II, Ib and Ic. However we have only two definite

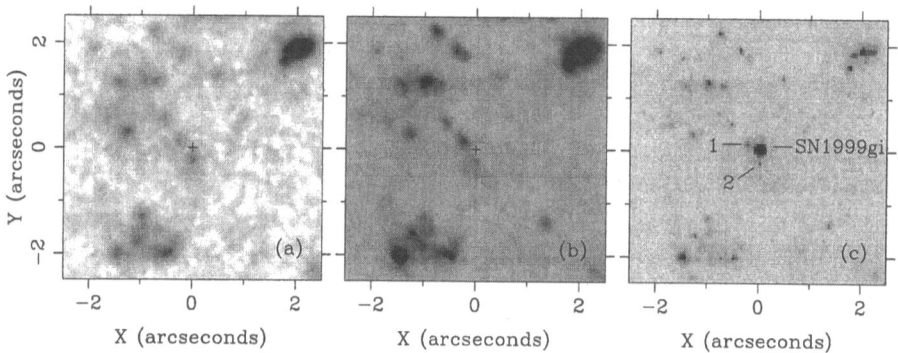

Fig. 1. The F300W (a) and F606W (b) images before the explosion of SN1999gi (marked with a cross) and the F555W (c) image of the SN after explosion. The pre-explosion images are from the WF4 chip, and the post-explosion image was centred on the PC1 hence it has better spatial resolution. There is no visible object at the position of SN1999gi in either pre-explosion image

detections of a SN progenitor before the explosion. It is well known that the progenitor of SN1987A in the LMC (50 kpc) was a hot blue supergiant of spectral type approximately (B3I; [1]). The optical UBVRI colours of the progenitor object of SN1993J in M81 (3.6 Mpc) are not consistent with a single stellar spectral type and the progenitor is likely to have been a G8-K0 Ia star plus excess UB-band flux from either a hot companion or the net flux from an unresolved OB association [2]. Neither progenitor is consistent with the canonical stellar evolution picture, in which core-collapse occurs while the massive star is an M-supergiant. However these were two peculiar SNe and their peculiarities may be due to the progenitor stars being in interacting binaries (e.g. [3,4]). We still don't understand the physical mechanisms which under-pin the different SNe types, and how these are related to the evolution of the progenitor star (e.g. see contributions by Heger, Nomoto and Maeda in these proceedings) and there is an understandable lack of observational data to constrain the last moments of stellar evolution.

2   Mass and Luminosity Limits from Pre-Explosion Archive Images

Nearby SNe such as SN 1987A and SN 1993J provide excellent opportunities to observe the progenitor star of a core-collapse SNe. Although the progenitor types were unexpected in each case, they were luminous supergiants and amongst the brightest stars in the host galaxies. However such fortuitous events will not occur more than once a decade or so, and to study larger numbers of progenitors one must look at more distant SNe. The very well maintained data archives of the HST, the CFH Telescope, the Isaac Newton Group of Telescopes and those of ESO contain a vast array of multi-colour images of late-type

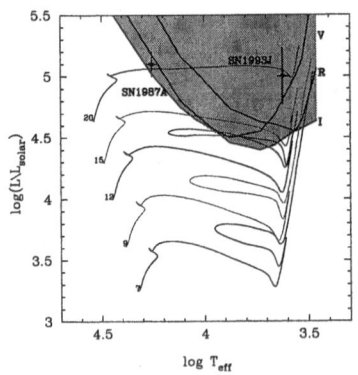

Fig. 2. SN1999em: a HRD with the detection limits converted to bolometric luminosity limits for each of three filters (VRI). The evolutionary tracks, without mass-loss, are from [9] (see also [14] for more details), and are fairly similar to the Geneva tracks for stars in the 11-20$M_\odot$ range. We suggest that the progenitor star must have definitely have been less than 12$M_\odot$ and possibly in the region 12±1$M_\odot$ (see Section 2)

galaxies within approximately 20 Mpc. At the spatial resolution of ground-based telescopes, the most luminous individual stars can be resolved and photometry obtained in galaxies within ∼8 Mpc. At the resolution of WFPC2 on board HST we can extend these measurement of individual massive stars to fainter intrinsic luminosities and distances out to ∼20 Mpc; the Cepheid Key Project is a clear demonstration of this [5]. Hence when a bright SN is discovered in a spiral galaxy within ∼20 Mpc there is now a reasonable chance that images have been taken of this galaxy either with HST or a ground based facility - allowing the exciting prospect of directly identifying the star which has exploded.

Three nearby type II-P SNe have occurred recently which have either HST or good seeing ground-based images taken before explosion. SN 1999em in NGC1637 (7.9 Mpc; [7,6]), SN 1999gi in NGC3184 (11.1 Mpc; [8]), and SN 2001du in NGC1365 (17.9 Mpc; [5]). The photometric and spectral evolution of SN 1999em and SN 1999gi have been studied in detail ([7,6,8]). They were both Type II-P, with very similar peak magnitudes ($M_V \simeq -16$), very similar ∼100 day plateaus. Both SN 1999gi and SN 2001du have pre-explosion HST images in the archive, and SN 1999em has a 0.7″ seeing image in the CFHT archive. Similar resolution images taken after explosion have allowed the SN position to be precisely determined on the pre-explosion frames (in the case of HST through a guaranteed target of opportunity program). However in each case *there is no detection of a progenitor star at the SN position*. Unfortunately the precursor objects are below the detection limits (Figs. 1 and 4). By measuring the sensitivity limits of the images, the bolometric luminosity limits of the progenitors (as a function of stellar effective temperature) can be determined, as in [10,9,11,12]. These can be plotted on an HR-diagram with stellar evolutionary tracks, which allows one to estimate the initial mass of the progenitor. This has been done for SN 1999em ([9] and Fig. 2), and for standard evolutionary tracks which go through the end

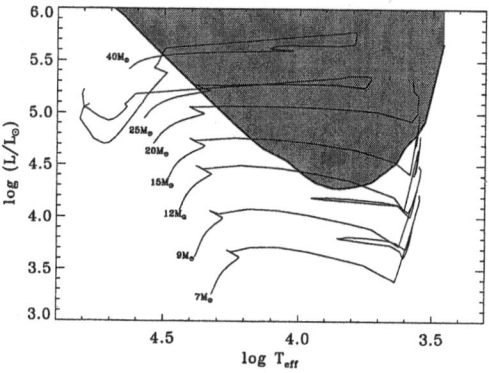

Fig. 3. SN1999gi: the Geneva evolutionary tracks [15,16] with the luminosity limit (from the F606W filter detection limit) plotted as a function of stellar effective temperature. The WFPC2 pre-explosion frames should be sensitive to all objects lying in the shaded region and the mass limit rises to $12^{+3}_{-3} M_\odot$

of helium burning, an upper mass limit of $< 12 M_\odot$ is derived for the *initial* mass of the progenitor. However the evolutionary calculations presented in [9] suggest stars below $11 M_\odot$ may experience second dredge-up, sending them to higher masses. The luminosity and temperature of red supergiants prior to explosion are critical parameters and we intend investigating the theoretical possibilities that occur during and after core C-burning in this phase in future work.

In [10] we originally derived a low upper mass limit of $9^{+3}_{-2} M_\odot$ for SN 1999gi, assuming a distance of 7.9 Mpc. However a more reliable distance using the expanding photosphere method has been derived by [8] as $11.1^{+2.0}_{-1.8}$ Mpc. Using this, they derive a less restrictive upper mass limit of $15^{+5}_{-3} M_\odot$. In Fig. 3 the results of [10] are recalculated using this distance, (and a reddening of E(B-V)=0.2), and a slightly finer grid of stellar types for the bolometric corrections. However with this larger distance, we can rule out a $20 M_\odot$ star as the track always remains in the detectable area (even when the errors are considered). The end point of the $15 M_\odot$ track also ends in the detectable zone, suggesting that we should see red supergiants of initial mass $15 M_\odot$ in the pre-explosion image. Taken at face value, the evolutionary tracks and the detection limits of the HST image would suggest an upper limit of $12 M_\odot$, with reasonable errors of ± 3. A value of $20 M_\odot$ is very unlikely, as this would have a magnitude $V_{606} = 23.5$, similar to the stars marked 1 and 2 in Fig. 1(c). The bolometric corrections and effective temperatures of M-supergiants, and the validity of detailed evolutionary models are critical in these derivations. But the upper mass of [8] seems too high, and $12 \pm 3 M_\odot$ would appear the most consistent result.

Images in the filters F336W, F555W, and F814W before the explosion of SN 2001du in NGC1365 exist in the HST WFPC2 archive. This was another Type II-P SN ([13,18]), although not as well monitored as SN 1999em or SN 1999gi.

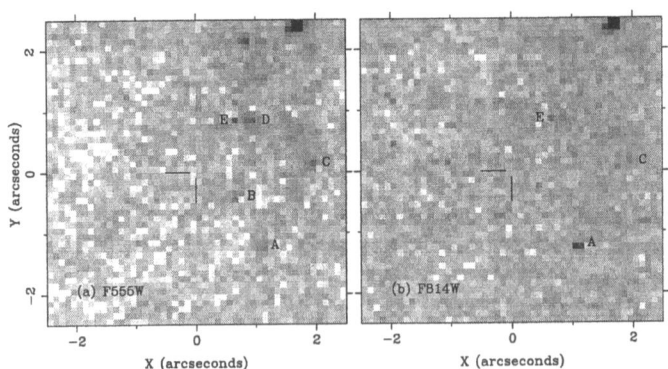

Fig. 4. SN2001du before explosion: the original F555W and F814W pre-explosion images ($5''$ square panels) from [12]. The position of the SN is marked at position (0,0) with the orthogonal lines. This position is determined from the centroid of SN in follow up WFPC images, and is accurate to better than $0.05''$. The stars detected closest to the position of SN2001du in the F555W frame are marked A-E

The target of opportunity programme (G09041) re-observed the position of the SN with WFPC2, for precise astrometric positioning. Again there is no sign of a progenitor at the position of the SN (see Fig. 4). Using the same method as above, an upper mass limit of $15M_\odot$ is determined.

Table 1. Summary of what we know about SN progenitors, taken from [9–12,17].

SN	Galaxy	Distance (Mpc)	Type	Z $(Z/Z_\odot)$	Progenitor Type	Mass
2002ap	NGC628	7.3	Ic	0.5	WR?	$<30M_\odot$
1997bs	NGC3627	11.4	IIn	1	?	$>20M_\odot$
1987A	LMC	0.05	II-pec	0.5	B3Ia	$20M_\odot$
1993J	M81	3.6	IIb	2	~K0Ia	$17M_\odot$
1980K	NGC6946	5.1	II-L	0.5	?	$<20M_\odot$
2001du	NGC1365	17.9	II-P	~ 1	M-type?	$<15M_\odot$
1999em	NGC1637	7.9	II-P	1-2	M-type?	$<12M_\odot$
1999gi	NGC3184	11.1	II-P	~2	M-type?	$<12M_\odot$

In summary, for three II-P SN we have very similar upper mass limits. In none of these cases do we find a very massive (e.g. similar to the progenitors of SN 1987A or SN 1993J) red or blue supergiant. In Table 1 the results of what we know about SN progenitors are summarized. All the available data are consistent with the three II-P events coming from moderate mass red supergiants. Both SN 1999em and SN 1999gi were faint x-ray sources ([19,20]), and the former was also detected in the radio. These studies have determined a mass-loss rate for the progenitor stars of $\sim 2 \times 10^{-6}M_\odot$ yr^{-1} (uncertain to within a factor 2, and assuming a stellar wind velocity of 10km s^{-1}), consistent with what one would expect for a cool red supergiant. The faint Type IIn SN 1997bs has a progenitor identified in NGC3627 [21], at approximately $M_V \simeq -8.1$. Only one observation through the filter (F606W) exists, but plotting the corresponding bolometric luminosity on a theoretical HRD suggests that the progenitor star must have had a mass at least $20M_\odot$. Prediscovery data of the recent type Ic SN 2002ap was presented by [11], and they found an upper mass limit of $30M_\odot$. It is likely that the progenitor was either a WR star with an initial mass of close to $30M_\odot$, or a star of initial mass $\sim 20 - 25M_\odot$ which was part of an interacting binary and stripped of its hydrogen and helium envelope via mass transfer.

While the data are understandably sparse, Table 1 suggests a picture may be emerging that most type II-P SNe come from moderate mass red supergiants. At higher masses, mass-loss becomes more important (as does its dependence on metallicity), and will inevitably lead to stripping of the outer envelopes of these stars. There will be a tendency to produce the IIb, Ib/c, as well as the II-n which are likely to arise through interaction of the SN ejecta with dense circumstellar material which itself may have been produced by the progenitor star.

3   Future Prospects

The chances of having suitable pre-explosion images available when a nearby SN is discovered are improving rapidly. A WFPC2 SNAP programme has just finished (GO9042) to enhance the number of late-type galaxies with high-quality archive images. Combining these with data available in the ground-based archives, [9] have estimated that on average ~ 2.4±2 SNe per year will have pre-explosion information and hence a project lasting 3-5 years should significantly improve our knowledge. This project has been greatly aided by the introduction of the ASTROVIRTEL[1] initiative which is a first step at creating tools for future Virtual Observatories. This is complemented with two HST GO approved proposals in Cycle 10 and Cycle 11 (9042 and 9353) that will provide the accurate astrometry of the SN for positioning on the pre-explosion images.

Acknowledgements

The support given by ASTROVIRTEL, a Project funded by the European Commission under FP5 Contract No. HPRI-CT-1999-00081, and financial support from PPARC is acknowledged. I thank G. Gilmore, C. Tout, P. Meikle, P. Mazzali, N. Trentham and E. Ramirez for many useful discussions, the SOC for accepting this contribution, and the LOC for a very enjoyable meeting.

References

1. Walborn N. et al., A&A, **219**, 229 (1989)
2. Aldering G., Humphreys R.M., Richmond M., AJ **107**, 662, (1994)
3. Podsiaklowski P., Hsu, J. J. L., Joss, P. C., Ross, R. R., Nature, **364**, 509, (1993)
4. Podsiaklowski P., PASP, **104**, 717, (1992)
5. Freedman W.L., et al., ApJ, **553**, 47, (2001)
6. Hamuy M., et al., ApJ, **558**, 615, (2002)
7. Leonard D., et al., PASP, **114**, 35, (2002)
8. Leonard D., et al., AJ, **24**, 2490, (2002)
9. Smartt S.J., Gilmore G.F., Tout C.A., Hodgkin S.T.,, ApJ, **565**, 2089, (2002)
10. Smartt S.J., et al., ApJ, **556**, L29, (2001)
11. Smartt S.J., et al., ApJ, **572**, L147, (2002)
12. Smartt S.J., et al., MNRAS, in prep., (2002)
13. Smartt S.J., Kilkenny D., Meikle P., IAUC No. 7704, (2001)
14. Pols O.R., Tout C.A., Eggleton P.P., Han Z., MNRAS, **274**, 964, (1995)
15. Meynet G., et al., A&ASS, **103**, 97, (1994)
16. Schaller G., Schaerer D., Meynet G., Maeder A., A&ASS, **96**, 269, (1992)
17. Thompson L.A., ApJ, **257**, L63 (1982)
18. Wang L., IAUC No. 7704, (2001)
19. Pooley D., et al., ApJ, **572**, 932 (2002)
20. Schlegel E.M., ApJ, **556**, L25 (2001)
21. Van Dyk S.D., et al., PASP, **112**, 1532 (2000)

---

[1] http://www.stecf.org/astrovirtel

Core-Collapse Supernova Progenitors in Hubble Space Telescope Images

Schuyler D. Van Dyk[1], Weidong Li[2], and Alexei V. Filippenko[2]

[1] IPAC/Caltech, 100-22, Pasadena, CA 91125, USA
[2] Astronomy Dept., Univ. of California, Berkeley, CA 94720-3411, USA

1 Introduction

Determining which stars give rise to supernovae (SNe) is key to SN research and stellar evolution studies. Without knowledge of SN progenitors, many of the conclusions and inferences made about the connection between SNe and important problems in astrophysics stand on precarious ground. The main obstacle is that a SN leaves few traces of the star that exploded.

Only five (SN 1961V, SN 1978K, SN 1987A, SN 1993J, SN 1997bs) of the more than 2000 historical SNe have had pre-explosion objects identified, and all of these were at least somewhat unusual. Both SN 1961V ([10,8,23]) and SN 1997bs [22] may not have been actual SNe (defined to be the catastrophic explosion of a star at the end of its life), but this is not yet certain.

SNe II and Ib/c arise from iron core collapse toward the end of a massive ($M \gtrsim 10 \ M_\odot$) star's life. Whereas it is generally agreed that SNe II must arise from H-rich supergiants, SNe Ib/c progenitors have not been unambiguously identified, but they must be stars that have lost most or all of their H envelopes. Two possibilities are Wolf-Rayet stars (see [4] and references therein) and massive interacting binary systems (e.g., [16,13,14]).

Evidence for the core-collapse nature of SNe II and Ib/c comes from both theoretical modelling (e.g., [26,27]) and from the few that have had progenitors directly identified. Evidence for massive progenitors has also been accrued from the environmental data for many other SNe, e.g., based on ground-based imaging ([18,19]). Recent studies ([2,20,21]) have exploited the superior spatial resolution afforded by the *Hubble Space Telescope* (*HST*) to resolve individual stars in SN environments and place constraints on the progenitor ages and masses.

Direct identification of additional core-collapse SN progenitors is essential. Our group ([21,22]) was able to directly identify the progenitor of SN 1997bs using *HST* archival images. The steadily increasing data volume in the *HST* archive and the remarkable success of modern search programs for nearby SNe, such as LOSS ([12,9]) and LOTOSS ([15,3]), make this possible. We [25] have attempted to isolate the progenitors of 16 core-collapse SNe (Table 1) using *HST* images and here we give a brief summary of the results.

Table 1. Core-Collapse SNe with *HST* data

SN	host	type	SN	host	type
1998Y	NGC 2415	II	2000C	NGC 2415	Ic
1999an	IC 755	II	2000ds	NGC 2768	Ib
1999br	NGC 4900	II-P	2000ew	NGC 3810	Ic
1999bu	NGC 3786	Ic	2001B	IC 391	Ib
1999bx	NGC 6745	II	2001ai	NGC 5278	Ic
1999dn	NGC 7714	Ib	2001ci	NGC 3079	Ic
1999ec	NGC 2207	Ib	2001du	NGC 1365	II-P
1999ev	NGC 4274	II	2001is	NGC 1961	Ib

2   Method

The crux of this work is determining at which location on the four WFPC2 chips the star should be. It is therefore of utmost importance to have high astrometric accuracy for all the images. Ideally, one could pinpoint the exact SN location by comparing a late-time image of the SN with a pre-SN image, which we did in three cases. However, even locating the fading SN in *HST* images is often quite difficult and requires high astrometric precision.

It is well known that positions based on the astrometric *HST* image header information are not very accurate – based on our experience, typically ~1″.5 *or worse*. We therefore apply an *independent* astrometric grid, using 2MASS as the basis, to both ground-based SN images and *HST* images, with astrometric solutions typically accurate to 0″.2–0″.3. This means we can potentially isolate the progenitor star location to within several WFPC2 pixels.

Once the SN site is located, photometry of the appropriate WFPC2 chip is performed using HSTphot ([6,7]). We can determine the magnitude and color of the candidate progenitor, although for most SNe, we can place only a 3σ limit on the magnitude and color.

3   Results

We have isolated candidate progenitors for five SNe, which, if verified, doubles the number of known progenitors. The candidate SNe II progenitors of SNe 1999br, 1999ev, and 2001du all have luminosities ($M_V^0 \approx -5.5$ to -6.9 mag) consistent with known Galactic red supergiants [11]. The SN 2001du progenitor color and the SN 1999bx progenitor color limit imply spectral types for supergiants that can only be as late as early M-type, further implying a more compact progenitor envelope or possible light contamination by a close binary companion.

The type Ib SN 2001B and SN 2001is candidate progenitors are very luminous, $M_V \approx -8$ to -9 mag, at or above the highest luminosities for known Galactic Wolf-Rayet stars. The brightness limits for the other SNe Ib/c are consistent both with the range of Wolf-Rayet brightnesses and with expectations for interacting binary models. The color limits are consistent with blue or yellow stars; for SNe 1999ec and 2001ci these limits indicate progenitors with spectral type A or earlier, consistent with the color range for Wolf-Rayet stars [17].

In general, we cannot yet place rigorous constraints on either the Wolf-Rayet or massive interacting binary models for SN Ib/c progenitors. The limitations of the data continue to be the restricted field of view, low signal-to-noise ratio, and poor color coverage. From their environments we find that SNe Ib/c seem to be more closely associated with massive-star regions than is true for SNe II (see [21]). Again, statistics are small, but this continues to suggest that at least some SN Ib/c progenitors may be more massive, in general, than SN II progenitors. With WFPC2 and ACS on board *HST* the amount of available archive data will continue to grow providing for larger SN samples in the future.

4   The SN 1993J Progenitor Revisited

From SINS project and archival *HST* WFPC2 images, we [24] identified four stars brighter than $V = 25$ mag within $2\!''\!.5$ of SN 1993J which contaminated the previous ground-based brightness estimates for the supernova progenitor ([1,5]). Correcting for the contamination, including the use of tests with fake stars (Fig. 1), we found that the progenitor's energy distribution is consistent

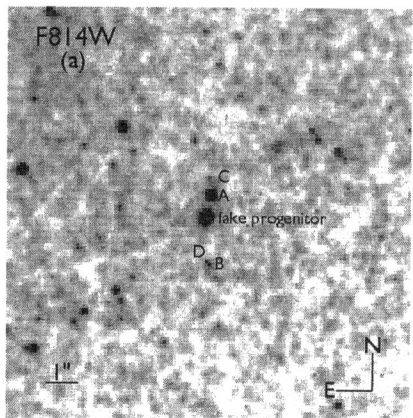
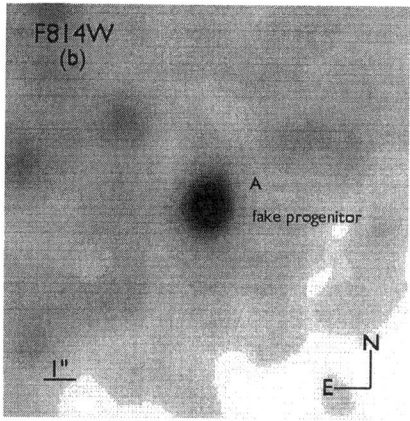

Fig. 1. (a) F814W (*I*-band) *HST* WFPC2 image from 2001 June, after subtraction of SN 1993J and the addition of a fake progenitor star with $I \approx 19.7$, $B - V \approx 0.8$, and $V - I \approx 1.9$ mag. The four contaminating stars, A–D, in the SN environment are indicated. **(b)** Same image as **(a)**, after convolution with a two-dimensional Gaussian with $\sigma = 0\!''\!.53$ (seeing $\approx 1\!''\!.25$ FWHM), to simulate the conditions experienced by pre-SN ground-based observers. Star A's position is indicated

with that of an early K-type supergiant with $M_V \approx -7.0$ mag and an initial mass of 13–22 $M_\odot$. The contamination is sufficient to account for the excess blue light seen pre-explosion from the ground; therefore, the progenitor did not necessarily have a blue companion (although a blue dwarf star with $M \lesssim 30$ $M_\odot$ would not have been detected). We also calculated model supergiant atmospheres with a range of plausible He abundances and found that the pre-SN colors are not strongly affected by the He abundance longward of 4000 Å. Abundances ranging between solar and 90% He (by number) are all consistent with the observations.

This publication makes use of data products from 2MASS, a joint project of the University of Massachusetts and IPAC/Caltech, funded by NASA and NSF; of the NASA/IPAC Extragalactic Database (NED), operated by JPL/Caltech for NASA; and of LEDA (http://leda.univ-lyon.fr). A.V.F.'s group at UC Berkeley is supported by NSF grant AST-9987438, the Sylvia and Jim Katzman Foundation, and NASA/HST grants AR-8754, AR-9529, and GO-8602.

References

1. G. Aldering, R.M. Humphreys, M.W. Richmond: AJ **107**, 662 (1994)
2. A.J. Barth, S.D. Van Dyk et al.: AJ **111**, 2047 (1996)
3. B. Beutler, W.D. Li et al.: IAUC 7906 (2002)
4. D. Branch, K. Nomoto, A.V. Filippenko: Comm. Ap. **XV**, 221 (1991)
5. J.G. Cohen, J. Darling, A. Porter: AJ **110**, 308 (1995)
6. A.E. Dolphin: PASP **112**, 1383 (2000a)
7. A.E. Dolphin: PASP **112**, 1397 (2000b)
8. A.V. Filippenko, A.J. Barth et al.: AJ **110**, 2261 (1995)
9. A.V. Filippenko, W.D. Li et al.: "The Lick Observatory Supernova Search with the Katzman Automatic Imaging Telescope." In *Small Telescope Astronomy on Global Scales* ed. B. Paczyński, W.-P. Chen, C. Lemme (ASP, San Francisco, 2001) pp. 121-130
10. R.W. Goodrich, G.S. Stringfellow et al.: ApJ **342**, 908 (1989)
11. R.M. Humphreys, K. Davidson: ApJ **232**, 409 (1979)
12. W.D. Li, M. Modjaz et al.: IAUC 7126 (1999)
13. K. Nomoto, A.V. Filippenko, T. Shigeyama: A&A **240**, L1 (1990)
14. Ph. Podsiadlowski, P.C. Joss, J.J.L. Hsu: ApJ **391**, 246 (1992)
15. M. Schwartz, W.D. Li et al.: IAUC 7514 (2000)
16. A. Uomoto: ApJ **310**, L35 (1986)
17. K.A. van der Hucht: New Astronomy Reviews **45**, 135 (2001)
18. S.D. Van Dyk: AJ **103**, 1788 (1992)
19. S.D. Van Dyk, M. Hamuy, A.V. Filippenko: AJ **111**, 2017 (1996)
20. S.D. Van Dyk, C. Y. Peng et al.: PASP **111**, 313 (1999a)
21. S.D. Van Dyk, C. Y. Peng et al.: AJ **118**, 2331 (1999b)
22. S.D. Van Dyk, C. Y. Peng et al.: PASP **112**, 1532 (2000)
23. S.D. Van Dyk, A.V. Filippenko, W.D. Li: PASP **114**, 700 (2002a)
24. S.D. Van Dyk, P.M. Garnavich et al.: PASP **114**, 1322 (2002b)
25. S.D. Van Dyk, W.D. Li, A.V. Filippenko: PASP, in press (2003)
26. S.E. Woosley, T.A. Weaver: ARA&A **24**, 205 (1986)
27. S.E. Woosley, T.A. Weaver: ApJS **101**, 181 (1995)

Part II

Core-Collapse Supernova Explosions

Core Collapse and Then?
The Route to Massive Star Explosions

Hans-Thomas Janka[1], Robert Buras[1], Konstantinos Kifonidis[1],
Tomek Plewa[2,3], and Markus Rampp[1]

[1] Max-Planck-Institut für Astrophysik, Karl-Schwarzschild-Str. 1, D-85741 Garching,
    Germany
[2] Dept. of Astronomy and Astrophysics and Center for Astrophysical Thermonuclear
    Flashes, The University of Chicago, Chicago, IL 60637, USA
[3] Nicolaus Copernicus Astronomical Center, Bartycka 18, 00716 Warsaw, Poland

Abstract. The rapidly growing base of observational data for supernova explosions
of massive stars demands theoretical explanations. Central to these is a self-consistent
model for the physical mechanism that provides the energy to start and drive the disruption of the star. We give arguments why the delayed neutrino-heating mechanism
should still be regarded as the standard paradigm to explain most explosions of massive
stars and show how large-scale and even global asymmetries can result as a natural
consequence of convective overturn in the neutrino-heating region behind the supernova shock. Since the explosion is a threshold phenomenon and depends sensitively on
the efficiency of the energy transfer by neutrinos, even relatively minor differences in
numerical simulations can matter on the secular timescale of the delayed mechanism.
To enhance this point, we present some results of recent one- and two-dimensional computations, which we have performed with a Boltzmann solver for the neutrino transport
and a state-of-the-art description of neutrino-matter interactions. Although our most
complete models fail to explode, the simulations demonstrate that one is encouragingly
close to the critical threshold because a modest variation of the neutrino transport in
combination with postshock convection leads to a weak neutrino-driven explosion with
properties that fulfill important requirements from observations.

1 Introduction

The primary energy source for powering supernovae of massive stars is the gravitational binding energy of the newly formed proto-neutron star or proto-black
hole (energy from nuclear reactions contributes at a minor level). To initiate
and drive the explosion, energy from some temporary storage, e.g. internal or
rotational energy of the compact remnant, must be transferred to the outer
stellar layers to be finally converted to kinetic energy of the ejecta. This might
be achieved by hydrodynamical shocks, by neutrinos or by magnetic fields as
mediators. Accordingly, one distinguishes between

(i) the (prompt) mechanism, which works on a dynamical timescale by the hy-
    drodynamical shock that is created at the moment of core bounce,
(ii) the delayed, neutrino-driven mechanism which starts the explosion on the
    secular timescale of neutrino-energy deposition behind the supernova shock,

(iii) and the magnetohydrodynamical (MHD) mechanism, which requires that initial seed magnetic fields are amplified to a dynamically relevant strength by differential rotation.

Intense radiation or relativistic outflows of charged particles might also play a role for very special conditions. They may, for example, originate from the vicinity of an accreting black hole that has formed after the collapse of a rotating stellar core. Relativistic jets as driving mechanism are currently discussed for stellar explosions that have been observed in association with gamma-ray bursts (see the contributions by S. Woosley and A. MacFadyen at this conference).

Depending on the mediator, the conditions for efficient energy transfer, the corresponding timescale and the tapped energy reservoir are different. A lot of work has been spent in the past 40 years on the search for a viable supernova mechanism and the study of the various theoretical suggestions. A brief review of these efforts and the current status of our knowledge can be found in Ref. [25].

The relevance of the different mechanisms for stellar explosions listed above depends on the (poorly known) physical conditions in collapsed stellar cores and on the properties of the progenitor stars. Some of the involved requirements are more likely to be fulfilled than others, some combinations of necessary conditions may be more frequent and more typical, while others may be realized only in rare cases and under very special, exceptional circumstances.

The neutrino-driven mechanism [54,2] involves a minimum of controversial assumptions and uncertain degrees of freedom in the physics of collapsing stars. It relies on the importance of neutrinos and their energetic dominance in the supernova core. After the detection of neutrinos in connection with SN 1987A and the overall confirmation of theoretical expectations for the neutrino emission, this can no longer be considered as a speculative assumption but is an established fact. Of course, this does not mean that such a minimal input is sufficient to understand the cause of supernova explosions and to explain all observable properties of supernovae. But at least it can be taken as a good reason to investigate how far one can advance with a minimum of imponderabilities.

2   Observational Facts

Progress in our understanding of the processes that lead to the explosion of massive stars is mainly based on elaborate numerical modeling, supplemented by theoretical analysis and constrained by a growing data base of observed properties of supernovae. The latter may carry imprints from the physical conditions very close to the center of the explosion. Observable features at very large radii, however, can be linked to the actual energy source of the explosion only indirectly through a variety of intermediate steps and processes. Any interpretation with respect to the mechanism that initiates the explosion therefore requires caution.

A viable model for the explosion mechanism of massive stars should ultimately be able to explain the observed explosion energies, nucleosynthetic yields (in particular of radioactive isotopes like ^{56}Ni, which are created near the mass cut), and the masses of the compact remnants (neutron stars or black holes) and their connection with the progenitor mass.

Recent evaluations of photometric and spectroscopic data for samples of well-observed Type-II plateau supernovae reveal a wide continuum of kinetic energies and ejected nickel masses. Faint, low-energy cases seem to be nickel-poor whereas bright, high-energy explosions tend to be nickel-rich and associated with more massive progenitors [14]. This direct correlation between stellar and explosion properties, however, is not apparent in an independent analysis by Nadyozhin [39] who speculates that more than one stellar parameter (rotation or magnetic fields besides the progenitor and core mass) might determine the explosion mechanism. A large range of nickel masses and explosion energies was also found for Type Ib/c supernovae [14]. Interpreting results obtained by the fitting of optical light curves and spectra, Nomoto et al. [41] came up with the proposal that explosions of stars with main sequence masses above 20–25 $M_\odot$ split up to a branch of extraordinarily bright and energetic events ("hypernovae") at the one extreme and a branch of faint, low-energy or even "failed" supernovae at the other. Stars with such large masses might collapse to black holes rather than neutron stars. The power of the explosion could depend on the amount of angular momentum in the collapsing core, which in turn can be sensitive to a number of effects such as stellar winds and mass loss, metallicity, magnetic fields, binarity or spiraling-in of a companion star in a binary system.

Anisotropic processes and large-scale mixing between the deep interior and the hydrogen layer had to be invoked in case of SN 1987A to explain the shape of the light curve, the unexpectedly early appearance of X-ray and γ-ray emission, and Doppler features of spectral lines (for a review, see [40]). More than ten years after the explosion, the expanding debris exhibits an axially symmetric deformation [50]. SN 1987A therefore seems to possess an intrinsic, global asymmetry. The same conclusion was drawn for other core-collapse supernovae (Type-II as well as Ib/c) based on the fact that their light is linearly polarized at a level around 1% with a tendency to increase at later phases when greater depths are observed [49,28]. This has been interpreted as evidence that the inner portions of the explosion, and hence the mechanism itself, are strongly non-spherical [20,53], possibly associated with a "jet-induced" explosion [50,26]. This is a very interesting and potentially relevant conjecture. It does, however, not necessarily constrain the nature of the physical process that mediates the energy transfer from the collapsed core of the star to the ejecta and thereby creates the asphericity.

Rotation plus magnetic fields were proposed as the "most obvious" way to break the spherical symmetry and to explain the global asphericity of core-collapse supernovae [52,1,51]. It was argued that current numerical calculations may be missing a major ingredient necessary to yield explosions. A proper treatment of rotation *and* magnetic fields may be necessary to fully understand when and how collapse leads to explosion. Of course, this might be true. But a confirmation or rejection will require computer models with ultimately the full physics.

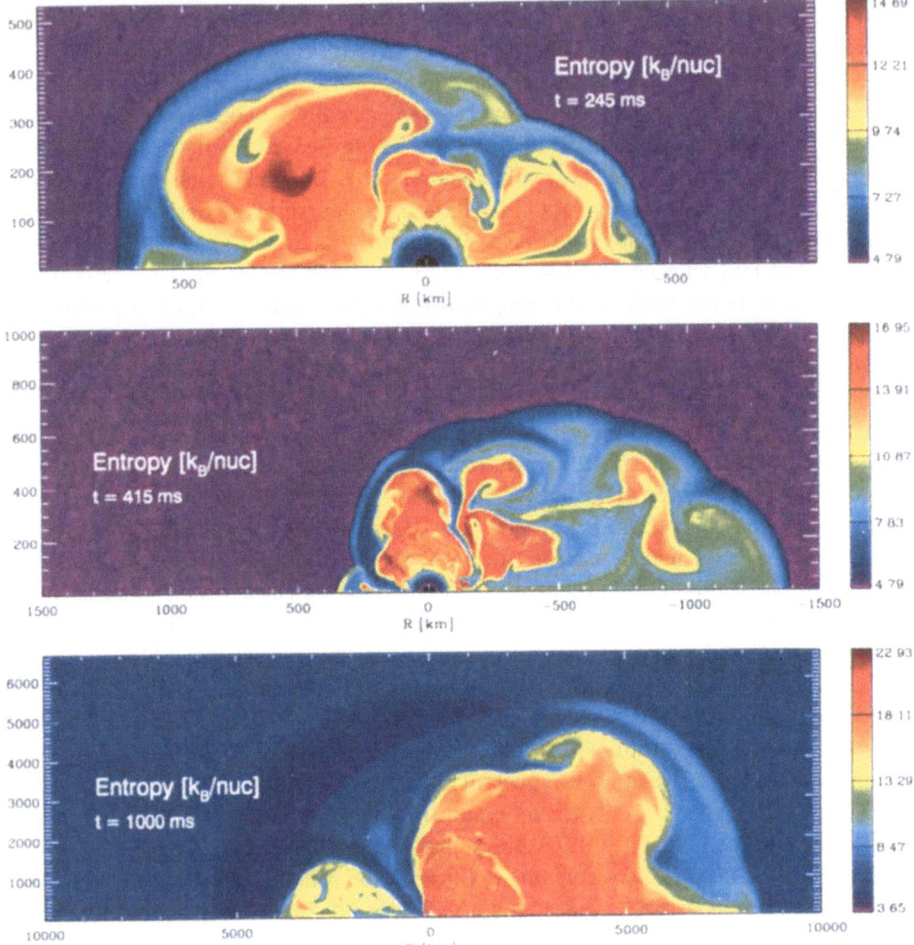

Fig. 1. Explosion that is driven by neutrino-energy deposition in combination with convective overturn in the region behind the supernova shock. The anisotropy of the neutrino- and shock-heated ejecta is growing in time and becomes very large due to an increasing contribution of the $m = 0$, $l = 1$ mode in the convective pattern. The snapshots (from top to bottom) show the entropy distribution (values between about 4 and 23 k_B per nucleon) at post-bounce times $t_{pb} = 245$ ms, 415 ms, and 1000 ms. Note that the radial scales of the figures differ. The neutron star is at the origin of the axially symmetric (2D) grid and plays the role of an isotropic neutrino "light bulb" [42]

It must be stressed, however, that current observations do not necessitate such conclusions and hydrodynamical simulations suggest other possible explanations. Strong convection in the neutrino-heating region behind the supernova shock can account for huge anisotropies of the inner supernova ejecta, even without invoking rotation. If the explosion occurs quickly, much power remains on smaller scales until the expansion sets in and the convective pattern gets frozen in. If, in contrast, the shock radius grows only very slowly and the explosion is delayed for several 100 ms after bounce, the convective flow can merge to increasingly larger structures. In two-dimensional (2D) hydrodynamic calculations including cooling and heating by neutrinos between the neutron star and the shock (with parameter choices for a central, *isotropic* neutrino "light bulb" which enabled explosions), Plewa et al. [42] found situations where the convective pattern revealed a contribution of the $l = 1$, $m = 0$ mode that was growing with time and was even dominant at about one second after bounce (Fig. 1). Herant [17] already speculated about such a possibility. Certainly three-dimensional (3D) calculations of the full sphere (and without the coordinate singularity on the axis of the spherical grid) are indispensable to convincingly demonstrate the existence of this phenomenon[1].

3   Do Neutrino-Driven Explosions Work?

Spherically symmetric simulations with the current input physics (neutrino interactions and the equation of state of dense matter) do not yield explosions by the neutrino-heating mechanism. There is no controversy about that. All computations are in agreement, independent of Newtonian or relativistic gravity and independent of the neutrino transport being treated in an approximate way by flux-limited diffusion methods (e.g., [38,37,5,6]) or very accurately by solving the frequency- and angle-dependent Boltzmann transport equation [43,35,31,30].

Whether neutrinos succeed in reviving the stalled shock depends on the efficiency of the energy transfer to the postshock layer, which in turn increases with the neutrino luminosity and the hardness of the neutrino spectrum. Wilson and collaborators [55,56,32,48] have obtained explosions in one-dimensional (1D) simulations for more than ten years now. In these models it is, however, *assumed* that neutron-finger convection in the hot neutron star boosts the neutrino luminosities. Moreover, Mayle et al. [32] used a special equation of state with a high abundance of pions in the nuclear matter, which again leads to higher neutrino fluxes from the neutron star and thus to enhanced energy-deposition behind the shock. Both assumptions are not generally accepted.

Two-dimensional [18,19,46,10,23,36,47] and 3D simulations [45,13] have shown that the neutrino-heating layer is unstable to convective overturn. The associated effects have a very helpful influence and can lead to explosions even in cases where spherical models fail. In the multi-dimensional situation downflows of cooler, low-entropy matter that has fallen through the shock, coexist with rising bubbles of high-entropy, neutrino-heated gas. On the one hand, the downflows take cool material close to the gain radius where it absorbs energy readily from the intense neutrino fluxes. On the other hand, the rising bubbles

---

[1] Another interesting possibility was pointed out by A. Mezzacappa at this conference. He showed results of 2D and 3D calculations performed in collaboration with J.M. Blondin [3], which revealed hydrodynamical instabilities growing to large-scale modes in the flow behind the accretion shock even in the absence of neutrino heating.

allow heated matter to expand and cool quickly, thus reducing the energy loss by the re-emission of neutrinos. They also increase the postshock pressure and hence push the shock farther out. This in turn enlarges the gain layer and thus the gas mass which can accumulate in the neutrino-heating region. It also means that the gas stays longer in the gain layer, in contrast to one-dimensional models where the matter behind the accretion shock has negative velocity and is quickly advected down to the cooling layer. When the gas arrives there, neutrino emission sets in and extracts again the energy which had been absorbed from neutrino heating shortly before. Due to the combination of all these effects postshock convection enhances the efficiency of the neutrino-heating mechanism. Therefore the multi-dimensional situation is *generically different* from the spherically symmetric case.

Nevertheless, the existence of convective overturn in the neutrino-heating layer does not guarantee explosions [23,36]. For insufficient neutrino heating the threshold to an explosion will not be overcome. Since neutrinos play a crucial role, an accurate description of the neutrino physics – transport and neutrino-matter interactions – is indispensable to obtain conclusive results about the viability of the neutrino-driven mechanism. All published multi-dimensional explosion models, however, have employed crude approximations or simplifications in the treatment of neutrinos.

4   A New Generation of 2D Supernova Simulations

In order to take a next step of improvement in supernova modelling, we have coupled a new Boltzmann code for the neutrino transport to the PROMETHEUS hydrodynamics program, which allows for spherically symmetric as well as multi-dimensional simulations [44]. Below we present some results of our first 2D supernova simulations with this new code, which we named MuDBaTH (**Mu**lti-**D**imensional **B**oltzmann **T**ransport and **H**ydrodynamics).

4.1   Technical Aspects and Input Physics

The Boltzmann solver scheme is described in much detail in Ref. [44]. The integro-differential character of the Boltzmann equation is tamed by applying a variable Eddington factor closure to the neutrino energy and momentum equations (and the simultaneously integrated first and second order moment equations for neutrino number). For this purpose the variable Eddington factor is determined from the solution of the Boltzmann equation, and the system of Boltzmann equation and its moment equations is iterated until convergence is achieved. Employing this scheme in multi-dimensional simulations in spherical coordinates, we solve the (one-dimensional) moment equations on the different angular bins of the numerical grid but calculate the variable Eddington factor only once on an angularly averaged stellar background. We point out here that it turned out to be necessary to go an important step beyond this simple "ray-by-ray" approach. Physical constraints, namely the conservation of lepton number

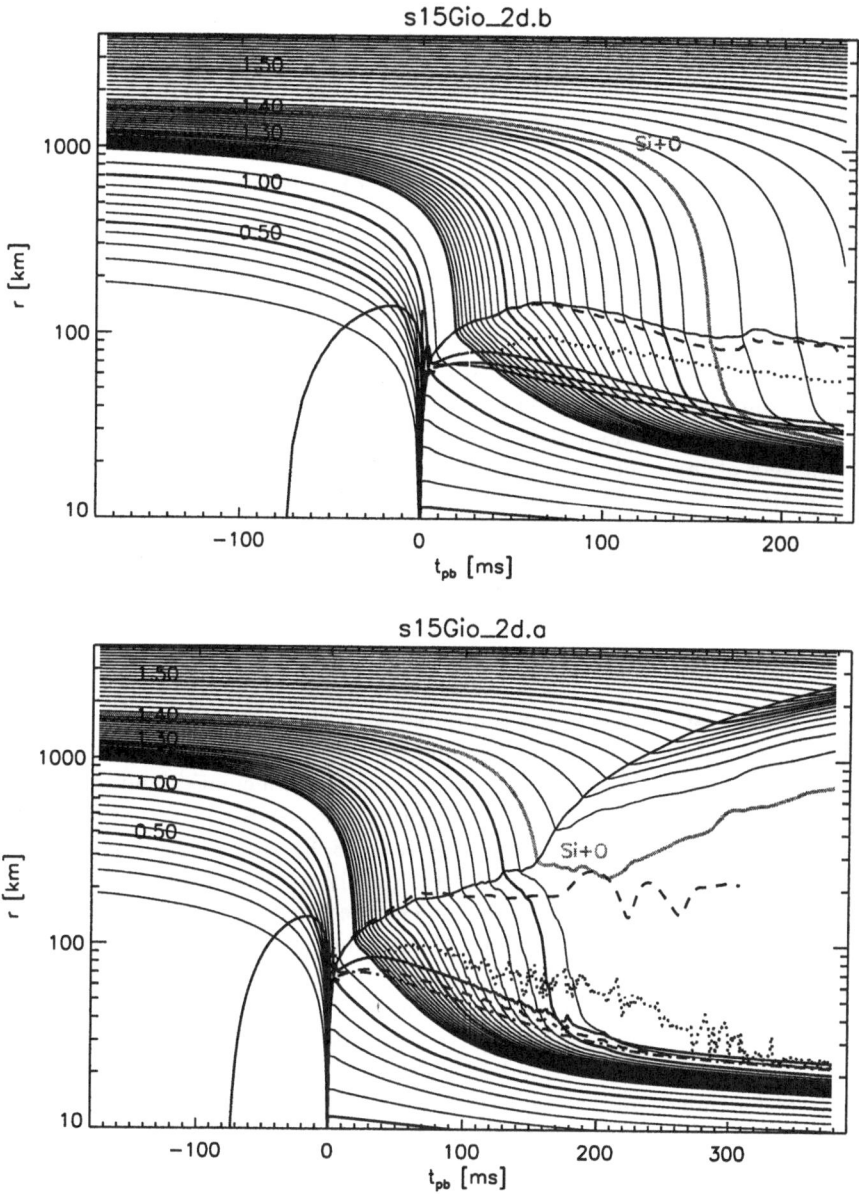

Fig. 2. Trajectories of mass shells (time being normalized to bounce) for the non-exploding (top) and the exploding 2D model. In the latter case one can see the shock starting a rapid expansion at about 150 ms after bounce. The dashed lines indicate the shock positions in the corresponding 1D simulations, where no explosions were obtained. The angle-averaged gain radius is given by the dotted line, and the neutrino-spheres of ν_e, $\bar{\nu}_e$ and the heavy-lepton neutrinos are also marked

and entropy within adiabatically moving fluid elements, and numerical requirements, i.e., the stability of regions which should not develop convection according to a mechanical stability analysis, make it necessary to take into account the
coupling of neighbouring rays at least by lateral advection terms and neutrino
pressure gradients [7].

General relativistic effects are treated only approximately in our code [44].
The current version contains a modification of the gravitational potential by
including correction terms due to pressure and energy of the stellar medium
and neutrinos, which are deduced from a comparison of the Newtonian and
relativistic equations of motion. The neutrino transport contains gravitational
redshift and time dilation, but ignores the distinction between coordinate radius
and proper radius. This simplification is necessary for coupling the transport
code to our basically Newtonian hydrodynamics.

As for the neutrino-matter interactions, we discriminate between two different sets of input physics. On the one hand we have calculated models with
conventional ("standard") neutrino opacities, i.e., a description of the neutrino
interactions which follows closely the one used by Bruenn and Mezzacappa and
collaborators [4,33,34]. It assumes nucleons to be uncorrelated, infinitely massive,
scattering targets for neutrinos. In these reference runs we have usually also added neutrino pair creation and annihilation by nucleon-nucleon bremsstrahlung
[15]. Details of our implementation of these neutrino processes can be found
in [44].

A second set of models was computed with an improved description of neutrino-matter interactions. Besides including nucleon thermal motions and recoil,
which means a detailed treatment of the reaction kinematics and allows for an
accurate evaluation of nucleon phase-space blocking effects, we take into account
nucleon-nucleon correlations (following Refs. [8,9]), the reduction of the nucleon
effective mass, and the possible quenching of the axial-vector coupling in nuclear
matter [11]. In addition, we have implemented weak-magnetism corrections as
described in Ref. [22]. The sample of neutrino processes was enlarged by also
including scatterings of muon and tau neutrinos and antineutrinos off electron
neutrinos and antineutrinos and pair annihilation reactions between neutrinos
of different flavors (i.e., $\nu_{\mu,\tau} + \bar{\nu}_{\mu,\tau} \longleftrightarrow \nu_e + \bar{\nu}_e$; [16]).

Our current supernova models are calculated with the nuclear equation of
state of Lattimer and Swesty [27], which we suitably extended to lower densities [44].

4.2   Models and Results

None of our spherically symmetric simulations, neither with the standard nor
with the improved description of neutrino opacities, has produced an explosion.
A compilation of a subset of our calculations, which we did for a 15 $M_\odot$ progenitor
star, Model s15s7b2, provided to us by S. Woosley, can be found in Refs. [24,7].
Here we discuss only two 2D runs (Models s15Gio_2d.a and s15Gio_2d.b), which
both were performed with our approximation of relativistic effects and the state-
of-the-art improvement of neutrino-matter interactions (cf. Sect. 4.1). We used a

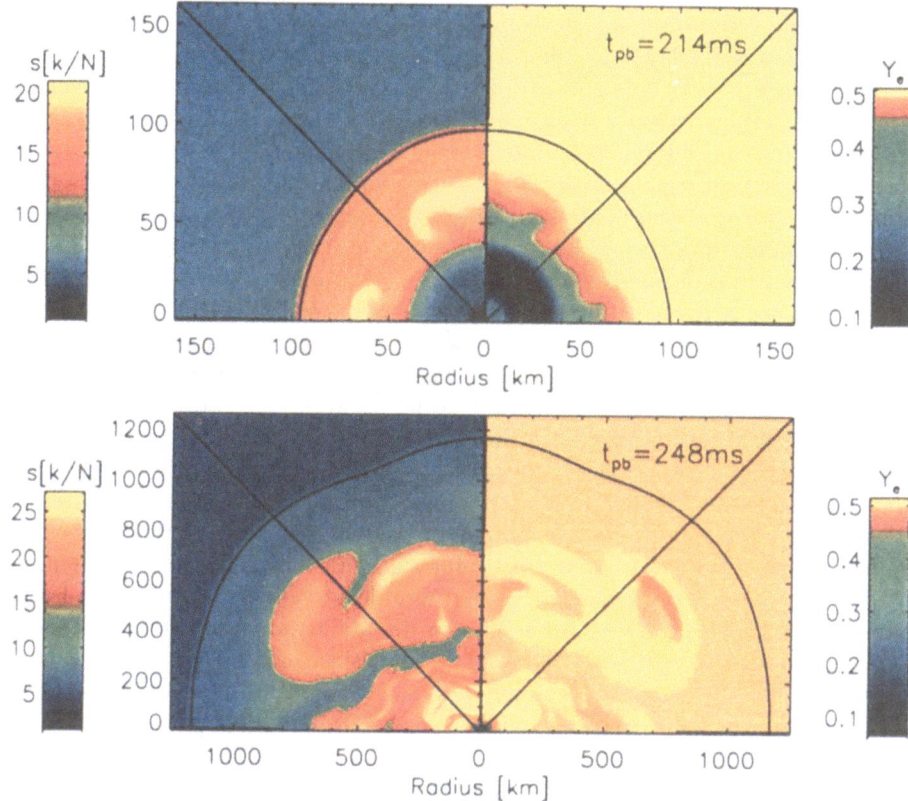

Fig. 3. Convection in the neutrino-heating region for the non-exploding 2D model (Model s15Gio_2d.b, top) and the exploding one (Model s15Gio_2d.a) at the post-bounce times indicated in the plots. The figures show the entropy distribution (left) and the electron fraction (proton-to-baryon ratio). A wedge of ±43.2° around the equatorial plane (marked by the diagonal solid lines) of the spherical coordinate grid was used for the computations

spherical coordinate grid with 32 equidistant zones within an angular wedge from −43.2° to +43.2° around the equatorial plane and assumed periodic conditions at the boundaries.

Both 2D simulations differ only in one important aspect: In Model s15Gio_2d.a the velocity dependent (Doppler shift and aberration) terms in the neutrino momentum equation (and the corresponding terms in the Boltzmann equation for the antisymmetric average of the specific intensity; see Ref. [44]) were omitted. These terms are formally of order v/c and are small for low velocities.

This simplification of the neutrino transport, however, has a remarkable consequence: The model with the most complete implementation of the transport equations, Model s15Gio_2d.b, fails to explode. In case of Model s15Gio_2d.a, however, the stalled shock is successfully revived by neutrino heating because very strong convection can develop in the gain region[2]. The time evolution of both models is displayed by the trajectories of mass shells in Fig. 2.

The reason for this dramatic difference is the following. Some of the velocity dependent terms (those in which derivatives with respect to the neutrino energy do not show up) in the neutrino momentum equation have a simple formal interpretation: In regions with mass infall (negative velocity) they effectively act like a reduction of the neutrino-medium interaction on the right hand side of this equation. The changes can be 10% or more for neutrino energies in the peak of the spectrum, depending also on time, radius, and the size of the postshock velocities. As a consequence, the neutrino flux streams more readily and the comoving-frame neutrino (energy) density is decreased. This is associated with somewhat larger neutrino losses in the cooling layer around the neutrinosphere and a significantly reduced neutrino heating between gain radius and shock.

Although the differences are moderate (10–30%, depending on the quantity) the accumulating effects during the first 80 ms after bounce clearly damp the shock expansion and finally lead to a dramatic shock recession after the initial phase of expansion. Before this happens postshock convection has not become strong enough to change the evolution. With the onset of contraction, the postshock velocities decrease (become more negative) quickly, neutrino-heated matter is rapidly advected inward below the gain radius and loses its energy by re-emission of neutrinos. The gain region shrinks to a very narrow layer, a fact which suppresses the convective activity later on. This is demonstrated by Fig. 3 where convection is weak in Model s15Gio_2d.b but very strong in Model s15Gio_2d.a.

Due to a combination of unfavorable effects and a continuously amplifying negative trend, Model s15Gio_2d.b remains below the explosion threshold while Model s15Gio_2d.a is just above that critical limit. In the vicinity of the threshold the long-time evolution of the collapsing stellar core depends very sensitively on "smaller details" of the neutrino transport.

5   Conclusions and Outlook

Our 2D models with a Boltzmann solver for the neutrino transport have considerably reduced the uncertainties associated with the treatment of the neutrino physics in previous multi-dimensional simulations. With the most complete implementation of the transport physics we could not obtain explosions. This result suggests that the neutrino-driven mechanism fails with the employed input physics, at least in case of the considered 15 $M_\odot$ star. We do not think that the remaining uncertainties in our simulations (mainly the approximate treatment of general relativistic effects) are likely to jeopardize this conclusion. A comparison with fully relativistic one-dimensional calculations (Liebendörfer, personal

---

[2] At the time of the conference, we had just this exploding 2D run and made a preliminary announcement of the success of this model. A later 2D computation with the full neutrino moment equations (Model s15Gio_2d.b) then turned out to produce a dud.

communication [29]) is very encouraging. Because of the remarkable similarity of the shock trajectories of different progenitors in spherical symmetry [30], it is likely that our negative conclusion is also valid for other pre-collapse configurations with a similar structure. Significant star-to-star variations of the progenitor properties with a non-monotonic dependence on the stellar mass [57], however, suggest that multi-dimensional core-collapse simulations of a larger sample of progenitors are needed before one can make final, more generally valid statements. The supernova problem is highly nonlinear and surprises may lurk behind every corner.

It would therefore be premature to conclude that the neutrino-driven mechanism fails and that not even postshock convection can alter this unquestioned outcome of all current spherical models. Besides studying other progenitors with multi-dimensional simulations, one should also investigate the effects of rotation and the influence of different high-density equations of state on the long-time post-bounce evolution and the neutrino-heating phase in a supernova. There is considerable uncertainty associated with the poorly known physics in the nuclear and supranuclear medium.

Our successfully exploding 2D model, Model s15Gio_2d.a, at least demonstrates that simulations which include the effects of postshock convection are rather close to an explosion. Therefore modest changes of the neutrino emission and transport seem to be already sufficient to push them beyond the critical threshold. The properties of the explosion in this case are very encouraging and may support one's belief in the basic viability of the delayed explosion mechanism. At 380 ms after bounce the shock has arrived at a radius of more than 2500 km and is expanding with about 10000 km/s. The explosion of this model does not seem to become very energetic. It is only $\sim 4 \times 10^{50}$ erg at that time, but still increasing. This may not be a serious problem if one recalls the large spread of energies of observed supernovae (Supernova 1999br, for example, is estimated to have an ejecta mass of $14\,M_\odot$ and an explosion energy of about 6×10^{50} erg [14]).

Since the explosion starts rather late (at ~ 150 ms post bounce), the proto-neutron star has accreted enough matter to have attained an initial baryonic mass of $1.4\,M_\odot$. Therefore our simulation does not exhibit the problem of previous successful multi-dimensional calculations which produced neutron stars with masses on the lower side of plausible values ($\sim 1.1\,M_\odot$). Also another problem of published explosion models (e.g., [19,10,23,12]) has disappeared: The ejecta mass with $Y_e \lesssim 0.47$ is less than $10^{-4}\,M_\odot$, thus fulfilling a constraint pointed out by Hoffman et. al. [21] for supernovae if they should not overproduce the $N = 50$ (closed neutron shell) nuclei, in particular ^{88}Sr, ^{89}Y and ^{90}Zr, relative to the Galactic abundances. Of course, final statements about explosion energy, ejecta composition, and the neutron star mass (which may grow by later fallback, especially when the explosion energy remains low) require to follow the explosion for a longer time.

Acknowledgements

We are grateful to K. Takahashi for providing routines to calculate the improved neutrino-nucleon interactions, and to C. Horowitz for correction formulae for the weak magnetism. We also thank M. Liebendörfer for making output data of his simulations available to us for comparisons. The Institute for Nuclear Theory at the University of Washington is acknowledged for its hospitality and the Department of Energy for support during a visit of the Summer Program on Neutron Stars, during which most of the work leading to Fig. 1 was done. HTJ, RB and MR are grateful for support by the Sonderforschungsbereich 375 on "Astroparticle Physics" of the Deutsche Forschungsgemeinschaft. TP was supported in part by the US Department of Energy under Grant No. B341495 to the Center of Astrophysical Thermonuclear Flashes at the University of Chicago, and in part by the grant 2.P03D.014.19 from the Polish Committee for Scientific Research. He performed his simulations on the CRAY SV1-1A at the Interdisciplinary Centre for Computational Modelling in Warsaw. The 2D simulations with Boltzmann neutrino transport were only possible because a node of the new IBM "Regatta" supercomputer was dedicated to this project by the Rechenzentrum Garching. Computations were also done on the NEC SX-5/3C of the Rechenzentrum Garching, and on the CRAY T90 and CRAY SV1ex of the John von Neumann Institute for Computing (NIC) in Jülich.

References

1. S. Akiyama, J.C. Wheeler, D.L. Meier, I. Lichtenstadt: Astrophys. J., in press (2002) (astro-ph/0208128)
2. H.A. Bethe, J.R. Wilson: Astrophys. J. **295**, 14 (1985)
3. J.M. Blondin, A. Mezzacappa, C. DeMarino: Preprint (2002)
4. S.W. Bruenn: Astrophys. J. Suppl. **58**, 771 (1985)
5. S.W. Bruenn: 'Numerical Simulations of Core Collapse Supernovae'. In: *Nuclear Physics in the Universe*, ed. by M.W. Guidry, M.R. Strayer (IOP, Bristol 1993) pp. 31–50
6. S.W. Bruenn, K.R. De Nisco, A. Mezzacappa: Astrophys. J. **560**, 326 (2001)
7. R. Buras, M. Rampp et al.: in preparation (2002)
8. A. Burrows, R.F. Sawyer: Phys. Rev. C. **58**, 554 (1998)
9. A. Burrows, R.F. Sawyer: Phys. Rev. C. **59**, 510 (1999)
10. A. Burrows, J. Hayes, B.A. Fryxell: Astrophys. J. **450**, 830 (1995)
11. G.W. Carter, M. Prakash: Physics Letters B **525**, 249 (2002)
12. C.L. Fryer: Astrophys. J. **522**, 413 (1999)
13. C.L. Fryer, M. S. Warren: Astrophys. J. Lett. **574**, L65 (2002)
14. M. Hamuy: Astrophys. J., in press (2002) (astro-ph/0209174)
15. S. Hannestad, G. Raffelt: Astrophys. J. **507**, 339 (1998)
16. R. Buras, H.-Th. Janka, M.-Th. Keil, G. Raffelt, M. Rampp: Astrophys. J., submitted (2002) (astro-ph/0205006)
17. M. Herant: Physics Rep. **256**, 117 (1995)
18. M. Herant, W. Benz, S.A. Colgate: Astrophys. J. **395**, 642 (1992)
19. M. Herant, W. Benz, W.R. Hix, C.L. Fryer, S.A. Colgate: Astroph. J. **435**, 339 (1994)

20. P. Höflich, J.C. Wheeler, L. Wang: Astrophys. J. **521**, 179 (1999)
21. R.D. Hoffman, S.E. Woosley, G.M. Fuller, B.S. Meyer: Astrophys. J. **460**, 478 (1996)
22. C.J. Horowitz: Phys. Rev. D **65**, 043001-1 (2002)
23. H.-Th. Janka, E. Müller: Astron. Astrophys. **306**, 167 (1996)
24. H.-Th. Janka, R. Buras, M. Rampp: 'The Mechanism of Core-Collapse Supernovae and the Ejection of Heavy Elements'. In: *Proceedings of the 7th Int. Symposium on Nuclei in the Cosmos, Fuji-Yoshida, Yamanashi, Japan, 8-12 July, 2002*, Nuclear Physics A, in press (2002)
25. H.-Th. Janka, R. Buras, K. Kifonidis, T. Plewa, M. Rampp: 'Explosion Mechanisms of Massive Stars'. In: *Core Collapse of Massive Stars*, ed. C.L. Fryer (Kluwer, Dordrecht) in preparation
26. A.M. Khokhlov, P. Höflich, E.S. Oran, J.C. Wheeler, L. Wang, A.Yu. Chtchelkanova: Astrophys. J. Lett. **524**, L107 (1999)
27. J.M. Lattimer, F.D. Swesty: Nucl. Phys. A **535**, 331 (1991)
28. D.C. Leonhard, A.V. Filippenko, D.R. Ardila, M.S. Brotherton: Astrophys. J. **553**, 86 (2001)
29. M. Liebendörfer et al.: in preparation (2002)
30. M. Liebendörfer, O.E.B. Messer, A. Mezzacappa, W.R. Hix, F.-K. Thielemann, K. Langanke: 'The Importance of Neutrino Opacities for the Accretion in Spherically Symmetric Supernova Models'. In: *Proc. 11th Workshop on Nuclear Astrophysics, Ringberg, Feb. 11-16, 2002*, Report MPA/P13, ed. by W. Hillebrandt, E. Müller (MPI für Astrophysik, Garching 2002) pp. 126–131
31. M. Liebendörfer, A. Mezzacappa, F. Thielemann, O.E. Messer, W.R. Hix, S.W. Bruenn: Phys. Rev. D. **63**, 3004 (2001)
32. R.W. Mayle, M. Tavani, J.R. Wilson: Astroph. J. **418**, 398 (1993)
33. A. Mezzacappa, S.W. Bruenn: Astrophys. J. **405**, 637 (1993)
34. A. Mezzacappa, S.W. Bruenn: Astrophys. J. **410**, 740 (1993)
35. A. Mezzacappa, M. Liebendörfer, O.E.B. Messer, W.R. Hix, F.-K. Thielemann, S.W. Bruenn: Phys. Rev. Lett. **86**, 1935 (2001)
36. A. Mezzacappa, A.C. Calder, S.W. Bruenn, J.M. Blondin, M.W. Guidry, M.R. Strayer, A.S. Umar: Astroph. J. **495**, 911 (1998)
37. E.S. Myra, S.A. Bludman: Astrophys. J. **340**, 384 (1989)
38. E.S. Myra, S.A. Bludman, Y. Hoffman, I. Lichenstadt, N. Sack, K.A. van Riper: Astrophys. J. **318**, 744 (1987)
39. D.K. Nadyozhin, Astron. Astrophys., submitted (2002) (Preprint MPA 1458)
40. K. Nomoto, T. Shigeyama, S. Kumagai, H. Yamaoka, T. Suzuki: 'Supernova 1987A: From Progenitor to Remnant'. In: *Supernovae, Les Houches Session LIV, July 31–Sept. 1, 1990*, ed. S.A. Bludman, R. Mochkovitch, J. Zinn-Justin (Elsevier/North-Holland, Amsterdam 1994) pp. 489–568
41. K. Nomoto, K. Maeda, H. Umeda, T. Ohkubo, J. Deng, P. Mazzali: 'Hypernovae and their Nucleosynthesis'. In: *A Massive Star Odyssey, from Main Sequence to Supernova, Proc. IAU Symposium 212*, ed. K.A. van der Hucht, A. Herrero, C. Esteban (ASP, San Francisco) in press (astro-ph/0209064)
42. T. Plewa, K. Kifonidis, H.-Th. Janka: in preparation (2002)
43. M. Rampp, H.-Th. Janka: Astrophys. J. Lett. **539**, L33 (2000)
44. M. Rampp, H.-Th. Janka: Astron. Astrophys., **396**, 361 (2002)
45. T. Shimizu, S. Yamada, K. Sato: Publ. Astron. Soc. Japan **45**, L53 (1993)
46. T. Shimizu, S. Yamada, K. Sato: Astrophys. J. Lett. **432**, L119 (1994)
47. T.M. Shimizu, T. Ebisuzaki, K. Sato, S. Yamada, Astrophys. J. **552**, 756 (2001)

48. T. Totani, K. Sato, H.E. Dalhed, J.R. Wilson: Astroph. J. **496**, 216 (1998)
49. L. Wang, D.A. Howell, P. Höflich, J.C. Wheeler: Astrophys. J. **550**, 1030 (2001)
50. L. Wang, J.C. Wheeler et al.: Astrophys. J. **579**, 671 (2002)
51. J.C. Wheeler: AAPT/AJP Resource Letter, American J. of Physics, in press (2002) (astro-ph/0209514)
52. J.C. Wheeler, D.L. Meier, J.R. Wilson: Astrophys. J. **568**, 807 (2002)
53. J.C. Wheeler, I. Yi, P. Höflich, L. Wang: Astrophys. J. **537**, 810 (2000)
54. J.R. Wilson: 'Supernovae and Post-Collapse Behavior'. In: *Numerical Astrophysics, Proc. Symposium in Honor of J.R. Wilson, Illinois, Oct. 1982*, ed. by J.M. Centrella, J.M. LeBlanc, R.L. Bowers, J.A. Wheeler (Jones and Bartlett, Boston 1985) pp. 422–434
55. J.R. Wilson, R. Mayle: Phys. Rep. **163**, 63 (1988)
56. J.R. Wilson, R. Mayle: Phys. Rep. **227**, 97 (1993)
57. S.E. Woosley, A. Heger, T.A. Weaver: Reviews of Modern Physics, submitted (2002)

The Mechanism of Core-Collapse
Supernova Explosions: A Status Report

Adam Burrows[1] and Todd A. Thompson[2,3]

[1] The University of Arizona, Tucson, AZ 85721, USA
[2] The University of California, Berkeley, CA 94720, USA
[3] Hubble Fellow

1 Introduction

Most massive stars ($8\ M_\odot \rightarrow 80\ M_\odot$) must transition at the ends of their lives into neutron stars or stellar-mass black holes. That they do so when their low-entropy cores reach the Chandrasekhar mass, gravitationally collapse, and launch a supernova explosion has been demonstrated both by direct observations (cf., the neutrinos from SN1987A) and by a host of compelling theoretical arguments. However, numerical simulations of the process of core collapse, bounce near nuclear densities, shock wave generation, and shock propagation have failed to recreate in detail the observed gravitational masses of known neutron stars, the expected nucleosynthesis pattern, and empirical supernova energies. All semi-realistic, one-dimensional (spherically-symmetric) simulations conducted to date fizzle into quasi-static accreting proto-black holes. What is more, they seem to do so convincingly, despite concerted attempts over the years to include all the known neutrino and nuclear physics or general relativity [3,20,4,5,18,16,17,22]. It is now fairly clear that the devil is not "in the details" and that 1D models don't explode. Something large and major, not something at the "10-20% level," would have to be missing to alter this conclusion.

Two-dimensional simulations conducted in the nineties [14,7,15,12] and more recent three-dimensional SPH simulations [13] do explode (when the corresponding simulations in one dimension do not), but these multi-D numerical experiments do not produce supernova explosions that satisfy all of the above observational constraints. Furthermore, they all employ some realization of a simple flux-limited, energy-integrated diffusion algorithm, in lieu of full neutrino transport (which is quite difficult in 2- or 3-D). It is thought by many that the apparent marginality of success demands that the transport be handled with a multi-group, multi-angle technique (and that the hydrodynamics be handled in full general relativity) before the multi-D explosion simulations are to be considered valid, or even indicative of the mechanism. Perhaps. However, this puts too much reliance on hardware and software, and not enough on physical understanding.

So, what is wrong? And how to fix it? In order to see where we are going we must first see where we have been.

2   One-Dimensional Simulations:
The Spherically-Symmetric Paradigm

The primary elements of the neutrino-driven mechanism were mapped out in the work of Colgate and White [11] and Arnett [1]. In the former, a cold (low-entropy), already deleptonized core was collapsed and after bounce at nuclear densities a neutrino luminosity which heated the outer mantle was turned on for 14 milliseconds. The neutrino transport was artificial, but it drove an explosion with 5×10^{52} ergs. In the latter, a higher-entropy core was collapsed, which bounced due to thermal nucleon pressure at a central density near a tenth of nuclear, and neutrino transport was handled more realistically with an energy-integrated diffusion approach. Neutrino diffusion was found to transport energy from the hot interior to the mantle on close to the dynamical timescale of the newly-generated shock wave. In Arnett's calculations, this kept the shock wave from stalling, an explosion ensued, and the total duration of the neutrino luminosity was less than 100 milliseconds, but longer than in Colgate and White. In both cases, neutrinos played the central role in launching the explosion and a neutron star remained, but the initial progenitor models, neutrino transport, neutrino physics, and equations of state were primitive by today's standards. The timescale for the cooling phase (detected for SN1987A) was off by a factor of 100-1000, the spectra were much harder than current simulations indicate, and "ν_μ" neutrinos were not included. Nevertheless, the core idea of the neutrino-driven mechanism, that neutrinos generated in the inner core and deposited in the outer mantle drive the explosion, was implicit in both papers.

The essential ingredient of the neutrino-driven mechanism is the transfer of gravitational binding energy from the core to the mantle of the newly-formed protoneutron star through the mediation of neutrinos liberated at the high temperatures and densities generated in its interior. An *efficient* coupling is required, that, if not achieved, will lead to a fizzle. With up-to-date nuclear equations of state, neutrino physics, progenitor models, and transport algorithms, the coupling efficiency in one-dimensional simulations has been determined to be inadequate both to forestall the stagnation of the bounce shock and to reignite it once it has stalled. Figure 1 depicts radius trajectories of mass zones versus time for one of our 1D multi-group, Feautrier/tangent-ray calculations (§2.2). The progress of the shock and its stagnation during the first 250 milliseconds after bounce are manifest. Wilson [26] and Bethe and Wilson [2] suggested that shock stagnation is only temporary, leading to the "delayed" scenario, but without the neutron-finger convective boost (an inherently multi-D effect which others have failed to reproduce) in the driving neutrino luminosity Wilson himself did not find explosions. In sum, embellishments with general relativity, the employment of multi-group, multi-angle Boltzmann solvers, and refinements in neutrino-matter interactions have not changed the conclusion that in spherical symmetry the coupling efficiency of the emergent neutrinos to the protoneutron star mantle is too small to lead to explosion, even after a pause [3,20,4,5,18,16,17,22]. Unfortunately, this conclusion seems to be independent of progenitor model.

2.1   Supernova Energetics Made Simple

It is important to note that one is not obliged to unbind the inner core ($\sim$10 kilometers) as well; the explosion is a phenomenon of the outer mantle at ten times the radius (50-200 kilometers). One consequence of this goes to the heart of a general confusion concerning supernova physics. Though the binding energy of a cold neutron star is $\sim$3 $\times$ 10^{53} ergs and the supernova explosion energy is near 10^{51} ergs, a comparison of these two numbers and the large ratio that results are not very relevant. More germane are the binding energy of the mantle (interior to the shock or, perhaps, exterior to the neutrinospheres) and the neutrino energy radiated during the delayed phase. These are both at most a few$\times$10^{52} ergs, not $\sim$3 $\times$ 10^{53} ergs, and the relevant ratio that illuminates the neutrino-driven supernova phenomenon is $\sim$10^{51} ergs divided by a few$\times$10^{52} ergs. This is $\sim$5-10%, not the oft-quoted 1%, a number which tends to overemphasize the sensitivity of the neutrino mechanism to neutrino and numerical details.

Furthermore, there is general confusion concerning what determines the supernova explosion energy. While a detailed understanding of the supernova mechanism is required to answer this question, one can still proffer a few observations. First is the simple discussion above. Five to ten percent of the neutrino energy coursing through the semi-transparent region is required, not one percent. Importantly, the optical depth to neutrino absorption in the gain region is of order $\sim$0.1. The product of the sum of the ν_e and $\bar{\nu}_e$ neutrino energy emissions in

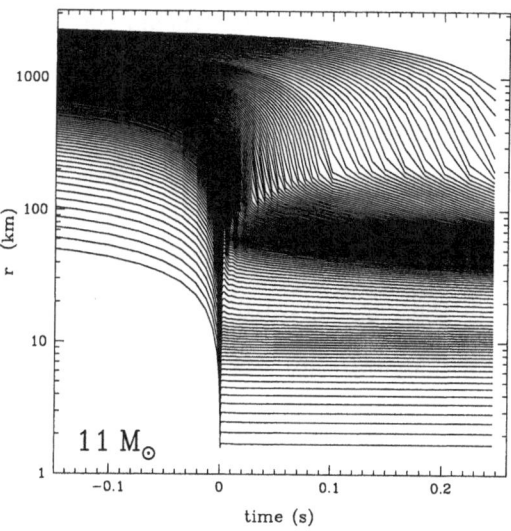

Fig. 1. Radius trajectories of selected mass zones in our 300-zone, 20 energy-group calculation of stellar collapse, bounce, and shock propagation for the Woosley and Weaver 11 M$_\odot$ progenitor [27]. In this simulation, as in all other 1D simulations, the shock does not revive

the first 100's of milliseconds and this optical depth gives a number near 10^{51} ergs. Furthermore, the binding energy of the progenitor mantle exterior to the iron core is of order a few$\times 10^{50}$ to a few$\times 10^{51}$ ergs and it is very approximately this binding energy, not that of a cold neutron star, that is relevant in setting the scale of the core-collapse supernova explosion energy. Given the power-law nature of the progenitor envelope structure, it is clear that this binding energy is related to the binding energy of the pre-collapse iron core (note that they both have a boundary given by the same GM/R), which at collapse is that of the Chandrasekhar core. The binding energy of the Chandrasekhar core is easily shown to be zero, modulo the rest mass of the electron times the number of baryons in a ~ 1.4 M$_\odot$ Chandrasekhar mass. (The Chandrasekhar mass/instability is tied to the onset of relativity for the electrons, itself contingent upon the electron rest mass). The result is $\sim 10^{51}$ ergs.

The core-collapse explosion energy is near the explosion energy for a Type Ia supernova because in a thermonuclear explosion the total energy yield is approximately the 0.5 MeV/baryon derived from carbon/oxygen burning to iron times the number of baryons burned in the explosion. The latter is $\geq$half the number of baryons in a Chandrasekhar mass. The result is $\sim 10^{51}$ ergs. This is the same number as for core-collapse supernovae because 1) in both cases we are dealing with the Chandrasekhar mass (corrected for electron captures, entropy, general relativity, and Coulomb effects) and 2) the electron mass and the per-baryon thermonuclear yield are each about 0.5 MeV.

While more detailed calculations are clearly necessary to do this correctly, the essential elements of supernova energetics are not terribly esoteric (if neutrino-driven), at least to within a factor of 5, and should not be viewed as such.

2.2   Some Recent Results
Using Our Feautrier/Tangent-Ray/ALI Method

One is firm in the general conclusion that 1D models with good physics and numerics do not explode because, coming from many different angles, various groups have now verified this. We here briefly describe our contribution to this activity.

It is thought that the angular distribution of neutrinos in the gain region needs to be calculated with precision, since neutrino energy deposition is proportional to neutrino energy density and, for given luminosities/fluxes, this is sensitive to ν_e and $\bar{\nu}_e$ angular distributions that can not be derived using flux-limited diffusion. Furthermore, the stiff dependence of the absorption cross sections on neutrino energy requires multi-group approaches. Hence, multi-group, multi-angle Boltzmann transport algorithms are to be preferred.

We have constructed such a code, using the Feautrier variables advocated in standard stellar atmospheres work [19] and the tangent-ray method to establish a dense angular grid [8]. Our transport solver calculates in the comoving frame, uses accelerated $\varLambda$ iteration (ALI) [9,10,24,21] to speed convergence of the solution, is implicit in time, second-order accurate in space, and iterates between the Boltzmann/transfer equation and the zeroth- and first-moment equations (in a method akin to the variable Eddington factor approach) until a converged global solution to the full transport equation is achieved. No ad hoc flux limiters or artificial closures are necessary. This iteration scheme is fast (convergence to a part in 10^6 in 2 to 10 steps) and automatically conserves energy in the transport sector. The Feautrier scheme can transition to the diffusion limit seamlessly and accurately and the tangent-ray method automatically adapts with the hydrodynamic grid as it moves. In constructing the tangent rays, we cast them from every outer zone to every inner zone. Hence, if there are 200 radial zones, the outer zone has 199 angular groups. Because of the spherical nature of the core-collapse problem and the need to accurately reproduce the angular distribution of the radiation field that transitions from the opaque (inner) to the transparent (outer) regions, such fine angular resolution is useful, though computationally demanding, as radiation becomes more and more forward-peaked. Figure 2 depicts a snapshot of the angular distribution of the ν_e neutrinos at various energies from 1 to 320 MeV, at a radius of 42 kilometers, 40 milliseconds after bounce. The positions of the angular bins are shown as dots in the lower hemisphere. The progressively more forward-peaked distribution at lower energies (less coupled) is clearly well-resolved. The code calculates the Feautrier variables (and, hence, the specific intensity) at every radial zone, for every energy group, for each neutrino species (we follow the standard "3"), at each timestep. We employ 20-40 energy groups either logarithmically or linearly spaced from 1 MeV to either 100 MeV (for $\bar{\nu}_e$ and "ν_μ") or 320 MeV (for ν_e).

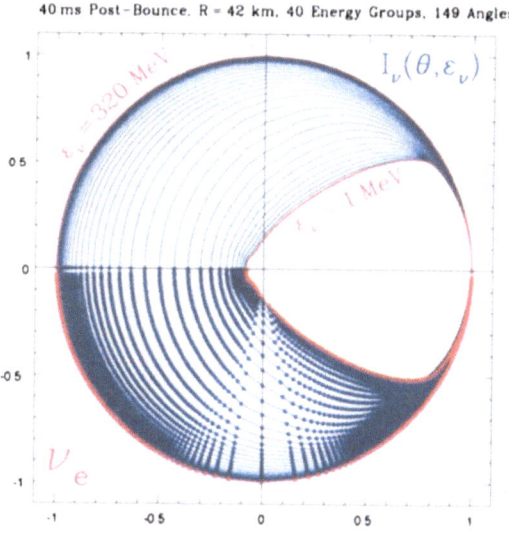

Fig. 2. A snapshot of the angular distribution of the ν_e specific intensity at a radius of 42 kilometers, 40 milliseconds after bounce, for energy groups from 1 to 320 MeV. The 11 M$_\odot$ progenitor of Woosley and Weaver [27] was used

The hydrodynamics is explicit in time and Lagrangean, uses a predictor/corrector method, employs artificial viscosity to handle shocks, and is Newtonian. We had originally used a PPM hydro scheme, but could not easily incorporate neutrino radiation pressure into the Riemann solver. We have constructed a dense equation of state table in the variables T, ρ, and Y_e that contains the 14 variables needed by the radiation hydrodynamics. We have constructed additional tables of the ν_i-electron scattering kernels and the e^+/e^- annihilation kernels [3,25]. Neutrino-electron redistribution/scattering is handled explicitly, which is a great time and memory saver. The radiation/matter couplings (both the energy and electron fraction updates) are handled implicitly and in operator-split fashion, using the ALI to facilitate convergence. Figure 3 provides snapshots we obtain of a representative evolution of the Y_e profile before and after trapping and bounce.

The minuses of our approach are that it is Newtonian, that the hydro is explicit, and that, being Lagrangean and not adaptive, the envelope must be pre-zoned densely in mass to maintain reasonable resolution of the shock at late times. Rampp and Janka [22,23] use an Eulerian grid and PPM hydro, remap between the comoving frame radiation solution and their static hydro grid, employ a variable Eddington factor/tangent-ray/Feautrier scheme as well, and operator split the neutrino-matter couplings. They cast tangent rays to a subset of interior Eulerian zones and thereby maintain a static, fixed set of angles. In addition, they have a simple method for approximately incorporating

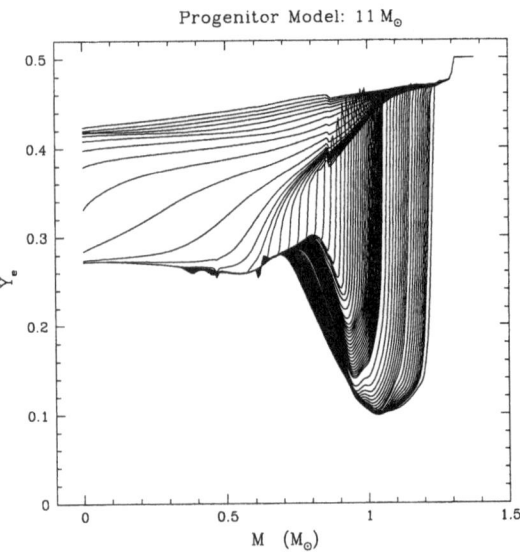

Fig. 3. A collection of snapshots of the Y_e profile versus interior mass for the Newtonian evolution of the Woosley and Weaver [27] 11 $M_\odot$ progenitor. A total of ∼250 milliseconds is depicted and the simulation was carried to ∼50 milliseconds after bounce

general relativity. There are lots of minor differences between our numerical implementations, but despite them our simulation results are quite similar [25].

Liebendörfer et al. [16,17] have assembled a very different radiation/hydro-dynamic code complex that has both strengths and weaknesses. Among their many strengths is that their code is fully implicit, adaptive (a great advantage), and general relativistic. Minor weaknesses are that they operator-split all the terms in the transport equations separately and use perforce (due to their implicit redistribution method) a small number of angles (6) in the S_n method they utilize to handle the angular distribution of the radiation field. S_n can not handle forward-peaked radiation fields, but for the core collapse problem, the gain region interior to a stalled shock (where the angular distribution is most problematic) is sufficiently compact that their S_n technique is more than adequate. Where it fails, in the outer regions, the neutrino/matter coupling is not so germane to the problem of the supernova mechanism. There are many other differences of implementation and approach, but, again, despite these differences their results for neutrino light curves, velocities, trapped electron fractions, entropies, densities, and the effect of neutrino-electron redistribution are both qualitatively and quantitatively similar to those of both ourselves [25] and Rampp and Janka [22,23].

Figure 4 portrays the evolution of the luminosity of the electron neutrinos with time up to 200 milliseconds after bounce for three progenitor models. The breakout peak, and the precursor peak just a few milliseconds earlier, are not only similar to one another, but they are very similar to the corresponding results of Rampp and Janka [22]. The peak luminosities of Liebendörfer et al. can be 20-30% higher. Due to the 20-$M_\odot$ model's thicker envelope, the subsequent accretion luminosity for this progenitor is ~60% higher than that for the two other progenitors 200 milliseconds after bounce. In none of our 1D simulations, carried out to as long as ~1 second after bounce and with the best neutrino and nuclear physics available, do the models explode. Moreover, we have incorporated, in approximate fashion, the possible effects of the neutrino oscillations, using the numbers derived from the solar and atmospheric neutrino experiments, and have seen no effect. This is not unexpected, since the "matter effect" severely suppresses oscillation between flavor neutrinos.

We are forced to conclude, on the basis of both our simulations and those of others, that the program to determine whether core-collapse supernovae explode in spherical symmetry (1D) when all the best physics and numerics are implemented has failed by a comfortable (?) margin. This leads us of necessity to multi-D effects.

3   Multi-Dimensional Simulations

The success of the 2D BHF [7] and Fryer et al. [12] simulations can be traced to the increase in the efficiency of the neutrino-matter coupling in the gain region interior to the shock due to the longer dwell time of convecting/absorbing parcels of matter. In 1D, accreted matter falls straight through the gain region.

Consequently, in 1D the increase in entropy due to neutrino energy deposition is modest. In 2- or 3-D, matter resides a bit longer in the gain region due to convection. As a result, the steady-state entropy of an average parcel of matter is higher. This translates into a larger gain region, that is on average less bound and more unstable to favorable changes in the global conditions of accretion and luminosity. In the language of Burrows and Goshy [6], the global instability condition of the protoneutron star is eased and changes in the control parameters (such as $\dot{M}$ and L_{ν_e}) can more easily result in explosion. Whether they do when better neutrino transfer and general relativity are implemented is unclear, but recently Janka and Rampp (JR, this volume) have seen a very weak explosion in a calculation using 2D hydro in a 27° wedge and their 1D Feautrier transport along rays. They obtain no explosion in 1D for the same inputs. In this 2D calculation, an average hydro solution in the wedge is used as input to a spherical calculation of the Eddington factors, and these are employed for all the individual Feautrier solutions along rays. This is better than the flux-limited solution along rays of BHF, incorporates better neutrino transport, has an approximate prescription for relativity, but seems to result in a weaker explosion. However, the JR algorithm is not yet true 2D transport and the calculations need to be done over a full 180° angular region. Nevertheless, it is clear from the JR calculation and those of BHF [7], Herant et al. [14], and Fryer et al. [12] that convection makes the protoneutron star mantle more unstable. What is not clear is whether this is enough.

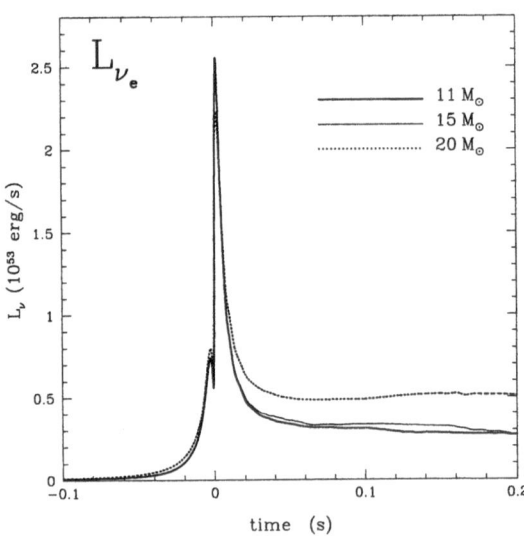

Fig. 4. The electron-neutrino luminosity versus time around bounce for three different progenitors. The primary focus of this plot is the prodigious shock breakout burst

Fryer and Warren [13] have recently demonstrated that their 3D SPH simulations explode quantitatively and qualitatively in much the same way as their 2D simulations, despite a general concern that 2D and 3D convection have different clump size spectra and cascade character. This seems not to matter much; it is the overall global behavior of the convecting region that matters and, according to them, the largest size and velocity scales. Small scales are a detail. The former are set by the size of the unstable region and are similar in both 2D and 3D. However, these calculations need to be verified with multi-group, multi-angle techniques, first in 2D (which has yet to be done), then in 3D (a major computational challenge for the future).

4   Coda

Multi-dimensional effects seem to be required to ensure or enable core-collapse supernova explosions. However, even though what the essential elements for explosion are is unclear, we believe that the solution, when found, will not be marginal. A number of groups are now embarked upon separate development paths to credible 2D (later 3D) radiation/hydro schemes (Arizona/Israel, ORNL, MPA). We, in collaboration with Eli Livne and Itamar Lichtenstadt, are developing a 2D, moving grid, discontinuous finite element (DFE), multi-group, and multi-angle (S_n) neutrino radiation/hydrodynamics code. With it, we hope soon to simulate multi-D collapse and to test out the numerous avenues now before us. Stay tuned.

Acknowledgements

Support for this work was provided by the Scientific Discovery through Advanced Computing (SciDAC) program of the DOE, grant number DE-FC02-01ER41184, and by NASA through Hubble Fellowship grant #HST-HF-01157.01-A awarded by the Space Telescope Science Institute, which is operated by the Association of Universities for Research in Astronomy, Inc., for NASA, under contract NAS 5-26555.

References

1. W.D. Arnett: Can. J. Phys. **45**, 1621 (1967)
2. H.A. Bethe & J.R. Wilson: ApJ **295**, 14 (1985)
3. S.W. Bruenn: ApJS **58**, 771 (1985)
4. S.W. Bruenn & A. Mezzacappa: Phys. Rev. D **56**, no. 12 (1997)
5. S.W. Bruenn, K.R. De Nisco, & A. Mezzacappa: ApJ **560**, 326 (2001)
6. A. Burrows & J. Goshy: ApJ **416**, L75 (1993)
7. A. Burrows, J. Hayes, & B.A. Fryxell: ApJ **450**, 830 (1995)
8. A. Burrows, T. Young, P. Pinto, R. Eastman, & T.A. Thompson: ApJ **539**, 865 (2000)
9. C.J. Cannon: J. Quant. Spectrosc. Rad. Transf. **13**, 627 (1973a)

10. C.J. Cannon: ApJ **185**, 621 (1973b)
11. S.A. Colgate and R.H. White: ApJ **143**, 626 (1966)
12. C.L. Fryer, W. Benz, M. Herant, & S. Colgate: ApJ **516**, 892 (1999)
13. C.L. Fryer & M. Warren: ApJL **574**, L65 (2002)
14. M. Herant, W. Benz, W.R. Hix, C.L. Fryer, and S.A. Colgate: ApJ **435**, 339 (1994)
15. H.-Th. Janka & E. Müller: A&A **306**, 167 (1996)
16. M. Liebendörfer, A. Mezzacappa, F.-K. Thielemann, O.E.B. Messer, W.R. Hix, & S.W. Bruenn: PRD **63**, 103004 (2001)
17. M. Liebendörfer, A. Mezzacappa, & F.-K. Thielemann: PRD **63**, 104003 (2001)
18. A. Mezzacappa, M. Liebendörfer, O.E.B. Messer, W.R. Hix, F.-K. Thielemann, & S.W. Bruenn: PRL **86**, 1935 (2001)
19. D. Mihalas: ApJ **238**, 1034 (1980)
20. E.S. Myra & A. Burrows: ApJ **364**, 22 (1990)
21. G.L. Olson, L.H. Auer, and J.R. Buchler: J. Quant. Spectrosc. Rad. Transf. **35**, 431 (1986)
22. M. Rampp & H.-Th. Janka: ApJL **539**, 33 (2000)
23. M. Rampp & H.-Th. Janka: A&A **396**, 361 (2002)
24. G.B. Scharmer: ApJ **249**, 720 (1981)
25. T.A. Thompson, A. Burrows, and P.A. Pinto: submitted to ApJ (2002)
26. J.R. Wilson: In *Numerical Astrophysics*, ed. J. Centrella, J. M. LeBlanc, R. L. Bowers, p. 422. Boston, (Jones & Bartlett Publ.) (1985)
27. S.E. Woosley & T.A. Weaver: ApJS **101**, 181 (1995)

A New Twist on the Core Collapse Supernova Mechanism?

Anthony Mezzacappa[1] and John M. Blondin[2]

[1] Physics Division, Oak Ridge National Laboratory, Oak Ridge, TN 37831-6354, USA
[2] Department of Physics, North Carolina State University, Raleigh, NC 27695-8202, USA

Abstract. We present results from the first numerical stability analysis of a stationary accretion shock in the core collapse supernova context during the critical shock reheating phase. We discuss the potential ramifications accretion shock instability may have for the supernova mechanism and supernova phenomenology.

1 Introduction

Rapidly increasing and ever more detailed observations of core collapse supernovae together with increasingly sophisticated stellar core collapse and postbounce simulations are uncovering a richness, a variety, and a veritable continuum of possibilities that future observational and modeling efforts must contend with (see, for example, [12]).

Past modeling efforts suggest that core collapse supernovae may be neutrino driven, MHD driven, or both [15,16,7,3,10,11,5,6], but uncertainties in the models prevent us from making firm conclusions at this point. Furthermore, in a scenario in which a supernova is neutrino driven, the magnetic fields in the proto-neutron star will likely have an impact on the dynamics of the explosion. Even small magnetic fields can be expected to qualitatively alter the fluid flow. Similarly, in a scenario in which a supernova is MHD-driven, the neutrino transport in no small part will dictate the dynamics of stellar core collapse, bounce, and the postbounce evolution, which in turn will create the environment in which the MHD-driven explosion occurs.

Ultimately, three-dimensional, general-relativistic radiation magnetohydrodynamics simulations with state of the art nuclear and weak interaction physics will be required to pin down the core collapse supernova mechanisms over the entire range of core collapse supernova progenitors. The neutrinos, fluid instabilities, core rotation, magnetic fields, and potentially other phenomena will act in a concerted way to drive these explosions, perhaps in different ways for different progenitor classes. To determine the "recipes for explosion" and to understand supernova phenomenology across these progenitor classes will require a systematic approach in which the dimensionality and the physics are layered to determine what is responsible for the explosions themselves and what is responsible for other observable characteristics of the explosions [14].

This systematic approach should begin with hydrodynamics-only simulations, provided they can be constructed in a meaningful way, in order to explore the complex, nonlinear hydrodynamics of stellar core collapse and bounce and the postbounce stellar core flow, particularly beneath the supernova shock wave and with a focus on the interaction of this flow with the shock. Along these lines, we have constructed models of stationary accretion flow that mimic the conditions in a postbounce stellar core during shock reheating and have discovered an instability that may have important ramifications for the supernova mechanism and phenomenology.

2   Accretion Shock Instability

The stability of accretion shocks has been considered in other contexts [9], but never in the context of the core collapse supernova problem during the neutrino heating phase and shock revival. Full details can be found in [2]. We present a few highlights below.

We consider an idealized adiabatic gas in one dimension accreting onto a star of mass M. We assume the infalling gas has had time to accelerate to free fall and that the free-fall velocity is highly supersonic. Below the standing accretion shock we assume radiative losses are negligible (we will discuss the appropriateness of this assumption later) and the gas is isentropic. The assumption of steady-state isentropic flow implies that there is a zero gradient in the entropy of the postshock gas, which in turn implies this flow is marginally stable to convection. This allows us to separate effects due to convection from other aspects of the multidimensional fluid flow. In the more general case where a negative entropy gradient would drive thermal convection [7,3,10,13,5], such convection could act as a seed for the instabilities we discuss here.

A reasonable fit to profiles from spherically symmetric stellar collapse and postbounce simulations [14] that include Boltzmann neutrino transport and a realistic equation of state is obtained with $\gamma = 1.25$, as shown in Fig. 1.

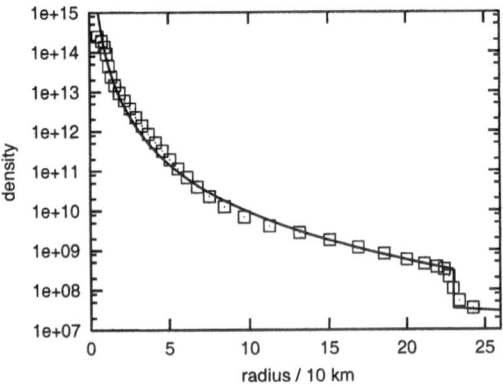

Fig. 1. For $\gamma = 1.25$, there is good agreement between the density profile in the spherically symmetric stationary accretion shock solution (solid line) and a postbounce profile taken from one of our simulations with neutrino transport and a realistic equation of state (squares)

If spherically symmetric perturbations are introduced, the accretion shock "rings" and eventually settles to its original configuration [2]. On the other hand, if non-spherically symmetric perturbations are introduced, the accretion shock becomes unstable. This is evident in Fig. 2. In this case the initial perturbation consists of two rings of overdense material that advect onto the shock from above (this is an axisymmetric simulation, and the figure shows one slice through the data). It is important to note that the instability is insensitive to the way the shock is initially perturbed. Once the shock is perturbed, the $l = 1, 2$ modes grow, become nonlinear, and do not saturate, leading to an oscillating, bipolar outflow [2].

The time sequences in Fig. 3 of both tangential velocity and pressure show the fundamental coupling at work in the accretion shock instability. Vorticity is introduced by the nonspherical shock and advects inward and is trapped. A low-pressure region at the base of the postshock flow associated with these trapped vortices becomes pronounced at $t \sim 240$ milliseconds. Pressure waves generated at the center when the vortices advect inward [as they collide or rebound off of the inner boundary (or the density cliff in a more realistic model)] propagate outward to further distort the shock, completing the loop and leading to the feedback that ultimately drives the instability. In Fig. 3, the radii and time are scaled.

In Fig. 4 we see that the shock radius increases dramatically when the turbulent energy, as measured by the (scaled) kinetic energy in the angular direction in the postshock flow, increases.

3 Discussion

The accretion shock instability described here adds another potential ingredient to core collapse supernova models and perhaps the explosion mechanism itself. Given sufficient conditions, this instability will develop and could alter the energetics of the explosion as well as other observables, such as the explosion morphology. As described in detail in [2], the accretion shock instability acts as a conduit between gravitational binding energy and outgoing kinetic energy, much like the neutrinos do in more realistic models. In the explosions obtained in the idealized case, *in the absence of neutrinos*, the gravitational binding and kinetic energies increase in magnitude, while the thermal energy decreases, and at the end of the simulation a significant fraction of the material is unbound. Regarding explosion morphology, the instability may be the underlying mechanism producing the polarization observed in core collapse supernovae [8]. In one two-dimensional model (with $\gamma = 4/3$; other cases have not yet been analyzed in this context), the outflow resulted in *time-independent* aspect ratios (i.e., the outflow became self similar after some point in the evolution) that were ~ 2 [2]. Aspect ratios of this size would provide at least one explanation for the spectro-polarimetry observations, although much work remains to be done to determine if an accretion shock instability occurs and causes such gross asymmetries in supernovae.

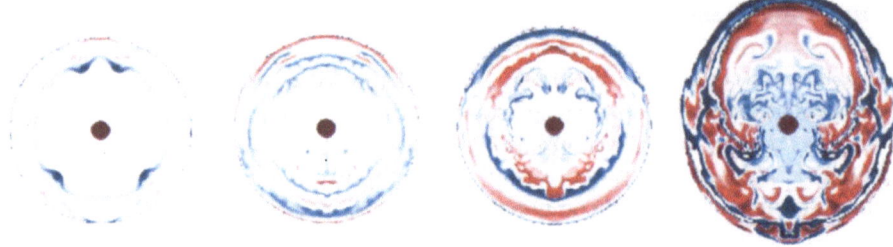

Fig. 2. In this sequence of two-dimensional entropy plots, both the growth of the accretion shock instability and the fact that the postshock region remains globally convectively stable – i.e., that entropy gradients are localized in radius and do not extend over a significant fraction of the postshock region – are evident

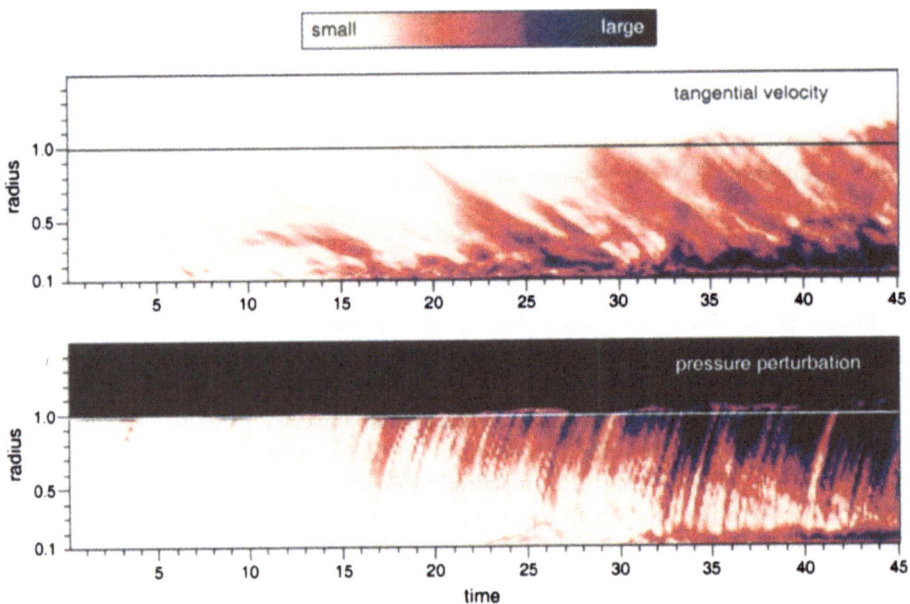

Fig. 3. Time sequences of tangential velocity and pressure show the feedback at work in the accretion shock instability. Tangential velocity introduced at the shock advects inward. Pressure waves caused by the advecting vortices propagate outward to further distort the shock. Deviations of the shock from its original radius are evident in both sequences. The radii are scaled to the initial shock radius, and one unit of our scaled time corresponds to ~ 6 ms in the more realistic model used in Fig. 1

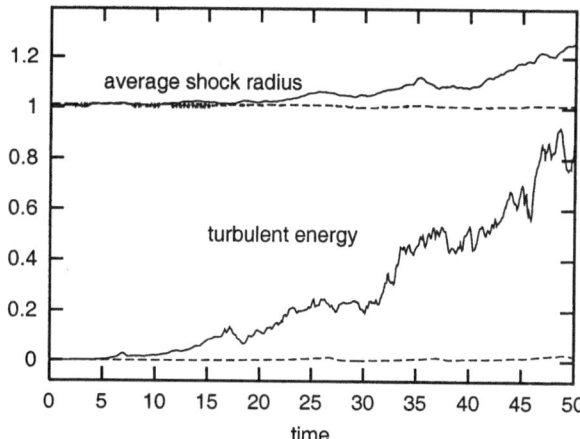

Fig. 4. The accretion shock begins to expand explosively when the turbulent energy beneath it begins to increase dramatically, as more and more vortices become trapped in the postshock flow. The dashed lines correspond to our two-dimensional simulation of the *unperturbed* spherically symmetric flow

We have presented the results of two-dimensional simulations in this paper. Three-dimensional simulations also exhibit the growth of $l = 1, 2$ modes and the same instability [1]. Therefore, the instability discussed here and its gross characteristics are not an artifact of the imposed axisymmetry and, more to the point, the imposed reflecting boundary conditions on the $\theta = 0$ axis used to numerically guarantee this axisymmetry. Indeed, while such boundary conditions may induce greater outflows on axis than off, they will certainly not lead to the unstable situation described here. Finally, in addition to our two- and three-dimensional simulations, linear stability analyses are planned. Linear stability analyses by Foglizzo [4] in a different context illuminated the existence of a vortical–acoustic feedback around accreting black holes.

Many questions must be answered before we can determine whether or not the instability described here plays a role in the dynamics of core collapse supernovae. Will neutrino cooling near the proto-neutron star surface dampen the feedback mechanism between the shock-induced vorticity and the resultant pressure waves by damping these waves? Will neutrino cooling below the shock drive the flow away from conditions in which an instability can develop – for example, by reducing the volume between the neutrinosphere and the shock? This seems to have occurred in the simulations documented in [13], where one can see the $l = 1, 2$ modes attempting to grow, but ultimately failing to do so as the shock recedes and the postshock volume shrinks. In light of these results, we investigated the growth of the instability in our model runs as (a) the adiabatic index was softened [which mimics to some extent the effects of radiative (neutrino) cooling] and (b) the ratio of the inner boundary radius to the shock radius was increased (a larger ratio would correspond to a smaller postshock volume in more

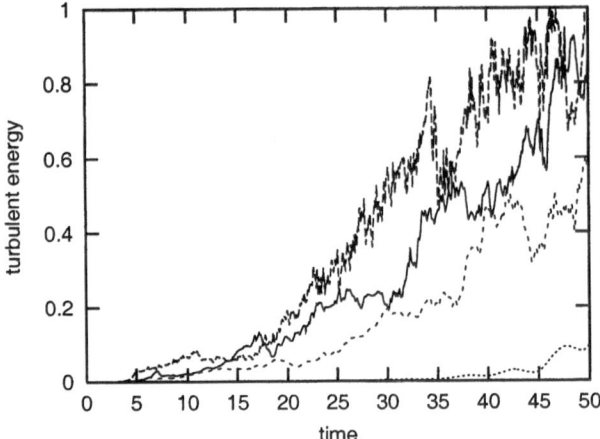

Fig. 5. The growth of the instability as a function of one model parameter is shown here by considering the growth of the interior turbulent energy for several different simulations with different inner boundary radii (from top to bottom: $R_i = 0.05$, 0.1, 0.2, and 0.4, where these radii are given as a fraction of the initial shock radius)

realistic models). Figure 5 shows the results from case (b). In our model problem, an instability still develops as the postshock volume decreases, but takes longer to develop with decreasing volume. Finally, in the $\gamma = 4/3$ case we also included a cooling layer at the base of the postshock region, as discussed in [9]. This was done to investigate the effects of cooling and to consider the impact of different inner boundary conditions on the development of the instability. In this case too, the instability developed. However, only two- and three-dimensional radiation hydrodynamics simulations can definitively address the questions posed here. We look forward to reporting on progress along these lines from the TeraScale Supernova Initiative (http://www.phy.ornl.gov/tsi/), whose goal it is to address these very questions, and others.

Acknowledgements

A.M. is supported at the Oak Ridge National Laboratory, managed by UT-Battelle, LLC, for the U.S. Department of Energy under contract DE-AC05-00OR22725. A.M. and J.M.B. are supported in part by a SciDAC grant from the U.S. DoE High Energy and Nuclear Physics Program.

References

1. J. M. Blondin and A. Mezzacappa: in preparation (2002)
2. J. M. Blondin, A. Mezzacappa, and C. DeMarino: Ap.J., in press (2002)
3. A. Burrows, J. Hayes, and B. A. Fryxell: Ap.J. **450**, 830 (1995)

4. T. Foglizzo: Astron. and Astrophys. **392**, 353 (2002)
5. C. L. Fryer and A. Heger: Ap.J. **541**, 1033 (2000)
6. C. L. Fryer and M. S. Warren: Ap.J. **574**, L65 (2002)
7. M. Herant, W. Benz, W. R. Hix, C. L. Fryer, and S. A. Colgate: Ap.J. **435**, 339 (1994)
8. P. Höflich, A. Khokhlov, and L. Wang: in Proc. of 20th Texas Symposium on Relativistic Astrophysics, eds. J. C. Wheeler and H. Martel (New York: American Institute of Physics) (2001)
9. J. C. Houck and R. A. Chevalier: Ap.J. **395**, 592 (1992)
10. H.-T. Janka and E. Müller: Astron. and Astrophys. **306**, 167 (1996)
11. A. I. MacFadyen and S. E. Woosley: Ap.J. **524**, 262 (1999)
12. P. A. Mazzali et al.: Ap.J. **572**, L61 (2002)
13. A. Mezzacappa, A. C. Calder, S. W. Bruenn, J. M. Blondin, M. W. Guidry, M. R. Strayer, and A. S. Umar: Ap.J. **495**, 911 (1998)
14. A. Mezzacappa, M. Liebendörfer, O. E. B. Messer, W. R. Hix, F.-K. Thielemann, and S. W. Bruenn: Phys. Rev. Lett. **86**, 1935 (2001)
15. E. M. D. Symbalisty: Ap.J. **285**, 729 (1984)
16. J. R. Wilson and R. Mayle: Phys. Rep. **227**, 97 (1993)

The Impact of Improved Weak Interaction Physics in Core-Collapse Supernova Simulations

O.E. Bronson Messer[1,2,3], Matthias Liebendörfer[1,2], W. Raphael Hix[1,2,3], Anthony Mezzacappa[2], and Stephen W. Bruenn[4]

[1] University of Tennessee, Knoxville TN 37996, USA
[2] Oak Ridge National Laboratory, Oak Ridge TN 37831, USA
[3] Joint Institute for Heavy Ion Research, Oak Ridge TN 37831, USA
[4] Florida Atlantic University, Boca Raton FL 33432, USA

Abstract. Recent core-collapse supernova simulations incorporating Boltzmann neutrino transport have highlighted the need for an improved set of "standard" microphysics to be employed in such simulations. Improved electron capture rates on nuclei, the inclusion of nucleon-nucleon bremsstrahlung, and the effects of weak magnetism and the strangeness content of the nucleon are examples of such improvements. We discuss the initial implementation of these updates in fully self-consistent radiation hydrodynamic core-collapse supernova simulations incorporating Boltzmann neutrino transport.

1 Motivation

The most recent generation of core-collapse supernova simulations employing state-of-the-art neutrino transport [19,16,9] make use of a set of weak interaction physics primarily due to [1]. The failure of these simulations to produce explosions in spherical symmetry suggests an update of the microphysical inputs is necessary to fully determine the sensitivity of the simulations to the description of the neutrino-matter interactions.

2 Improved Electron Capture Rates During Core Collapse

Heger et al. [5](HLMPW) have repeated the evolution calculations of Woosley & Weaver [21](WW95) for initial progenitor masses of 15 $M_\odot$, 25 $M_\odot$, and 40$M_\odot$, replacing the weak interaction rates for electron and positron captures and β^- and β^+ decays. The WW95 models used the electron capture rates of [2] and older sets of beta decay rates [12,4]. HLMPW have updated both with a new set of shell model weak interaction rates for electron capture, positron capture, and β^- and β^+ decays [8,18]. The most noticeable effect of these changes is a marked increase in the electron fraction (Y_e) throughout the iron core before collapse. Because the final size of the homologous core, and therefore the shock formation radius, is proportional to the square of the trapped lepton fraction ($Y_l{}^2$) at core bounce [22], the persistence of these initial differences in Y_e throughout collapse might have a discernible effect on the shock energetics.

In an initial attempt to determine the influence of these improved rates on iron core collapse, we have performed full radiation hydrodynamic collapse simulations using the neutrino radiation hydrodynamics code AGILE-BOLTZTRAN [14,17,10,11].

We observe no difference in initial shock formation position between the two sets of progenitor models when our set of standard physics is used [13]. In these cases, capture rates on nuclei are quickly shut off when $A > 65$, and the capture rate on free nucleons dominates. The steep dependence of the free proton fraction on changes in the electron fraction assures convergence to the same Y_e profile inside the homologous core. We have also performed simulations in which the electron capture rate on nuclei was turned off as well as simulations in which the capture rate on free nucleons was increased by a factor of 10. Neither parameterization had any discernible effect on the shock formation radius.

In an attempt to investigate the impact of including the updated electron capture rates during core collapse we have also performed collapse simulations with the HLMPW 15 $M_\odot$ core with a parameterized form of our usual electron capture rates. We write the neutrino emissivity from nuclei as [1,15]:

$$j_{nuc} = \frac{2}{7} \frac{(2\pi)^4 G_F^2 g_A^2}{\pi h^4 c^4} \frac{\rho X_H}{m_B A} N_p(Z) N_h(N)(E + Q)^2 \left[1 - \left(\frac{M_e}{E + Q}\right)^2\right]^{1/2} F_e(E + Q), (1)$$

where

$$N_p(Z) = \begin{cases} 0 & Z < 20 \\ Z - 20 & 20 < Z < 28 \\ 8 & Z > 28 \end{cases} \quad N_h(N) = \begin{cases} 6 & N < 34 \\ 40 - N & 34 < N < 40 \\ 0 & N > 40. \end{cases} \quad (2)$$

To parameterize this rate we replace the product $N_p(Z)N_h(N)$ with a constant (0.1, 1, 10, or 100). This parameterization has a significant effect on the formation of the bounce shock. Setting the capture parameter equal to 10 moves the shock formation point inward more than 0.1 $M_\odot$ (0.64 $M_\odot$ versus 0.52 $M_\odot$). The most important feature of the parameterization is the persistence of captures on nuclei throughout collapse. The precise value of the parameter is of secondary importance. The fundamentally different behavior of the parametrized rate as compared to the standard rate is shown in Fig. 1.

3   Nucleon-Nucleon Bremsstrahlung

It has recently been shown that nucleon-nucleon bremsstrahlung can be the dominant production mechanism for μ and τ neutrinos and antineutrinos in dense regions of the protoneutron star [3,20]. We have included this process in recent test simulations of core collapse and shock breakout using AGILE-BOLTZTRAN (with reduced angular resolution). The impact on the $\nu_{\mu,\tau}$ luminosity is shown in Fig. 2. During collapse, the bremsstrahlung process leads to markedly increased production, as the only other production mechanism, electron-positron pair annihilation, is essentially inoperative. After shock breakout the average neutrino energy is reduced by the inclusion of bremsstrahlung ($\approx 10\%$ throughout most

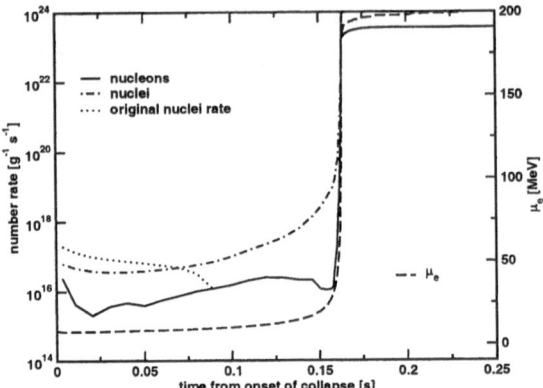

Fig. 1. The number rate of electron captures during core collapse as a function of time. The electron chemical potential is also shown

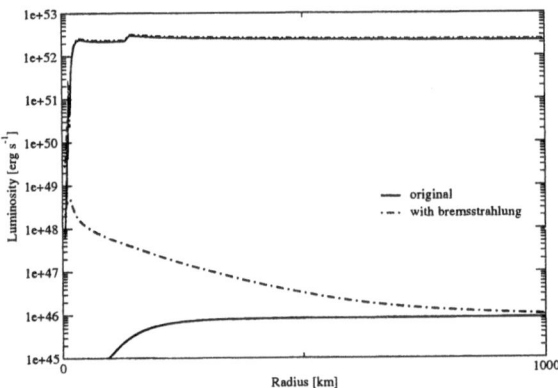

Fig. 2. $\nu_{\mu,\tau}$ luminosities for test runs with and without incorporation of nucleon-nucleon bremsstrahlung. The lower set of curves corresponds to a central density of 6.470×10^{13} g/cm^3, roughly 2 ms before core bounce. The upper curves are just after shock breakout

of the core), but the increased flux induced by the softened spectra more than compensates for this reduction, yielding a slightly higher luminosity ($\approx 8\%$ above the shock).

4   Weak Magnetism

Horowitz [6] gives a simple prescription for the inclusion of weak magnetism effects in neutrino transport calculations. Fig. 3 compares neutrino luminosities at 1000 km for simulations incorporating the corrections of [6] for weak magnetism and corrections for the strangeness content of the nucleon [7] to results using our standard set of microphysics. Antineutrino luminosities are most affected, being ≈10% higher when both weak magnetism and strangeness corrections are included. Weak magnetism effects alone produce a ≈6-7% enhancement. Neutrino luminosities behave in much the same way, but the maximum change in luminosity for each flavor is only about half that for the antineutrinos (≈5%).

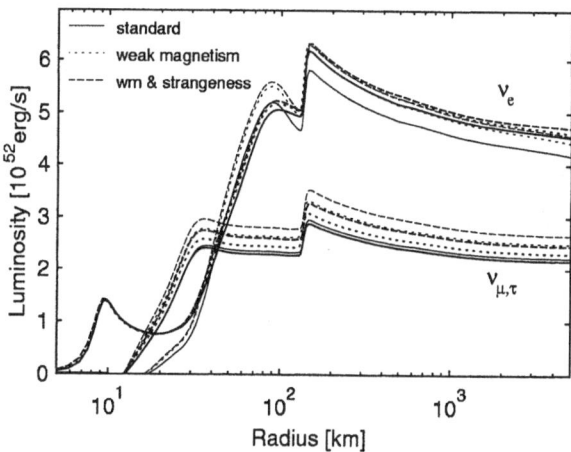

Fig. 3. Neutrino (*heavy lines*) and antineutrino (*thinner lines*) luminosities at 100 ms post-bounce with standard opacities, including weak magnetism effects, and including both weak magnetism effects and the effect of the strangeness content of the nucleon

Acknowledgements

We acknowledge support from the U.S. Department of Energy Scientific Discovery through Advanced Computing Program, the Joint Institute for Heavy Ion Research, the National Science Foundation under contract AST-987713, and NASA under contract NAG5-3903. Oak Ridge National Laboratory is managed by UT-Battelle, LLC, for the U.S. Department of Energy under contract DE-AC05-00OR22725. This research used resources of the National Energy Research Scientific Computing Center, which is supported by the Office of Science of the U.S. Department of Energy under Contract No. DE-AC03-76SF00098.

References

1. S. W. Bruenn: ApJS **58**, 771 (1985)
2. G. M. Fuller, W. A. Fowler, M. J. Newman: ApJS **42**, 447 (1980)
3. S. Hannestad, G. Raffelt: ApJ **507**, 339 (1998)
4. C. J. Hansen: Neutrino Emission from Dense Stellar Interiors. PhD Thesis, Yale University, New Haven (1966)

5. A. Heger, K. Langanke, G. Martínez-Pinedo, S.E. Woosley: Phys. Rev. Lett. **86**, 1678 (2001)
6. C. Horowitz: Phys. Rev. D **65**, 043001 (2002)
7. E. Kolbe, K. Langanke, S. Krewald, F.-K. Thielemann: ApJL **401**, L89, (1992)
8. K. Langanke, G. Martinez-Pinedo: Nucl. Phys. A. **673**, 481 (2000)
9. M. Liebendörfer, A. Mezzacappa, F.-K. Thielemann, O. E. B. Messer, W. R. Hix, S. W. Bruenn: Phys. Rev. D **63**, 103004, (2001)
10. M. Liebendörfer, S. Rosswog, F.-K. Thielemann: ApJS **141**, 229 (2002)
11. M. Liebendörfer, O. E. B. Messer, A. Mezzacappa, S. W. Bruenn, C. Y. Cardall, F.-K. Thielemann: ApJS, submitted (2002)
12. T. J. Mazurek: Thermonuclear Supernova in Binary Helium Dwarfs. PhD thesis, Yeshiva University, New York (1973)
13. O. E. B. Messer, W. R. Hix, M. Liebendörfer, A. Mezzacappa: in preparation
14. A. Mezzacappa, S. W. Bruenn: ApJ **405**, 637 (1993)
15. A. Mezzacappa, S. W. Bruenn: ApJ **405**, 669 (1993)
16. A. Mezzacappa, M. Liebendörfer, O. E. B. Messer, W. R. Hix, F.-K. Thielemann, S. W. Bruenn: Phys. Rev. Lett. **86**, 1935 (2001)
17. A. Mezzacappa, O. E. B. Messer: J. Comp. and App. Mathematics **109**, 281 (1999)
18. T. Oda, M. Hino, K. Muto, M. Takahara, K. Sato: Atomic Data & Nucl. Data Tables **56**, 231 (1994)
19. M. Rampp, H.-T. Janka: ApJL **539**, L33 (2000)
20. T. A. Thompson, A. Burrows, J. E. Horvath: Phys. Rev. C **62**, 035802 (2000)
21. S. E. Woosley, T. A. Weaver: ApJS **101**, 181 (1995)
22. A. Yahil: ApJ **265**, 1047 (1983)

Supernova Explosion Physics

Wolfgang Kundt

Institut für Astrophysik der Universität, Auf dem Hügel 71, D-53121 Bonn, Germany

Abstract. Quite likely, all supernovae are core-collapse supernovae. When the progenitor star's burnt-out core contracts under its own gravity – on the time scale of seconds – angular-momentum conservation raises its spin energy as $1/r^2$, towards some $10^{52.5}$ erg, whilst neutron-degeneracy pressure halts the collapse at a neutron star's radius, some 10^6 cm.

Magnetic-flux winding will then tap the core's large spin energy – on the time scale of $\lesssim 30$s – bringing the spin period P into the range of neutron-star birth periods – ms $< P < 10$ s – and transferring the excess angular momentum to the overlying mantle. Subsequent reconnection of the huge toroidal magnetic fields creates a magnetized relativistic cavity, both leptons and hadrons, with particle energies up to 10^{20} eV, ready to launch the envelope (via adiabatic expansion, through some 10^7 in radius). Magnetic Rayleigh-Taylor instabilities tear and squeeze the ejected shell into a large number ($> 10^4$) of filamentary fragments, like a splinter bomb.

1 Constraints on Supernovae

When in 1962, Shklovskii [10] proposed to model supernovae (SNe) by Sedov-Taylor waves, he had seen photographs of hydrogen-bomb explosions in air, cf. [11]. But Sedov-Taylor waves (STWs) apply only to thin-walled bombs, because they assume that only a negligible fraction of the explosion energy is stored in the blown-out material of the former container (which material hardly sweeps, because it has a small filling factor). And STWs apply only to a short initial stage after the explosion during which photons are trapped inside the ejected shell, as they do not conserve radial momentum. *Thick-walled bombs*, on the other hand, do not sweep, have much larger ranges, have no reverse shock, and have (non-spherical) morphologies that are controlled by the progenitor's circumstellar medium (which is impacted, heated, and caused to radiate). As an illustrative example of this splinter morphology, I consider the youngest known SN in the Galaxy, centered on the Becklin-Neugebauer Kleinmann-Low IR object in Orion [8]; it looks like fireworks.

How massive are SN progenitors? In 1985, Adrian Blaauw [1] compared the number of pulsars in a 0.5 Kpc-environment of the Sun with the number of expected *progenitor stars*, and concluded that all stars with mass $\gtrsim 6$ $M_\odot$ are needed for replenishment. This estimate involves a reliable judgement of incompletenesses in our catalogues, which amounts to more than a factor of 3 in the case of pulsars [6]. Moreover, there will be at least one binary X-ray source for every pulsar (involving a neutron star: its elder brother in a binary system), so that we need twice as many progenitors as those of mass $\gtrsim 6\ M_\odot$, approaching a 'critical mass' of $\gtrsim 5\ M_\odot$ for neutron-star formation; see also [12].

This estimate is much lower than often traded *progenitor masses* – in excess of 10 $M_\odot$ – masses which should form a small percentage ($< 10\%$) among all SNe, consistent with the 'empty fields' found by Stephen Smartt (this volume), and independently implied by the change of SN light curves from optically thick to thin – after $\lesssim 10^2$d – according to

$$\Delta M = t(8\pi mE/\sigma)^{1/2} = 3M_\odot(t/10^2d)(m/m_p)^{1/2}, \tag{1}$$

where ΔM is the ejected shell mass, t = transition time (from thick to thin), m = mean particle mass, E = kinetic SN energy and σ = continuum cross section [4,6]. In the sequel, therefore, I shall focus consideration on progenitor stars of some 6 $M_\odot$. A comparison with the complementary (in progenitor mass) nova phenomenon is recommended [2].

This low mass estimate is at variance with what was found for the progenitor of SN 1987A – though not with mass estimates from its light curve, Fig. 1e, – and at variance with mass estimates based on the assumption that *SN light curves* were powered by radioactive decay of ^{56}Ni to ^{56}Fe via ^{56}Co. The latter assumption has, however, failed quantitatively by factors of several in most well-studied cases, cf. [9]. It ignores three other energy inputs into the light curve which may be dominant: (i) overtaking crashes of filamentary ejecta during launch, i.e. a tapping of the (huge) kinetic-energy reservoir, (ii) friction on the (overtaking) relativistic piston, and (iii) enhanced convective cooling of the central neutron star, via volcanoes.

2   Core-Collapse Supernovae

One reason of why numerical simulations have not yet been successful in describing a (core-collapse) SN, after some 30 years, is the non-trivial transfer of the liberated core-collapse energy to the overlying progenitor star's mantle, from some 10^6cm to some 10^{13}cm, through a factor of order 10^7 in radius. Adiabatic expansion of the piston decreases its internal energy as $1/r^2$ for a non-relativistic gas or plasma, or as $1/r$ for a relativistic plasma. Here neutrinos do not qualify as a piston because for them, the mantle is essentially transparent; and photons do not qualify because they mix with matter, resulting in a non-relativistic plasma for typical SN energetics [4–7]. For a SN, the only *relativistic pistons* I can think of are magnetic fields and their decay product, a relativistic cavity.

Figure 1a shall serve as a plausible model of the progenitor's density profile. Some day, when the degenerate core (of mass $\gtrsim 1\ M_\odot$) has burnt most of its nuclear fuel and cooled sufficiently, it will collapse under its own gravity, on the free-fall time scale

$$t_{ff} \approx (2r/g)^{1/2} \lesssim 4\ \text{s}, \tag{2}$$

until neutron-degeneracy pressure halts its collapse, and supports it as a (hot) new-born neutron star (or black hole, a highly unlikely possibility: [6]). The onset

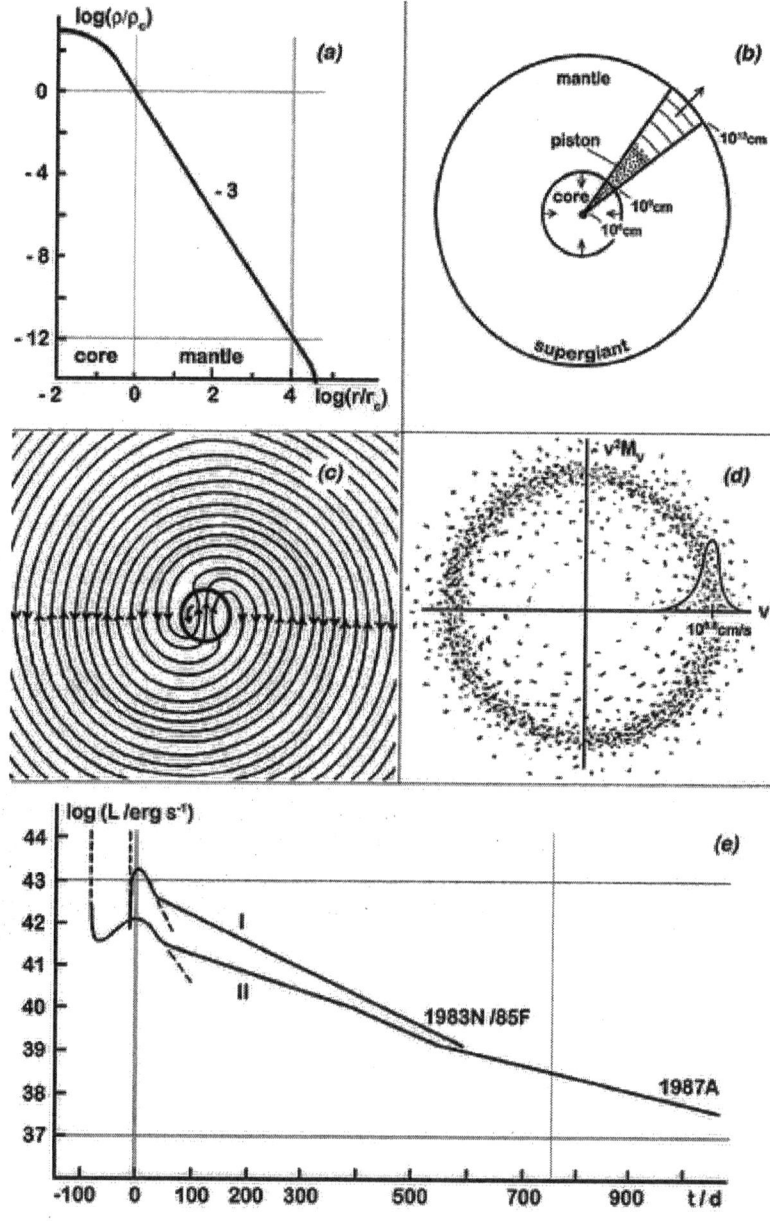

Fig. 1. Sketch of assumptions and results made and obtained in this communication. (a) Assumed mass profile of the SN progenitor star. (b) Section through the ejection geometry. (c) Idealized flux-winding geometry around the collapsed core. (d) Symbolical splinter geometry of the ejected shell. (e) Two observed bolometric SN light curves, for types I and II respectively. The broken initial parts are hypothesized; the broken parts beyond the (second) maximum plot the underlying photon continuum

of this *core collapse* may have to be synchronized between antipodal regions at sound speed, thereby stretching the collapse time to

$$t_{core} \approx r/c_s \approx 10 \text{ s } T_8^{-1/2}, \tag{3}$$

where $T_8 := T/10^8 \text{K}$ is the core's temperature in units of 10^8K.

Crucial for a SN is the spin of the core, which will be almost conserved under collapse, whence $\Omega \sim r^{-2}$ and $E_{rot} \sim r^{-2}$ grow by factors of 10^6. The rotational energy E_{rot} is bounded by centrifugal stability, at $\Omega \lesssim 10^4 \text{s}^{-1}$: $E_{rot} \lesssim 10^{52.7} \text{erg}$. Observed neutron stars require $10^{44.3} < E_{rot}/\text{erg} \lesssim 10^{52}$ at birth, corresponding to $10^7 > P_{core}/\text{s} \gtrsim 10^{3.2}$ before collapse, consistent with evolutionary estimates (for stars of 6 $M_\odot$). I.e. the highest possible *spin energy* (of the collapsed core) is just 5 times larger than E_{rot} of a neutron star of (extreme) spin period 1.56 ms, and some 50 times larger than kinetic SN energies; it can launch a SN at a transfer efficiency of 2% and, at the same time, give birth to (even) the fastest observed pulsars.

Inside a highly conductive plasma, the collapsed core is expected to wind its protruding *magnetic flux* around itself in proportion to time, [3] and Fig. 1c, some $10^{2.8}$ times per second, to a maximum given by energy balance:

$$B_\phi \lesssim (3I/R^3)^{1/2}\Omega = 10^{17} \text{ G } \Omega_4 \tag{4}$$

for a moment of inertia $I = 10^{45} \text{g cm}^2$; i.e. magnetic fields are expected to stay below 10^{17}G. At the same time, the spin is reduced by the magnetic torque, on the time scale

$$\Omega/\dot{\Omega} = 3I\Omega/R^3 B_r B_\phi = 10^{1.5} \text{ s } (B_\phi/B_r)_4^{-1} \tag{5}$$

for $\Omega = 10^4/\text{s}$, $B_r = 10^{13}\text{G}$. The growth of B_ϕ will be limited by *reconnection*, because layers of opposite field orientation get into contact, and likewise by the rapid reduction of differential angular velocities all the way back to $|B_\phi| \lesssim |B_r|$. During sonic reconnection, particles will be created whose energies will range up to

$$\Delta W = e \int (\beta \times B) dx \leq 10^{20} \text{ eV } \beta_{-1} B_{16}(dx)_{2.5} , \tag{6}$$

[7], from extremely relativistic $e^\pm$ through pions and $p, \bar{p}$ all the way up to whatever particle resonances are permitted. This freshly created UHE *relativistic cavity* – the *piston* – can eject the overlying mantle under adiabatic expansion, Fig. 1b, as will be elaborated next.

The *ejection speed* v_{ej} of the mantle, under adiabatic expansion of the relativistic piston, follows from ram-pressure balance: $\rho_{mantle} v_{ej}^2 = p_{piston} = E_{piston}/3V_{piston}$, whence

$$v_{ej} = (E_{piston}/3\rho_{mantle}V_{piston})^{1/2} = c \, r_6^{-1/2} \tag{7}$$

for a piston of energy 10^{52}erg and volume $10^{18.5}\text{cm}^3$ pushing a plasma of mass density $\rho = 10^{12}\text{g/cm}^3$. I.e. at radius $r = 10^6\text{cm}$, v_{ej} equals the speed of light c, and at $r = 10^9$cm – the former core's radius – it equals 10^9cm/s. Beyond this distance, the piston's energy stays almost constant – as the sum of the energy of the relativistic plasma plus that of the pushed material – and so does the product ρV for a stellar mass profile as in Fig. 1a, so that the ejection will take place at mildly decreasing speed, converging towards the observed bulk SN speed of $10^{8.8}$cm/s. The ejection time results as

$$t_{ej} = R_{prog}/v_{ej} = 10^4 \text{ s } R_{13}, \tag{8}$$

of order hour. During the ejection, calculated gyration radii of the relativistic particles through the screening magnetic fields stay small, i.e. they penetrate only a few permille of the mantle material.

According to shock dynamics, the pushed mantle material is transiently heated to *kinetic temperatures*

$$T_{shock} = (2mv^2/k)(\kappa - 1)(\kappa + 1)^2 = 10^{9.4} \text{ K } (m/m_p)v_9^2 \tag{9}$$

for an adiabatic index $\kappa = 5/3$, which kinetic energy is quickly shared with the photon bath, via collisional thermalization, resulting in true *temperatures*

$$T = [(f/32\pi^2)(\Delta M/m)(kT_{shock}c/\sigma_{SB}R^2\Delta R)]^{1/4} \approx 10^{6.5} \text{ K }, \tag{10}$$

where f $= 4/(\kappa - 1)$ counts the number of degrees of freedom of the (ionized) mantle gas. Once this heat flash breaks through the photosphere, individual surface elements will emit a *UV flash* of this temperature, during the photospheric crossing time of

$$\Delta r/v_{ej} = \text{ ms } (\Delta r)_6 \tag{11}$$

– where the photospheric depth Δr results from $\Delta r = \Delta N/n = 1/n\sigma = 10^6$cm n_{16}, $(\Delta N := \text{column density})$ – corresponding to local radiative spikes of integrated luminosity

$$L = 4\pi R^2 \sigma_{SB} T^4 = 10^{49} \text{ erg s}^{-1} R_{13}^2 T_{6.5}^4. \tag{12}$$

These spikes reach a distant observer with a range of delays of order $R/c = 10^{2.5\pm0.5}$s, i.e. on a $10^{5.5\pm1}$ times longer timescale than emitted locally, so that a UV flash will have a luminosity of $\lesssim 10^{44}$erg/s. Its time-integrated energy amounts to some 10^{46}erg, a tiny fraction of the whole radiated SN energy, of order $10^{49.5\pm0.5}$erg. Nevertheless, this UV flash was clearly observed for SN 1987A, in the shape of Napoleon's hat [9].

As the star's mantle material is some 10^9 times heavier than its accelerating piston, the ejection must be starkly Rayleigh-Taylor unstable, hence tear the mantle. This conclusion follows more quantitatively from the *filling factor* f_{th} of the thermal component, assumed homogeneous, in pressure equilibrium with the relativistic gas:

$$f_{th} = p_{therm}(hom)/p_{rel} = 2nkT/(E_{rot}/3V) = T_{9.5}\Delta M_{(0.5)}, \tag{13}$$

i.e. is of order 10^{-3} for a shell of mass 3 $M_\odot$ at temperature $10^{6.5}$K, and even smaller locally for an r-dependent estimate. The ejected shell gets torn and squeezed into thousands of small-filling-factor magnetized filaments.

3 Neutron-Star Spin Periods at Birth

If all SNe are of the core-collapse type – as I have argued earlier – and if 'recycling' of neutron stars has been an erroneous assumption [5–7], then the observed spin-period distribution does not differ much from the *spin-period distribution* at birth, whose two-humpedness should reflect the magnetic core structure of the progenitors. Does the $\gtrsim$ms hump derive from SNe of type I, and the $\gtrsim$s hump from type II? As the globular clusters house (so far) only PSRs with $P \lesssim 2.5$ ms, their neutron-star progenitors should then all belong to type I. This somewhat naïve-sounding dichotomy may have to be refined in the future; but a better insight into the spin and magnetic-field structure of the progenitors has clearly the potential of yielding the answer.

Acknowledgements

It is a pleasure to thank Hans Baumann and Gernot Thuma for the manuscript, and Olga Kostromina and Günter Lay for its electronic conversion.

References

1. A. Blaauw: 'The Progenitors of the Local Pulsar Population'. In: *Birth and Evolution of Massive Stars and Stellar Groups*, eds. Boland and van Woerden (Reidel, Dordrecht 1985) pp. 211–224
2. R.D. Gehrz, J.W. Truran, R.E. Williams, S. Starrfield: PASP **110**, 3–26 (1998)
3. W. Kundt: Nature **261**, 673–674 (1976)
4. W. Kundt: 'Supernova Structure and Light Curves'. In: *Supernova Shells and their Birth Events*, ed. by W. Kundt, Lecture Notes in Physics **316** (Springer, Heidelberg 1988), pp. 165–183
5. W. Kundt: 'Supernova Explosions and their Ejected Shells'. In: *Neutron Stars and their Birth Events*, ed. W. Kundt, NATO ASI C **300** (Kluwer, Dordrecht 1990), pp. 40–50
6. W. Kundt: Fundamentals of Cosmic Physics **20**, 1–119 (1998)
7. W. Kundt: *Astrophysics, A Primer*, (Springer, Heidelberg 2001), pp. 38–40
8. W. Kundt, A. Yar: Astrophys. Space Sci. **254**, 1–12 (1997)
9. R. McCray: Ann. Rev. Astron. Astrophys. **31**, 175–216 (1993)
10. I.S. Shklovskii: Soviet Astronomy **6**, 162–166 (1962)
11. S. Starrfield, S.N. Shore: Scientific American **272**, p.61 (1995)
12. S. Van den Bergh, G.A. Tammann: Ann. Rev. Astron. Astrophys. **29**, 363–407 (1991)

The Neutrino Burst from Supernovae and Neutrino Oscillations

Katsuhiko Sato[1,2], Keitaro Takahashi[1], and Shin'ichiro Ando[1]

[1] Department of Physics, School of Science, the University of Tokyo,
   113-0033 Hongo, Bunkyo-ku, Tokyo, Japan
[2] Research Center for the Early Universe, School of Science,
   the University of Tokyo, 113-0033 Hongo, Bunkyo-ku, Tokyo, Japan

Abstract. Core-collapse driven supernovae are the most luminous neutrino source in the universe. If neutrinos have finite masses and convert into each other, the time profile and energy spectrum of the neutrino burst from supernovae are greatly modified. We review the neutrino conversion in a supernova mantle, and how the burst will be detected by SK (Super-Kamiokande) and SNO (Sudbury Neutrino Observatory) if a supernova appears at the Galactic Center. We show that various neutrino oscillation models can be discriminated by the combination of the SK and SNO detections of a future galactic supernova. We also discuss effects of neutrino oscillation on the supernova relic neutrino observations.

1 Introduction

Recently the SK (Super-Kamiokande) collaboration showed that neutrinos have finite masses and oscillate each other from the observations of solar neutrinos [9] and atmospheric neutrinos [10]. More recently the SNO (Sudbury Neutrino Observatory) collaboration confirmed the neutrino oscillation by combining the SK data with SNO data [1], and clearly showed that electron type neutrinos are converted into muon type or tau type neutrinos from the solar neutrino observation by detecting neutral current events [2]. Although the neutrino oscillation is confirmed, the values of the mass squared differences and mixing angles are not firmly established. For the observed ν_e suppression of solar neutrinos, four solutions were proposed: large mixing angle (LMA), small mixing angle (SMA), low Δm^2 (LOW), and vacuum oscillation (VO). Although SMA solution has been almost ruled out and LMA solution is the most probable model, the other models have not yet been ruled out. For θ_{13}, the mixing angle between mass eigenstate ν_1, ν_3, only an upper bound is known from reactor experiments [4] and combined three generation analysis [8]. Also the nature of neutrino mass hierarchy (normal or inverted) is still a matter of controversy.

   In 1987, Supernova 1987A appeared in Large Magellanic cloud. Huge water Čherenkov counters embedded deeply in the Earth, Kamiokande [11] and IMB [5], detected the neutrino burst from the SN1987A. Core-collapse driven supernovae are the most powerful neutrino source in the universe. It is very natural to consider to use the neutrino burst from supernova as a tool to probe neutrino properties. Fortunately huge water Čherenkov counters, SK and SNO, are running at present. If a supernova appears at the Galactic Center, more than 10,000 events will be detected by SK, and a few hundreds events by SNO, as discussed in the following sections. There have been some studies on future supernova neutrino detection taking neutrino oscillation into account. Dighe and Smirnov [6] estimated qualitatively the effects of neutrino oscillation in a collapse-driven supernova on the neutronization peak, the distortion of energy spectra, and the Earth matter effects. They concluded that it is possible to identify the solar neutrino solution and to probe the mixing angle θ_{13}. Dutta et al. [7] showed numerically that the events involving oxygen targets increase dramatically when there is neutrino mixing. Takahashi et al. [16,17,14,15] investigated the effects of three flavor oscillation on the supernova neutrino spectra, taking into account the constraints on the neutrino mixing and masses imposed by solutions consistent with the solar and atmospheric neutrino problems in detail. They proposed a method to discriminate quantitatively the solutions of the solar neutrino problem.

Detection of the cosmological neutrino background originated by past supernovae is one of the interesting targets for neutrino astronomy. Detectability of these supernova relic neutrinos has been investigated by many people. Recently Ando et al. [3] investigated this problem in detail taking into account the neutrino oscillation. In this talk, we present summary of our recent work.

2   Expected Event Rates in SuperKamiokande and SNO

SuperKamiokande is a water Čherenkov detector with 32,000 ton pure water based at Kamioka in Japan. The relevant interactions of neutrinos with water are shown, for example, in [18]. In these interactions, the $\bar{\nu}_e p$ CC interaction has the largest contribution to the detected events at SK. Hence the energy spectrum detected at SK (including all the reactions) is almost the same as the spectrum derived from the interaction only. Figure 1 shows energy spectrum positrons and electrons expected to be detected at SuperKamiokande. As the neutrino burst model, the Lawrence Livermore group simulation [18] was adopted. The Woosley and Weaver presupernova model with $15 M_\odot$ [19] is used to calculate the time evolution of neutrino wave functions. The distance is assumed as $d = 10$kpc.

Sudbury Neutrino Observatory(SNO) is a water Čherenkov detector based at Sudbury, Ontario. SNO is unique in its use of 1000 tons of heavy water, by which both the charged-current and neutral-current interactions can be detected. The charged current interactions are detected when electrons emit Čherenkov light. These reactions produce electrons and positrons whose energies are sensitive to the neutrino energy, and hence the energy spectra of electrons and positrons give us the information on the original neutrino flux. Figure 2 shows the energy spectrum of positrons and electrons, produced by the two CC interactions expected at SNO. As can be seen in Figs. 1 and 2, when there is neutrino oscillation, neutrino spectra are harder than those in absence of neutrino oscillation. This is because average energies of ν_e and $\bar{\nu}_e$ are smaller than those of ν_x and neutrino oscillation produces high energy ν_e and $\bar{\nu}_e$ which was originally ν_x. Neutrino oscillation makes ν_e and $\bar{\nu}_e$ spectra harder. Therefore, the ratio of high-energy

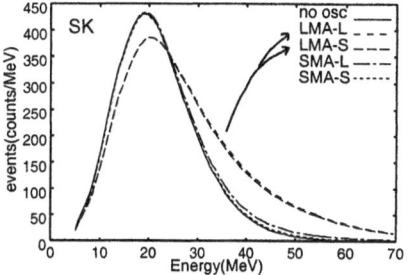

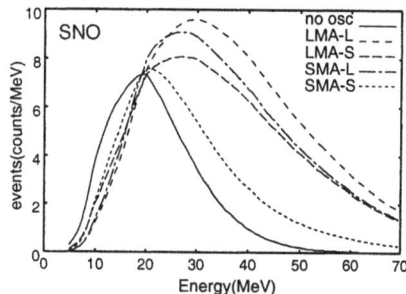

Fig. 1. Energy spectrum of positrons and electrons expected to be detected at SuperKamiokande. Solid, dashed, long-dashed, dash-dot-dash, and dotted lines correspond to no oscillation, model LMA-L, LMA-S, SMA-L, and SMA-S, respectively. The symbols -L and -S represent $\theta_{13} = 0.043$ and 1.0×10^{-6}, respectively

Fig. 2. Energy spectrum of positrons and electrons expected to be detected at SNO taking only CC events into account. Solid, dashed, long-dashed, dash-dot-dash, and dotted lines correspond to no oscillation, model LMA-L, LMA-S, SMA-L, and SMA-S, respectively

events to low-energy events will be a good measure of neutrino oscillation effects. We calculated the following ratio of events at both detectors:

$$R_{SK} \equiv \frac{\text{number of events at } 30 < E < 70\text{MeV}}{\text{number of events at } 5 < E < 20\text{MeV}} \tag{1}$$

$$R_{SNO} \equiv \frac{\text{number of events at } 25 < E < 70\text{MeV}}{\text{number of events at } 5 < E < 20\text{MeV}} \tag{2}$$

The plots of R_{SK} vs. R_{SNO} are shown in Fig. 3. The error bars include only statistical errors. At first glance, it seems to be possible to distinguish all the models including the no oscillation case (for detailed discussion, see [16,17,14,15]).

3   Supernova Relic Neutrinos and Neutrino Oscillation

It is generally believed that the core-collapse supernova explosions have traced the star formation history in the universe and have emitted a great number of neutrinos, which should make a diffuse background. This supernova relic neutrino (SRN) background is one of the targets of the currently working large neutrino detectors, SK and SNO. Comparing the predicted SRN spectrum with the observations by these detectors provides us with potentially valuable information on the nature of neutrinos as well as the star formation history in the universe. This SRN background has been discussed in a number of previous papers (see, papers cited in [3]).

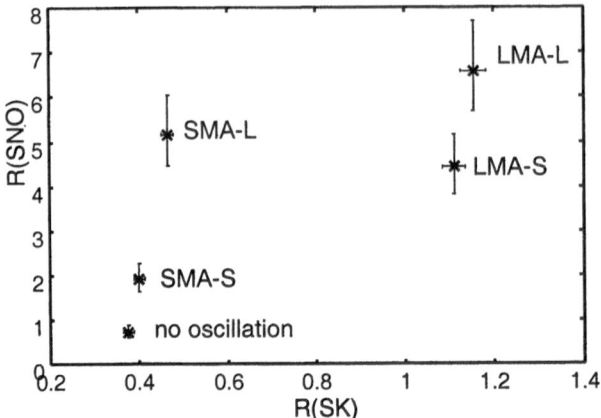

Fig. 3. The plot of R_{SK} vs. R_{SNO} for all the models. The error-bars represent the statistical errors

Ando et al. [3] investigated the SRN flux and the event rate at SK for various neutrino oscillation models taking into account of a realistic neutrino spectrum [18]. If neutrino oscillation occurs, $\bar{\nu}_{\mu,\tau}$'s are converted into $\bar{\nu}_e$'s which are mainly detected at SK detector. Because $\bar{\nu}_{\mu,\tau}$'s interact with matter only through the neutral-current reactions in supernovae, they are weakly coupled with matter compared to $\bar{\nu}_e$. Thus the neutrino sphere of $\bar{\nu}_{\mu,\tau}$'s is deeper in the core than $\bar{\nu}_e$'s and their temperatures are higher than $\bar{\nu}_e$'s. Therefore neutrino oscillation enhances the mean $\bar{\nu}_e$'s energy and enhances the event rate at the SK detector. In Fig. 4, the event rate of various neutrino oscillation models is shown assuming the supernova rate model estimated by observations of the star formation rate [13]. A clear difference of the LMA model from SMA or no oscillation models can be seen, especially at the high energy tail. This property results from the flux dependence on oscillation models. As a result, if we can detect SRN events above $\sim$ 10 MeV, we can discriminate the LMA model from the SMA or no oscillation models in any supernova rate models and any sets of cosmological parameters.

There are several background events which hinder the detection of SRN. These includes atmospheric and solar neutrinos, anti-neutrinos from nuclear reactors, spallation products, and decay electrons from invisible muons. We should find the energy region which is not contaminated by these background events and then calculate the detectable event rate of SRN. By careful examination of these events, we found that there is a narrow energy window from 10 MeV to 27 MeV, which is free from solar, atmospheric, and reactor neutrinos. However, according to Kaplinghat et al. [12] electrons or positrons from invisible muons are the largest background in the energy window from 19 to 35 MeV. In Fig. 5, the SRN event rate is compared with the invisible muon events. From this figure it is shown that SRN events can be seen only below about 12 MeV. In practice,

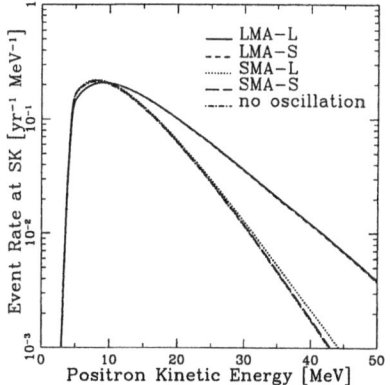

Fig. 4. Event rate of $\bar{\nu}_e$'s at SK for the neutrino oscillation models

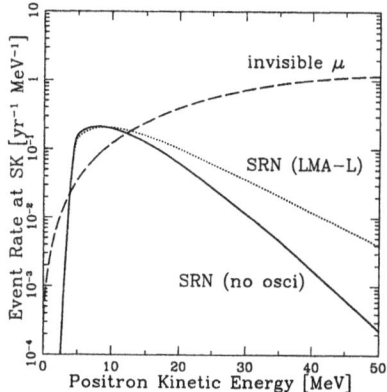

Fig. 5. Event rate at SK detector of SRN and invisible μ decay products. Two oscillation models are shown (no oscillation and LMA-L)

there is another serious background, i.e., spallation products induced by cosmic ray muons. Ultra high energy cosmic ray muons spall oxygens in the detector, and radioactive decay processes of these spalled nuclei occur. The event rate of the spallation background is several hundred per day per 22.5 kton. Although most of them can be rejected by the information of preceding muons, a small fraction cannot. (Roughly, this spallation product produces about 200,000 events per a year. Because the expected SRN event rate is less than 1 per a year, we should reject all these 200,000 events. For future detectors this problem is also quite difficult to solve.) Thus, this makes a serious background at the energy range below the maximum energy of beta spectrum of spallation products, 16 MeV. It looks there is no energy window of SRN, but we can detect the SRN events by subtracting the other background events from the total detected events. Consider the energy range $17 < (T_e/\text{MeV}) < 25$, where T_e is positron kinetic energy. This range corresponds to $19 < (E_{\bar{\nu}_e}/\text{MeV}) < 27$ by the simple relation, $E_{\bar{\nu}_e} = T_e + 1.8\text{MeV}$. There are two advantages in using this energy region. First, the SRN event rate is rather large, and second, the background (invisible muon) event rate is fairly well known by SK observation. The SRN event rate at SK in this energy range is $0.4 - 0.8 \text{ yr}^{-1}$. In contrast, the event rate of the invisible muon over the same energy range is 3.4 yr^{-1}. When SRN event rate is larger than the statistical error of background event rate, we can conclude that the SRN is detectable as a distortion of the expected invisible muon background event. Unfortunately, only one year observation does not provide any useful information about SRN. However, we can expect that ten-year observation provides several statistically meaningful results. The statistical error of invisible muon events in ten years is $\sqrt{34} = 5.8$, which is smaller than the event rate of LMA models and is larger than that of SMA and no oscillation models. Then these neutrino oscillation models can be distinguished by the observation of the event rate of invisible muon events. (If there is a discrepancy from expected event rate, this is due to SRN events and LMA models are favored.)

In future, it is expected that the next generation of water Čherenkov detectors have a much larger volume than that of SK. For example the HyperKamiokande project is now under consideration. HyperKamiokande is planed to be a water Čherenkov detector whose mass is about 1,000,000 tons (about 20 times larger than SK), and its location is near the SK detector. We expect that the SRN event becomes about 10 per one year for this detector, and statistically sufficient discussion of SRN is possible even using only one year data.

Acknowledgements

We would like to thank the Super Kamiokande people including Y. Totsuka, Y. Suzuki,A. Suzuki, M. Nakahata, and Y. Fukuda for useful discussions and comments. This work was supported in part by grants-in-aid for scientific research provided by the Ministry of Education, Science and Culture of Japan through Research grant no. S14102004 and 14079202.

References

1. Q.R. Ahmad et al., SNO Collaboration, Phys. Rev. Lett. **87**, 071301 (2001)
2. Q.R. Ahmad et al., SNO Collaboration, Phys. Rev. Lett. **89**, 011301 (2002)
3. S. Ando, K. Sato, and T. Totani, Astropart. Phys. (2002),(astro-ph/0202450).
4. M. Apollonio et al., Phys. Lett. B **466**, 415 (1999)
5. R.M. Bionta et al., Phys. Rev. Lett. **58**, 1494 (1987)
6. A.S. Dighe, A.Yu. Smirnov, Phys. Rev. D **62**, 033007 (2000)
7. G. Dutta, D. Indumathi, M.V.N. Murthy, and G. Rajasekaran, Phys. Rev. D **61**, 013009 (1999)
8. G.L. Fogli et al., hep-ph/0104221; O. Yasuda and H. Minakata, hep-ph/9602386; H. Minakata and O. Yasuda, Phys. Rev. D **56**, 1692 (1997)
9. S. Fukuda et al., Phys. Rev. Lett. **86**, 5656 (2001)
10. Y. Fukuda et al., Phys. Rev. Lett. **82**, 2644 (1999); K. Scholberg, hep-ex/9905016.
11. K. Hirata et al., Phys. Rev. Lett. **58**, 1490 (1987)
12. M. Kaplinghat, G. Steigman and T.P. Walker, Phys. Rev. D **62**, 043001 (2000)
13. C. Porciani and P. Madau, Astrophys. J. **548**, 522 (2001)
14. K. Takahashi and K. Sato, Phys. Rev. D, (2002). (hep-ph/0110105)
15. K. Takahashi and K. Sato, (hep-ph/0205070)
16. K. Takahashi, M. Watanabe, K. Sato and T. Totani, Phys. Rev. D **64**, 093004 (2001).
17. K. Takahashi, M. Watanabe and K. Sato, Phys. Lett. B **510**, 189 (2001)
18. T. Totani, K. Sato, H.E. Dalhed, and J.R. Wilson, Astrophys. J. **496**, 216 (1998)
19. S.E. Woosley and T.A. Weaver, ApJ. Suppl. **101**, 181 (1995)

Gamma-Ray Bursts and Jet-Powered Supernovae

S.E. Woosley[1], Weiqun Zhang[1], and A. Heger[2]

[1] Department of Astronomy and Astrophysics, University of California,
Santa Cruz, CA 95064, USA
[2] Department of Astronomy and Astrophysics, Enrico Fermi Institute,
The University of Chicago, 5640 S. Ellis Ave, Chicago, IL 60637, USA

Abstract. The last five years have seen growing challenges to the traditional paradigm of a core collapse supernova powered by the neutrino emission of a young proto-neutron star. Chief among these challenges are gamma-ray bursts (GRBs) and the supernovae that seem to accompany them. Here we review some recent – and not so recent – models for GRBs and supernovae in which strong magnetic fields, rotation, or accretion into a black hole play a role. The conditions for these energetic explosions are special and, at this point, there is no compelling reason to invoke them in the general case. That is, 99% of supernovae may still operate in the traditional fashion.

1 Introduction

The demise of spherically symmetric models for supernovae can be traced to 1987. Though we certainly already understood that stars (and pulsars) rotated and had magnetic fields and that this might affect the explosion [22], that neutrino powered convection had to be included in any realistic model [49] and that instabilities would be encountered as the shock moved out [9], it was the clear evidence for mixing on a large scale in SN 1987A [4] that drove us inexorably to multi-dimensional models. The migration was facilitated by developments in computer hardware and software that made multi-dimensional calculations practical. Still hope remained that, globally, things would still be pretty spherically symmetric. In particular, the shock wave coming out from the neutron star, though bounding regions that bubbled and mixed, was roughly spherical.

Events of similar significance happened in 1997, when it became clear that GRBs are located at cosmological distances [10,44], and again in 1998 when a supernova, SN 1998bw, was discovered in conjunction with a GRB. The supernova had very peculiar properties and, if modeled in one dimension (surely a gross approximation), had a kinetic energy in excess of 10^{52} erg [51,19]. Even more dramatic was the discovery that a significant fraction of that energy was contained in relativistic ejecta [21].

During the next four years evidence accumulated both for supernovae associated with GRBs [6,36,7] and for unusually energetic supernovae (see talk by Mazzali et al.). The term "hypernovae" [30] was often used to describe these exceptional explosions, and has lately come to denote almost any unusual supernova with inferred high energy (along the line of sight) or broad lines. Here we will avoid the term, which is not associated with any particular model, and speak of the specific *mechanisms* that might be responsible for exploding stars with great energy, gross asymmetry, and/or relativistic mass ejection. Obviously energy and asymmetry are not independent. A grossly asymmetric explosion may appear anomalously energetic – in terms of broad lines for example – along one line of sight and not another.

In this regard, GRBs themselves may be an extreme case of a continuous distribution of events ranging from nearly spherical supernovae with kinetic energies of order 10^{51} erg to events like GRB 990123 with an inferred equivalent isotropic energy of over 10^{54} erg. But is it energy or asymmetry? Recent analysis of the afterglows of GRBs [11] has shown that the total energies in GRBs are really remarkably clustered around 10^{51} erg, even for 990123, and that their exceptional brilliance is a consequence of having focused some appreciable fraction of that energy into a narrow, relativistic jet ($\Gamma \sim 200$) moving in our direction. Other observations have also shown the association of GRBs with star forming regions inside galaxies [8]. Taken together, a picture is emerging that at least some massive stars die while producing relativistic jets.

2   Jet-Powered Supernovae (JetSN) and Pulsar-Powered Supernovae

2.1   Rotation

All modern models for GRBs and JetSN invoke rapid rotation, either of a neutron star or of a disk around a black hole. For typical equations of state, a neutron star with radius 10 km and period ~ 5 ms has $\sim 10^{51}$ erg of rotational kinetic energy, and this is an upper bound on the final period that is needed, especially since most of the action occurs when the radius is 30 km, not 10 km.

The evolution of massive stars including the transport of angular momentum by magnetic [14] and non-magnetic processes [15] has been considered until core collapse in various papers by Heger, Woosley, Spruit, and Langer. To summarize, common red supergiants, the progenitors of most supernovae, end up producing neutron stars with rotation rates near break up when magnetic fields are omitted, and around 10 ms when current estimates of magnetic torques [39] are included. The 10 ms value accounts for angular momentum loss due to neutrinos flowing out of the neutron star, but does not include possible braking by a neutrino-powered magnetic stellar wind or by the propeller mechanism operating in conjunction with fallback [15].

For GRB progenitors a bare helium core is more appropriate. A helium star born (e.g. from a merger) with equatorial rotation 10% of Keplerian and low metallicity can retain enough angular momentum to form a centrifugally supported disk around a central (Kerr) black hole of $\sim 3 \, M_\odot$ provided that magnetic fields are left out of the calculation [14,16]. However, when an approximate treatment of angular momentum transport by magnetic fields is included [39] along with mass loss, the resulting rotation become too low to form centrifugally supported disks in the inner part of the core [16].

Admittedly our knowledge of magnetic torques inside evolved massive stars is still uncertain, but these results suggest that: a) magnetic field torques during the pre-supernova evolution are an important consideration, and b) within uncertainties, all current models may be allowed, but may require special circumstances. This may be why GRBs only occur in about 1% of supernovae (based on estimates of GRB beaming and the supernova rate in the universe).

Since nature is continuous, however, we may also expect many supernovae in which rotation plays an important role (i.e. the inferred pulsar rotation rate is faster than 5 ms), but no GRB is produced.

2.2   Pulsar-Powered Supernovae

Shortly after pulsars were discovered and their rapid rotation rates inferred, it was suggested that they might power supernovae [29]. If energies of $> 10^{50}$ erg must be rapidly dissipated by means other than neutrinos or gravity waves, it is unavoidable that a pulsar will influence supernova dynamics, leading, for example, to additional mixing. However, pulsars as the cause of common supernova explosions encounters at least two objections. First, the accretion rate shortly after neutron star formation is ~ 0.1 to 1 $M_\odot$ s^{-1}. The Alfven radius for this accretion rate is then [2]

$$r_A = 1.3 \times 10^4 \text{ cm } \mu_{30}^{4/7} \dot{M}_{32}^{-2/7} \tag{1}$$

with μ_{30} the magnetic moment in G cm^{-3} (10^{30} is approximately the value for B $\sim 10^{12}$ G) and $\dot{M}_{32}$, the accretion rate in units 10^{32} g s^{-1}. For the Alfven radius to be greater than the neutron star radius, ~ 10 km, with an accretion rate of 0.3 $M_\odot$ s^{-1} the magnetic moment must exceed 5×10^{33} and the B field must exceed 5×10^{15} G. When the explosion is developing, the protoneutron star radius is actually more like 30 km and the necessary magnetic moment about 10 times greater. This implies, baring ultrastrong magnetic fields, that no pulsar will be able to function during the critical epoch when the accretion rate is high and the probability of black hole formation large.

Second is the issue of ^{56}Ni nucleosynthesis. A shock like the one produced in neutrino powered explosions will raise a significant quantity of material to temperatures greater than 5×10^9 K and thus make iron group elements [52]. To do so the shock must receive its energy in a time short compared with that needed to cross the region where the nickel is made, about 4000 km. This is, at most, a few tenths of a second. If a pulsar does not deposit at least 10^{51} erg in this brief interval, very little nickel will be made to power the light curve. Such short braking times again require very large magnetic fields and rotation rates. While it may be that the occasional neutron star is born with these extreme properties (see below), we do not think it happens in most supernovae.

2.3   MHD Jets and Explosions

Another supernova model with us for over 30 years invokes powerful bi-polar outflows energized by magnetic wind up and instabilities in a differentially ro-

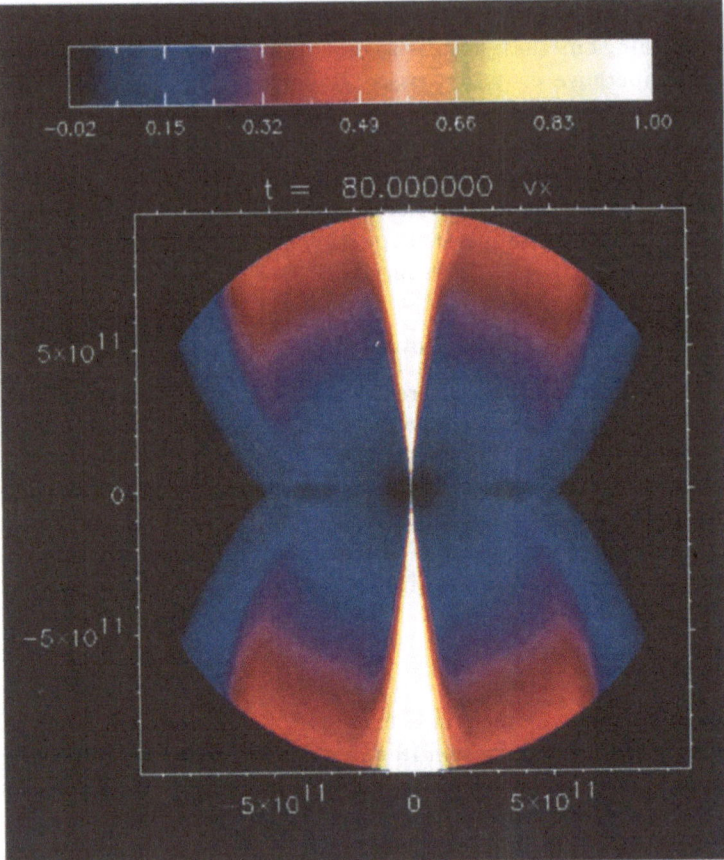

Fig. 1. A massive Wolf-Rayet star being exploded by the passage of relativistic jets along its axes [54]. The jet was initiated at 2000 km in a Wolf-Rayet star with radius 700,000 km and had a Lorentz factor of 10 for the first 10 seconds which slowly declined to 2 at 1000 s. The energy input was 5×10^{50} erg s^{-1} (per jet) for 10 s declining to 10^{47} erg s^{-1} at 1000 s. The initial ratio of internal energy to kinetic energy in the jet was 20 and the opening angle, 20 degrees (which was quickly reduced by hydrodynamical focusing). The picture shows radial velocity 80 s after the initiation of the jet

tating proto-neutron star [22,5,27]. Generically these outflows are referred to as LeBlanc-Wilson jets. Their creation again requires very large magnetic fields and rotation rates, once regarded as unrealistic. Interest in this variety of model has been rekindled however [47,48,3], both by the observation of jets in GRBs and by promising models for soft gamma-ray repeaters and anomalous S-ray pulsars that invoke magnetic fields up to 10^{15} G [41,42].

Granted that such neutron stars exist and may be born rotating rapidly, a robust supernova model does not necessarily follow. A jet is not a particularly efficient way to explode a star. Even one introduced with a significant opening angle is rapidly collimated by its passage through the star [53] and collides with only a small fraction of the mass (Fig. 1). Lateral shocks move around the star, but a lot of the matter falls back, enough that it may be difficult to preserve the neutron star. The jet may also produce very little ^{56}Ni, not enough to explain the light curves of Type Ib and Ic supernovae. The velocities of the resulting supernova will be highly asymmetric with very high values along the axis. This will give great variation in the properties of ordinary supernovae seen at different angles. Such variations are not seen.

This is not to say that a model for supernovae in which rotation and magnetic fields play a major role is ruled out. The magnetic torque on a spinning protoneutron star, $\tau = dL/dt$ with L the angular momentum, is approximately $B_r B_\phi R^3$, suggesting that an angular momentum of $I\omega \sim 10^{48}(I/10^{45})(\omega/10^3)$ erg s could be braked in a few seconds if the wound up poloidal field, B_ϕ, and radial field, B_r exceeded 10^{15} gauss. This would lead to the rapid dissipation of $\sim 10^{51}$ erg, possibly by Alfven waves ($\sim r^2(\delta B)^2 v_A$ with $v_A \sim 10^{10}$ cm s^{-1}, the Alfven speed), long wavelength electromagnetic waves [43] or magnetic reconnection. Larger rotation rates and stronger fields could provide greater energies. Neutrino energy deposition and the overturn it causes might aid in producing the necessary B_r. Further work is needed here, especially on the idea that neutrino energy deposition and MHD models for supernovae are not exclusive.

3   Models with Black Holes

Models for supernovae in which a large part of the energy comes from an accreting black hole are newcomers to the scene, motivated chiefly by a need to explain GRBs. However, it is recognized that these same models may have broader applicability and, in less extreme versions or in stars that still retain their hydrogen envelope, might power supernovae. Such supernovae would probably retain unusual properties such as gross asymmetry or high energy.

3.1   Supranovae

It has been suggested by Vietri & Stella [45,46] and others that GRBs might result from the delayed implosion of rapidly rotating neutron stars to black holes. The neutron star forms in a traditional (neutrino-powered) supernova, but is "supramassive" in the sense that without rotation, it would collapse, but with rapid rotation, collapse is delayed until angular momentum is lost. The momentum can be lost by gravitational radiation and by magnetic field torques. Vietri and Stella assume that the usual pulsar formula holds and, for a field of 10^{12} gauss, a delay of order years (depending on the field and mass) is expected, but other parameters might give a shorter delay. When the centrifugal support becomes sufficiently weak, the star experiences a period of runaway deformation and gravitational radiation before collapsing into a black hole. It is assumed that $\sim 0.1 M_\odot$ is left behind in a disk which accretes and powers the burst explosion.

As a GRB model, the supranova has several advantages. It, as well as the collapsar model discussed later, predicts an association of GRBs with massive stars and supernovae. Moreover it produces a large amount of material enriched in heavy elements located sufficiently far from the GRB as not to obscure it. The irradiation of this material by the burst or afterglow can produce X-ray emission lines as have been reported in several bursts [31,32,35]. However, the supranova model also has some difficulties [26]. It may also take fine tuning to produce a GRB days to years after the neutron star is born. Shapiro [38] has shown that neutron stars requiring differential rotation for their support will collapse in only a few minutes. The requirement of rigid rotation reduces the range of masses that can be supported by rotation to, at most, ~20% above the non-rotating limit [38,37].

3.2 Collapsars

Basic collapsars Generically, a collapsar is a rotating massive star whose central core collapses to a black hole surrounded by an accretion disk [50,23]. Accretion of at least a solar mass through this disk produces outflows that are further collimated by passage through the stellar mantle. These flows attain high Lorentz factor as they emerge from the stellar surface and, after traversing many stellar radii, produce a GRB and its afterglows by internal and external shocks. The passage of the jet through the star also gives a very asymmetric supernova of order 10^{51} erg [53].

There are three ways to make a collapsar and each is likely to have different observational characteristics.

- A standard (Type I) collapsar is one where the black hole forms promptly in a helium core of approximately 15 to 40 $M_\odot$. There never is a successful outgoing shock after the iron core first collapses. A massive, hot proto-neutron star briefly forms and radiates neutrinos, but the neutrino flux is inadequate to halt the accretion. Such an occurrence seems likely in helium cores of mass over ~ 15$M_\odot$ because of their large binding energy [52] and the rapid accretion that characterizes the first second after core collapse [12].

- A variation on this theme is the "Type II collapsar" wherein the black hole forms after some delay – typically a minute to an hour, owing to the fallback of material that initially moves outwards, but fails to achieve escape velocity [25]. Such an occurrence is again favored by massive helium cores. Unfortunately the long time scale associated with the fall back may be, on the average, too long for typical long, soft bursts. Their accretion disks are also not hot enough to be neutrino dominated and this may affect the accretion efficiency [28] and therefore the energy available to make jets.

- A third variety of collapsar occurs for extremely massive metal-deficient stars that probably existed only in the early universe [1,13]. For non-rotating stars with helium core masses above 133 $M_\odot$ (main sequence mass 260 $M_\odot$), it is known that a black hole forms after the pair instability is encountered [17]. It is widely suspected that such massive stars existed in abundance in the first generation after the Big Bang at red shifts $\sim$5 – 20. For *rotating* stars the mass limit for black hole formation will be raised. The black hole that forms here, about 100 $M_\odot$, is more massive, than the several $M_\odot$ characteristic of Type I and II collapsars, but the accretion rate is also much higher, $\sim$10 $M_\odot$ s^{-1}, and the energy released may also be much greater. The time scale is also much longer.

For both Type I and II collapsars it is also essential that the star loses its hydrogen envelope before death. No jet can penetrate the envelope in less than the light crossing time, typically 100 s for a blue supergiant and 1000 s for a red one. After running into $1/\Gamma$ of its rest mass, a ballistic jet loses its energy.

Because of space limitations, we will not review details of the collapsar model here, but refer the reader to the published literature especially [23,53,54]. We will emphasize however two recent developments of great interest: nucleosynthesis in collapsar disks and the prediction by the collapsar model of other forms of high energy transients, especially cosmological X-ray flashes and events like GRB 980425/SN 1998bw.

^{56}Ni production and the r-process Lacking a hydrogen envelope, the supernova that accompanies a GRB made by a collapsar will be Type Ib or Ic with an optical luminosity given entirely by the yield of ^{56}Ni. In Type I collapsars however, the material that would have become ^{56}Ni falls into the black hole. The jet itself subtends a small solid angle and carries a small, albeit very energetic mass. It cannot propagate outwards until the mass flux inwards at the pole has declined, i.e., the density has gone down. This makes it hard for the jet itself to synthesize much ^{56}Ni. How then is the supernova visible?

It is believed that the ^{56}Ni in collapsars is made not by the jet, but by the disk wind [23,28]. In the parlance of Narayan et al. [28], it could be that at late times (after $\sim$10 s), a neutrino-dominated accretion disk (NDAF) switches to a convection dominated accretion disk (CDAF) with a large fraction of the mass flow being ejected. MacFadyen and Woosley [23] even found considerable mass outflow from NDAFs. We postulate that a certain fraction of the accreting matter – composed initially of nucleons or iron group elements – is ejected at high velocity ($\sim$0.1 c) by the accretion disk.

But will the material be ^{56}Ni? Nucleosynthesis in collapsar disks has been explored recently by Pruet and colleagues at LLNL [33]. They find that the composition flowing out from the disk and in the jet is very sensitive to both the accretion rate and assumed viscosity of the disk. For an "α-disk" with $\alpha \approx 0.1$ or less and accretion rates 0.1 $M_\odot$ s^{-1} and more the composition will not be ^{56}Ni, but more neutron-rich isotopes of iron, or even r-process nuclei. For accretion rates around 0.01 $M_\odot$ s^{-1} the composition will be *proton*-rich ($Y_e \approx 0.51$), though still dominated by ^{56}Ni. Interestingly typical accretion rates for Type I collapsars are $\sim$ 0.05 $M_\odot$ s^{-1} (less at later times) and ^{56}Ni synthesis is possible. For Type II collapsars the accretion rate is lower and the disk is proton-rich. Lower values of α shift the nucleosynthesis to low Y_e and for $\alpha = 0.01$ or less, Type I collapsar disks make no ^{56}Ni.

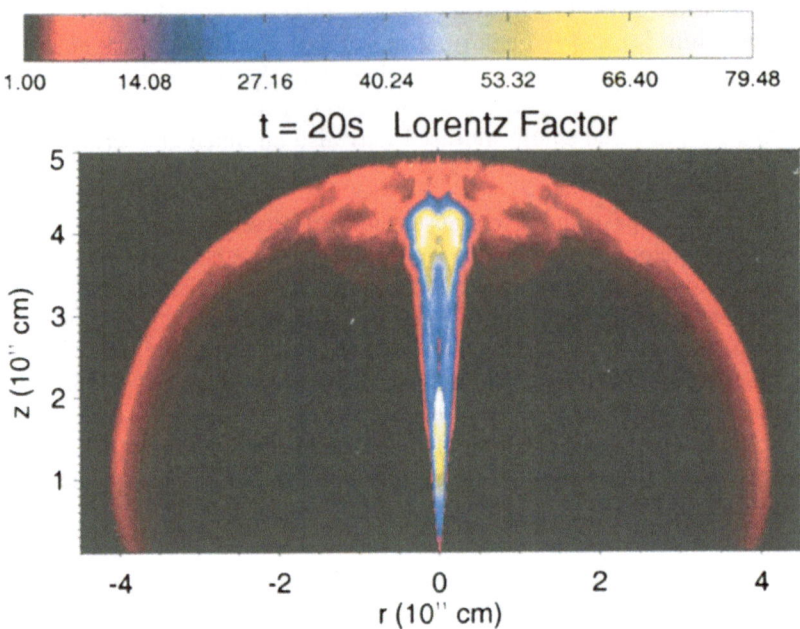

Fig. 2. The break out of a relativistic jet and its cocoon 22 seconds after the jet's initiation in the star [54]

X-ray flashes and supernovae The collapsar model was originally intended as an explanation for GRBs but time, additional calculations and observations suggest it has broader implications. These essentially hinge on the answer to the question "If a GRB from a collapsar is only seen by observers in about 0.3% of the sky, what do other observers see?" Clearly these will be the most common events. Additionally, one may inquire what happens when a collapsar occurs in a star still having a hydrogen envelope [25], if the parameters are such that high Lorentz factor is not achieved or the jet engine turns off before the jet emerges from the star [24].

In the equatorial plane of a collapsar – the common case – probably little more is seen than an extraordinary supernova. In fact the supernova may not even appear exceptionally energetic because the high velocities are all along the rotational axis (Fig. 1). Off axis though, in a collapsar that made a GRB, one will see X-ray flashes made by the explosion of the jet cocoon as it breaks out of the star [34,54]. The cocoon contains about $10^{50} - 10^{51}$ erg [53] and has Lorentz factor $\Gamma \sim 5 - 10$ (Fig. 2). By way of an external shock with the pre-explosive wind of the stellar progenitor, this material can produce a bright transient visible out to ~30 degrees from each axis. Even though it has lower energy per solid angle than the GRB jet (which is concentrated within about 5 degrees), relativistic beaming compensates to make the observable fluence comparable. That is, the GRB beams its emission to $1/\Gamma \sim 0.005$ radians = 1/4 degree while the X-ray flash (XRF) is beamed to perhaps 10 degrees. The duration of such events depends on the Lorentz factor and and the pre-explosive mass loss, but could be from tens of seconds to minutes.

These properties mesh well with the recently discovered class of cosmological XRFs [18,20] which share many of the properties of long-duration GRBs (duration, frequency of occurrence, isotropy on the sky, non-thermal spectrum, non-recurring), but have no hard emission above about 10 keV. If our speculations are correct, every (long-soft) GRB should have an underlying XRF that may even be visible as a precursor to the GRB. We also predict supernovae in association with XRFs and these might be looked for [40].

Acknowledgements

This research has been supported at UCSC by the NSF (AST 02-06111), NASA (NAG5-12036, MIT-292701), and the SciDAC Program of the DOE (DE-FC02-01ER41176). At Chicago, work was supported by the DOE ASCI Program (B347885). AH is supported in part by the Department of Energy under grant B341495 to the Center for Astrophysical Thermonuclear Flashes at the University of Chicago.

References

1. Abel, T., Bryan, G.L., & Norman, M.L. 2002, Science, 295, 93
2. Alpar, M.A. 2001, ApJ, 554, 1245
3. Ardeljan, N.V., Bisnovatyi-Kogan, G.S., & Moiseenko, S.G. 2001, Ap&SS, 276, 295
4. Arnett, W.D., Bahcall, J.N., Kirshner, R.P., & Woosley, S.E., 1989, ARA&A, **27**, 629.
5. Bisnovatyi-Kogan, G.S. 1970, Astron. Zh., 47, 813
6. Bloom, J.S., Kulkarni, S.R., Djorgovski, S.G., Eichelberger, A.C., Cote, P., Blakeslee, J.P., Odewahn, S.C., Harrison, F.A., et al. 1999, Nature, 401, 453
7. Bloom, J.S., Kulkarni, S.R., Price, P.A., Reichart, D., Galama, T.J., Schmidt, B.P., Frail, D.A., Berger, E. 2002, ApJ, 572, L45
8. Bloom, J.S., Kulkarni, S.R., & Djorgovski, S.G. 2002, AJ, 123, 1111
9. Chevalier, R.A., & Klein, R.I. 1978, ApJ, **219**, 994
10. Costa, E., Frontera, F., Heise, J., Feroci, M., et al. 1997, Nature, 387, 783
11. Frail, D.A., Kulkarni, S.R., Sari, R., Djorgovski, S.G., Bloom, J.S., Galama, T.J., Reichart, D.E., Berger, E., et al. 2001, ApJ, 562, L55
12. Fryer, C.L. 1999, ApJ, 522, 413
13. Fryer, C.L., Woosley, S.E., & Heger, A. 2001, ApJ, 550, 372
14. Heger, A., Langer, N., & Woosley, S.E. 2000, ApJ, 528, 368
15. Heger, A., Woosley, S.E., & Spruit, H. 2002, ApJ, in preparation
16. Heger, A., & Woosley, S.E. 2002, in Proc. Woods Hole GRB meeting, ed. Roland Vanderspek, in press (astro-ph/0206005)
17. Heger, A., & Woosley, S.E. 2002, ApJ, 567, 532
18. Heise, J., in't Zand, J., Kippen, R.M., Woods, P.M. 2001, GRBs in the Afterglow Era, eds. Costa, Frontera, & Hjorth, ESO Astrophysics Symposia, (Springer), 16
19. Iwamoto, K., Mazzali, P.A., Nomoto, K., Umeda, H. et al. 1998, Nature 395, 672

20. Kippen, R.M., Woods, P.M., Heise J., in't Zand, J.J.M. 2002, in Proc. Woods Hole GRB meeting, ed. Roland Vanderspek, in press (astro-ph/0203114)
21. Kulkarni, S.R., Frail, D.A., Wieringa, M.H., Ekers, R.D., et al. 1998, Nature, 395, 663
22. LeBlanc, J.M., & Wilson, J.R. 1970, ApJ, 161, 541
23. MacFadyen, A., Woosley, S.E. 1999 ApJ, 524, 262
24. MacFadyen, A. 2000, PhD Thesis, UCSC
25. MacFadyen, A., Woosley, S.E., & Heger, A. 2001, ApJ, 550, 410
26. McLaughlin, G.C., Wijers, R.A.M.J., & Brown, G.E., & Bethe, H.A. 2002, 567, 454
27. Meier, D.L., Epstein, R.I., Arnett, W.D., & Schramm, D.N. 1976, ApJ, 204, 869
28. Narayan, R., Piran, T., & Kumar, P. 2001, ApJ, 557, 949
29. Ostriker, J.P. & Gunn, J.E. 1971, ApJ, 164, L95O
30. Paczynski, B. 1998, ApJ, 494, L45
31. Piro, L., Costa, E., Feroci, M., Frontera, F., Amati, L., dal Fiume, D., Antonelli, L.A., et al. 1999, ApJ, 514, L73.
32. Piro, L., Garmire, G., Garcia, M., Stratta, G., Costa, E., Feroci, M., Meszaros, P., Vietri, M., et al. 2000, Science, 290, 955
33. Pruet, J., Woosley, S.E., & Hoffman, R.D. 2002, ApJ submitted, astroph-0209412
34. Ramirez-Ruiz, E., Celotti, A. & Rees, M.J. 2002, MNRAS, in press (astro-ph/0205108)
35. Reeves, J.N., Watson, D., Osborne, J.P., Pounds, K.A., O'Brien, P.T., Short, A.D.T., Turner, M.J.L., Watson, M.G., et al, 2002, Nature, 416, 512.
36. Reichart, D. 1999, ApJL, 521, 111
37. Salgado, M., Bonazzola, S., Gourgoulhon, E., & Haensel, P. 1994, A&AS, 108, 455
38. Shapiro, S.L. 2000, ApJ, 544, 397
39. Spruit, H.C. 2002, A&A, 381, 923
40. Soderberg, A.M., Price, P.A., Fox, D.W., Kulkarni, S.R. et al. 2002, GCN 1554, http://gcn.gsfc.nasa.gov/gcn3/1554.gcn3
41. Thompson, C., & Duncan, R.C. 1995, MNRAS, 275, 255
42. Thompson, C., & Duncan, R.C. 1996, ApJ, 473, 322
43. Usov, V. 1992, Nature, 357, 472
44. van Paradijs, J., Groot, P.J., Galama, T., Kouveliotou, C., et al. 1997, Nature, 386, 686
45. Vietri, M., & Stella, L. 1998, ApJ, 507, L45
46. Vietri, M., & Stella, L. 1999, ApJ, 527, L43
47. Wheeler, J.C., Yi, I., Höflich, P., & Wang, L. 2000, ApJ, 537, 810
48. Wheeler, J.C., Meier, D.L., & Wilson, J.R. 2002, ApJ, 568, 807
49. Wilson, J.R., Mayle, R., Woosley, S.E., & Weaver, T.A. 1986, Proc 12th Texas Symp. Rel. Ap., Ann. N.Y. Acad. Sci., 470, 267
50. Woosley, S.E. 1993, ApJ, 405, 273.
51. Woosley, S.E., Eastman, R.G., Schmidt, B.P. 1999, ApJ, 516, 788
52. Woosley, S.E., Heger, A., & Weaver, T.A. 2002, RMP, 74, 1015
53. Zhang, W., Woosley, S.E., & MacFadyen, A. 2002, ApJ, in press
54. Zhang, W., Woosley, S.E., & Heger, A. 2002, ApJ, in preparation

Supernova Explosions
from Accretion Disk Winds

A.I. MacFadyen

Theoretical Astrophysics, California Institute of Technology,
Pasadena, CA 91125, USA

Abstract. Winds blown from accretion disks formed inside massive rotating stars may result in stellar explosions observable as Type Ibc and Type II supernovae. A key feature of the winds is their ability to produce the radioactive ^{56}Ni necessary to power a supernova light curve. The wind strength depends on accretion disk cooling by neutrino emission and photo-disintegration of bound nuclei. These cooling processes depend on the angular momentum of the stellar progenitor via the virial temperature at the Kepler radius where the disk forms. The production of an observable supernova counterpart to a Gamma-Ray Burst (GRB) may therefore depend on the angular momentum of the stellar progenitor. Stars with low angular momentum may produce a GRB without making an observable supernova. Stars with large angular momentum may make extremely bright and energetic supernovae like SN 1998bw. Stars with an intermediate range of angular may simultaneously produce a supernova and a GRB.

1 Introduction

Observational evidence continues to establish the association of long gamma-ray bursts with active star forming regions of galaxies, e.g. [4]. In addition, there are indications from the optical light curves of several (long) GRB afterglows that supernova components may be directly observed as emission from the decelerating relativistic ejecta fades [2,13,3]. These observations continue to support the collapsar models for GRBs in which the core of a massive rotating star collapses to a black hole and rapidly accretes [14,7]. It is therefore of interest to try to understand under what conditions a star which makes a GRB will also make an observable supernova.

Collapsars [7,9] form dense accretion disks ($\rho \gtrsim 10^9$ g cm^{-3}) which are extremely optically thick to photons ($\tau_\gamma \sim 10^{19}$). As the stellar gas spirals through the disk, photons are trapped and accrete with the gas. This is in distinction from "thin" accretion disks in which photons are assumed to escape to infinity carrying away the locally dissipated energy. Since photons are trapped, viscous dissipation of orbital energy increases the disk entropy, pressure gradients are important for the force balance and the disk is "thick". Such non-radiating accretion flows are capable of ejecting gas away from the black hole [5]. Accretion in these disks is inefficient with significant fractions of the gas supplied at large radii being ejected from the system.

An important feature of collapsar disks is the realization at sufficiently high accretion rates ($\dot{M} \sim 0.1 M_\odot s^{-1}$) or temperatures ($T \sim 10^{10}$ K) and densities

$(\rho \sim 10^9 \text{ g cm}^{-3})$ at which the loss of thermal energy to neutrino emission and photodisintegration of heavy nuclei allows for accretion with a range of efficiency.

2   Winds

MacFadyen & Woosley [7] showed that collapsar disks eject comparable amounts of material in a wind as is accreted by the central black hole. The fraction of accreted gas depends on the efficiency of neutrino cooling at removing entropy from the accreting gas. The remainder is ejected from the black hole as an outflowing wind. Recent semi-analytic work [10] has mapped the parameter space of inefficient neutrino-cooled accretion in agreement with detailed calculations of [7] for limited parameters.

Of particular interest in the case of collapsars is the chemical composition of the wind. Collapsar disks are hot enough to completely photodisintegrate heavy nuclei to free nucleons (neutrons and protons). Recent simulations [8] and [7] show expulsion of free nucleon gas in the wind. This is important for two reasons: 1) energetics: free nucleons combining to iron group nuclei (e.g. ^{56}Ni) release 8 MeV/nucleon or 1.5×10^{52} erg per solar mass of recombined material. 2) observability: this ejected material provides a long term energy supply to the explosion (through radioactive decay of ^{56}Ni) enabling the gas to shine on time scales of months.

3   Light Curve

Models of the light curve of the energetic and peculiar Type Ibc supernova SN1998bw require large quantities of ^{56}Ni (M(^{56}Ni) $\sim 0.5 M_\odot$) [6,15]. Conventional models require large explosion energies to produce sufficient nickel and fit the light curve. In addition abnormally high expansion velocities were inferred from the unusual spectrum indicating a large explosion energy ($\sim 10^{52}$ erg). Several groups have also interpreted deviations from power law decay of GRB optical transients as supernovae light curve components and have matched them with appropriately shifted SN1998bw light curves [2].

Since supernovae are invoked to interpret these observations it is important to note that a stellar explosion (e.g. jets piercing a star) is not necessarily a supernova. Supernovae, as an observable phenomenon, require a persistent source of energy input to power a light curve for long times (weeks to months). It is necessary to make ^{56}Ni in the explosion so that radioactive decay (to Cobalt to Iron) injects energy into the gas so that it can shine. Lacking a persistent source of energy input, a stellar explosion would be unobservable via electromagnetic radiation. Explosion energy released in the optically thick star would simply be converted to expansion kinetic energy with little or no light emitted.

In conventional core collapse supernovae some nickel is thought to be produced via explosive nuclear burning behind the explosion shock. However, current models for these "delayed" supernova explosions have trouble producing the 10^{51} erg for a normal supernova (in fact, some current models fail to get any explosion at all!) and are unlikely to be capable of producing the higher energies required for SN 1998bw.

4 Angular Momentum

As we have seen, neutrino cooling and photodisintegration of heavy nuclei are crucial for allowing gas to accrete efficiently. The neutrino cooling depends sensitively on temperature (e.g. $Q_\nu^- \propto T^6$ for neutrino losses due to pair capture on free nucleons) and therefore on the radius where the disk forms. This radius is, in turn, dependent on the angular momentum of the accreting gas with the disk first forming at the Kepler radius $R_{kep} \equiv j^2/GM = 2.5 \times 10^7 j_{17}^2 M_3^{-1}$ cm, where j_{17} is the specific angular momentum of the accreting gas in units of 10^{17} cm^2 s^{-1} and M_3 is the mass of the central black hole in units of three solar masses. The virial temperature for gas falling to its Kepler radius $T_{vir} = GMm_p/3k_B R_{kep} = 3.3 \times 10^{10} M_3^2 j_{17}^{-2}$ K, where m_p is the proton mass and k_B is the Boltzmann constant. In terms of gravitational radii, $R_G \equiv GM/c^2$, this temperature is $T_{vir} = m_p c^2/3k_B r = 1.8 \times 10^{12} r^{-1}$ K (assuming a Newtonian potential), where $r \equiv R/R_G$. We see that gas with $j_{17} \approx 1$ is heated to above 10^{10} K so that it is fully photodisintegrated to free neutrons and protons from its original composition of Silicon, Oxygen and Helium. This means that capture of electron-positron pairs onto the free neutrons and protons serves as an efficient neutrino emission process which cools the gas and helps it to accrete efficiently. Gas with $j_{17} \gtrsim 2.6$, however, heats to less than 5×10^9 K. At these lower temperatures the heavy nuclei fail to photodisintegrate and pair capture neutrino cooling is suppressed. This gas is therefore poorly cooled and subject to being driven from the disk.

It is worth noting that gas with $1 \lesssim j_{17} \lesssim 2.6$ is partially photodisintegrated. Photodisintegration acts as a loss for thermal energy for the gas and thus is effectively a cooling process, robbing about 10^{19} erg of thermal energy from every gram of photodisintegrated nuclei. This process helps the gas to accrete and provides free nucleons which enhance the neutrino cooling.

The above discussion assumed $M_3 = 1$ though the scaling with M_3 is apparent.

5 The Afterburner

Interestingly, energy lost to photodisintegration becomes available again if the gas is ejected from the disk and begins to reassemble.

Effectively, gravitational energy is temporarily stored in the freeing of nucleons from the heavy nuclei. These nucleons are volatile in the sense that the have a huge nuclear energy source if they manage to escape the energetic photons trapped in the (optically thick) accretion disk. Accretion physics may provide the nucleons with opportunity to escape the disk's nuclei-disintegrating photon bath. Once free they can quickly recover nuclear binding energy by reassembling into iron group elements. This process can be explosive since the nuclei may

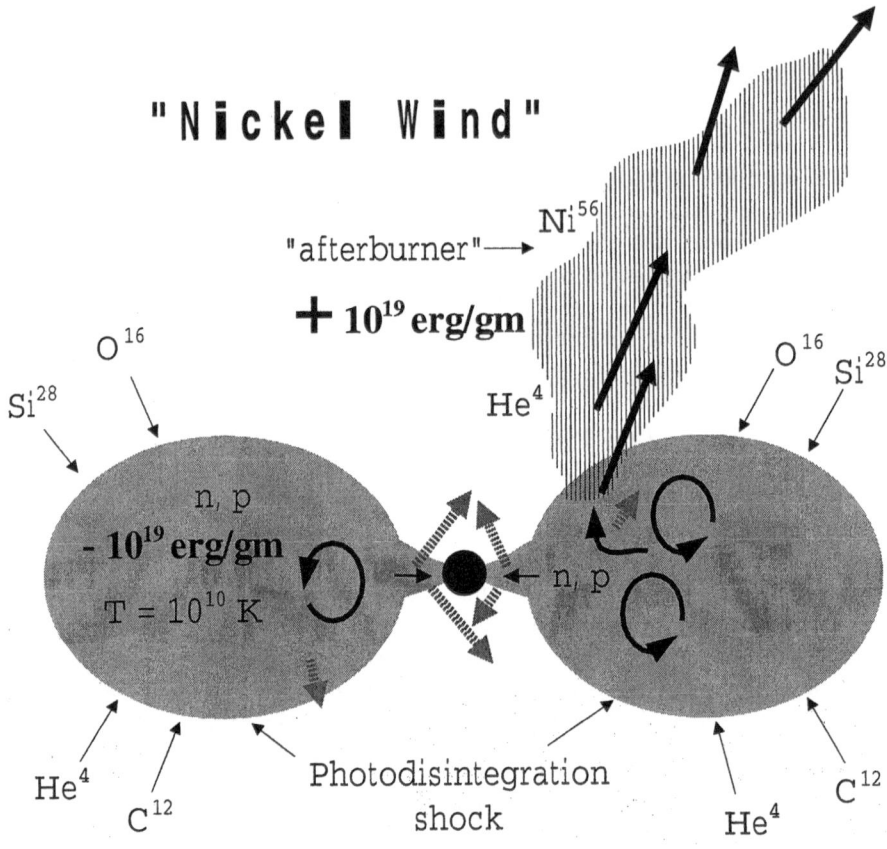

Fig. 1. Cross section of a collapsar accretion disk feeding a stellar mass black hole (center). The disk is embedded in a collapsing star that is falling onto the disk at rates above $0.1M_\odot$ s^{-1}. A wind (the striped region in the upper right) is blown from the collapsar disk at speeds of up to $\sim 40,000$ km s^{-1}. The wind is composed of free neutrons and protons which can recombine to iron group elements injecting 1.5×10^{51} erg per 0.1 $M_\odot$ of reassembled nucleons. ^{56}Ni in the wind can power a long term "supernova" light curve via radioactive decay of nickel and cobalt. The black solid arrows indicate the velocity of the gas flow while the thick dashed lines represent neutrino emission. The wind is shown only in the upper right quadrant for clarity but is in reality present in all four quadrants

recombine in seconds compared to millions of years it took them to assemble (burn) the first time around during the slower pre-explosion nuclear burning stages. In fact the reassembly, plus the kinetic luminosity of the disk wind, may power extremely energetic explosions. SN1998bw may be an example.

6 GRB with Supernova

The collapsar model relies on rapid accretion into the central black hole to power relativistic jets which pierce the star and make a GRB and afterglow via internal and external shocks. It is notable that for an interesting range of angular momentum the accretion of the star simultaneously feeds the black hole rapidly and powers a wind [7].

There are several interesting regimes determined by the angular momentum present in the collapsing star:

The following values of angular momentum correspond to important transition radii in the accretion flow:

j_{isco} angular momentum of the innermost stable circular orbit. This is the minimum angular momentum needed to form a disk around a black hole.

j_ν angular momentum of gas that falls deep enough in the gravitational potential to photodisintegrate the heavy nuclei to free nucleons activating pair capture neutrino emission as an efficient coolant.

j_γ angular momentum of gas that falls deep enough to cool partially. Some gas accretes and some is expelled in a wind. The relative amount depends on the exact value of j

1. $j_{isco} < j < j_\nu$ - efficient neutrino cooling allows rapid accretion into black hole with plenty of power potentially going into jets with little or no outflows expected. This kind of star would not be expected to produce a bright supernova since little or no ^{56}Ni is expected to be present in the exploding star. A possible caveat is that there is some nickel production via explosion burning in the lateral jet shock but the temperature is low in this region and not much mass is involved.

2. $j_\nu < j < j_\gamma$ - some gas accretes and some is ejected in a wind rates can be comparable depending on j. This can make both a GRB and a "supernova".

3. $j > j_\gamma$ Gas doesn't cool efficiently so doesn't feed the hole rapidly. Not good for making an accretion powered GRB. Outflows may result with some recombination nickel possible if some gas is heated above 5×10^9 K by a combination of virialization and viscous dissipation. Of interest here is explosive burning of centrifugally supported oxygen.

A less interesting regime is $j < j_{lso}$ for which the gas falls directly to the innermost stable circular orbit without forming an accretion disk.

Note that electromagnetic extraction of black hole spin energy is a possible source of jet energy even for "slowly" accreting black holes. Convective motions may even be favorable for building up large magnetic fields needed to extract the hole spin energy.

7   Photodisintegration

As stellar gas collapses onto the collapsar accretion disk, adiabatic compression and shocks can raise the temperature sufficiently to photodisintegrate the gas to alpha particles and free nucleons. The destruction of the heavy nuclei is an energy sink for the gas which is helpful in allowing accretion to occur. Photodisintegration of heavy nuclei (e.g., Silicon, Oxygen in the collapsing core) costs ηc^2 per unit mass where $\eta \approx 0.007$ (or 8 MeV/nucleon) for complete disintegration to free nucleons. Gas falling in a gravitational potential can dissipate its accretion energy by swapping nuclear binding energy for gravitational binding energy. A measure of where this occurs is where the gravitational binding energy equals the nuclear binding energy $\delta\phi = GM/r = \eta c^2$. If gas near the equator falls to its keplerian radius the other half goes into the kinetic energy of keplerian rotation so $\delta\phi = 1/2GM/r$

Measuring radius in gravitational radii $r \equiv \hat{r} r_g$ where $r_g \equiv GM/c^2$ we can define a photodisintegration radius $\hat{r}_{pd} \equiv 1/2\eta$. We thus expect photodisintegration to "cool" accreting gas when it falls to a radius of $\sim 70\, r_g$.

8   Viscosity

The above scenario assumes significant viscosity in the disk gas corresponding to a Shakura-Sunyaev alpha viscosity parameter $\gtrsim 0.1$. The temperature and density of the disk wind and hence the nucleosynthesis depend on the disk viscosity. Observations of the supernova powered by a collapsar wind may therefore help to constrain the viscosity of the collapsar accretion disk.

Recent calculations of the neutron abundance in one-dimensional accretion disk models relevant to GRBs [11] indicate that the inner parts of collapsar accretion disks may be too neutron rich to produce significant quantities of ^{56}Ni [12,1]. Outflows from these inner regions may instead be of interest for rare nucleosynthesis like the r-process. However, much of the mass loss from GRB accretion disks (collapsars and otherwise) may come from the outer regions of the disks where electron capture has not significantly neutronized the material composing the outflow.

Collapsar disks may have more than one active wind blowing region: 1. the outer disk where low densities imply non-degenerate electrons and little neutron excess ($Y_e \sim 0.5$). Radioactive ^{56}Ni may result from nucleosynthesis taking place in as this wind expands. 2. The innermost disk where the disk becomes optically thick to neutrino emission and is again poorly cooled. This innermost disk wind is significantly neutronized and will not produce ^{56}Ni. However, it is an interesting site for the r-process because of the large neutron fraction and large entropies attained from viscous dissipation.

In between these two disk regions, the disk is (partially) neutrino-cooled and most (but not necessarily all) of the gas can accrete. This region of the disks transports the fraction of gas received from the outer, i.e. the gas not ejected in the wind, plus gas falling onto the neutrino-cooled region of the disk.

References

1. A. M. Beloborodov, arXiv:astro-ph/0210522.
2. Bloom, J. S., Kulkarni, S. R., Djorgovski, S. G., Eichelberger, A. C., Cote, P., Blakeslee, J. P., Odewahn, S. C., Harrison, F. A., Frail, D. A., Filippenko, A. V., Leonard, D. C., Riess, A. G., Spinrad, H., Stern, D., Bunker, A., Dey, A., Grossan, B., Perlmutter, S., Knop, R. A., Hook, I. M., and Feroci, M., Nature, **401**, 453–456 (1999).
3. Bloom, J. S., Kulkarni, S. R., Price, P. A., Reichart, D., Galama, T. J., Schmidt, B. P., Frail, D. A., Berger, E. 2002, ApJ, **572**, L45
4. Fruchter, A. 2002, American Physical Society, April Meeting, Jointly Sponsored with the High Energy Astrophysics Division (HEAD) of the American Astronomical Society April 20 - 23, 2002 Albuquerque Convention Center Albuquerque, New Mexico Meeting ID: APR02, abstract #Y2.004, 2004
5. Hawley, J. F., Balbus, S. A., and Stone, J. M., ApJ, **554**, L49–L52 (2001).
6. Iwamoto, K., Mazzali, P. A., Nomoto, K., Umeda, H., Nakamura, T., Patat, F., Danziger, I. J., Young, T. R., Suzuki, T., Shigeyama, T., Augusteijn, T., Doublier, V., Gonzalez, J.-F., Boehnhardt, H., Brewer, J., Hainaut, O. R., Lidman, C., Leibundgut, B., Cappellaro, E., Turatto, M., Galama, T. J., Vreeswijk, P. M., Kouveliotou, C., van Paradijs, J., Pian, E., Palazzi, E., and Frontera, F., Nature, **395**, 672–674 (1998).
7. MacFadyen, A. I., and Woosley, S. E., ApJ, **524**, 262–289 (1999).
8. MacFadyen, A. I., and Woosley, S. E., ApJ, in preparation (2002).
9. MacFadyen, A. I., Woosley, S. E., and Heger, A., ApJ, **550**, 410–425 (2001).
10. Narayan, R., Piran, T., and Kumar, P., ApJ, **557**, 949–957 (2001).
11. Popham, R., Woosley, S. E., & Fryer, C. 1999, ApJ, 518, 356
12. J. Pruet, S. E. Woosley and R. D. Hoffman (astro-ph/0209412)
13. Reichart, D. 1999, ApJL, 521, 111
14. Woosley, S. E. 1993, ApJ, 405, 273
15. Woosley, S. E., Eastman, R. G., and Schmidt, B. P., ApJ, **516**, 788–796 (1999).

Formation and Evolution of Hypernova Progenitors in Massive Binary Systems

Paul C. Joss and J. Alex Becker

Massachusetts Institute of Technology, Cambridge, MA 02139, USA

Abstract. If γ-ray bursts are produced by hypernovae, a problem that must be confronted is how the core of the hypernova progenitor retains or acquires sufficient angular momentum to produce the requisite axisymmetric collapse. Physical processes during the evolution of an isolated massive star will tend to extract any initial angular momentum from the stellar core, rendering it difficult for such a star to become a hypernova. However, a substantial fraction of massive stars are members of binary systems. Tidal locking, mass transfer, or stellar merger in an evolved massive binary may lead to the transfer of orbital angular momentum to the core of one of the stars (or the merged star), sufficient to produce the progenitor of a hypernova. We have developed a new stellar evolution code that includes rotation and the mixing of nuclear species and angular momentum due to dynamical and secular shear instabilities, convective motions, and gravity waves. Preliminary results indicate that over a wide range of initial conditions, mass transfer or merger during the course of the evolution of a massive binary results in the formation of an evolved massive star with a rapidly rotating core, providing the initial conditions necessary for a hypernova event.

1 Introduction

One approach to reducing the otherwise prodigious energy requirements of γ-ray bursts at cosmological distances is to assume that the underlying explosions release at least a portion of their energy in collimated beams rather than isotropically. Paczynski's [1] hypernova/microquasar mechanism (see also the related collapsar hypothesis [2]) invokes the supernova explosion of a rotating star to provide the requisite energy: the central portions of the star's rapidly rotating core collapse to form a Kerr black hole surrounded by a torus of material from regions near the core's rotational equator. The subsequent accretion of material from the torus onto the black hole, or magnetic coupling between the torus and the hole [3], might then generate beams of energetic radiation along the rotation axis, resulting in the emission of a γ-ray burst for observers that lie within the beams. A similar mechanism for emission from active galactic nuclei was first proposed by Blandford & Znajek [4].

A crucial ingredient of this mechanism is a massive stellar core containing a large amount of angular momentum at the time of its collapse. The rapidly rotating core of a massive main-sequence star is prone to lose much of its initial angular momentum as it evolves. The outer envelope of a massive star loses angular momentum via a stellar wind, and the differential rotation that arises between the envelope and the core can result in the transfer of angular momentum from the latter to the former via a variety of mechanisms, including the turbulence arising from shear instability and the action of a magnetic field [5]. (For another point of view see Heger et al. [12], who present calculations which suggest that, under certain circumstances, the stellar core may retain a marginally sufficient amount of angular momentum to produce a hypernova.)

In a close binary system, an evolved star may acquire a substantial amount of rotational angular momentum that was originally stored as orbital angular momentum of the binary. The angular momentum transfer may be accomplished by several mechanisms, including tidal locking (as proposed by Paczynski [1]), the accretion of matter from the companion [6], or the merger of the two stars during a common-envelope phase [6]. We here examine the accretion/merger scenario, and show that an initially non-rotating massive star in a close binary system may be able to accrete and transport inward a sufficient amount of angular momentum to bring the stellar core into rapid rotation by the time of its collapse. In §2 the numerical models of the evolution are described, emphasizing the treatment of rotation and mixing of angular momentum and nuclear species. §§3 and 4 discuss the physical mechanisms responsible for the transport of angular momentum in our models. §5 presents some representative computational results, and we briefly describe our conclusions in §6.

2   Computational Methods

The calculations reported herein were performed using an updated Henyey-type stellar evolution code originally developed by Kippenhahn et al. [7] and augmented by Podsiadlowski et al. [8]. For the present work this code has been extended to calculate the evolution of stars that can accrete matter and angular momentum from a binary companion or the common envelope of a merging binary. The equilibrium structure calculations have been updated to include the effects of rotation, and the time evolution calculations have been extended to include the redistribution of angular momentum and nuclear species due to rotationally induced fluid instabilities. Both angular momentum and chemical composition are mixed in convectively unstable regions, while only angular momentum is transported in regions subject to instabilities other than convection.

We include the effects of rotation upon the mechanical structure of the star by using the method of Kippenhahn [11] as implemented by Endal & Sofia [13], which assumes that rotation is constant on surfaces of the combined gravitational and centripetal pseudo-potential (the "shellular" approximation).

The transfer of angular momentum and mixing of nuclear species by fluid instabilities was modeled by a coupled set of nonlinear diffusion equations, with the strength of mixing parameterized by effective eddy viscosity and eddy diffusivity coefficients, respectively (see [14] for a similar treatment). These equations are solved in a fully implicit finite-differenced form via Newton-Raphson iteration, subject to (1) reflective boundary conditions at the stellar center and (2) surface boundary conditions that depend on the ratio of the surface equatorial rotational angular velocity to the local Keplerian angular velocity, ω_k, as well as the currently active accretion process, if any (see below). The calculations described herein were simplified in that eddy diffusivities were assumed to be zero (no mixing of nuclear species) in regions of fluid instability and radiative stability; only the transport of angular momentum was allowed for in such regions. The eddy viscosity coefficients in these regions were assigned as described below. More complete calculations that include the diffusion of nuclear species in regions of fluid instability are now under way and will be reported elsewhere.

Mixing in a convectively unstable region (determined by the Schwarzschild criterion) was modeled by assigning eddy viscosity and diffusivity coefficients in such regions that correspond to mixing on a dynamical time scale. This has the effect of establishing chemical homogeneity and solid body rotation in convective regions virtually instantaneously relative to the stellar evolutionary time scale.

The accreted matter has the same initial composition as the zero-age main-sequence (ZAMS) models of the binary stellar components, and is assumed to fall softly onto the stellar surface (vertical component of velocity small compared to the local escape speed) with no modification to the pressure of the envelope. The surface boundary conditions depend on the current evolutionary stage of the model. For spin-up during accretion (when the equatorial angular velocity, ω, of the stellar envelope is less than ω_k), reflective boundary conditions are applied. During this phase the matter added to the outermost zone of the model is assumed to carry specific angular momentum, j_k, corresponding to angular velocity ω_k. When the angular velocity of the envelope reaches ω_k, the reflective condition on ω is replaced by one that fixes its value in the envelope at ω_k, while the accreted matter continues to carry specific angular momentum j_k. (In practice, this condition can result in the loss of angular momentum from the envelope of the model; this angular momentum is presumably carried outward through an accretion disk). At all other (non-accreting) stages of evolution, reflective conditions are applied at both the stellar surface and the stellar center. In these stages, the stellar envelope is held to subcritical rotation velocities by the loss of angular momentum to a rotationally enhanced stellar wind.

The ZAMS chemical composition for all models was taken to be $X = 0.70$ and $Z = 0.02$. The mass loss rate due to a stellar wind was computed according to the parameterization of Nieuwenhuijzen et al. [9], modified to account for rotational enhancement by the prescription of Friend & Abbott [10].

3   Fluid Instabilities

Turbulence-induced mixing and momentum transfer occur on several time scales.

The most vigorous transfer is that due to hydrodynamic instabilities, including convection and dynamical shear, and occurs on a characteristic time that is comparable to the dynamical time scale. Secular shear instability can also produce mixing, but on a thermal time scale. Both dynamical and secular shear instabilities are treated in the present models as described in Endal & Sofia

[14], with the effective eddy viscosity determined from the applicable time scale (dynamical or thermal) and characteristic size of the region of instability.

Other mixing mechanisms commonly considered in rotating stars include baroclinic instabilities and Eddington-Sweet currents. The differential rotation induced by the accretion of large amounts of angular momentum results in shear instabilities which are expected to overwhelm these others, owing primarily to their relatively long development time scales. Hence they are not considered in the present models.

4   Angular Momentum Transport by Gravity Waves

In certain cases the molecular weight gradient that builds up at the edge of the stellar core during central hydrogen burning contributes to a density stratification that is highly stable against shear. Hence, for the purpose of inward angular momentum transport by shear instabilities, the core is effectively decoupled from the stellar envelope.

An alternative mechanism for transporting angular momentum across this barrier is via Reynolds stresses associated with propagating internal gravity waves [16]. These waves are a ubiquitous phenomenon in gravitationally stratified fluids such as the terrestrial atmosphere and oceans (see e.g. Plumb [17]; Fritts et al. [18]); they can transport substantial amounts of momentum between the regions where they are excited and those in which they eventually dissipate [15].

Angular momentum transport by internal gravity waves generated at the interface between the convective and radiative regions of the sun has recently been invoked as a possible mechanism for the spindown of the solar core [19–21]. These authors argue that gravity waves generated by convective motions in the solar surface layers, which are isotropically distributed in direction of propagation and hence carry no net angular momentum, can nevertheless redistribute angular momentum in the solar core. This redistribution may, in turn, induce shear instabilities that can extract sufficient angular momentum to spin down the core in $\sim 10^7$ years. In contrast to this case, gravity waves generated via shear flow, such as the time-varying corrugations at the bottom of a rapidly rotating stellar convection zone above a slowly rotating radiative layer, will have a spectrum that is anisotropic with respect to the direction of propagation [18]. Such a spectrum will transport angular momentum into the radiative region in which they propagate. In particular, if these waves are trapped in the stellar core, they will eventually dissipate, depositing angular momentum in the core.

In one of the models presented below, we assume that gravity waves are generated at the interface between the convective hydrogen-burning shell and the radiative core following an early post main-sequence mass transfer event. If sufficient angular momentum has been transported inward to the burning shell, it will be rotating rapidly relative to the core and the mechanism described above will be active. The resulting net flux of angular momentum into the core is estimated by scaling arguments similar to those of Fritts et al. [18]. In order to model this effect, we calculate an effective eddy viscosity coefficient at the core-shell interface. For simplicity this coefficient, once calculated, is held constant for the remainder of the hydrostatic evolution of the star. Similarly, the details of angular momentum redistribution are not followed for the core; instead, we assume that the core rotates as a solid body throughout the core helium burning phase of the star.

5   Results

5.1   Late Main Sequence Accretion – $26.2 M_\odot$ ZAMS

Figures 1 and 2 display the HR diagram and interior angular velocity distribution, respectively, for a prototypical evolutionary sequence in which accretion of matter and angular momentum by the secondary star commences during its late main-sequence evolution. During the earliest stages of accretion the angular momentum deposited in the envelope of the star is transported inward through layers subject to shear instability, giving rise to a distribution of $\omega(m)$ in the envelope that increases monotonically with increasing m, where m is the mass enclosed within a pseudo-potential surface (curve 2 in Fig. 2). After further accretion, the distribution of angular velocity in the outermost layers develops a "no-slip" level, where $d\omega/dr = 0$ (r being the distance from the stellar center on the star's rotational equator) and through which no further angular momentum may pass. The existence of a no-slip level persists through the remainder of the accretion phase and into the post-accretion evolution; its presence acts to shield the angular momentum content of the bulk of the stellar interior from loss through the stellar surface layers. The no-slip level eventually dissipates, with the angular velocity distribution becoming a monotonically decreasing function of m in the outer layers of the star, through which angular momentum is transported outward to the stellar surface. Some of this angular momentum is eventually lost to a rotationally enhanced wind (see §2). However, owing to the relatively short amount of time over which this portion of the evolution occurs, the effect of these losses on the final rotational state of the stellar core is small.

Since the adverse density gradient at the edge of the radiative core is relatively mild when the inwardly transported angular momentum first impinges upon it (compared to that in the post-main-sequence model considered in the next subsection), the shear that develops is sufficient to excite instability with consequent angular momentum penetration into the core. In the subsequent post-main-sequence evolution, the core and adjacent hydrogen-burning shell are rapidly brought into near solid-body rotation, aided by the establishment of convection in the core once helium burning is ignited. Transport of angular momentum into the core by gravity waves is both inefficient and unnecessary in this case. The increase in the magnitude and non-uniformity of the core's angular velocity distribution during late stages of evolution (curves 6 and 7 in Fig. 2) are primarily a result of the contraction of the core prior to the onset of carbon burning.

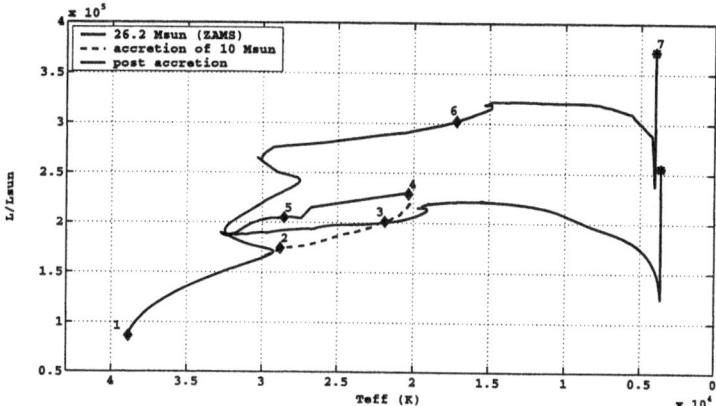

Fig. 1. HR diagram illustrating the evolution of an initially nonrotating star with a ZAMS mass $26.2\,M_\odot$, which accretes $10\,M_\odot$ at a rate of $10^{-3}\,M_\odot\,\mathrm{yr}^{-1}$ from a close-binary companion near the end of its main-sequence evolution, under the assumption of no transport of angular momentum by gravity waves. Note that the star has lost $\sim 1.2\,M_\odot$ in a stellar wind during its main-sequence evolution prior to the accretion phase, leading to a stellar mass of 25 and $35\,M_\odot$ just prior to and just following the accretion episode, respectively. The lower solid line shows the post-main-sequence evolution of an isolated, non-rotating $26.2\,M_\odot$ star until core carbon ignition, which is marked on the diagram by a terminating asterisk. The dashed and upper solid lines represent evolution during the accretion event (of duration 10^4 years) and following accretion until core carbon ignition, respectively. Loss due to a rotationally enhanced stellar wind reduces the final stellar mass to $\sim 25\,M_\odot$. Numbered diamonds indicate points in the evolution for which ω is plotted in Fig. 2

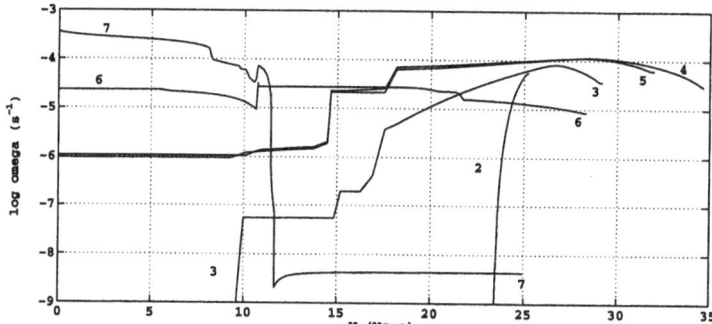

Fig. 2. Rotational angular velocity, ω, as a function of enclosed mass for seven points along the evolutionary track, shown in Fig. 1, for the late-main-sequence accretion scenario. Gravity wave transport of angular momentum was not included in the calculations and is is not, in fact, necessary for core spin-up, as is evident from curves 4 – 7. Contraction of the core just prior to carbon ignition gives rise to most of the increase in ω within the core at these late evolutionary stages

5.2   Early Post Main Sequence Accretion – 26.2$M_\odot$ ZAMS

Figures 3 and 4 illustrate the evolution of a 26.2$M_\odot$ secondary which accretes 10$M_\odot$ from the primary just after hydrogen burning has ceased in the core and begun in a surrounding shell. Once the inward moving wave of angular momentum reaches the hydrogen burning shell, it is redistributed nearly instantaneously via convection, causing the shell to rotate as a solid body. The density gradient at the edge of the core in this case is large enough to stabilize

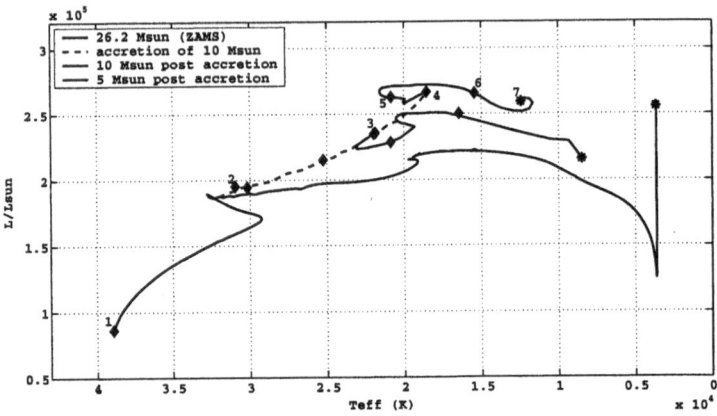

Fig. 3. HR diagram illustrating the evolution of an initially nonrotating star with ZAMS mass 26.2 $M_\odot$, which accretes either 5 $M_\odot$ or 10 $M_\odot$ at a rate of $10^{-3} M_\odot$ yr^{-1} from a close-binary companion near the beginning of its hydrogen shell burning phase (early post-main-sequence accretion). Transport of angular momentum by gravity waves is assumed to be operative at the inner edge of the hydrogen-burning shell. The middle solid line represents evolution after 5 $M_\odot$ has been accreted and up to the time of carbon ignition. The notation is otherwise the same as in Fig. 1

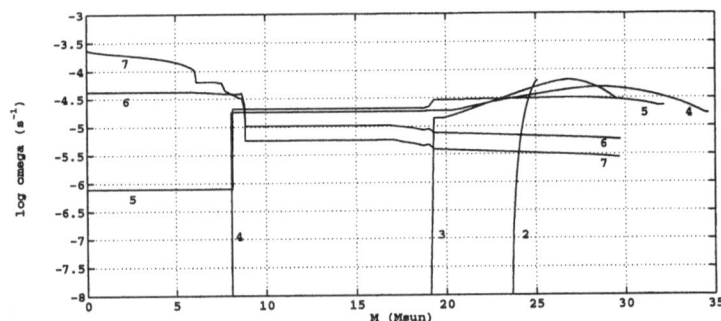

Fig. 4. Angular velocity, ω, as a function of enclosed mass for the early post-main-sequence case illustrated in Fig. 3. Gravity wave transport of angular momentum is assumed to be operative at the inner edge of the hydrogen burning shell, resulting in the spin up of the core evident in curve 5

the shear flow between the rotating shell and nonrotating, radiative core; angular momentum penetrates no further. The flow of the radiative fluid of the core over the convectively generated corrugations of the overlying hydrogen burning shell does, however, efficiently generate gravity waves, which are effective in transporting angular momentum across this barrier, and which spin up the core. Through convection and other mechanisms not followed in detail in these calculations, the core is eventually brought into a state of solid body rotation, after which evolution proceeds as described in the previous case.

6   Conclusions

We have carried out binary stellar evolution calculations of the accretion of matter and angular momentum by a massive late main-sequence or early post-main-sequence star from its companion, and the subsequent transport of the accreted angular momentum by a combination of convection, shear-induced turbulence, and/or gravity waves. Our results indicate that the stellar core, prior to its collapse, can acquire a mean specific angular momentum of at least $\sim 10^{16}$ cm^2 s^{-1}, which should be sufficient to lead to a hypernova event (cf. MacFadyen & Woosley [22]). For the early post-main-sequence case, the accretion event resulted in a more compact star at the time of core collapse, with a radius ~ 25 times smaller than the same star if it had evolved in isolation (see Fig. 3; Podsiadlowski et al. [8]). This compaction would greatly facilitate the escape of beamed γ-rays during a hypernova event (see e.g. Heger & Woosley [23]). However, such compaction was absent in the pre-core collapse evolution of the late main-sequence case, rendering this case a less promising scenario for γ-ray burst progenitors.

Accretion at an earlier evolutionary stage is not of interest in the present context, since the result would simply be a rejuvenated ZAMS star that would need to retain a large amount of angular momentum throughout its main-sequence, as well as post-main-sequence, evolution. Similarly, accretion during later post-main-sequence stages is of lower interest, because (1) the density stratification at the core-envelope interface would be more extreme, rendering it more difficult for the accreted angular momentum to penetrate the core, (2) the time between accretion and core collapse would be reduced, further complicating the problem of transporting sufficient angular momentum into the core prior to its collapse, and (3) the distension of the stellar envelope during late evolutionary phases would, as noted above, inhibit the emission of γ-rays even if a hypernova event were to ensue.

As noted by Podsiadlowski et al. [8], the behavior of the secondary star during common-envelope evolution is similar to that of a secondary experiencing quasi-hydrostatic accretion in a mass-transfer binary system that does not undergo a common-envelope merger. Hence, our models serve as good surrogates for secondary stars of the same mass that gain mass during common-envelope evolution at the same evolutionary epoch, provided that the primary star within the common envelope dissolves before reaching the surface of the secondary during the spiral-in process. Interestingly, Ivanova & Podsiadlowski [24] (see also these

Proceedings) have recently examined the consequences of the residual primary in a common-envelope system penetrating well into the envelope of the secondary before dissolving. They found that this scenario can result in a stream of material impinging on the core of the secondary, introducing a large amount of angular momentum just outside the core, and can also result in the star being relatively compact at the time of core collapse. In the context of the present work, such a scenario could be modeled by altering the point of injection of mass and angular momentum from the stellar surface to a point within its envelope. Since this is closer to the core, the injected angular momentum may well penetrate the core more readily than in the cases reported here, yielding another promising candidate for a hypernova progenitor created within a massive binary.

We are grateful to Philipp Podsiadlowski for stimulating discussions. P.C.J. is pleased to acknowledge a travel grant from the U.S. National Science Foundation, which enabled him to attend this workshop.

References

1. B. Paczynski: Ap.J. **494**, L45 (1998)
2. S.E. Woosley, N. Langer, T.A. Weaver: Ap.J. **411**, 823 (1993)
3. M.H.P.M. van Putten: Phys. Reports **345**, 1 (2001)
4. R.D. Blandford, R.L. Znajek: MNRAS **179**, 433 (1977)
5. H. Spruit, E.S. Phinney: Nature **393**, 139 (1998)
6. G.E. Brown, et al.: New Astr. **5**, 191 (2000)
7. R. Kippenhahn, A. Weigert, E. Hofmeister: Meth. Comp. Phys. **7**, 129 (1967)
8. Ph. Podsiadlowski, P.C. Joss: Ap.J. **391**, 246 (1992)
9. H. Nieuwenhuijzen, C. de Jager: A&A **231**, 134 (1990)
10. D.B. Friend, D.C. Abbott: Ap.J. **311**, 701 (1986)
11. R. Kippenhahn, H.C. Thomas: IAU Colloq. **4**, *Stellar Rotation*, ed. by A. Slettebak (Dordrecht: Reidel), 20 (1970)
12. A. Heger, N. Langer, S.E. Woosley: Ap.J. **528**, 368 (2000)
13. A.S. Endal, S. Sofia: Ap.J. **210**, 184 (1976)
14. A.S. Endal, S. Sofia: Ap.J. **220**, 279 (1978)
15. J.S. Turner: *Buoyancy Effects in Fluids.* (Cambridge Univ. Press: Cambridge 1980)
16. W.H. Press: Ap.J. **245**, 286 (1981)
17. R.A. Plumb, J. Atmos. Sci. **34**, 1847 (1977)
18. D.C. Fritts, S.L. Vadas, O. Andreassen: A&A **333**, 343 (1998)
19. J.P. Zahn, S. Talon, J. Matias: A&A **322**, 320 (1997)
20. S. Talon, P. Kumar, J.P. Zahn: Ap.J. **574**, 175 (2002)
21. P. Kumar, E.J. Quataert: Ap.J. **475**, 143 (1997)
22. A.I. MacFadyen, S.E. Woosley: Ap.J. **524**, 262 (1999)
23. A. Heger, S.E. Woosley: preprint, astro-ph/0206005 (2002)
24. N. Ivanova, Ph. Podsiadlowski: Astrop. & Space Sci. **281**, 191 (2002)

Part III

Progenitors of Thermonuclear Supernovae

Type Ia Supernovae: Progenitors and Diversities

Ken'ichi Nomoto[1], Tatsuhiro Uenishi[1], Chiaki Kobayashi[2], Hideyuki Umeda[1], Takuya Ohkubo[1], Izumi Hachisu[3], and Mariko Kato[4]

[1] Department of Astronomy, School of Science, University of Tokyo, Japan
[2] Max-Planck-Institut für Astrophysik, Garching, Germany
[3] Department of Earth Science & Astronomy, University of Tokyo, Japan
[4] Department of Astronomy, Keio University, Japan

Abstract. A key question for supernova cosmology is whether the peak luminosities of Type Ia supernovae (SNe Ia) are sufficiently free from the effects of cosmic and galactic evolution. To answer this question, we review the currently popular scenario of SN Ia progenitors, i.e. the single degenerate scenario for the Chandrasekhar mass white dwarf (WD) models. We identify the progenitor's evolution with two channels: (1) the WD+RG (red-giant) and (2) the WD+MS (near main-sequence He-rich star) channels. The strong wind from accreting WDs plays a key role, which yields important age and metallicity effects on the evolution.

We suggest that the variation of the carbon mass fraction $X(C)$ in the C+O WD (or the variation of the initial WD mass) causes the diversity of SN Ia brightness. This model can explain the observed dependence of SNe Ia brightness on the galaxy types. We then predict how SN Ia brightness evolves along the redshift (with changing metallicity and age) for elliptical and spiral galaxies. Such evolutionary effects along the redshift can be corrected as has been made for local SNe Ia.

We also touch on several related issues: (1) the abundance pattern of stars in dwarf spheroidal galaxies in relation to the metallicity effect on SNe Ia, (2) effects of angular momentum brought into the WD in relation to the diversities and the fate of double degenerates, and (3) possible presence of helium in the peculiar SN Ia 2000cx in relation to the sub-Chandrasekhar mass model.

1 Introduction

Relatively uniform light curves and spectral evolution of Type Ia supernovae (SNe Ia) have led to the use of SNe Ia as a "standard candle" to determine cosmological parameters. Whether a statistically significant value of the cosmological constant can be obtained depends on whether the peak luminosities of SNe Ia are sufficiently free from the effects of cosmic and galactic evolutions [20].

SNe Ia have been widely believed to be a thermonuclear explosion of a mass-accreting white dwarf (WD). However, the immediate progenitor binary systems have not been clearly identified yet [3]. In order to address the above questions regarding the nature of high-redshift SNe Ia, we need to identify the progenitors systems and examine the "evolutionary" effects (or environmental effects) on those systems [48].

Here we review several issues such as Chandra vs. sub-Chandra, double degenerates vs. single degenerate (AIC vs. SN Ia), C/O ratio, and rotation. For complete discussion, see earlier reviews on SN Ia progenitors [1,2,11,22,31–33].

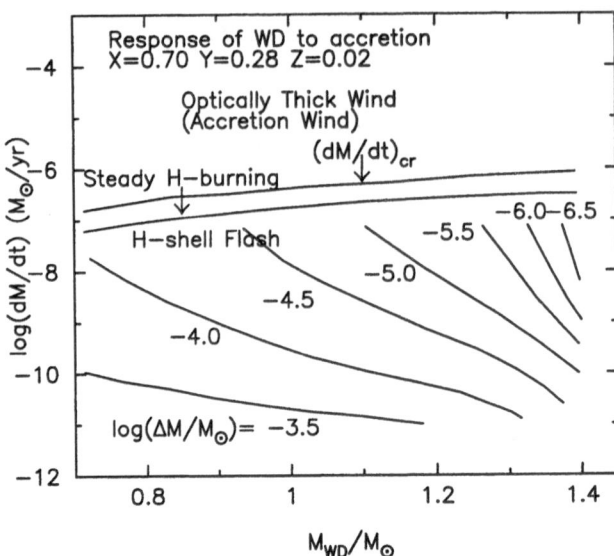

Fig. 1. The nature of hydrogen burning on accreting WD as functions of the WD mass and the accretion rate [25]. Above the steady burning regime, the WD blows an optically thick wind [5]

2 Evolution of Progenitor Systems

Chandra vs. Sub-Chandra: There exist two models proposed as progenitors of SNe Ia: 1) the Chandrasekhar mass model, in which a mass-accreting carbon-oxygen (C+O) WD grows its mass M_{WD} up to the critical mass $M_{Ia} \simeq 1.37 - 1.38 M_\odot$ near the Chandrasekhar mass and explodes as an SN Ia (e.g. [30,31]), and 2) the sub-Chandrasekhar mass model, in which an accreted layer of helium atop a C+O WD ignites off-center for a WD mass well below the Chandrasekhar mass (e.g. [1]). The early time spectra of the majority of SNe Ia are in excellent agreement with the synthetic spectra of the Chandrasekhar mass models, while the spectra of the sub-Chandrasekhar mass models are too blue to be comparable with the observations [12,34]. However, the peculiar SN Ia 2000cx might be the sub-Chandrasekhar mass explosion as will be discussed in §6.

Double degenerates vs. single degenerate: For the evolution of accreting WDs toward the Chandrasekhar mass, two scenarios have been proposed: 1) a double degenerate (DD) scenario, i.e. merging of double C+O WDs [14,50], and 2) a single degenerate (SD) scenario, i.e. accretion of hydrogen-rich matter via mass transfer from a binary companion (e.g. [25,31]). The issue of DD vs. SD is still debated, although theoretical modeling has indicated that the merging of WDs leads to the accretion-induced collapse (AIC) rather than SN Ia explosion [40,42]. Whether the effect of rotation on accretion changes the conclusion on AIC vs. SN Ia will be discussed in §5.2.

2.1 White Dwarf Winds

In the SD Chandrasekhar mass model, a WD explodes as a SN Ia only when its rate of the mass accretion ($\dot{M}$) is in a certain narrow range (Fig. 1 [25,5]). In particular, if $\dot{M}$ exceeds the critical rate $\dot{M}_b$, the accreted matter extends to form a common envelope [29]. Here $\dot{M}_b$ is the rate at which steady burning can process the accreted hydrogen into helium as $\dot{M}_b \approx 0.75 \times 10^{-6} \left(\frac{M_{WD}}{M_\odot} - 0.40 \right) M_\odot$ yr^{-1}.

For $\dot{M} \gtrsim \dot{M}_b$, the strong peak of Fe opacity [15] drives the radiation-driven wind from the WD [7]. If the wind is sufficiently strong, the WD can avoid the formation of a common envelope and steady hydrogen burning increases its mass continuously at a rate $\dot{M}_b$ by blowing the extra mass away in a wind. For $\dot{M} \lesssim 0.5\ \dot{M}_b$, hydrogen shell burning becomes unstable to trigger weak shell flashes but still burns a large fraction of accreted hydrogen. Recurrent novae appear in the upper-right region of Fig. 1. In this way, strong winds from the accreting WD play a key role to increase the WD mass to M_{Ia}.

2.2 Progenitor Binary Systems

For the actual binary systems which grow M_{WD} to M_{Ia}, the following two systems are appropriate.

WD+RG system (Symbiotic system): This is a symbiotic binary system consisting of a WD and a low mass red-giant (RG) [7]. The immediate progenitor binaries in this symbiotic channel to SNe Ia may be observed as symbiotic stars, luminous supersoft X-ray sources, or recurrent novae like T CrB or RS Oph, depending on the wind status [5].

WD+MS system (Super-soft system): This is a system consisting of a mass-accreting WD and a lobe-filling, more massive, slightly evolved main-sequence or sub-giant star. In this scenario, a C+O WD is originated, not from an AGB star with a C+O core, but from a red-giant star with a helium core of $\sim 0.8 - 2.0 M_\odot$. The helium star, which is formed after the first common envelope evolution, evolves to form a C+O WD of $\sim 0.8 - 1.1 M_\odot$ with transferring a part of the helium envelope onto the secondary main-sequence star [6].

A part of the progenitor systems are identified as the luminous supersoft X-ray sources [49] during steady H-burning (but without wind to avoid extinction), or the recurrent novae like U Sco if H-burning is weakly unstable [5]. Actually these objects are characterized by the accretion of helium-rich matter.

Realization frequency: The rate of SNe Ia originating from these channels in our Galaxy is estimated with equation (1) of [14]. The realization frequencies of SNe Ia through the WD+RG and WD+MS channels are estimated as ~ 0.0017 yr^{-1} (WD+RG) and ~ 0.001 yr^{-1} (WD+MS), respectively. The total SN Ia rate of the WD+MS/WD+RG systems becomes ~ 0.003 yr^{-1}, which is close enough to the inferred rate of our Galaxy.

2.3   Metallicity Dependence of Type Ia Supernovae

The optically thick winds are driven by a strong peak of OPAL opacity due to iron lines. Thus the wind velocity v_w is lower for lower [Fe/H]. The SN Ia regions are much smaller for lower metallicity, and very few SN Ia occur at [Fe/H] ≤ -1.1 in this model. It is possible to test such metallicity effects on SNe Ia with the chemical evolution of galaxies.

In the one-zone uniform model for the chemical evolution of the solar neighborhood, the heavy elements in the metal-poor stars originate from the mixture of the SN II ejecta of various progenitor masses. The abundances averaged over the progenitor masses of SNe II predict [O/Fe] ~ 0.45 (e.g. [44]). Later SNe Ia start ejecting mostly Fe, so that [O/Fe] decreases to ~ 0 around [Fe/H] ~ 0. The low-metallicity inhibition of SNe Ia predicts that the decrease in [O/Fe] starts at [Fe/H] ~ -1. Such an evolution of [O/Fe] well explains the observations [18,19].

We should note that some anomalous stars have [O/Fe] ~ 0 at [Fe/H] $\lesssim -1$. The presence of such stars, however, is not in conflict with the metallicity dependence of SNe Ia, but can be understood as follows: The formation of such anomalous stars (and the diversity of [O/Fe] in general) indicates that the interstellar materials were not uniformly mixed but contaminated by only a few SNe II (or even single SN II) ejecta. The Fe and O abundances produced by a single SN II vary depending mainly on the mass of the progenitor. Relatively smaller mass SNe II ($13 - 15 M_\odot$) and higher explosion energies tend to produce [O/Fe] ~ 0 [44,46]. Those metal-poor stars with [O/Fe] ~ 0 may be born from the interstellar medium polluted by such SNe II. Alternatively, such stars were born in nearby dwarf spheroidal galaxies and captured by our Galaxy (see below).

2.4   Abundances in Dwarf Spheroidal Galaxies

The chemical abundances of individual stars in local dwarf spheroidal galaxies (dSph) have been measured (e.g. [43] for recent observations and references therein). These stars have [Fe/H] < -1 but the ratios between the α elements and Fe are found to be as low as [α/Fe] ~ 0 being significantly lower than the metal-poor halo-stars in our Galaxy. If the low [α/Fe] was due to the Fe-enrichment by SNe Ia, it implies that SNe Ia appeared in low metallicity environment, which may be difficult to explain with the metallicity dependent SN Ia model.

However, it is possible to produce such low [α/Fe] with core-collapse SNe. As discussed above, yields from 13-15 $M_\odot$ stars have low (even sub-solar) [α/Fe]. Thus if IMF in those dwarf spheroidal galaxies is steep enough or the upper limit mass of core-collapse SNe is truncated around 20 $M_\odot$ [43], the contribution of 13-15 $M_\odot$ stars could be significantly larger than in our Galaxy. Also it is possible that some fraction of AGB stars suffers less mass loss because of low metallicity, thus growing degenerate C+O cores to the Chandrasekhar mass before losing entire envelopes. If this is the case, those metal-poor AGB stars undergo SN Ia-like explosions (SN I+1/2). If the mass range is very narrow, say ~ 7.5-8.0 $M_\odot$, SNe I+1/2 do not affect [O/Fe] for our Galaxy if integrated over the Salpeter IMF up to 50 $M_\odot$ [28]. However, if the IMF is steeper or the upper-mass limit is lower, contributions of SNe I+1/2 and 13-15 $M_\odot$ stars are larger, thus reducing [α/Fe].

The abundance patterns in dSph as well as some Damped Lyman-α systems can provide important constraint on the progenitor mass ranges of SNe Ia, SNe I+1/2, and SNe II. Also the appearance of SNe Ia in more metal-rich regions of dSph may not be impossible in view of spatial inhomogeneity.

3   The Origin of Diversity of Type Ia Supernovae

There are some observational indications that SNe Ia depend on the type of the host galaxies. The most luminous SNe Ia seem to occur only in spiral galaxies, while both spiral and elliptical galaxies are hosts for dimmer SNe Ia. Thus the mean peak brightness is dimmer in ellipticals than in spiral galaxies [9]. Also the SNe Ia rate per unit luminosity at the present epoch is almost twice as high in spirals as in ellipticals [4].

Umeda et al. [48] suggested that the variation of the C/O ratio is the main cause of the variation of SNe Ia brightness, with larger C/O ratio yielding brighter SNe Ia. Here we will show that the C/O ratio depends indeed on the metallicity and age of the companion of the WD, and that the model can explain most of the observational trends discussed above. We then make some predictions about the brightness of SN Ia at higher redshift.

3.1   C/O Ratio in White Dwarf Progenitors

The C/O ratio in C+O WDs depends primarily on the main-sequence mass of the WD progenitor and on metallicity. The most important metallicity effect is that the radiative opacity is smaller for lower Z. Therefore, a star with lower Z is brighter, thus having a shorter lifetime than a star with the same mass but higher Z. In this sense, the effect of reducing metallicity for these stars is almost equivalent to increasing a stellar mass.

For stars with larger masses and/or smaller Z, the luminosity is higher at the same evolutionary phase. With a higher nuclear energy generation rate, these stars have larger convective cores during H and He burning, thus forming larger He and C-O cores.

According to the evolutionary calculations for $3-9$ $M_\odot$ stars [47], the C/O ratio and its distribution are determined in the following evolutionary stages of the close binary.

(1) At the end of central He burning in the $3-9$ $M_\odot$ primary star, C/O< 1 in the convective core. The mass of the core is larger for more massive stars.

(2) After central He exhaustion, the outer C+O layer grows via He shell burning, where C/O$\gtrsim$ 1 [47].

(3a) If the primary star becomes a red giant (case C evolution), it then undergoes the second dredge-up, forming a thin He layer, and enters the AGB phase. The C+O core mass, $M_{\rm CO}$, at this phase is larger for more massive stars. For a larger $M_{\rm CO}$ the total carbon mass fraction is smaller.

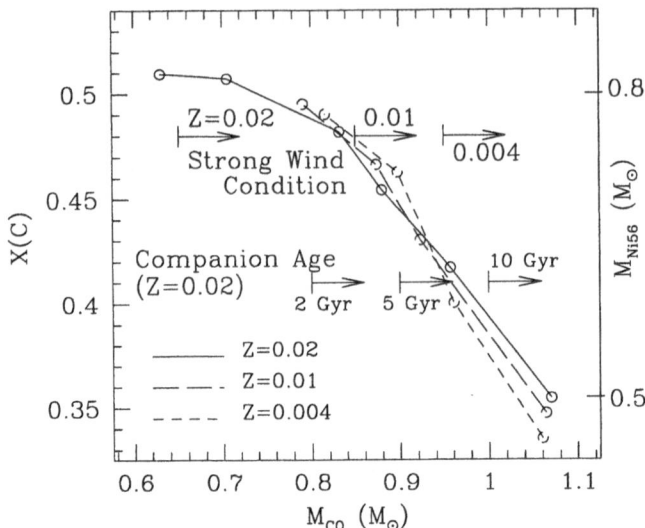

Fig. 2. The total ^{12}C mass fraction included in the convective core of mass, $M = 1.14 M_\odot$, just before the SN Ia explosion as a function of the C+O core mass before the onset of mass accretion, M_{CO}. The lower bounds of M_{CO} obtained from the age effects and the conditions for strong wind to blow are also shown by arrows [48]

(3b) When it enters the AGB phase, the star greatly expands and is assumed here to undergo Roche lobe overflow (or a super-wind phase) and to form a C+O WD. Thus the initial mass of the WD, $M_{WD,0}$, in the close binary at the beginning of mass accretion is approximately equal to M_{CO}.

(4a) If the primary star becomes a He star (case BB evolution), the second dredge-up in (3a) corresponds to the expansion of the He envelope.

(4b) The ensuing Roche lobe overflow again leads to a WD of $M_{WD,0} = M_{CO}$.

(5) After the onset of mass accretion, the WD mass grows through steady H burning and weak He shell flashes, as described in the WD wind model. The composition of the growing C+O layer is assumed to be C/O=1.

(6) The WD grows in mass and ignites carbon when its mass reaches $M_{Ia} = 1.367 M_\odot$, as in the model C6 [30]. Carbon burning grows into a deflagration for a central temperature of 8×10^8 K and a central density of 1.47×10^9 g cm^{-3}. At this stage, the convective core extends to $M_r = 1.14 M_\odot$ and the material is mixed almost uniformly (model C6).

Figure 2 shows the carbon mass fraction $X(C)$ in the convective core of this pre-explosive WD, as a function of metallicity (Z) and initial mass of the WD before the onset of mass accretion, M_{CO}. We note:

1) $X(C)$ is smaller for larger $M_{CO} \simeq M_{WD,0}$.

2) The dependence of $X(C)$ on metallicity is small when plotted against M_{CO}, even though the relation between M_{CO} and the initial stellar mass depends sensitively on Z [47].

3.2   Brightness of Type Ia Supernovae and the C/O Ratio

In the Chandrasekhar mass models for SNe Ia, the brightness of SNe Ia is determined mainly by the mass of ^{56}Ni synthesized (M_{Ni56}). Observational data suggest that M_{Ni56} for most SNe Ia lies in the range $M_{Ni56} \sim 0.4 - 0.8 M_\odot$ (e.g. [24]).

Here we postulate that M_{Ni56} and consequently the brightness of a SN Ia increase as the progenitors' C/O ratio increases (and thus $M_{WD,0}$ decreases). As illustrated in Fig. 2, the range of $M_{Ni56} \sim 0.5 - 0.8 M_\odot$ is the result of an $X(C)$ range $0.35 - 0.5$, which is the range of $X(C)$ values of our progenitor models. The $X(C) - M_{Ni56} - M_{WD,0}$ relation we adopt is still only a working hypothesis, which needs to be proven from studies of the turbulent flame during explosion (e.g. [11]).

3.3   Metallicity and Age Effects

Assuming the relation between M_{Ni56} and $X(C)$ given in Fig. 2, the model predicts the absence of brighter SNe Ia in lower metallicity environments.

In this model, the age of the progenitor system also constrains the range of $X(C)$ in SNe Ia. In the SD scenario, the lifetime of the binary system is essentially the main-sequence lifetime of the companion star, which depends on its initial mass M_2. In order for the WD mass to reach M_{Ia}, the donor star should transfer enough material at the appropriate accretion rates. The donors of successful SN Ia cases are divided into two categories: one is composed of slightly evolved main-sequence stars with $M_2 \sim 1.7 - 3.6 M_\odot$ (for $Z=0.02$), and the other of red-giant stars with $M_2 \sim 0.8 - 3.1 M_\odot$ (for $Z=0.02$) [17].

If the progenitor system is older than 2 Gyr, it should be a system with a donor star of $M_2 < 1.7 M_\odot$ in the red-giant branch. Systems with $M_2 > 1.7 M_\odot$ become SNe Ia in a time shorter than 2 Gyr. Likewise, for a given age of the progenitor system, M_2 must be smaller than a limiting mass. This constraint on M_2 can be translated into the presence of a minimum M_{CO} for a given age, as follows: For a smaller M_2, i.e. for the older system, the total mass which can be transferred from the donor to the WD is smaller. In order for M_{WD} to reach M_{Ia}, therefore, the initial mass of the WD, $M_{WD,0} \simeq M_{CO}$, should be larger. This implies that the older system should have larger minimum M_{CO} as indicated in Fig. 2. Using the $X(C)$-M_{CO} and M_{Ni56}-$X(C)$ relations (Fig. 2), we conclude that WDs in older progenitor systems have a smaller $X(C)$, and thus produce dimmer SNe Ia.

3.4   Morphology of the Host Galaxies

Among the observational indications which can be compared with our model is the possible dependence of the SN brightness on the morphology of the host galaxies. Hamuy et al. [9] found that the most luminous SNe Ia occur in spiral galaxies, while both spiral and elliptical galaxies are hosts to dimmer SNe Ia. Hence, the mean peak brightness is lower in elliptical than in spiral galaxies.

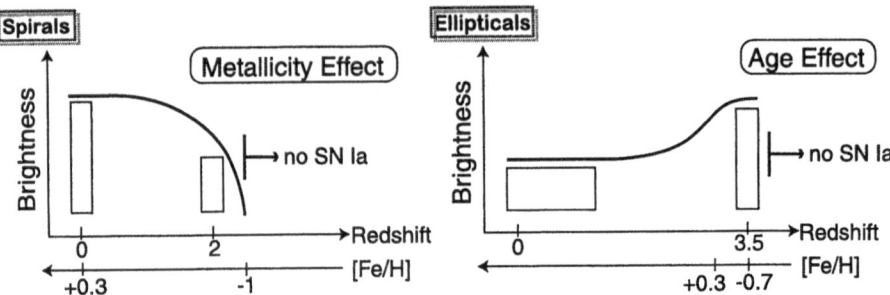

Fig. 3. Illustration of the predicted variation in SN Ia brightness with redshift

In our model, this property is simply understood as the effect of the different age of the companion. In spiral galaxies, star formation occurs continuously up to the present time. Hence, both WD+MS and WD+RG systems can produce SNe Ia. In elliptical galaxies, on the other hand, star formation has long ended, typically more than 10 Gyr ago. Hence, WD+MS systems can no longer produce SNe Ia. Since a WD with smaller M_{CO} is assumed to produce a brighter SN Ia (larger M_{Ni56}), our model predicts that dimmer SNe Ia occur both in spirals and in ellipticals, while brighter ones occur only in spirals. The mean brightness is smaller for ellipticals and the total SN Ia rate per unit luminosity is larger in spirals than in ellipticals. These properties are consistent with observations.

3.5   Evolution of Type Ia Supernovae at High Redshift

Our model predicts that when the progenitors belong to an old population, or to a low metal environment, the number of very bright SNe Ia is small, so that the variation in brightness is also smaller, which is shown in Fig. 3. In spiral galaxies, the metallicity is significantly smaller at redshifts $z \gtrsim 1$, and thus both the mean brightness of SNe Ia and its range tend to be smaller (Fig. 3). At $z \gtrsim 2$ SNe Ia would not occur in spirals at all because the metallicity is too low. In elliptical galaxies, on the other hand, the metallicity at redshifts $z \sim 1-3$ is not very different from the present value. However, the age of the galaxies at $z \simeq 1$ is only about 5 Gyr, so that the mean brightness of SNe Ia and its range tend to be larger at $z \gtrsim 1$ than in the present ellipticals because of the age effect.

We note that the variation of $X(C)$ is larger in metal-rich nearby spirals than in high redshift galaxies. Therefore, if $X(C)$ is the main parameter responsible for the diversity of SNe Ia, and if the light curve shape (LCS) method is confirmed by the nearby SNe Ia data [38,8], the LCS method can also be used to determine the absolute magnitude of high redshift SNe Ia.

Here we consider the metallicity effects only on the C/O ratio; this is just to shift the main-sequence mass - $M_{WD,0}$ relation, thus resulting in no important evolutionary effect. However, some other metallicity effects could give rise to evolution of SNe Ia between high and low redshifts (i.e. between low and high metallicities) (see, e.g. [33]).

4   Rotation

The accreting WD gains angular momentum from the rotating disk in addition to the mass, thus rotating faster and faster. Rapid rotation of a WD affects its limiting mass and final structure. The effects of rotation on the evolution of accreting WDs have been studied mostly with 1D approximate models (e.g. [53,36,41]).

4.1   Diversity and Rotation

Uenishi et al. [45] have calculated the axisymmetric structure of *uniformly* rotating WDs and followed the evolutionary sequence in the total angular momentum J and the WD mass M (Fig. 4). Here accreting gas is supposed to obey the Keplerian rotation law and gives its angular momentum to the WD. The almost straight lines show the evolutionary tracks of the accreting WDs starting from the initial masses of 0.6 $M_\odot$, 0.8 $M_\odot$, and 1.07 $M_\odot$, respectively. After gaining $\sim 0.1\ M_\odot$, the WDs reach the critical rotation at the upper edge of the $J - M$ diagram where the equatorial centrifugal force is equal to the gravity.

For WDs which have reached the critical rotation, Paczyński [35] and Pooham & Narayan [37] found a solution that permits the WD to accrete without becoming secularly unstable. In such a solution, the angular momentum is transported backwards from the WD to the disk while the mass of the WD increases. Then the WD evolves along the upper envelope in the J-M plane and eventually explodes at the upper-right corner. (Along the right edge of the diagram, the central density of WDs is 2×10^9 g cm^{-3} where carbon is ignited.) Then all the rapidly rotating WDs explode with the same (M, J). In other words, *uniform* rotation leads to the uniformity rather than the diversity of SNe Ia.

However, rotation could produce the diversity of SNe Ia in the following way. The uniform rotation is realized if the timescale of angular momentum transport in the WDs is much shorter than the accretion time scale. If the angular momentum transport is much slower, then the initially non-rotating part of the WD may remain non-rotating, while the accreted outer part of the WD rotates fast and reaches the critical rotation. When the WD reaches the carbon ignition at the limiting mass, the angular momentum brought into the WD is larger (smaller) for smaller (larger) M_0. The WD with smaller M_0 grows to larger $M_{\rm fin}$ at the explosion. Although $M_{\rm fin} \sim 1.45\ M_\odot$ are not so different from each other, larger $M_{\rm fin}$ could produce brighter SNe Ia.

Also the distribution of the angular velocity is different, i.e. the non-rotating core is smaller (larger) for smaller (larger) M_0. Such a difference might affect the flame propagation and lead to different amount of ^{56}Ni.

If the WD with smaller M_0 produce a brighter SN Ia with the effect of rotation, the upper limit of SNe Ia brightness is higher in spirals than in ellipticals. SNe Ia in spirals can originate from WDs with a wide range of M_0, thus having a larger dispersion of brightness than those in ellipticals. This could explain the observation that the most luminous SNe Ia appear to be observed only in spirals, while dimmer SNe Ia are observed in both spirals and ellipticals.

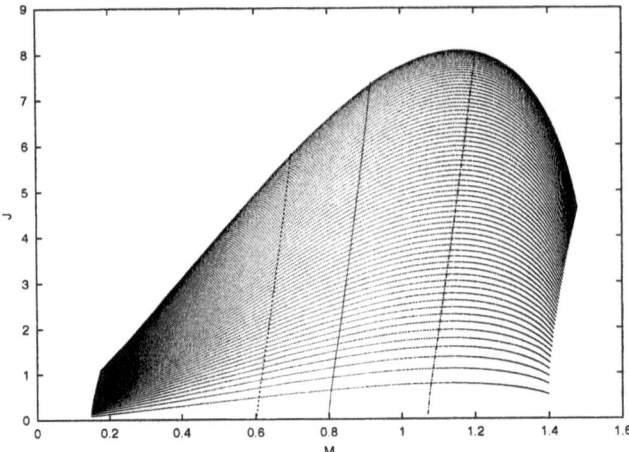

Fig. 4. The evolutionary track in the total angular momentum J (10^{49} erg sec) and the WD mass M ($M_\odot$) of uniformly rotating WDs starting from the initial masses of 0.6 $M_\odot$, 0.8 $M_\odot$, and 1.07 $M_\odot$ [45]

4.2   Rotation and Rapid Accretion in Merging White Dwarfs

Nomoto & Iben [27] and Saio & Nomoto [40] have simulated the merging of double WDs in 1D and shown that the rapidly accreting WDs undergo off-center carbon ignition if $\dot{M} \gtrsim 2 \times 10^{-6}$ $M_\odot$ yr^{-1} because of compressional heating. Afterwards the carbon flame propagates inward through the center and converts C+O into O+Ne+Mg. Then the final outcome is most likely an accretion-induced collapse rather than SNe Ia.

Recently Piersanti et al. [36] calculated the evolution of WDs with rotation in 1D approximation and argued that the lifting effect of rotation reduces compressional heating, and the WD could avoid off-center carbon ignition and reach the central carbon ignition to produce SNe Ia. However, they did not follow the accretion after the critical rotation is reached because the backward transport of angular momentum to the disk was not taken into account.

Saio & Nomoto [41] have also calculated the accretion of C+O onto the C+O WD with rotation for various timescale of angular momentum transport in 1D approximation. The outermost layer of the accreting WD quickly reaches the critical rotation as in Fig. 4. Afterwards, the angular momentum is transported backward to the disk and accretion continues. For $\dot{M} \gtrsim 4 \times 10^{-6}$ $M_\odot$ yr^{-1}, off-center carbon burning is ignited prior to the central C-ignition. Thus the lifting effect of rotation increases the critical accretion rate for the occurrence of off-center C-ignition by a factor of ~ 2 compared with the non-rotating case, but the basic conclusion is the same as for the non-rotating case, i.e. the accretion-induced collapse is the most likely outcome in the DD scenario [40].

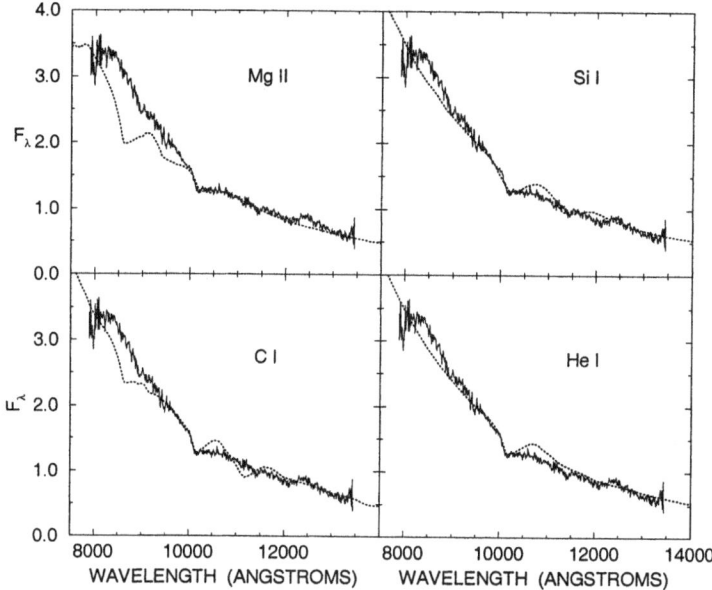

Fig. 5. The average of two infrared spectra of SN 2000cx obtained by Rudy et al. [39] at Lick Observatory in July 2000 in the J band (solid lines) is compared with four synthetic spectra (dotted lines), each of which contains lines of only one ion [10]

5   SN 2000cx

Supernova 2000cx was a very well observed SN Ia that in certain respects resembled the peculiar, "powerful" SN 1991T, but also showed some photometric and spectroscopic characteristics that are unprecedented among well observed SNe Ia (Li et al. [21]; hereafter L01).

5.1   Infrared Spectra

Rudy et al. ([39]; hereafter R02) obtained infrared spectra covering the range 0.8 to 2.5 μm, 6 and 5 days before optical maximum. They attributed an absorption feature near 1.0 μm to Mg II λ10926. Hatano et al. [10] have used the spectrum code SYNOW and found that the absorption feature is most likely due to He I λ10830. The presence of helium would have important implications for models.

In Fig. 5 the infrared spectrum [39] is compared with four synthetic spectra, each containing lines of only one ion. The synthetic spectra have $T_{bb} = T_{exc} = 12,000$ K and $v_{phot} = 22,000$ km s^{-1} (Mg II), 16,000 km s^{-1} (C I), and 20,000 km s^{-1} (Si I and He I).

Mg II λ10926 has been considered as a possible identification for the absorption feature by R02. However, the upper left panel of Fig. 5 shows that when λ10926 is strong enough to account for the 1.0 μm absorption, other Mg II lines produce unwanted absorptions near 8600 Å (λ9226) and 9400 Å (λ9632). In

LTE $\lambda 9226$ has a larger optical depth than $\lambda 10926$ for any reasonable excitation temperature. It is unlikely that Mg II is responsible for the 1.0 μm feature.

The lower left panel of Fig. 5 shows that C I, with $\lambda 10695$ producing the 1.0 μm absorption, has a similar problem of unwanted features, due to $\lambda 9055$ and $\lambda 11755$. The upper right panel shows that a group of Si I lines (mainly a multiplet at $\lambda 10790$) fit the 1.0 μm feature without severe problems; a feature due to $\lambda 12047$ is from the same lower level and is stronger, but one might argue that it is possibly present, blended with other lines, in the observed spectrum. However, unless the level of ionization in SN 2000cx was much lower at the epoch of the infrared spectrum than it was at the time of the first optical spectrum, just a few days later, Si I lines would not be expected. Finally, the lower right panel of Fig. 5 shows that He I $\lambda 10830$, with a optical depth at the photosphere of 0.9, can account for the 1.0 μm feature without causing any obvious problems. By this process of elimination, Hatano et al. [10] find that the most likely identification of the 1.0 μm feature in SN 2000cx is He I $\lambda 10830$. The evidence for helium in SN 2000cx, although not conclusive, is strong enough to be taken seriously.

5.2   Sub-Chandrasekhar Mass Model

Both L01 and R02 suggest a delayed detonation as a favorable model for SN 2000cx. In particular, R02 argued that the 1.0 μm feature is due to Mg II and the presence of such high velocity Mg as $\gtrsim$ 20,000 km s^{-1} supports the delayed detonation model. However, this argument does not hold if the 1.0 μm feature is due to He I. Rather the delayed detonation models do not contain such high velocity He (e.g. [16]).

The presence of He, instead, may suggest a sub-Chandrasekhar model for SN 2000cx. In the sub-Chandrasekhar mass models, the He detonation in the outer He layer synthesize mostly ^{56}Ni but a significant amount of He is also present due to strong α-rich freezeout [26,23,51,32]. The velocities of He are $\sim$ 11,000 - 14,000 km s^{-1} near the bottom of the He layer and $\sim$ 25,000 - 30,000 km s^{-1} near the surface, depending on M_{WD} and the mass of the He layer (M_{He}). The He velocities and the ^{56}Ni mass of SN 2000cx could be consistent with the model with relatively large M_{WD} and M_{He}. In the outer detonated layers, little Si is produced, which is also consistent with the weak Si features of SN 2000cx.

The presence of high velocity He with ^{56}Ni (and little Si) is also seen in the late detonation model W7DHE of the Chandrasekhar mass white dwarf [52]. The explosion of this type would occur for slower accretion than the sub-Chandrasekhar mass explosion [25].

References

1. Arnett, W. D. 1996, Nucleosynthesis and Supernovae (Princeton Univ. Press)
2. Branch, D. 1998, ARA&A, 1998, 36, 17
3. Branch, D., Livio, M., Yungelson, L., Boffi, F., Baron, E. 1995, PASP, 107, 717
4. Cappellaro, E. et al. 1997, A&A, 322, 431

5. Hachisu, I., & Kato, M., 2001, ApJ, 558, 323
6. Hachisu, I., Kato, M., Nomoto, K., & Umeda, H. 1999a, ApJ, 519, 314
7. Hachisu, I., Kato, M., & Nomoto, K. 1999b, ApJ, 522, 487
8. Hamuy, M., et al. 1995, AJ, 109, 1
9. Hamuy, M., Phillips, M.M., Schommer, R., & Suntzeff, N.B., 1996, AJ, 112, 2391
10. Hatano, K., Branch, D., Nomoto, K., Baron, E. 2002, in preparation
11. Hillebrandt W., & Niemeyer J.C. 2000, ARA&A, 38, 191
12. Höflich, P., & Khokhlov, A., 1996, ApJ, 457, 500
13. Höflich, P., Wheeler, J. C., & Thielemann, F. -K., 1998, ApJ, 495, 617
14. Iben, I. Jr., & Tutukov, A. V. 1984, ApJS, 54, 335
15. Iglesias, C. A., & Rogers, F. 1993, ApJ, 412, 752
16. Iwamoto, K., Brachwitz, F., Nomoto, K., et al. 1999, ApJS, 125, 439
17. Kobayashi, C., & Nomoto, K. 2002, in this volume
18. Kobayashi, C., Tsujimoto, T., Nomoto, K., et al. 1998, ApJ, 503, L155
19. Kobayashi, C., Tsujimoto, T., & Nomoto, K. 2000, ApJ, 539, 26
20. Leibundgut, B. 2001, ARA&A, 39, 67
21. Li, W. D. et al. 2001, PASP, 113, 1178 (L01)
22. Livio, M. 2000, in Type Ia Supernovae, ed. J. Niemeyer (Univ. Chicago press) 33
23. Livne, E., & Arnett, W. D. 1995, ApJ, 452, 62
24. Mazzali, P. A. & Lucy, L. B. 1998, MNRAS, 295, 428
25. Nomoto, K. 1982a, ApJ, 253, 798
26. Nomoto, K. 1982b, ApJ, 257, 780
27. Nomoto, K., & Iben, I. Jr. 1985, ApJ, 297, 531
28. Nomoto, K., Shigeyama, T., & Tsujimoto, T. 1991, in IAU Symp. 145, Evolution
    of Stars: The Photospheric Abundance Connection, ed. G. Michaud (Kluwer) 21
29. Nomoto, K., Nariai, K., & Sugimoto, D. 1979, PASJ, 31, 287
30. Nomoto, K., Thielemann, F. -K., & Yokoi, K., 1984, ApJ, 286, 644
31. Nomoto, K., Yamaoka, H., Shigeyama, T., Kumagai, S., Tsujimoto, T. 1994, in
    Supernovae, Les Houches Session LIV, ed. S. Bludman (North-Holland), 199
32. Nomoto, K. et al. 1997a, Science, 276, 1378; 1997b, in Thermonuclear Supernovae,
    eds. P.Ruiz-Lapuente et al. (Dordrecht: Kluwer), 349
33. Nomoto, K. 2000, in Type Ia Supernovae, ed. J. Niemeyer (Univ Chicago) 63
34. Nugent, P., Baron, E., Branch, D., Fisher, A., Hauschildt, P. 1997, ApJ, 485, 812
35. Paczyński, B. 1990, ApJ, 370, 597
36. Piersanti, L., Gagliardi, S., Iben, I.,Jr., Tornambe, A. 2002, ApJ, in press
37. Popham, R., & Narayan, R. 1990, ApJ, 370, 604
38. Riess, A. G., Press, W. H., & Kirshner, R. P. 1995, ApJ, 438, L17
39. Rudy, R.J., Lynch, D.K., Mazuk, S., et al. 2002, ApJ, 565, 413 (R02)
40. Saio, H., & Nomoto, K. 1998, ApJ, 500, 388; 1985, A&A, 150, L21
41. Saio, H., & Nomoto, K. 2002, in preparation
42. Segretain, L., Chabrier, G., & Mochkovitch, R. 1997, ApJ, 481, 355
43. Shetrone, M., Venn, K., Tolstoy, E., et al. 2002, A&A, in press (astro-ph/0211167)
44. Thielemann, F.-K., Nomoto, K., & Hashimoto, M. 1996, ApJ, 460, 408
45. Uenishi, T., Nomoto, K., & Hachisu, I. 2002, in New Trends in Theoretical and
    Observational Cosmology, ed. K. Sato (Tokyo: Universal Acad. Press), 129
46. Umeda, H., & Nomoto, K. 2002, ApJ, 565, 385
47. Umeda, H., Nomoto, K., Yamaoka, H., & Wanajo, S. 1999a, ApJ, 513, 861
48. Umeda, H., Nomoto, K., Kobayashi, C., Hachisu, I, Kato, M. 1999b, ApJ, 522, L43
49. van den Heuvel, E.P.J., Bhattacharya, D., Nomoto, K., et al. 1992, A&A, 262, 97
50. Webbink, R. F. 1984, ApJ, 277, 355
51. Woosley, S. E., & Weaver, T. A. 1994, ApJ, 423, 371
52. Yamaoka, H., Nomoto, K., Shigeyama, T., Thielemann, F.-K. 1992, ApJ, 393, L55
53. Yoon, S.-C., Langer, N., & Scheithauer, S. 2002, A&A, submitted

Are Cataclysmic Variables the Progenitors of Thermonuclear Supernovae?

Sumner Starrfield

Department of Physics and Astronomy
Arizona State University
Tempe, AZ 85287-1504, USA

Abstract. In 1973, Whelan and Iben suggested that binary stars, similar in structure to Cataclysmic Variables but not called that by them, were the progenitors of SN Ia's. Their evidence included the fact that thermonuclear supernovae appear to come from compact objects with similar properties. In addition, they appear late in the evolution of our galaxy when compared to the appearance of SN II and they also appear in elliptical galaxies which presumably have an older population. Since that time a large number of different types of binary stars have been proposed as the progenitors but no viable solution has appeared. I will discuss the properties of the various types of CV's, including the Super Soft X-ray Sources, and present some of the pro's and cons's for them actually being the progenitors.

1 Introduction

Exploding stars participate in the cycle of Galactic chemical evolution in which metal enriched gas ejected at high speeds is a source of heavy elements for the Interstellar Medium (ISM). In addition, some of the material ejected in these explosions condenses into grains which survive the formation of the Solar System and studies of pre-solar grains are now providing information on both their source and their formation processes (Zinner 1998; Amari et al. 2001). While AGB stars are expected to produce most of the lower mass and s-process nuclei, supernovae and novae produce higher mass nuclei and some exotic isotopes. In addition to the nuclear processed material returned to the ISM by exploding stars, they also insert a large amount of kinetic energy into the gas which enhances mixing throughout the galaxy. In addition to the intrinsic interest in the properties and evolution of these supernovae, SN Ia are thought to be responsible for the iron in our Galaxy, and they have been used to determine that the expansion of the Universe may be accelerating. Excellent reviews of the properties, spectral developments, and implications of thermonuclear supernovae can be found in Filippenko (1997) and Leibundgut (2000, 2001).

In contrast to massive stars which are the progenitors of SN II, the progenitors of SNe Ia are thought to be white dwarfs (WDs) in binary systems (Whelan & Iben 1973). Both the short duration of the peak in the light curve and the small amount of ejected mass suggests that the explosion occurs on a compact object. The declining light is powered by the radioactive decay of ^{56}Ni, which also implies little ejected mass as compared to that of a SN II. Since SNe Ia spectra show no evidence for hydrogen or helium in their ejecta, the compact object must be highly evolved (and a great deal of mass loss has occurred) so that the best candidate is a WD. Since a WD with a mass smaller than the Chandrasekhar mass ($\sim 1.4 M_\odot$) is stable, it must accrete mass from a companion until explosive conditions occur. Moreover, the companion of the WD should also be evolved since it cannot be transferring material rich in hydrogen and helium. Otherwise, these elements would be entrained by the explosion and appear prominently in the spectrum.

It is unfortunate that the class of supernova in which observational cosmology is putting so much effort still lacks an agreed upon progenitor. Clearly, until we have identified the progenitor we can only guess at the effects of metallicity on either the characteristics of the explosion or the evolutionary processes that culminate in the progenitor.

2   General Structure

Virtually every type of binary containing a WD has been proposed as a progenitor. One such system is the cataclysmic variable (CV) which is a close binary (orbital periods of hours or less) that consists of a large cool star which fills its Roche Lobe and orbits a WD. Unfortunately, there are few if any systems where the masses of the WDs are known to high accuracy. This is because there are so many sources of light in these systems that they confuse the spectroscopic determinations of WD masses. The cool star loses mass through the inner Lagrangian point and that material spirals into an accretion disk prior to falling onto the surface of the WD. The literature on accretion disks is extensive and will not be reviewed here. However, it is important to point out that the disk must be viscous in order for the gas to ultimately fall onto the surface of the star. In addition, the viscosity in the disk must far exceed the molecular viscosity or the amount of angular momentum transferred outward and mass transferred inward will be far smaller than observed. Complicating the studies of accretion disk viscosity, is the realization that it is not constant but probably depends on the disk density and temperature.

3   Types of Cataclysmic Variables

Before I begin the discussion in this section, it is important for the reader to keep in mind an extremely important observational result (Filippenko 1997; Leibundgut 2000, 2001). **There is no evidence, whatsoever, for the presence of hydrogen or helium in the ejecta of SN Ia.** This strongly implies that the object on which the explosion takes place must also be depleted in hydrogen because the opacity of hydrogen is large at the temperatures and densities determined for supernova ejecta and, if present, it would show up in the spectrum (Baron 2001, private communication). In fact, Baron and collaborators have been able to set upper limits of $0.1 M_\odot$ or less in the SN Ia that they have studied. In addition, once a supernova starts to show the slightest hint of hydrogen in its spectrum, it is reclassified as a Ib or Ic (as we heard at this meeting in the talk by Hamuy et al.).

This means that the Ia's are an extremely uniform class of objects with no hydrogen in their spectrum. Therefore, without any loss in generality we can assume that Ia's do not contain any hydrogen in their ejecta. Helium has a lower opacity and is more problematical. Clearly, it would be useful if someone did the necessary calculations to determine upper limits for helium in Ia ejecta. What this means in the search for the progenitors is that we must find objects that show as little hydrogen in their spectrum as possible and still appear to be able to evolve to a Ia explosion.

3.1 Dwarf Novae

All classes of CV's have the same general structure, although the evolutionary state of the secondary can differ between classes. Dwarf novae are so named because they go through low amplitude ($\sim$5 mag) outbursts every few months and the outbursts last a few days. There is such a variety in behavior, however, that no one system can be considered typical. Unlike the Classical or Recurrent Novae, their outbursts are thought to be caused by episodes of rapid mass accretion onto the WD and the rise in light is caused by the release of gravitational potential energy in the infalling material. If the mass transfer rate into the accretion disk is low, then the accretion disk will store material until it reaches a certain critical density. At this point the viscosity rises rapidly and drives gas inward and angular momentum outward.

The detailed observational studies of these systems suggest a wide range in WD mass with many of the systems probably having low mass WDs ($M_* <$ $1.0 M_\odot$). However, the measured mass transfer rates from the secondary in Dwarf Nova systems are small ($\leq 10^{-10} M_\odot$ yr^{-1}) implying very long times to accrete sufficient material to reach the Chandrasekhar limit if no Classical Nova event intervenes and ejects all the accreted material plus core material. One last problem with these systems is that the spectroscopic studies show that the accretion disks are hydrogen rich, as are the secondary stars, so that if an explosion took place it would show the presence of hydrogen.

3.2 Symbiotic Variables and Symbiotic Novae

This class was suggested as the progenitors of SN Ia by Hachisu et al. (1999). Unfortunately, there is no way that these systems can either explode or resemble a SN Ia. Unlike the other CV classes, the secondary is a red giant and the orbital periods are many months to years. For many years, these systems were thought to be single red giants with a strange atmospheric structure. It was the IUE studies of these variables, which indicated the presence of the hot, compact source in the system, that demonstrated their binary nature. In some cases, moreover, it does not appear that the secondary fills its lobe and mass transfer must occur via wind accretion which reduces the rate of mass transfer onto the WD. Nevertheless,

there are systems in which the WD does accrete enough material to experience
an outburst and these are called Symbiotic Novae. The outbursts take so long
(AG Peg has been in outburst since about 1850) that they must be occurring on
low mass WDs. In some cases, spectroscopic studies imply masses below $0.5M_\odot$
(Kenyon & Mikolajewska 1995). With such a low mass and low accretion rate,
the likelihood of such a system reaching the conditions for accretion induced
collapse seem unlikely.

Nevertheless, there are Symbiotic Novae in which the outburst has been observed to reoccur and the behavior of the outburst implies an explosion on a
massive WD. These systems are RS Oph, T CrB, V745 Sco, and V3890 Sgr
and they are also considered Recurrent Novae. However, the most important
argument against these systems being the progenitors is that the explosion will
occur inside the extended envelope of a hydrogen rich red giant. It does not seem
possible to hide the presence of hydrogen and helium in such a case.

3.3   Classical Novae

Classical novae, another member of the CV class, have also been suggested as
the progenitors. In order for the nova explosion to occur and to be as violent
as observed, these systems must contain a high mass WD accreting material at
rates from $\sim 10^{-8}M_\odot$ yr^{-1} to $\sim 10^{-10}M_\odot$ yr^{-1}. In addition, observations show
that the ejected material is enriched either in carbon and oxygen or oxygen,
neon, and magnesium which must come from the core of the WD (Starrfield
1989; Starrfield et al. 1998, S98). Their binary structure is very similar to that
of the Dwarf Novae but the mass transfer rates are higher. Why this is the case
is not understood. Unfortunately, the analyses of the ejected material show that
it contains sufficient core material for the WD to be losing mass as a result of
the outburst. Therefore, they cannot grow to the Chandrasekhar Limit. Also,
as for the above systems, there is too much hydrogen and helium present for an
explosion of the WD to spectroscopically resemble a SN Ia.

3.4   Super Soft X-Ray Sources

Another class of proposed progenitors are the recently discovered Super Soft X-
ray Sources (SSS). These systems were first identified as an astrophysical class
by studies with ROSAT (van den Heuvel et al. 1992; Greiner 1996; Kahabka &
van den Heuvel 1997). Most members of this class are binaries with one star
probably a WD and the other star thought to be transferring mass onto the
WD at high rates. In contrast to classical or dwarf novae, where the rates of
infall are sufficiently low to allow a thermonuclear explosion to occur, in the
SSS the infalling material is thought to be burning at the same rate that it is
being accreted (called steady burning). Since little or no mass is thought to be
lost (although jets or jet-like features are identified in the spectra: Cowley et al.
1998), the mass of the WD is supposed to be growing toward the Chandrasekhar
limit. Unfortunately, there are both observational and theoretical problems with

this scenario (Leibundgut 2000) and, again, there is too much hydrogen and helium present to be hidden during the explosion.

4   Double Degenerates – One Solution?

A final suggestion is that of the CO double-degenerates in which both the donor and the accretor are WDs (Cameron & Iben 1986). A number of such systems have been found in the past few years making this a tenable suggestion for some Ia explosions (Maxted & Marsh 1999). The final evolution of such a system would be the rapid accretion of nearly pure carbon-oxygen enriched material by a carbon-oxygen WD. Evolving over the Chandrasekhar mass would produce a SN Ia explosion (Saio & Nomoto 1998). It seems likely that such an explosion should occur in systems with a wide range of total mass and, thereby, exhibit a wide range in nickel mass and outburst characteristics.

5   Recurrent Novae – Another Solution?

Given all the negative statements about the CVs above, nevertheless, there is one class that could be the progenitors of some of the SNe Ia. These are the Recurrent Novae such as U Sco, V394 CrA, and LMC 1990 #2 in which the orbital periods are short (a day or less) and there is no evidence for a giant in the system (Starrfield et al. 1985; 1988; Shore et al. 1991). This is in contrast to those Recurrent Novae which contain a cool giant and the orbital periods are long (T CrB, RS Oph, V745 Sco, V3890 Sgr).

These systems are designated as recurrent because they have been observed to experience more than one explosion during the lifetime of an astronomer. Moreover, modeling has shown that in order to obtain recurrence times as short as 10 years (such as for U Sco), the outburst must occur on a massive WD ($M_* \sim 1.35 M_\odot$ or larger), and the mass accretion rate onto the WD must be high which requires an evolved secondary (Starrfield et al. 1985, 1988). Of great importance to SNe Ia, the material being transferred in these systems is extremely depleted in hydrogen. In fact, the accretion disk does not show any hydrogen lines and the amount of hydrogen in the ejecta is less than the amount of helium (Shore et al. 1991, 1996). This implies that the secondary star is the evolved core of a red giant that has lost most, if not all, of its hydrogen envelope. Another important result of combined observational plus theoretical studies of these systems is that the mass of the WD is growing as a result of the TNR. In fact, less than 10% of the accreted mass is ejected during the outburst, and the ejected gases show no evidence of core material having been mixed up into the accreted material. The major problems with this scenario are that there appear to be too few of these systems to agree with Ia rates, and there is helium present in the ejecta. Nevertheless, the problems with this proposal are fewer than problems with most other suggestions and, in addition, in the last two years we have discovered two new Recurrent Novae (CI Aql and IM Nor) suggesting that the number of systems is larger than currently thought.

6   Summary

I have briefly reviewed the various types of CV's that have been proposed as progenitors of SN Ia at one time or another. The evidence against most of them actually evolving to an explosion seems insurmountable. The only systems that have any hope of evolving to an explosion are the double degenerates and the Recurrent Novae with short orbital periods and minimal hydrogen present in the system (U Sco, V394 CrA, LMC 1990 # 2). These systems appear to be the only CV's in which the WD is growing as a result of accretion. The small amount of hydrogen seen in the ejected material would probably be hidden in a SN Ia explosion.

Acknowledgements

This work was partially supported by NSF and NASA grants to Arizona State University.

References

1. Amari, S., et al. 2001, ApJ, 551, 1065.
2. Cameron, A.G.W., Iben, I. 1986, ApJ, 305, 228
3. Cowley, A., Schmidtke, P., Crampton, D., Hutchings, J. 1998, ApJ, 504, 854
4. Filippenko, A. V. 1997, ARAA, 35, 309
5. Greiner, J. 1996, Supersoft X-ray Sources, Springer-Verlag, Berlin
6. Hachisu, I., Kato, M., Nomoto, K. 1999, ApJ, 522, 487.
7. Kahabka, P., van den Heuvel, E.P.J. 1997, ARAA, 35, 69
8. Kenyon, S.J., Mikolajewska, J. 1995, AJ, 110, 391
9. Leibundgut, B. 2000, A&A Reviews, 10, 179
10. Leibundgut, B. 2001, ARAA, 39, 67
11. Maxted, P.F.L., Marsh, T.R. 1999, MNRAS, 307, 122
12. Saio, H., Nomoto, K. 1998, ApJ, 500, 388
13. Shore, S.N., Kenyon, S., Starrfield, S., Sonneborn, G. 1996, ApJ, 456, 717.
14. Shore, S.N., Sonneborn, G., Starrfield, S., Hamuy, M., Williams, R.E., Cassatella, A., Drechsel, H. 1991, ApJ, 370, 193
15. Starrfield, S., 1989, in *The Classical Novae*, ed. M. Bode and A. Evans, (New York: Wiley), 39
16. Starrfield, S., Sparks, W.M., Shaviv, G. 1988, ApJL, 326, L35
17. Starrfield, S., Sparks, W.M., Truran, J.W. 1985, ApJ, 291, 136
18. Starrfield, S., Truran, J.W., Wiescher, M.C. and Sparks, W.M. 1998, MNRAS, 296, 502. (S98)
19. van den Heuvel, E.P.J., Bhattacharya, D., Nomoto, K., Rappaport, S.A. 1992, A&A, 262, 97
20. Whelan, J., Iben, I. 1973, ApJ, 186, 1007
21. Zinner, E. 1998, AREPS, 26, 147

Search for Double Degenerate Progenitors of Supernovae Type Ia with SPY[*] [**]

R. Napiwotzki[1], N. Christlieb[2], H. Drechsel[1], H.-J. Hagen[2], U. Heber[1],
D. Homeier[4], C. Karl[1], D. Koester[3], B. Leibundgut[5], T.R. Marsh[6],
S. Moehler[3], G. Nelemans[7], E.-M. Pauli[1], D. Reimers[2], A. Renzini[5], and
L. Yungelson[8]

[1] Dr. Remeis-Sternwarte, Astronom. Institut, Universität Erlangen-Nürnberg,
Sternwartstr. 7, 96049 Bamberg, Germany
[2] Hamburger Sternwarte, Universität Hamburg, Gojenbergsweg 112, 21029 Hamburg,
Germany
[3] Institut für Theoretische Physik und Astrophysik, Universität Kiel, 24098 Kiel,
Germany
[4] Department of Physics & Astronomy, University of Georgia, Athens,
GA 30602-2451, USA
[5] European Southern Observatory, Karl-Schwarzschild-Str. 2, 85748 Garching,
Germany
[6] University of Southampton, Department of Physics & Astronomy, Highfield,
Southampton SO17 1BJ, UK
[7] Institute of Astronomy, Madingley Road, Cambridge CB3 0HA, UK
[8] Institute of Astronomy of the Russian Academy of Sciences, 48 Pyatnitskaya Str.,
109017 Moscow, Russia

Abstract. We report on a large survey for double degenerate (DD) binaries as potential progenitors of type Ia supernovae with the UVES spectrograph at the ESO VLT (ESO **SN** Ia **P**rogenitor surve**Y** – SPY). About 560 white dwarfs were checked for radial velocity variations until now. Ninety new DDs have been discovered, including short period systems with masses close to the Chandrasekhar mass.

1 The Project

Supernovae of type Ia (SN Ia) play an outstanding role for our understanding of galactic evolution and the determination of the extragalactic distance scale. However, the nature of their progenitors is still unknown (e.g. [7]). There is general consensus that the event is due to the thermonuclear explosion of a white dwarf when the Chandrasekhar mass ($1.4M_\odot$) is reached, but the nature of the progenitor system remains unclear. Two main options exist: the merging

---

[*] Based on data obtained at the Paranal Observatory of the European Southern Observatory for programs 165.H-0588, 167.D-0407, 266.D-5658, 268.D-5739, 68.D-0483, 69.D-0534

[**] Based on observations collected at the German-Spanish Astronomical Center, operated by the Max-Planck-Institut für Astronomie Heidelberg jointly with the Spanish National Commission for Astronomy

of two WDs in the so called double degenerate (DD) scenario [2], or mass transfer from a red giant/subgiant in the so-called single degenerate (SD) scenario [15].

We know that most stars end up as white dwarf remnants and that a major fraction of stars are in binary systems, hence DDs must be common among WDs. What we do not know is whether there exist enough DDs able to merge in less than one Hubble time and produce a SN Ia event. In the SD case, we know that white dwarfs accreting from a non-degenerate companion do exist, but we do not know whether such systems exist in sufficient number to account for the observed SN Ia frequency, nor whether the WD grows enough to reach the critical mass for ignition. There is also evidence from a class of sub-luminous SN Ia that both routes might be significant [1].

On the theoretical side three possible outcomes of the merger of a super-Chandrasekhar mass DD are discussed in the literature: a SN Ia explosion leaving no remnant [2,13], an accretion induced collapse producing a neutron star without a SN Ia explosion [14], or the formation of a neutron star during a SN Ia event [5]. On the observational side several systematic radial velocity (RV) searches for DDs have been undertaken starting in the mid 1980's checking a total of ≈ 200 white dwarfs RV for variations (cf. [8] and references therein), but have failed to reveal any massive, short-period DD progenitor of SN Ia. This negative result cast some doubt on the DD scenario. However, it is not unexpected, as theoretical simulations suggest that only a few percent of all DDs are potential SN Ia progenitors [3,12].

In order to perform a definitive test of the DD scenario we have embarked on a large spectroscopic survey of ≥ 1000 white dwarfs (ESO SN Ia Progenitor surveY – SPY). SPY will overcome the main limitation of all efforts so far to detect DDs that are plausible SN Ia precursors: the samples of surveyed objects were too small. In SPY spectra are taken with the high-resolution UV-Visual Echelle Spectrograph (UVES) at the UT2 telescope (Kueyen) of the ESO VLT in service mode. Our instrument setup provides nearly complete spectral coverage from $3200\,\text{Å}$ to $6650\,\text{Å}$ with a resolution $R = 18500$ ($0.36\,\text{Å}$ at Hα). Due to the nature of the project, two spectra at different, "random" epochs separated by at least one day are observed.

ESO provides a data reduction pipeline for UVES, which forms the basis for our first selection of DD candidates. A careful re-reduction of the spectra is in progress. Differing from previous surveys we use a correlation procedure to determine RV shifts of the observed spectra (cf. [9]). We routinely measure RVs with an accuracy of $\approx 2\,\text{km}\,\text{s}^{-1}$ or better, therefore running only a very small risk of missing a merger precursor, which has orbital velocities of $150\,\text{km}\,\text{s}^{-1}$ or higher.

2   Results

We have analyzed spectra of 558 white dwarfs and pre-white dwarfs taken during the first two years of the SPY project and detected 90 new DDs, 13 are double-lined systems (only 6 were known before). Results are summarized in Table 1.

Our observations have already increased the DD sample by a factor of six. After completion, a final sample of ≈200 DDs is expected. SPY is the first RV survey which performs a systematic investigation of both classes of white dwarfs: DAs *and* non-DAs. Previous surveys were restricted to DA white dwarfs, because the sharp NLTE core of Hα allows a very accurate RV determination. This feature is not present in the non-DA (DB, DO) spectra, but the use of several helium-lines enables us to reach a similar accuracy. Our results in Table 1 indicate that the DD frequency among DA and non-DA white dwarfs are similar.

Table 1. Fraction of RV variable stars in the current SPY sample for different spectral classes. WD+dM denotes systems for which a previously unknown cool companion is evident from the red spectrum (not included in the DA/non-DA entries).

Spectral type	total	RV variable	detection rate
All DDs	558	90	16%
non-DA (DB,DO,DZ)	72	10	14%
WD+dM	30	14	47%

Follow-up observations of this sample are mandatory to determine periods and white dwarf parameters and find potential SN Ia progenitors among the candidates. Good statistics of a large DD sample will also set stringent constraints on the evolution of close binaries, which will dramatically improve our understanding of this phase of stellar evolution. Starting in 2001 follow-up observations have been carried out with VLT and NTT of ESO as well as with the 3.5 m telescope of the Calar Alto observatory/Spain and the INT [9–11,4]). Our sample includes many short period binaries, several with masses closer to the Chandrasekhar limit than any system known before. During our follow-up observations we have detected a very promising potential SN Ia precursor candidate discussed below.

Our follow-up observations concentrated on candidates with high RV variations, indicating short periods. Double-lined systems (with both white dwarfs visible in the spectrum) are of special interest, because these binaries allow the determination of individual masses for both components.

Exemplary for other double-lined systems we discuss here the DA+DA system HE 1414−0848 [11]. The orbital period of $P = 12^h25^m44^s$ and semi-amplitudes of $127\,\mathrm{km\,s^{-1}}$ and $96\,\mathrm{km\,s^{-1}}$ are derived for the individual components. RV curves for both components are displayed in Fig. 1. The ratio of velocity amplitudes is directly related to the mass ratio of both components. Additional information comes from the mass dependent gravitational redshift. The difference in gravitational redshift corresponds to the apparent difference of "systemic velocities" of both components, as derived from the RV curves (Fig. 1). Only one set of individual white dwarf masses fulfills the constraints given by both the amplitude ratio and redshift difference (for a given mass-radius relation). We estimate the masses of the individual components with this method to be

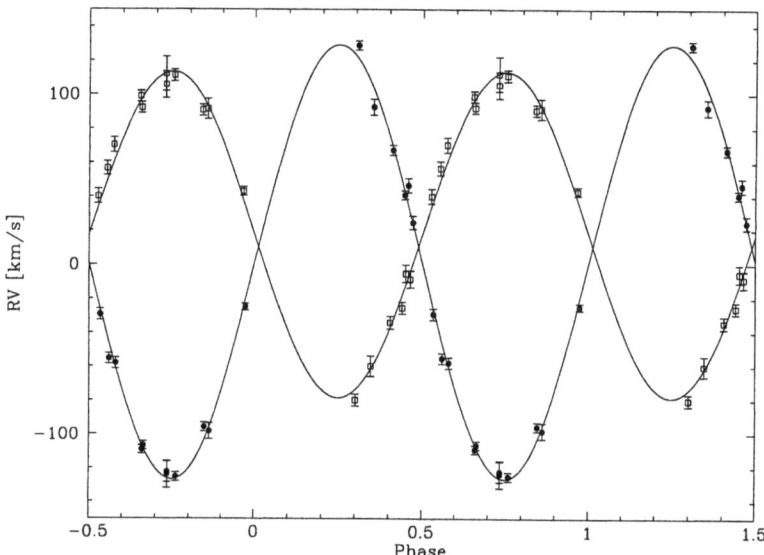

Fig. 1. Measured RVs as a function of orbital phase and fitted sine curves for HE 1414-0848. Filled circles/open rectangles indicate the less/more massive components A and B. Note the difference of the "systemic velocities" γ_0 between both components caused by gravitational redshift

$0.55 M_\odot$ and $0.71 M_\odot$, respectively. This translates into $\log g = 7.92$ and 8.16, respectively.

Another estimate of the white dwarf parameters is available from a model atmosphere analysis of the combined spectrum. We have developed a new tool (FITSB2), which performs a spectral analysis of both components of a double-lined system. The fit is performed on all available spectra, covering different orbital phases simultaneously. We fitted temperatures and gravities of both components of HE 1414−0848 (the mass ratio fixed at the accurate value derived from the RV curve). The results are $T_{\mathrm{eff}}/\log g = 8380\,\mathrm{K}/7.83$ and $10900\,\mathrm{K}/8.14$ for A and B (Fig. 2), which are in good agreement with the $\log g$ values predicted from the analysis of the RV curve. The total mass of the HE 1414−0848 system is $1.26 M_\odot$, only 10% below the Chandrasekhar limit. The system will merge due to loss of angular momentum via gravitational wave radiation after two Hubble times.

Our follow-up observations yielded parameters for several DDs. We have completed the analysis for three double-lined systems summarized in Table 2. The mass sum of the HE 2209−1444 system is as high as that of HE 1414−0848, but the HE 2209−1444 system is closer and will merge within 4 Gyrs. On the other hand the WD 1349+144 system consists of two low mass white dwarfs and has a rather long period and will merge only after more than 100 Hubble times.

Our sample includes also another short period ($P = 7^{\mathrm{h}}12^{\mathrm{min}}$) system, which will merge after 4 Gyrs and has probably a system mass above the Chandrasekhar

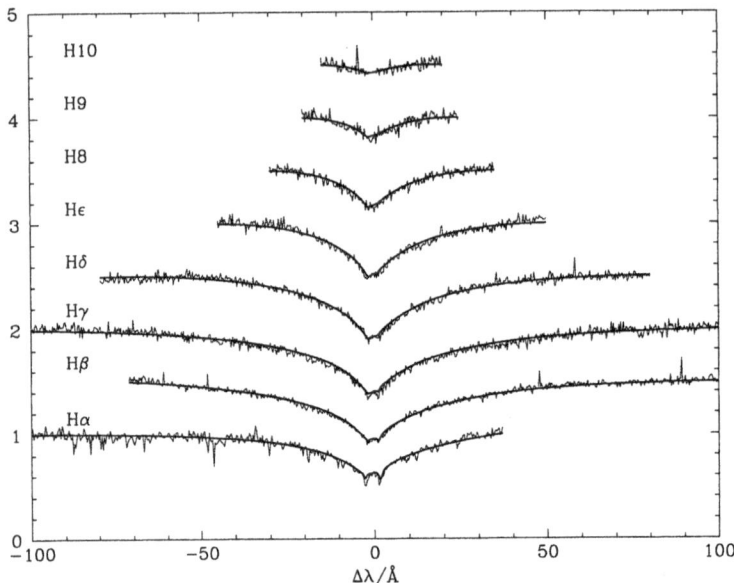

Fig. 2. Model atmosphere fit of the Balmer series of HE 1414−0848 with FITSB2. This is only a sample fit. All available spectra, covering different orbital phases, were used, which allows an unambiguous parameter determination

Table 2. New double-lined DDs from SPY (data taken from [11] and [4]). The table details the mass sum ($M_1 + M_2$) of the systems, the orbital periods (P), and the time (T_{merge}) until merging due to gravitational wave radiation.

system	$M_1 + M_2$ ($M_\odot$)	P (h)	T_{merge} (Gyrs)
HE 1414−0848	1.26	12.4	25
HE 2209−1444	1.26	6.6	4
WD 1349+144	0.88	53.0	2000

limit. Thus this system is a very promising potential SN Ia precursor candidate. However, the Hα line core of the secondary in this system is broad and shallow compared to typical DA line cores, which makes the determination of a precise RV curve and therefore of the masses difficult. Some additional data are necessary to verify our RV curve solution. Results will be reported elsewhere.

3 Conclusions

SPY has now finished a major fraction of the survey. Our analysis of the data from the first 24 months has already quadrupled the number of white dwarfs checked for RV variability (from 200 to 760) and increased the number of known DDs from 18 to 108 compared to the results of the last 20 years. Our sample includes many short period systems (Table 1; [10,11,4]), several with masses closer to the Chandrasekhar limit than any system known before, greatly improving the statistics of DDs. We expect this survey to produce a final sample of ≈200 DDs.

This will also provide a census of the final binary configurations, hence an important test for the theory of close binary star evolution after mass and angular momentum losses through winds and common envelope phases, which are very difficult to model. An empirical calibration provides the most promising approach. A large sample of binary white dwarfs covering a wide range in parameter space is the most important ingredient for this task. We have started a project to exploit the information provided in the SPY sample of DDs.

Our ongoing follow-up observations already revealed the existence of three short period systems with masses close to the Chandrasekhar limit, which will merge within 4 Gyrs to two Hubble times. Even if it will finally turn out that the mass of our most promising SN Ia progenitor candidate system is slightly below the Chandrasekhar limit, our results already allow a qualitative evaluation of the DD channel. Since the formation of a system slightly below the Chandrasekhar limit is not very different from the formation of a system above this limit, the presence of these three systems alone provides evidence that potential DD progenitors of SN Ia do exist.

References

1. D.A. Howell: ApJ 554, L193 (2001)
2. I. Iben Jr., A.V. Tutukov: ApJS 54, 335 (1984)
3. I. Iben Jr., A.V. Tutukov, L.R. Yungelson: ApJ 475, 291 (1997)
4. C. Karl, R. Napiwotzki, U. Heber, et al.: 'Double Degenerates from the Supernova Ia Progenitor Survey', in: *White Dwarfs, Proc. XIII European Conference on White Dwarfs*, eds. D. de Martino, R. Kalytis, R. Silvotti, J.E. Solheim, (Kluwer, Dordrecht), in press (astro-ph/0210004)
5. A.R. King, J.E. Pringle, D.T. Wickramasinghe: MNRAS 320, L45 (2001)
6. D. Koester, R. Napiwotzki, N. Christlieb, et al.: A&A 378, 556 (2001)
7. M. Livio: 'The Progenitors of Type Ia Supernovae', in *Type Ia Supernovae: Theory and Cosmology*, eds. J.C. Niemeyer & J.W. Truran, (Cambridge Univ. Press, Cambridge 2000) p. 33
8. T.R. Marsh: NewAR 44, 119 (2000)
9. R. Napiwotzki, N. Christlieb, H. Drechsel, et al.: AN 322, 401 (2001)
10. R. Napiwotzki, H. Edelmann, U. Heber, et al.: A&A 378, L17 (2001)
11. R. Napiwotzki, D. Koester, G. Nelemans, et al.: A&A 386, 957 (2002)
12. G. Nelemans, L.R. Yungelson, S.F. Portegies Zwart, F. Verbunt: A&A 365, 491 (2001)
13. L. Piersanti: PASP 114, 471 (2002)
14. H. Saio, K. Nomoto: A&A 150, L21 (1985)
15. J. Whelan, I. Iben Jr: ApJ 186, 1007 (1973)

Search for the Companions of Galactic SNe Ia

Pilar Ruiz–Lapuente[1,2], Fernando Comeron[3], Stephen Smartt[4], Robert Kurucz[5], Javier Mendez[1,6], Ramon Canal[1], Alex Filippenko[7], and Ryan Chornock[7]

[1] Department of Astronomy, University of Barcelona, Spain
[2] Max–Planck Institut für Astrophysik, Garching, Germany
[3] European Southern Observatory, Garching, Germany
[4] Institute of Astronomy, Cambridge, UK
[5] Harvard–Smithsonian Center for Astrophysics, Cambridge, USA
[6] Isaac Newton Group, La Palma, Spain
[7] Department of Astronomy, University of California at Berkeley, USA

Abstract. The central regions of the remnants of Galactic SNe Ia have been examined for the presence of companion stars of the exploded supernovae. We present the results of this survey for the historical SN 1572 and SN 1006. The spectra of the stars are modeled to obtain Teff, log g and the metallicity. Radial velocities are obtained with an accuracy of 5–10 km s^{-1}. Implications for the nature of the companion star in SNeIa follow.

1 Introduction

Type Ia supernovae have long been recognized as close binary systems where one of the stars, a carbon–oxygen WD (C+O WD), undergoes a thermonuclear runaway after reaching the explosive conditions at its center. While the remnants and the ejecta left by the explosion have been studied in great detail, little is known about companion stars. Motivated by the lack of definitive constraints on the progenitors of SNe Ia, a programme was started to find the moving companion of the exploding WD (Ruiz–Lapuente 1997; Canal, Mendez & Ruiz–Lapuente 2001).

Several possibilities for the companion have been proposed. It could be a white dwarf, giant, subgiant, or a main sequence star. Each of these possibilities and in particular the expected peculiar velocities are discussed below.

White Dwarf. A double–degenerate (DD) system, where the mass donor is a second WD, should not leave any companion. The companion is destroyed in the mass–transfer process. Although first estimates of the efficiency in producing SNe Ia disfavor this option, as reported in this conference (Napiwotzki et al. 2002), more double degenerate systems able to merge in less than a Hubble time are being discovered.

Subgiant or Giant. Close binaries consisting of WDs with subgiant or giant companions (also called Algol–like systems) can produce the growth in mass of the WD until it reaches the explosive condition (Whelan & Iben 1973; Hachisu, Kato & Nomoto 1996, 1999; Hachisu et al. 1999). The transfer takes place when the subgiant or giant fills its Roche lobe due to its thermonuclear evolution and pours material onto the WD. The material transferred to the WD is H, and depending on the accretion rate the WD will either grow in mass up to the Chandrasekhar limit or ignite a detonation in an outer shell. Once the WD explodes, the ejecta hit the companion. That produces several effects: the system is disrupted, the companion moves with the orbital velocity it had before the explosion, plus the kick velocity it acquires from the impact of the ejecta. The interaction of the ejecta with the companion will also strip away part of its mass. In addition, the bound remnant of the companion will have its hydrostatic and thermal equilibrium altered (Marietta, Burrows & Fryxell 1999; Canal, Méndez & Ruiz–Lapuente 2001). Hydrostatic equilibrium will resume on a very short time scale, but thermal equilibrium will only be regained on a longer, global Kelvin–Helmholtz time scale (see also the earlier calculations done by Colgate 1970; Wheeler, Lecar & McKee 1975; Fryxell & Arnett 1981; Taam & Fryxell 1985; Livne, Tuchman & Wheeler 1992). That means the companion will exhibit an increase in radius, increased surface temperature, and an increased luminosity, which should be observable in the most recent Galactic Type Ia supernovae. Depending on the separation between the two stars at the time of explosion, the companion will acquire different velocities. As a result of the orbital motion that the star had before the explosion and the kick velocity, we expect the star to be moving at velocities of a few hundred km s^{-1} (see Table 1). That is one order of magnitude larger than the systemic velocities of typical stars at the same location in our Galaxy. The motion will be an identifying signature of the companion.

Main Sequence Star. The final possibility for a SN progenitor which has been proposed consists of a WD plus a main–sequence companion, i.e. a cataclysmic variable. Orbital shrinkage is driven in those systems by magnetic braking plus the emission of gravitational wave radiation (see Ruiz–Lapuente, Canal, & Burkert 1997 and references therein). In this case, the moving companion remaining after the explosion will be fainter than in the giant/subgiant case, but typical velocities will be higher.

Several considerations lead to the selection of SNIa remnants for the detection of the companion star within this project. First, the X-ray shell morphology should be spherically symmetric with a well defined center of the remnant. The distance and the age of the SNeIa remnants should be in an adequate range to allow a reasonable search radius (see Table 1). SNe of Type Ia such as Tycho (SN 1572) and SN 1006, which have shell morphologies with high spherical symmetry preserved up to 1000 yr after the explosion seem to be the most natural targets for this study.

Table 1. Typical apparent magnitudes, proper motions, radial velocities, and maximum angular distance from explosion site (after 10^3 yr) of SNeIa companions

Companion type and distance	m_V	π (arcsec yr^{-1})	v_r (km s^{-1})	θ (arcmin)
Main sequence				
1 kpc	15.1	0.067	320	1.6
5 kpc	18.6	0.013	320	0.3
Subgiant				
1 kpc	12.6	0.038	180	0.9
5 kpc	16.1	0.008	180	0.2
Red giant				
1 kpc	10.5	0.015	70	0.4
5 kpc	14.0	0.003	70	0.1

2   SN 1572

SN 1572 (Tycho Brahe's supernova) is close to the Galactic plane ($b = +1.4\,^0$). The field of the supernova is 3.8' in radius. In SN 1572 the radio shell is very regular and early radio and optical expansion measurements have found similar results on the ejecta expansion rate. The distance inferred by the expansion of the radio shell and by other methods lies between 2.25 and 4.5 kpc (Strom, Goss & Shaver 1982).

An estimate of the center is possible with an uncertainty less than 10 % of the radio shell. Therefore, it seemed a good strategy to complete observations down to a magnitude limit $m_R \sim 23$ of the stars within 0.7' of the center. In Fig. 1 we show the spectra of some of the stars near the center of the remnant. A red giant is very near the geometrical center of SN 1572. Other stars in the vicinity range from supergiants to WDs.

We have obtained spectra with high enough resolution to allow detection of motions in the radial direction. Our spectra correspond to different epochs and this allows an additional check of variation of the velocities in those directions. Spectra were obtained with ISIS and UES at the William Herschel Telescope and with ESI at the Keck Telescope (see Table 2 for the observations).

Table 2. Observations of SN 1572

Run	Rd (')[1]	m_R [2]	Telescope [3]	R	Spec Range (A)	stellar types
(1)	0.7	14	WHT (UES)	50,000	4000–7100	red giant
(2)	0.7	23	WHT (ISIS)	15,000	4600-7500	red giants to WD
(3)	0.7	23	Keck (ESI)	7000	4000–10000	as above

[1] Radius of the search
[2] Limiting magnitude
[3] Telescopes(Instrumentation)

Table 3. Observations of SN 1006

Run	Rd (')[1]	m_R [2]	Telescope [3]	R	Spec range (A)	stellar types
(1)	5	13	NTT (EMMI)	10,000	3950–7660	red giants
(2)	5	15	VLT (UVES)	50,000	3500–9000	all types

[1] Radius of the search
[2] Limiting magnitude
[3] Telescopes(Instrumentation)

3   SN 1006

SN 1006 is at a Galactic latitude $b = +14.6^0$, about ~ 550 pc above the Galactic plane. The field of this supernova extends 15' in radius. An examination of all the centers given in the literature up to now (Winkler & Long 1997; Reynolds and Gilmore 1986; van den Bergh 1976) and our inspection of the geometry of the SNIa, suggest that within 1', the center of SN 1006 should be at $\alpha = 15\ 02\ 55$, $\delta = -41\ 55\ 12$ (J2000) as given by Allen et al. (2001). The distance estimates to this SNR are in the range between 1.5 and 2.5 kpc. Recently Winkler, Gupta & Long (2002) measure a distance of 2.17 ± 0.08 kpc to SN 1006 from the expansion rate of the remnant as derived from the optical filaments. Given the predictions for the movement of the companion, at a distance between 1.5 and 2.5 kpc, a search around 5' at the best determined center of the remnant should find the companion of this SNIa.

Our search goes down to mag 15 in R. At the distance to this remnant, this means to reach stars of solar luminosity. Thus, an exhaustive test of the hypothesis of supergiant, giant and main sequence stars is obtained in this way (see Table 3 for a summary of the observations).

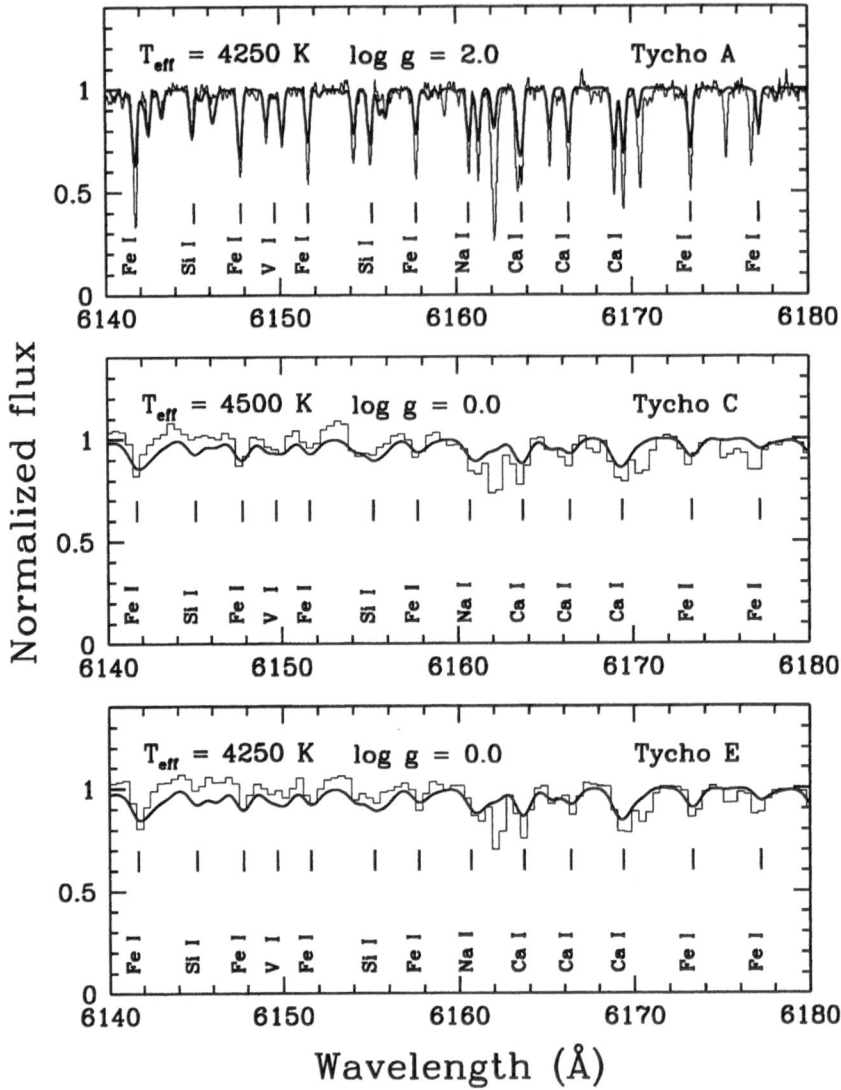

Fig. 1. Calculated synthetic spectra compared with the observed spectra of the SN companion candidates near the center of SN 1572. These are the closest red giants and supergiants to the center of the explosion, their surface gravity goes from log g=2 for Tycho A to log g=0 for C and E. The effective temperatures for those stars are similar. They are in the range 4200–4500 K. Model atmospheres with solar chemical abundances give a good account of the spectra. The overall spectral comparison allow us to exclude overabundances of the Fe–peak elements. Moreover, the stars show no enhancement of iron–peak elements versus intermediate–mass elements in the spectra. Synthetic spectra are shown with bold continuous lines

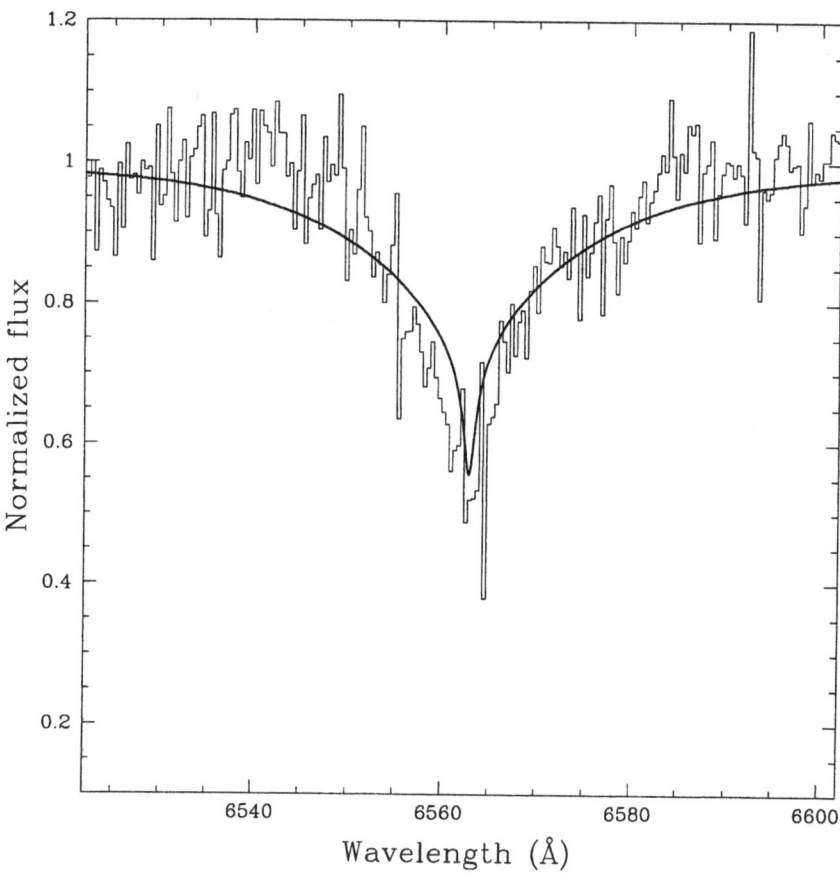

Fig. 2. DA WD near the center of Tycho remnant. A preliminary analysis suggests log g=6 and Teff=25,000 K

4   Modeling

We have compared calculated LTE spectra with the observed spectra of the SN companion candidates. We have used Kurucz's grids of model atmospheres (Kurucz 1993), a spectral synthesis code based on the Uppsala Synthetic Spectrum Package (Gustafsson et al. 1975), and the atomic data from the linelists in Kurucz. The profiles of the Balmer lines of H, calculated separately for the model atmospheres of the grids, have also been used in the spectral fits. The comparison of the synthetic spectra with the observations is shown in Fig. 1 for some of the stars in our sample of candidates to SN companion. Identifications of the most significant lines are given. Good agreement is achieved assuming solar abundances (Anders & Grevesse 1989). Moreover, none of the closest stars to the center show signs of any spectroscopic anomaly. Radial velocities have also been measured from the wavelength shifts of several lines in each observed spectrum: they are in the range –40 to –60 km s^{-1}, which is perfectly attributable to Galactic rotation alone.

The low–velocity tail of the ejecta of a thermonuclear (Type Ia) supernova is made of Fe–peak elements (Fe, Co, Ni, Mn, Cr, V, Ti), as it comes from the material at the center of the exploding star and it thus reached the highest temperatures ($\sim 10^{10}$ K), being then processed to Nuclear Statistical Equilibrium (Thielemann 1989). Since this material is the most likely to have contaminated the surface of a binary companion (see papers cited in section 1), especial care has been given to the determination of the abundances of those elements in the atmospheres of the stars in our sample. An overabundance of those elements in relation to intermediate–mass elements is expected.

The giants at the core do not show any signs of contamination and are consistent with solar metallicities (or slightly lower than solar metallicities). This result disfavors the subgiant possibility and that of a red giant closely bound to the WD prior to the moment of explosion.

Strong constraints can be placed also on the main sequence star candidates. No main sequence star moving at high radial velocity has been detected in SN 1572. An examination over a more extended radius around the center is being done.

For SN 1006 we have found radial velocities in the range 30–120 km s^{-1} for the giant stars within the search radius. Stars show a larger dispersion in radial velocity than in the field of the remnant of SN 1572. A full account of the observations, the measurement of the reddening, the overall spectral modeling and a discussion on the distance to those candidates will be presented elsewhere.

This paper is based on observations obtained at the WHT, operated by the Isaac Newton Group of Telescopes, funded by the PPARC at the Spanish Observatorio del Roque de los Muchachos of the Instituto de Astrofisica de Canarias; at W.M. Keck Observatory, which is operated as a scientific partnership among the California Institute of Technology, the University of California and the National Aeronautics and Space Administration (the Observatory was made possible by the generous financial support of the W.M.Keck Foundation); and the NTT and VLT at the La Silla and Paranal Observatories of ESO for programs 67.D-0348(A) and 69.D-0397(A). We express our gratitude to the ESO User Support Group for providing help and conducting service observations and pipeline reductions of the data taken at Paranal. P.R.L would like to thank as well the assistance provided by Carlos Abia with the Uppsala Synthetic Spectrum Package.

References

1. E. Anders & N. Grevesse: Geochim. Cosmochim. Acta 53, 197 (1989)
2. W.D. Arnett: *Supernovae and Nucleosynthesis* (Princeton Univ. Press, Princeton) (1996)
3. R. Canal, J. Mendez, & P. Ruiz–Lapuente: ApJ 550, L53 (2001)
4. S.A. Colgate: Nature 225, 247 (1970)
5. B.A. Fryxell & W.D. Arnett: ApJ 243, 994 (1981)
6. B.A. Gustafsson et al.: A&A 42, 407 (1975)
7. I. Hachisu, M.N. Kato & K. Nomoto: ApJ, 470, 97 (1996)
8. I. Hachisu, M.N. Kato, K. Nomoto & H. Umeda: ApJ, 519, 314 (1999)
9. I. Hachisu, M.N. Kato & K. Nomoto: ApJ, 522, 487 (1999)
10. R.L. Kurucz: *ATLAS9 Stellar Atmospheres Programs, Grids of Model Atmospheres and Line Data* (CD–Rom, Smithsonian Astrophysical Observatory, Cambridge) (1993)
11. E. Livne, Y Tuchman, J.C. Wheeler: ApJ 399, 665, (1992)
12. E. Marietta, A. Burrows & B. Fryxell: ApJS 128, 615 (2000)
13. R. Napiwotzki et al. (SPY consortium): A&A 386, 957 (2002)
14. S.P. Reynolds & D.M. Gilmore: AJ 92, 1138 (1986)
15. P. Ruiz–Lapuente: Science 276, 1813 (1997)
16. P. Ruiz–Lapuente, R. Canal & A. Burkert: in *Thermonuclear Supernovae*, Ed. P. Ruiz–Lapuente, R. Canal & J. Isern (Kluwer, Dordrecht), 205 (1997)
17. R.G. Strom, W.M. Goss, P.A. Shaver: MNRAS 200, 473 (1982)
18. R.E. Taam & B.A. Fryxell: ApJ 294, 303 (1985)
19. F.K. Thielemann: in *Nuclear Astrophysics*, Ed. M. Lozano, M.I. Gallardo & J.M. Arias (Springer–Verlag, Berlin), 106 (1989)
20. S. van den Bergh: ApJ 208, L17 (1976)
21. J.C. Wheeler, M. Lecar & C. F. McKee: ApJ 200, 145 (1975)
22. J. Whelan & I. J. Iben: ApJ 186, 1007 (1973)
23. P.F. Winkler & K.S. Long: ApJ 491, 829 (1997)
24. P.F. Winkler, G. Gupta & K.S. Long: astro–ph/0208415 (2002)

Part IV

Models of Thermonuclear Supernova Explosions

Small Steps Toward Realistic Explosion Models of Type Ia Supernovae

Jens Niemeyer, Martin Reinecke, Claudia Travaglio, and Wolfgang Hillebrandt

Max-Planck-Institut für Astrophysik, Karl-Schwarzschild-Strasse 1,
D-85748 Garching, Germany

Abstract. We describe recent progress in multidimensional simulations of exploding Chandrasekhar mass white dwarfs, interpreted as the canonical model for type Ia supernovae. We particularly stress the importance of the initial ignition surface, chosen here as a distribution of 30 spherical, randomly located bubbles near the center of the star. In addition, we present the first results for nucleosynthetic yields obtained by post-processing a multidimensional explosion model.

1 Introduction

If there were only two messages a theorist took home from the talks reviewing Type Ia supernova (SN Ia) observations at this conference, they would probably be: 1) the hunting season for SNe Ia is open, and 2) a few years from now, there won't be a single self-respecting observatory that is not involved in at least one of the many SN Ia surveys about to be started. Clearly, theorists have some catching up to do.

The need for a solid theoretical framework of how SNe Ia evolve and explode, and how they tell their story in the light emitted by the explosion products, is growing along with the quality and quantity of observational data. Theorists are tackling this problem in a twofold way. For the last three decades, one-dimensional spherically symmetric models have been used to study the various channels that may give rise to a successful SN Ia in terms of the predicted spectra, light curves, and nucleosynthesis. Much of this work was centered on the Chandrasekhar-mass (M_{ch}) scenario wherein a C+O white dwarf accretes H or He from its binary companion (see [1] for details) and ignites explosive carbon burning just before it reaches a total mass of $M_{ch} \sim 1.4$ $M_{\odot}$. The subsequent explosion produces enough ^{56}Ni (~ 0.5 $M_{\odot}$) and intermediate mass elements to reproduce "normal" SN Ia spectra and light curves, provided that the amount of C+O burned at any given density is suitably chosen. In 1D models, this can easily be achieved by parameterizing the thermonuclear flame speed and, if desired, the density at which a spontaneous transition to supersonic burning (detonation) occurs. Alternative scenarios, including sub-M_{ch} explosions and merging white dwarfs (double degenerates), have met with mixed success but may be needed to explain some of the "weirder" events (see e.g. [2,3] for details and references).

More recently, it has become possible to perform three-dimensional simulations of (a part of) an exploding M_{ch}-white dwarf [4–6]. The principal difficulty in these models is the fact that the hydrodynamically unstable, and consequently turbulent, flame front develops structures on more scales than can be resolved on any computer now or in the foreseeable future. It is circumvented by employing subgrid-scale (SGS) models for the unresolved scales that provide an educated guess of the effective turbulent flame speed on the scale of the computational grid [7]. This quantity is then used to propagate a model flame front, meant to represent the turbulent flame brush on smaller scales, which turns nuclear fuel into ashes. Much effort has been spent lately on making this flame algorithm more robust [8,9].

It is important to stress that despite the need for SGS modelling, 3D simulations reach a qualitatively different level of predictive power than 1D models. The amount of material burned at a given density cannot be fine-tuned by varying the SGS parameters or the grid resolution [4]. Once the parameters and resolution are fixed, the only remaining degree of freedom is the way the explosion is ignited. Exactly how strongly the outcome depends on the ignition conditions is currently under investigation; an update is given in Sec. (2). Furthermore, we have begun analyzing the detailed nucleosynthetic yields of our explosion models by post-processing the ejecta, using the density and temperature history of passively advected tracer particles. First preliminary results are presented in Sec. (3). For details regarding our computational methods and earlier results we refer to recent publications [4,5].

2   Recent Three-Dimensional SN Ia Calculations

2.1   Mode of Ignition

So far, very little is known about the number, size and location of quietly burning "hot spots" on whose surface the thermonuclear runaway eventually sets in and triggers the SN Ia explosion. This is a consequence of the complicated physical processes taking place in the white dwarf's core during the convective smoldering phase prior to ignition lasting for ~1000 years. The long time scales combined with the relatively slow convective motions make numerical simulations of this phase a daunting task which has not been undertaken in its full complexity so far.

Theoretical considerations and simplified simulations carried out by Garcia-Senz & Woosley [10] suggest that fast burning starts on the surface of many small bubbles ($r \leq 5$ km) within 100 km of the star's center. We try to model this scenario as closely as possible in our calculations.

For an accurate description of the ignition conditions mentioned above, extremely high central resolutions would be required. With the given resources it was possible to achieve a central resolution of 3.33 km, using a grid consisting of 768^3 zones. In the simulated octant of our model b30_3d_768, 30 bubbles with a radius of 10 km were distributed randomly. The bubble locations were drawn from a Gaussian probability distribution with a dispersion of $\sigma = 75$ km. Bubbles located more than 2.5 σ away from the center were rejected. A particular realization of these prescriptions is shown in Fig. 1.

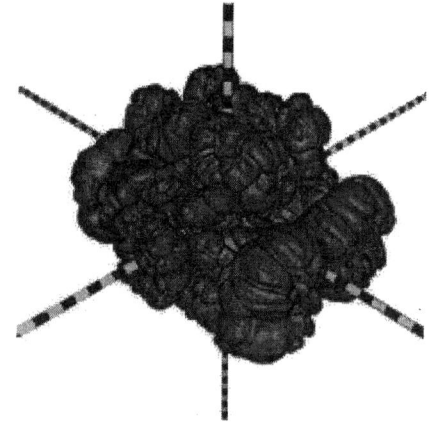

Fig. 1. Initially burned material, representing 30 ignition "bubbles" distributed close to the center of the white dwarf

Fig. 2. Snapshot of the flame surface 0.6 s after ignition

2.2 Results

Owing to the small volume of the bubbles, the initial hydrostatic equilibrium is only slightly disturbed. During the first stages the energy release is therefore lower than in previous simulations. Only after the total flame surface has grown considerably (mostly by deformation of the bubbles), vigorous burning sets in (Fig. 2). The final energy level is reached slightly earlier than in previous simulations, and is comparable to the older multi-bubble calculation b9_3d_512; for an energy comparison of all 3D simulations performed so far see Fig. 3.

In addition to the total energy release, the mass fraction of unburned material in the central region of the remnant appears to be a good criterion for judging the validity of our simulations, because a high amount of C/O in this region would most likely produce a characteristic signature in the late-time spectra which has not yet been observed. In this respect the results of our earlier calculations were not very encouraging since the ashes rose towards the surface in large structures and left nearly pure fuel in the center. Using many initial bubbles, however, seems to alleviate this problem insofar as the statistical isotropy of the initial flame at least delays the development of large-scale turnover motions. As a consequence, C/O is lower than 20% in the central 0.2 $M_\odot$ after 0.9 s for model b30_3d_768.

3   Nucleosynthesis in Multidimensional SNe Ia

The evolution of the elements in the range between C and Ni is dominated by two alternative explosive stellar sources, i.e. by the combined contribution of SNe II and SNe Ia. From the theoretical point of view, the general nucleosynthesis outcome of SNe Ia is dominated by Fe-group elements, involving also sizable

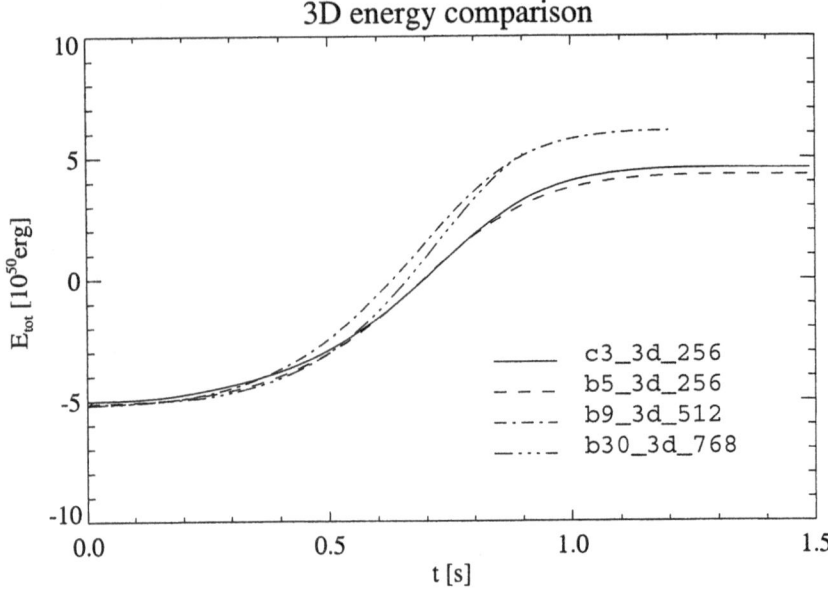

Fig. 3. Comparison of 3D explosion models with varying grid resolution. The lines represent the total energy as a function of time after ignition

fractions of Si–Ca and minor amounts of unburned or pure C-burning products (e.g. C, O, Ne, Mg).

The multidimensional SN Ia simulations described in this work employed a minimal nuclear reaction network, sufficient for a good approximation of the thermonuclear energy release. Although the predicted chemical composition agrees well with the expectations (see e.g. [5]), we present here the preliminary results of a more detailed study of the nuclear abundances in the ejecta obtained by post-processing the output of the model b5_3d_256. Since the multidimensional hydrodynamical scheme applied is of Eulerian type (i.e. the grid doesn't move with the fluid), in order to record temperature and density evolution as a function of time (necessary input for the nucleosynthesis calculations) we homogeneously distributed 8000 marker particles and followed their T and ρ evolution. We then calculated the nucleosynthesis experienced by each marker and computed the total yield as a sum over all the markers including the decays of unstable isotopes.

The nuclear reaction network employed in computing the explosive nucleosynthesis contains 259 nuclear species ranging from neutrons, protons, and α-particles to germanium (F. Thielemann, priv. comm.). The network reaction rates include experimental as well as theoretical rates [11]. Weak interaction rates applied in the network are from Fuller, Fowler & Newman [12] and Langanke & Martinez-Pinedo [13].

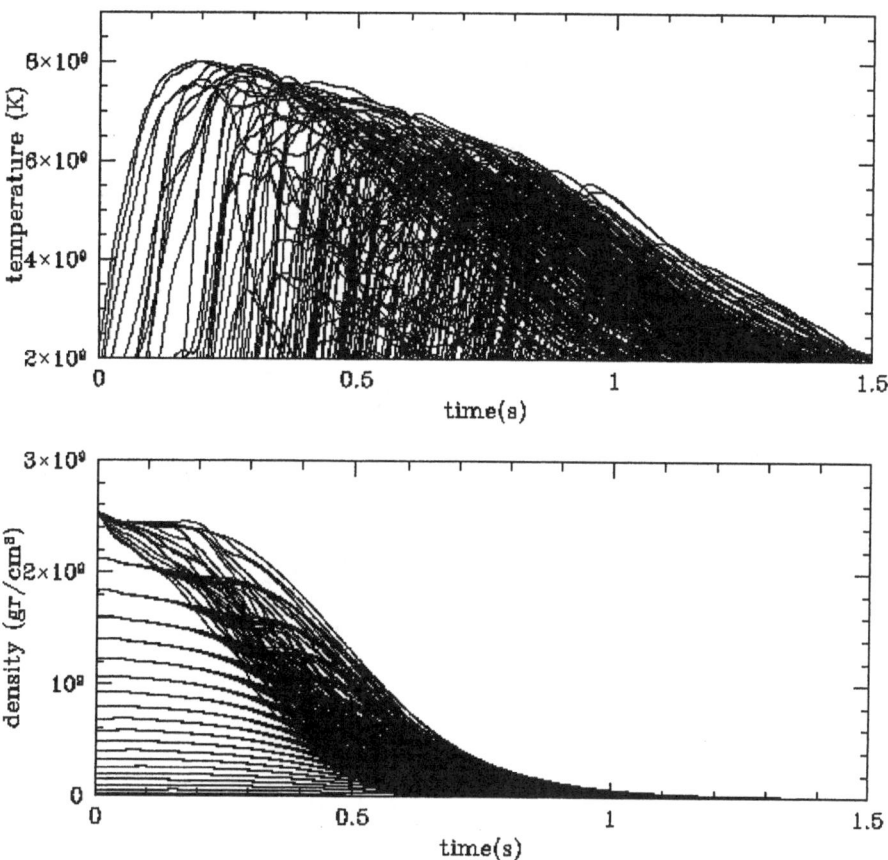

Fig. 4. Temperature (*upper panel*) and density (*lower panel*) distribution as a function of time in the marker particles

For this work, the initial mixture we used consists of (mass fraction) 0.475 $M_\odot$ of ^{12}C, 0.5 $M_\odot$ of ^{16}O, and 0.025 $M_\odot$ of ^{22}Ne. When the flame passes through the fuel, C, O and Ne are converted to ash, with different compositions depending on the initial T and ρ. In Fig. 4 we show the typical burning conditions of the 3D model, T vs. time (*upper panel*) and ρ vs. time (*lower panel*). Each line in the figure corresponds to a marker particle. Nuclear statistical equilibrium (NSE) conditions are assumed in the marker particles with $T > 6 \ 10^9$ K. At such temperatures ($T > 6 \cdot 10^9$ K) a mixture of ^{56}Ni and α-particles in NSE is synthesized. Below that temperature burning only produces intermediate mass elements. Once the temperature drops $T < 2 \cdot 10^9$ K, no burning takes place during the short timescale ($\simeq 1.5$ s) of the explosion ("unburned" material).

The resulting yields are shown in Fig. 5. They are very preliminary and should be considered with care. However, with a few exceptions, isotopic abundances are within the expected range. For comparison, we also plotted the yields for

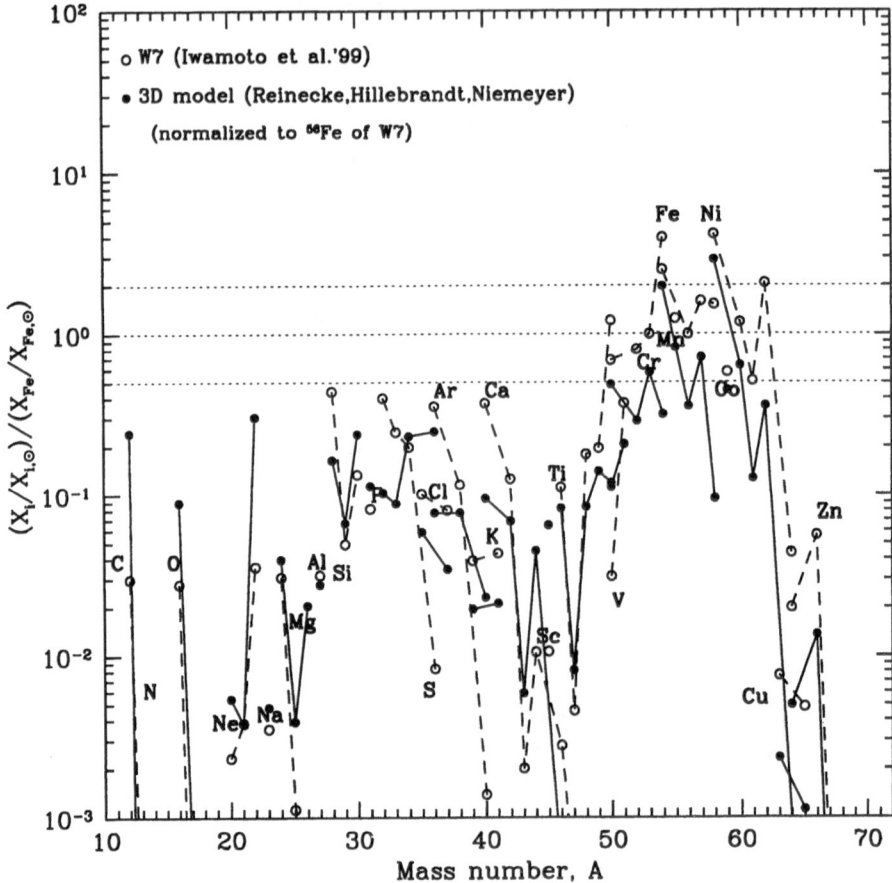

Fig. 5. Isotopic abundances (with respect to solar ratio) obtained for the 3D model b5_3d_256. For comparison, the abundances of the W7 model of [14] are also shown

the canonical SN Ia model W7 [14]. Exceptions include the overproduction of ^{48,50}Ti, ^{54}Fe and ^{58}Ni, as well as the high abundances of *unburned* C and O (i.e. $\sim 20\%$ of the total mass). This may have an interesting impact on the role of SNIa in the context of Galactic chemical evolution, in particular for C. Again, these results should not yet be taken at face value: work is in progress to compute models with higher resolution (cf. Sec. 2), different white dwarf properties, as well as with a higher number of markers.

4 Summary

We have presented recent progress in modelling type Ia supernovae (SNe Ia) with highly resolved three-dimensional simulations of exploding Chandrasekhar mass C+O white dwarfs. The simulations employ a modern flame tracking/capturing algorithm and a subgrid-scale model to compute the turbulent flame speed based on large scale velocity fluctuations. Our current research focusses on the sensitivity of the explosion on the ignition conditions and on the detailed nucleosynthetic yields that they predict. While our results are certainly far from conclusive, one trend appears to be robust: their potential to reproduce observations of normal SNe Ia grows with computational resolution, most of all by allowing more finely resolved ignition conditions. Furthermore, the preliminary nuclear yields are consistent with expectations. We can therefore say with some confidence (and without parameterization) that the Chandrasekhar mass scenario with a pure turbulent deflagration is a viable candidate for SN Ia explosions. Whether or not a delayed detonation is needed to explain the observed spectra will be answered in future work, but Occam's razor suggests that it is not.

Needless to say, much work lies ahead of us, in particular regarding a better understanding of the evolution toward ignition and radiation transport in multiple dimensions. Clearly, the most direct test of explosion models and nucleosynthesis is provided by observed light curves and spectra. Sorokina & Blinnikov [15] have used our 3D-model nucleosynthesis result, averaged over spherical shells, to compute color light curves in the UBVI-bands. Their first results are very promising.

References

1. K. Nomoto: these proceedings
2. W. Hillebrandt, J.C. Niemeyer: Ann. Rev. Astron. Astrophys. 38, 191 (2000)
3. J.C. Niemeyer, M. Reinecke, W. Hillebrandt: astro-ph/0203369 (2002)
4. M. Reinecke, W. Hillebrandt, J.C. Niemeyer: Astron. Astrophys. 386, 936 (2002)
5. M. Reinecke, W. Hillebrandt, J.C. Niemeyer: Astron. Astrophys. 391, 1167 (2002)
6. A. Khokhlov: astro-ph/0008463 (2001)
7. J.C. Niemeyer, W. Hillebrandt: Astrophys. J. 452, 769 (1995)
8. M. Reinecke, W. Hillebrandt, J.C. Niemeyer, R. Klein, A. Gröbl: Astron. Astrophys. 347, 724 (1999)
9. F. Röpke, J.C. Niemeyer, W. Hillebrandt: Astrophys. J. (subm.) (2002)
10. D. Garcia-Senz, S.E. Woosley: Astrophys. J. 454, 895 (1995)
11. T. Rauscher, F.-K. Thielemann: At. Data Nucl. Data Tabl. 75, 1 (2000)
12. G.M. Fuller, W.A. Fowler, M.J. Newman: Astrophys. J. 293, 1 (1985)
13. K. Langanke, G. Martinez-Pinedo: Nucl. Phys. A673, 481 (2000)
14. K. Iwamoto, F. Brachwitz, K.I. Nomoto, N. Kishimoto, H. Umeda, W.R. Hix, F.K. Thielemann: Astrophys. J. Suppl. Ser. 125, 439 (1999)
15. E. Sorokina, S. Blinnikov: these proceedings

Influence of Geometry in the Delayed Detonation Model of SNIa

Domingo García-Senz[1,2] and Eduardo Bravo[1,2]

[1] Dpt. Física i Eng. Nuclear, Universitat Politècnica de Catalunya, Jordi Girona 1-3, Mòdul B5, 08034 Barcelona, Spain
[2] Institut d'Estudis Espacials de Catalunya, Gran Capità 2-4, 08034 Barcelona, Spain

Abstract. We present several hydrodynamical simulations of thermonuclear supernovae dealing with multiple delayed detonations. The calculations were carried out in three dimensions, making possible to study the influence of geometry of the flame front in two aspects. First, the evolution of its fractal dimension during the deflagration phase has been followed until a critical value is reached such that the deflagration may turn into a detonation. Second, as the resulting detonation could probably be scattered through the flame, the effect of its initial location on the detonation propagation, final energetics and nucleosynthesis has been explored.

1 Flame Acceleration: From Very Subsonic to (Maybe) Supersonic Combustion Rate

The study about how a white dwarf is disrupted by a thermonuclear explosion, giving rise to a Type Ia Supernovae, is a hot topic in theoretical astrophysics. Taking as starting point a carbon-oxygen white dwarf near the Chandrasekhar-mass limit current models must face two crucial questions: where the initial carbon runaway starts? and, once switched on, how does combustion propagate through the volume of the white dwarf? The first question is related with the conditions settled in the convective core prior to the explosion and will not be addressed here. The second question has deserved a lot of work during many years. In the last decade there has been a considerable theoretical advance owing to the increasing feasibility to carry out calculations in more than one dimension and also because many useful ideas have been borrowed and adapted from the terrestrial combustion research. The central point in any modelization is how to accelerate the combustion from its subsonic laminar value at the beginning ($\simeq 0.01c_s$, with c_s the local sound speed) to a maximum value, around $0.3c_s$, before the expansion quenches the nuclear energy input. It is today widely accepted that hydrodynamical instabilities are the physical agent responsible for combustion acceleration by increasing enormously the effective heat exchange area between fuel and ashes.

To some extent flame propagation becomes a problem of geometry. The ratio between the actual corrugated flame surface, S_{eff}, and the equivalent spherical surface encompassing the same volume can be written as:

$$\frac{S_{eff}}{4\pi r_b^2} \simeq \left(\frac{v_{max}}{v_l}\right)^{(D-2)k} \tag{1}$$

where r_b is the mean radius of the burned zone, v_l is the microscopic laminar velocity of the flame [1], D is the fractal dimension of the effective surface and k is a scaling exponent which links the maximum and minimum scale lengths with the corresponding velocity fluctuations at these scales. For example, for Kolmogorov turbulence k=3 and the minimum length scale l_G, called the Gibson length in combustion theory is $l_G \simeq r_b(v_l/v_{max})^3$. An interesting limiting case takes place for $D = 2 + 1/k$, then the effective combustion velocity becomes independent of v_l. This behaviour has been observed in terrestrial experiments dealing with turbulent chemical flames in the corrugated regime where $D = 7/3$. In the context of supernovae there are more physical mechanism affecting the flame surface other than turbulence but the above statement still approximately holds [2]. This result opens a way to model the explosion by simply substituting the laminar velocity v_l by another one, v_b, high enough to rise l_G to the level of resolution of the hydrocode. The adoption of such *baseline velocity* v_b, instead of v_l is equivalent to have a subgrid below the level of resolution of the code (see Sect. 2).

The above picture gives a combustion front which moves very subsonically at the beginning, when instabilities are not so strong, but rapidly accelerates when plenty of instabilities develop. However, in order to get the adequate explosion energetics and synthesize a substantial amount of intermediate-mass elements the effective front velocity must reach quite a high value, $v_{eff} \geq 10^3$ km s^{-1} once the density at the front location declines below $\rho = 5\ 10^7$ g cm^{-3}. Multidimensional models based on pure deflagrations have traditionally had some difficulty to obtain such precise effective velocity profile, especially at low densities (but see J. Niemeyer, this volume). One way to cure this drawback is to turn the deflagration into a supersonic detonation at late times, as many one-dimensional models have shown [3]. Even though the physical mechanism driving such a transition has not been found yet it could be related to two possible causes: 1) spontaneous burning of a critical mass of carbon when combustion enters the distributed regime [4,5], 2) an effective velocity exceeding the maximum value for a Chapman-Jouguet (CJ) deflagration to be stable [4,6]. Although it is not evident that the CJ limit, strictly defined for planar fronts, can be extrapolated to different geometries using the effective velocity instead of the laminar one we have adopted this second possible mechanism in order to set a practical criterion to select those points of the flame which turn into a detonation.

Up to now there are very few multidimensional calculations dealing with the delayed detonation scenario [6,7]. Two dimensional simulations [7] are forced to detonate carbon in a single region located near the singularity axis. Full three-dimensional calculations are, however, desirable because of their ability to handle multipoint ignitions and better represent the important previous deflagration phase. When calculated in three dimensions the detonation trajectories can be significatively altered by the irregular distribution of fuel and ashes set in the deflagration phase and by the detonation-detonation and detonation-deflagration wave interactions. As these effects will affect the outcome of the explosion it is worth to make a first exploration of this scenario.

2   Method of Calculation and Models

The simulations were carried out using the SPH technique adapted to handle both steady thermal waves (flames) and shocks (detonations) [8]. We included realistic physics: EOS consisting of relativistic partially degenerated electrons with pair corrections, ions as ideal gas plus Coulomb corrections and radiation. The nuclear part is a small nuclear network of 9 nuclei from helium to nickel, wherever the temperature exceeded $5 \cdot 10^9$ K the material is isochorically processed to the nuclear statistical equilibrium, electron captures were allowed in this regime. Detailed nucleosynthesis was calculated by postprocessing the output of the hydrodynamics of each model.

The advance of the flame in the deflagration stage was simulated by expanding the flame thickness (much smaller than 1 cm!) to the actual resolution of the code (several dozens of km) by adequately rescaling the actual microscopic conductivity and nuclear energy rate in the energy equation. In addition, ellipsoidal kernels were used in the detonation phase in order to increase the resolution in this regime. Typically an improvement between 2-3 is achieved.

The initial model is an isothermal white dwarf of 1.36 $M_\odot$ and $\rho_c = 1.8\ 10^9$ g cm^{-3} in hydrostatic equilibrium. The first stages of the explosion were followed by using a one-dimensional Lagrangian code until the central density declined to $\rho_c = 1.4\ 10^9$ g cm^{-3}. At this point the structure was mapped to a 3D distribution of 250,000 particles, and the velocity field around the flame contour perturbed by a sample of 20 sinusoidal functions of different amplitude and wavelength. Afterwards the evolution was followed with the SPH code.

During the progression of the calculation the geometrical features of the flame front were tracked by calculating the main scaling parameters of the surface such as l_{min}, l_{max} and D, using the method described in [8]. The value of l_{min} can be used to set the baseline velocity v_b. In principle an optimum value can be found by solving the equation $l_{min}(v_b, r) = h(r)$, where h is the smoothing length parameter. Even though v_b is a time-dependent local quantity we have found that, on average, a constant value of $v_b \simeq 60 - 70$ km s^{-1} is enough to move l_{min} (or equivalently the Gibson length, l_G) to the coarse resolution of the SPH code.

2.1   Deflagration Phase

The spherical symmetry was kept during the first half second of the evolution, the flame moving with v_b, almost the laminar value near the center. Afterwards the typical fingers of the Rayleigh-Taylor instability in the linear regime developed. When t=0.7 s the effective velocity and the nuclear rate input began to increase (Figs. 1,2). After one second the front displayed the mushroom like shape characteristic of the non-linear regime. At approximately t=1.1 s the nuclear energy input reached a maximum of $1.3 \cdot 10^{51}$ ergs s^{-1}, the effective velocity of the front being $\simeq 0.12c_s$. A maximum value of $v_{\text{eff}} = 0.14c_s$ was reached at t=1.41 s. From here on the effective velocity declined owing to the expansion. The main features at the end of this phase are summarized in Table 1.

Table 1. Main features at the end of the deflagration phase

t	M_b	Total energy	M_{Ni}	M_{C+O}	M_{Si}
seconds	$(M_\odot)$	(10^{51}erg)	$(M_\odot)$	$(M_\odot)$	$(M_\odot)$
1.54	0.52	0.18	0.27	0.65	0.07

The evolution of the fractal dimension is also shown in Fig. 1. After an initial period where the fractal dimension of the flame surface remained very close to two it rapidly increased for $t > 0.7$ s. At that time the averaged density of the flame dropped below $5 \ 10^8$ g cm^{-3}. After t=1.1 s the fractal dimension rose slowly until a value $D \simeq 2.6$, similar to that associated with the RT instability in the non-linear regime, $D_{RT} = 2.5$.

2.2 Detonation Phase

When $< \rho_{flame} > = 2 \cdot 10^7$ g cm^{-3} the flame surface was rather complex, displaying many regions with high fractal dimension. In fact, a multifractal analysis showed that $\simeq 20\%$ of the mass within the flame had $D > 2.5$. A high value of D means that the local effective velocity of the front is higher than the average value of v_{eff} making possible the transition to a detonation by the second mechanism commented in Sect. 1. A rough estimation of the minimum fractal dimension needed to exceed the Chapman-Jouguet limit was given in [6] where it was assumed that turbulence ($D_{turb} = 7/3$) dominates in those scale lengths not directly resolved by the hydrocode. According to that analysis, a fractal dimension $D > 2.5$ when $\rho \simeq 2 \cdot 10^7$ g cm^{-3} could make the transition possible. We took this criteria as a procedure to pick up the points of the flame prone to turn into a detonation in Model A of Table 2, our main calculation. In models B, C, D we explore the influence of the altitude in the outcome of the model by artificially detonating all the mass within the flame at different radii.

Model A gave an acceptable SNIa model, as can be seen in Table 2 and Fig. 2b. The resulting explosion energy was a little weak but not bad and the ^{56}Ni amount is enough to power the light curve. The final abundance of intermediate-mass elements was right although they showed a large spread in velocity space as can be seen in Fig. 2. Nevertheless there remained 0.34 M$_\odot$ of unburnt carbon and oxygen chiefly due to both, the shielding effect of the ashes

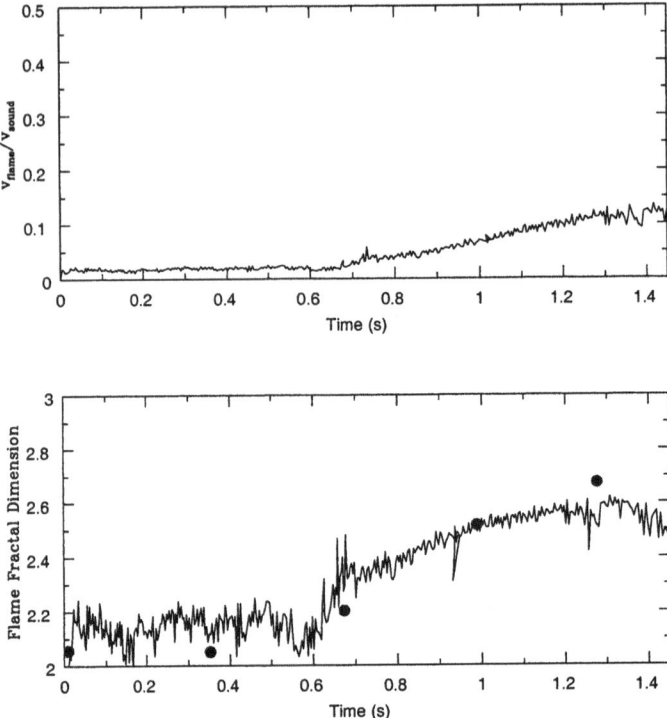

Fig. 1. Evolution during the deflagration phase. (**a**) Effective velocity of the flame in units of the local sound speed. (**b**) Evolution of the fractal dimension of the flame in two *independent* approximations: correlation dimension calculated as in [8] (big dots) and derived from the burning mass rate given by the SPH (continuum line). Both calculations are consistent with each other

produced during the deflagration phase and the degradation of the resolution when the detonation propagates in the low-density regions. The detonation altitude largely influences the outcome of the explosion as shown in Table 2 and Fig. 3. Models B and C were not totally unsatisfactory but Model D was clearly bad, the explosion was too weak and an unacceptable amount of unburnt $^{12}C + ^{16}O$ was left even at the center of the white dwarf (Fig. 3).

3   Conclusions

The explosion of a Chandrasekhar-mass white dwarf has been simulated in three dimensions. The adopted explosion mechanism was a deflagration followed by multiple detonations when the mean density of the flame front declined below $\rho = 2 \cdot 10^7$ g cm^{-3}. In addition the dependence of the main observables of the explosion against the geometrical location of the initial detonating spots has

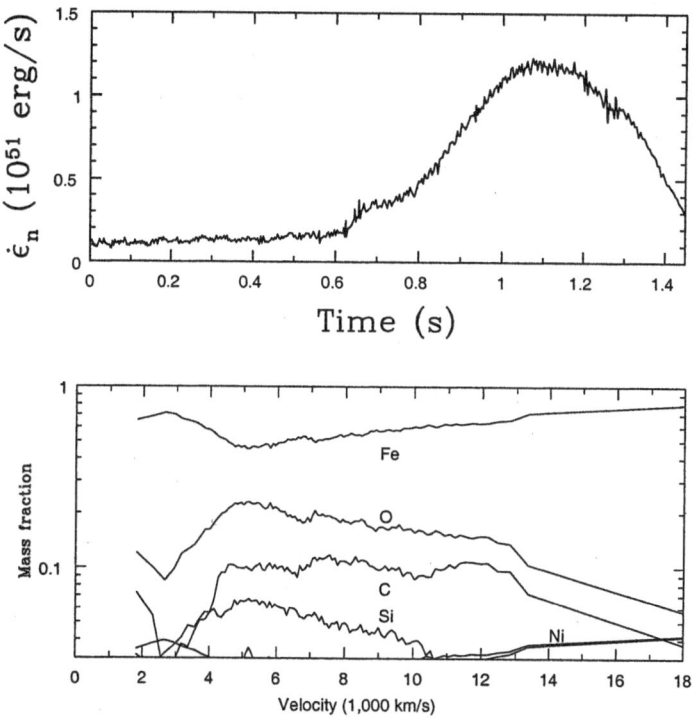

Fig. 2. (a) Rate of released nuclear energy in the deflagration phase. (b) Distribution of $^{56}Fe(^{56}Ni)$, ^{28}Si and $^{12}C,^{16}O$ in velocity space

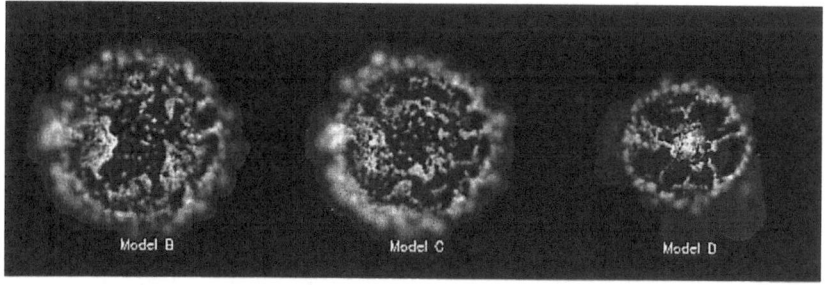

Fig. 3. Final abundance of $^{12}C + ^{16}O$ for models B,C, and D in a meridian slice

Table 2. Calculated models

Model	Detonation criteria at $\langle\rho_{flame}\rangle = 2\ 10^7$ g.cm^{-3}	E_{kin} $(10^{51}\,erg)$	M_{Ni} $(M_\odot)$	M_{C+O} $(M_\odot)$	M_{Si} $(M_\odot)$
A	$D_{fractal} > 2.5$	0.75	0.54	0.34	0.16
B	$r < 1.3\ 10^8$ cm	0.48	0.43	0.48	0.1
C	$1.7\ 10^8\,cm < r < 2.0\ 10^8\,cm$	0.51	0.42	0.45	0.14
D	$2.4\ 10^8\,cm < r < 2.8\ 10^8\,cm$	0.33	0.34	0.57	0.09

been explored. Our best model (model A of Table 2) is an acceptable model for Type Ia Supernovae albeit the explosion was a little weak. A relevant feature is that ^{56}Ni as well as intermediate-mass elements showed a large dispersion in velocity space. A negative point was the large amount of unburnt carbon and oxygen, which were left in isolated pockets. The sensitivity of the outcome on the initial detonation distribution is strong, as discussed in Sect. 2. This fact relies on the irregular distribution of fuel and ashes settled by the initial deflagration and may provide a natural way to explain the Type Ia supernovae diversity.

There is still a lot of work needed to confirm or reject the delayed detonation scenario. In particular, asynchronous multipoint detonations with more resolution must be attempted to shape the model. On the other hand there are some expectatives that deflagration models dealing with active turbulence could also give the right models for thermonuclear supernovae without invoking detonations (J.Niemeyer and Bravo and García-Senz, this volume). For the time being it is necessary to investigate both mechanisms.

This work has been benefited from the MCYT grants EPS98-1348 and AYA2000-1785 and by the DGES grant PB98-1183-C03-02.

References

1. F.X. Timmes and S.E. Woosley: ApJ **396**, 649 (1992)
2. A.M. Khokhlov: ApJ **449**, 695 (1995)
3. A.M. Khokhlov: A&A **245**, 114 (1991)
4. J. Niemeyer and S.E. Woosley: ApJ **475**, 740 (1997)
5. A.M. Khokhlov, E.S. Oran and J.C. Wheeler: ApJ **478**, 678 (1997)
6. E.Bravo, D. García-Senz: 'Asymmetrical delayed detonation from a 3D hydrosimulation of a white dwarf explosion'. In *Cosmic Chemical Evolution. Proceedings of the 187th Symposium of the IAU, Kyoto*, ed. by K. Nomoto and J.W.Truran (Dordrecht: Kluwer Academic Publishers, 2002), p220
7. E. Livne: ApJ **527**, L97 (1999)
8. D.García-Senz, E. Bravo., N. Serichol: ApJSS **115**, 119 (1998)

Thermonuclear Supernovae:
Is Deflagration Triggered by Floating Bubbles?

Eduardo Bravo[1,2] and Domingo García-Senz[1,2]

[1] Dpt. Física i Eng. Nuclear, Universitat Politècnica de Catalunya, Diagonal 647, 08028 Barcelona, Spain
[2] Institut d'Estudis Espacials de Catalunya, Gran Capità 2-4, 08034 Barcelona, Spain

Abstract. In recent years, it has become clear from multidimensional simulations that the outcome of deflagrations depends strongly on the initial configuration of the flame. We have studied under which conditions this configuration could consist of a number of scattered, isolated, hot bubbles. Afterwards, we have calculated the evolution of deflagrations starting from different numbers of bubbles. We have found that starting from 30 bubbles a mild explosion is produced ($M(^{56}Ni) = 0.56M_\odot$), while starting from 10 bubbles the star becomes only marginally unbound ($K = 0.05$ foes).

1 The Initial Configuration of Deflagrations in White Dwarfs and Its Outcomes

The explosion mechanism(s) responsible for thermonuclear supernovae (SNIa) is still not well known. Although there have been recently some claims that the delayed detonation mechanism lacks a physical background, there are still unexplored mechanisms by which the transition from deflagration to detonation could occur (see García-Senz & Bravo, this volume). Here we concentrate on the other possibility, i.e. a pure deflagration that would process about a solar mass, synthesizing of the order of $0.5M_\odot$ of ^{56}Ni in order to make a typical SNIa. Multidimensional calculations of deflagrations bear the advantage over one-dimensional models that the energy generation rate becomes eventually independent from the local value of the flame speed ([5,4], see also Niemeyer, this volume). In these calculations, the acceleration of the fuel consumption rate is due to the deformation of the flame surface, which is well accounted for by 3D hydrocodes.

In this picture, the main free parameter is the initial configuration of the flame. There have been a few [3] simulations of the transition from the hydrostatic phase to the hydrodynamic one, but a self consistent multidimensional initial model is still far from being available. However, a plausible ignition scenario was suggested in [2]: the nearly-simultaneous runaway at several different spots (from here on, bubbles) in the central region of the white dwarf. In this paper, we explore the dependence of the outcome of the explosion on the initial number of bubbles. First, we address the statistics of the initial distribution of bubble radii and, then, we present the results of a couple of hydrodynamical simulations performed with an SPH code ([1] and García-Senz & Bravo, this volume).

1.1   Statistical Approach to the Initial Distribution of Igniting Bubbles

Our statistical approach is based on the following assumptions: **A)** As a result of convection there appear a number of hot spots, characterized by the central (peak) temperature, T_0, and the thermal profile (which we take here exponential, with characteristic exponent R_0). **B)** Each hot spot evolves adiabatically in place in a time given by its ignition timescale. **C)** The peak temperatures of the hot spots can be characterized by a statistical continuous function, $\theta\,(T_0)$.

Depending on the thermal gradient inside each hot spot our results can be split into two different regimes: **1)** If the thermal profile is shallow enough, bubbles grow due to spontaneous flame propagation, **2)** otherwise, bubbles grow conductively. In the first case, the distribution function of bubble radii is a time function given by

$$\frac{dN}{dR_b} = \theta\,(T_0)\,\frac{A^{1/B}\cdot 10^9}{R_0}\,\frac{\tau_i\,(T_{0b})\exp\left(BR_b/R_0\right)}{\{t + \tau_i\,(T_{0b})\,[1 - \exp\left(BR_b/R_0\right)]\}^{1+1/B}}\,, \tag{1}$$

where $\tau_i\,(T_{0b}) \sim A\,(T_{0b}/10^9\,\mathrm{K})^{-B}$ is the ignition timescale, and $T_{0b} = 10^9$ K. In the second case, the distribution function becomes

$$\frac{dN}{dR_b} = \theta\,(T_0)\,\frac{A^{1/B}}{Bv_{cond}}\left(t - \frac{R_b}{v_{cond}}\right)^{1+1/B}\,. \tag{2}$$

In the first case, and for any reasonable choice of the function θ, the resulting distribution has a sharp peak at a determined value of R_b so that the initial configuration is composed of an arbitrary number of *equal size bubbles*. In contrast, in the second case the bubble radii distribution is continuous up to a maximum radius, which results in a initial distribution of *unequal size bubbles*.

1.2   SPH Simulations of Deflagrations Triggered by Floating Bubbles

Given the constraints that the value of R_0 must satisfy in order to get an initial configuration composed of identical bubbles, we estimate this case as quite improbable. However, for simplicity, here we have adopted this configuration as the starting point of our simulations. We have followed the evolution of the explosion

Table 1. Model parameters and results

Model	N_b	R_b	M_b	E_{kin}	M_{Ni}	M_{C+O}	M_{Si}
		$(10^6\,cm)$		$(10^{51}\,erg)$	$(M_\odot)$	$(M_\odot)$	$(M_\odot)$
B30U	30	6.7	2.8%	0.44	0.56	0.65	0.05
B10U	10	9.7	3%	0.05	0.24	0.88	0.04

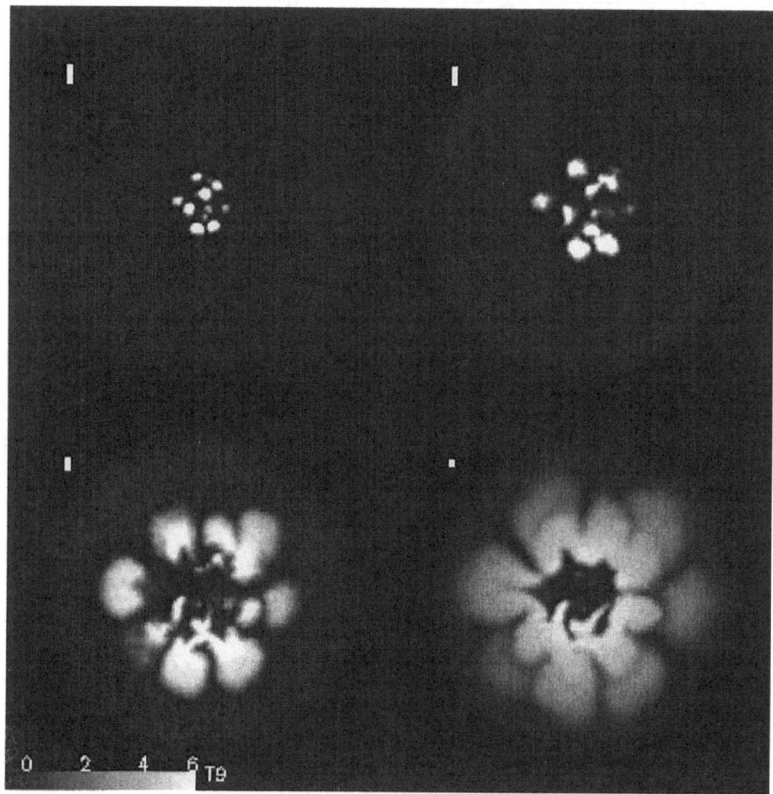

Fig. 1. Snapshots of the temperature distribution in a meridian plane at times 0, 0.27, 0.58, and 0.81 s for model B30U. The temperature scale is shown at the bottom left of the image, while the length scale is shown at the top left of each snapshot (the length of the vertical bar is equivalent to 200 km)

of a Chandrasekhar mass white dwarf of initial density $1.9 \cdot 10^9$ g/cm^3 starting from two different numbers of equal size bubbles, as detailed in Table 1 (N_b is the initial number of bubbles, R_b its initial radius, M_b the mass incinerated initially, and the other symbols have their usual meanings). The calculation was performed in 3D with the above mentioned SPH code using 250,000 particles, with a central resolution of 20 km. It is important to emphasize that we did not impose any artificial symmetry conditions and, thus, our model has no artificial characteristic length other than the own resolution of the code. Actually, the imposition of artificial symmetry conditions (for instance, symmetry planes) in any 3D hydrodynamic calculation biases the development of large-scale structures or even of the small-scale ones if they are close enough to a symmetry plane.

The results of our calculations can be found in Table 1 and in Figs. 1 and 2a. The outcome depends strongly on the initial number of bubbles. The deflagration is powered by the evolution and interaction of the bubbles: growth, buoyancy

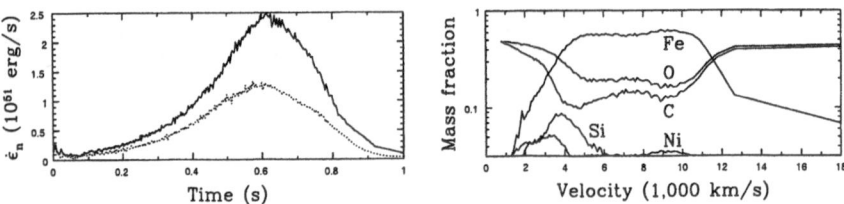

Fig. 2. Results of the hydrodynamical calculation. (**a**) Nuclear energy generation rate as a function of time for models B30U (*solid line*) and B10U (*dotted line*). (**b**) Final distribution of elements in velocity space for model B30U

Fig. 3. Spatial distribution of the main chemical species at the end of the calculation of model B30U, in the same meridian plane as in Fig. 1

(second snapshot in Fig. 1), and merging (third snapshot). The maximum acceleration of combustion is obtained when the bubbles interact with each other, feeding a rich spectrum of scale lengths to the hydrodynamic instabilities (third snapshot in Fig. 1, see also Fig. 2a). It is not a surprise that this interaction is favoured in the presence of a large number of bubbles. The distribution of nuclei in velocity and space for model B30U is shown in Figs. 2b and 3. Our results agree qualitatively with those obtained in [5].

In summary, our best model (B30U) fails to give the magnitudes adequate for a typical SNIa. A possible cause is a poor representation of the subsonic flame at low densities (a common problem in most SNIa hydrocodes). However, a thorough exploration of the parameter space of initial conditions is in order. This work has been supported by the MCYT grants EPS98–1348 and AYA2000–1785 and by the DGES grant PB98-1183–C03-02.

References

1. D. García-Senz, E. Bravo, N. Serichol: ApJSS **115**, 119 (1998)
2. D. García-Senz, S.E. Woosley: ApJ **454**, 895 (1995)
3. P. Höflich, J. Stein: ApJ **568**, 779 (2002)
4. A.M. Khokhlov: astro-ph/0008463 (2000)
5. M. Reinecke, W. Hillebrandt, J.C. Niemeyer: astro-ph/0206459 (2002)

Part V

Supernova Spectra and Light Curves

Optical Spectra and Light Curves of Supernovae

Alexei V. Filippenko

Department of Astronomy, University of California, Berkeley, CA 94720-3411, USA

Abstract. I review recent optical observations of supernovae (SNe) conducted by my group. The Lick Observatory Supernova Search with the 0.76-m Katzman Automatic Imaging Telescope is currently the world's most successful search for nearby SNe. We also use this telescope to obtain multicolor light curves of SNe. One of the more interesting SNe we discovered is SN 2000cx, which differs from all previously observed SNe Ia. Another very strange SN Ia that we studied is SN 2002cx, many of whose properties are opposite those of SN 2000cx. Extensive data on the SNe II-P SN 1999em and SN 1999gi were used to derive distances with the expanding photosphere method. Results from spectropolarimetry suggest that the deeper we peer into the ejecta of core-collapse SNe, the greater the asphericity. We are using *Hubble Space Telescope* data to identify, or set limits on, the progenitors of core-collapse SNe.

1 The Lick Observatory Supernova Search (LOSS)

In 1989, my team began to work on developing a robotic telescope for CCD imaging of relatively faint objects. The history of the project is discussed in several papers (e.g. Filippenko *et al.* 2001; Richmond, Treffers, & Filippenko 1993), and several prototypes were used over the years. In 1996, we achieved first light with our present instrument, the 0.76-m Katzman Automatic Imaging Telescope (KAIT) at Lick Observatory on Mt. Hamilton, California. It took the better part of another year to eliminate most of the remaining bugs in the system, and useful scientific results started appearing in 1997. Absolutely vital contributions to the programming and to the observing strategy were made by Dr. Weidong Li, who joined my group in 1997.

KAIT is a fully robotic instrument whose control system checks the weather, opens the dome, points to the desired objects, acquires guide stars (in the case of long exposures), exposes, stores the data, and manipulates the data automatically, all without human intervention. We reach a limit of ~ 19 mag (4σ) in 25-s unfiltered, unguided exposures, while 5-min guided exposures yield $R \approx 20$ mag. KAIT acquires well-sampled, long-term light curves of SNe and other variable or ephemeral objects – projects that are difficult to conduct at other observatories having a large number of users with different interests.

One of our main goals is to discover nearby SNe to be used for a variety of studies. Special emphasis is placed on finding them well before maximum brightness. Although the original sample of our Lick Observatory Supernova Search (LOSS; Li *et al.* 2000; Filippenko *et al.* 2001) had only about 5000 galaxies, in the year 2000 we increased the sample to ~ 14,000 galaxies (most with red-shift $\lesssim 10,000$ km s^{-1}), separated into three subsets (observing baselines of 2 days for about 100 galaxies, 3–6 days for ~ 3000 galaxies, and 7–14 days for ~ 11,000 galaxies). We are able to observe ~ 1000 galaxies per night in unfiltered mode. Our software automatically subtracts new images from old ones (after registering, scaling to account for clouds, convolving to match the point-spread-functions, etc.), and identifies SN candidates (Fig. 1) which are subsequently examined and reported to the Central Bureau for Astronomical Telegrams by numerous undergraduate research assistants in my group, working with Weidong.

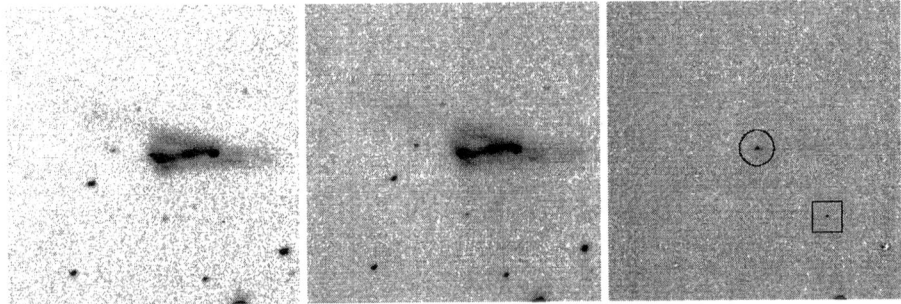

Fig. 1. KAIT images of NGC 523 before *(left)* and after *(center)* SN 2001en appeared. The difference image *(right)* shows the SN (circled), along with a cosmic ray (boxed)

A Web page describing LOSS is at http://astro.berkeley.edu/~bait/kait.html. LOSS found its first supernova in 1997 – SN 1997bs, which might not even be a "genuine" SN (Van Dyk *et al.* 2000). In 1998, mostly during the second half of the year, LOSS discovered 20 SNe, thereby breaking the previous single-year record of 15 held by the Beijing Astronomical Observatory Supernova Search. In 1999, LOSS doubled this with 40 SNe. In 2000, LOSS found 38 SNe, even though we spent a significant fraction of the observing time expanding the data-base of monitored galaxies rather than searching for SNe. With this expanded database, LOSS discovered 68 SNe in 2001 and 52 SNe in 2002 through July. We discovered SN 2000A and SN 2001A, and hence the first supernova of the new millennium, regardless of one's definition of the turn of the millennium! During the past few years, KAIT has discovered *about half* of all nearby SNe reported world-wide, from all searches combined – and through July 2002 it accounted for well over half (52/86) of them. Thus, KAIT/LOSS is currently the world's most effective search engine for nearby SNe.

To further increase the sample of SNe, my group recently decided to co-ordinate with Michael Schwartz of the Tenagra Observatory (0.60 m and 0.75 m telescopes), using largely complementary sets of galaxies (the overlap between galaxy lists is larger during those months when either observatory often has bad weather). Our "Lick Observatory and Tenagra Observatory Supernova Search" (LOTOSS; Schwartz *et al.* 2000) should discover almost all of the nearby SNe over the accessible areas of the sky. Already, the Tenagra telescopes have discovered several SNe (e.g. Schwartz & Li 2001, 2002). Also, this strategy leaves KAIT with more time to conduct multicolor follow-up photometry of SNe.

At Lick and Keck Observatories, we spectroscopically confirm and classify nearly all of the SNe that other observers haven't already classified. Thus, the sample suffers from fewer biases than most. The distribution of types through SN 2002dy is 93 SNe Ia, 76 SNe II, 12 SNe IIn, 1 SN IIb, 12 SNe Ib, 16 SNe Ic, and 6 unknown. We have started to determine the Hubble types of the host galaxies of the SNe (van den Bergh, Li, & Filippenko 2002), as a first step in the calculation of rates of various types of SNe. Already, our observations and Monte-Carlo simulations have shown that the rate of spectroscopically peculiar SNe Ia is considerably larger than had previously been thought (Li et $al.$ 2001a).

Follow-up observations for the discovered SNe are emphasized during the course of LOSS. Our goal is to build up a multicolor database for nearby SNe. Because of the early discoveries of most LOSS SNe, our light curves usually have good coverage from pre-maximum brightening to post-maximum decline. Moreover, all LOSS SNe are automatically monitored in unfiltered mode as a byproduct of our search; these can sometimes be useful for other studies (e.g. Matheson et $al.$ 2001). The positions of SNe in KAIT early-time images were used to identify the same SNe at very late times in $Hubble$ $Space$ $Telescope$ images (Li et $al.$ 2002), allowing us to determine the late-time decline rates.

LOSS also discovers novae in nearby galaxies (e.g. M31), cataclysmic variable stars, and occasionally comets (e.g. Li 1998). Although it records many asteroids, we don't conduct follow-up observations, so most of them are lost.

2   SN 2000cx: A Very Weird SN Ia

High-quality observations of nearby SNe Ia provide valuable information about their progenitor evolution and the relevant physics. Analyses of samples of well-observed, nearby SNe Ia enable observers to study the differences among SNe Ia, empirical correlations, and possible environmental effects (e.g. Hamuy et $al.$ 2000). It is thus important to expand the sample of well-observed nearby SNe Ia. Thus, a substantial fraction of KAIT's time is devoted to follow-up photometry of bright SNe Ia.

Moreover, studies of high-redshift SNe Ia have revealed a surprising cosmological result, that the expansion of the Universe is currently accelerating, perhaps due to a nonzero cosmological constant (e.g. Riess et $al.$ 1998, 2001; Perlmutter et $al.$ 1999). This result, however, is based on the assumption that there are no significant differences between SNe Ia at high redshift and their low-redshift counterparts. In particular, we rely on the luminosity/light-curve correlation, as quantified in a number of ways (e.g. Riess et $al.$ 1998; Phillips et $al.$ 1999; Perlmutter et $al.$ 1999), to "standardize" the luminosities of different SNe Ia. But what if some SNe Ia don't conform with this correlation? If there are more of them at high redshifts than at low redshifts, systematic errors may creep into the analysis. We need to find and investigate such objects at low redshifts.

SN 2000cx in the S0 galaxy NGC 524 is a case in point. It was discovered and confirmed by LOSS in July 2000, and at became the brightest SN of the year 2000. A follow-up program of multicolor photometry and spectroscopy was established at Lick Observatory; photometry of SN 2000cx was also gathered at the Wise Observatory in Israel. The results are described in detail by Li *et al.* (2001b); here I summarize the main points.

A very peculiar object, SN 200cx is indeed unique among all known SNe Ia. The light curves cannot be fit well by any of the fitting techniques currently available (e.g. MLCS and the stretch method); see Fig. 2. There is an apparent asymmetry in the rising and declining parts of the B-band light curve, while there is a unique "shoulder"-like evolution in the V-band light curve. The R-band and I-band light curves have relatively weak second maxima. In all $BVRI$ passbands the late-time decline rates are relatively large compared to other SNe Ia.

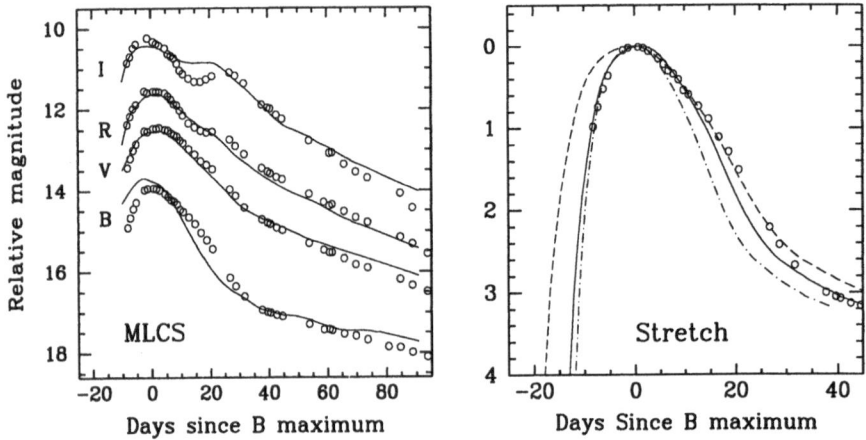

Fig. 2. The MLCS fit (Riess *et al.* 1998; *left panel*) and the stretch method fit (Perlmutter *et al.* 1999; *right panel*) for SN 2000cx. The MLCS fit is the worst we had ever seen through 2000. For the stretch method fit, the solid line is the fit to all the data points from $t = -8$ to 32 days, the dash-dotted line uses only the premaximum datapoints, and the dashed line only the postmaximum datapoints. The three fits give very different stretch factors. From Li *et al.* (2001b)

SN 2000cx has the reddest $(B - V)_0$ color before $t \approx 7$ days among several SNe Ia, and it subsequently has a peculiar plateau phase where $(B-V)_0$ remains at 0.3 mag until $t = 15$ days. The late-time $(B - V)_0$ evolution of SN 2000cx is found to be rather blue, and is inconsistent with the fit proposed by Lira (1995) and Phillips *et al.* (1999). SN 2000cx also has very blue $(V - R)_0$ and $(V - I)_0$ colors compared with other SNe Ia.

Our earliest spectrum of SN 2000cx ($t = -3$ days) reveals remarkable resemblance to those of SN 1991T-like objects, with prominent Fe III lines and weak Si II lines. As in the case of SN 1991T, Si II lines strengthened around the time of maximum brightness. However, the subsequent spectral evolution of SN 2000cx is quite different from that of SN 1991T. The Fe III and Si II lines remain strong, and the Fe II lines remain weak, in the spectra of SN 2000cx until $t \approx 20$ days, indicating that the excitation stages of iron-peak elements change relatively slowly in SN 2000cx compared with other SNe Ia, and suggesting that the photosphere of SN 2000cx stays hot for a long time. Both iron-peak and intermediate-mass elements are found to be moving at very high velocities in SN 2000cx. The V_{exp} measured from the Si II $\lambda6355$ line shows a peculiar (nearly constant) evolution.

We find that the delayed detonation model DD3 (Woosley & Weaver 1994) investigated by Pinto & Eastman (2001) accounts for the observations of SN 2000cx rather well. This model suggests that SN 2000cx is similar to SN 1991T, but with a larger ^{56}Ni production and a higher kinetic energy (i.e. greater expansion velocity for the ejecta). We emphasize that because of uncertainties in the current theoretical models for SNe Ia, various views should be considered. For example, the big difference between SN 2000cx and SN 1991T in their $V, R,$ and I light curves may suggest that they are two very different objects.

3   SN 2002cx: An Even Weirder SN Ia

But SN 2000cx is not the end of the story, when it comes to peculiar SNe Ia. A more recent, even stranger SN Ia was SN 2002cx, many of whose properties are the opposite of those of SN 2000cx! SN 2002cx was discovered in May 2002 by Wood-Vasey et al. (2002) with the Oschin 1.2-m telescope at Palomar Observatory in unfiltered images. It host galaxy is CGCG 044-035, at a redshift of $cz = 7184$ km s^{-1} (determined from H II region emission lines). An optical spectrum (Matheson et al. 2002) identified the SN as a peculiar SN 1991T-like event at about a week before maximum brightness, but the object is very *underluminous* (instead of somewhat overluminous) compared with normal SNe Ia. The Si II $\lambda6355$ and Ca II H & K lines are extremely weak or absent, but the Fe III lines at 4300 Å and 5000 Å are present and indicate very low expansion velocity – only half that of normal SNe Ia.

Recognizing the uniqueness of SN 2002cx shortly after its discovery, we established a follow-up program of multicolor photometry at Lick Observatory. Spectra of the SN were obtained with the Fred L. Whipple Observatory 1.5-m telescope and also with the Keck 10-m telescopes. Li et al. (2003) discuss the results in detail; here I provide only a brief summary.

Besides being subluminous by ~ 2 mag at all optical wavelengths relative to normal SNe Ia (implying that only a small amount of ^{56}Ni was produced), SN 2002cx has peculiar photometric evolution (Fig. 3). In the B band it has a decline rate of $\Delta m_{15}(B) = 1.29 \pm 0.11$ mag, similar to those of SN 1994D and SN 1999ac, but it is less luminous by ~ 1.4 mag than SN 1994D and SN 1999ac. The R band has a broad peak, and the I band has a unique plateau that lasts until about 20 days after B maximum. The late-time decline is rather slow in all $BVRI$ bands. The $(B-V)$ color evolution is nearly normal, but the $(V-R)$ and $(V-I)$ colors are very red.

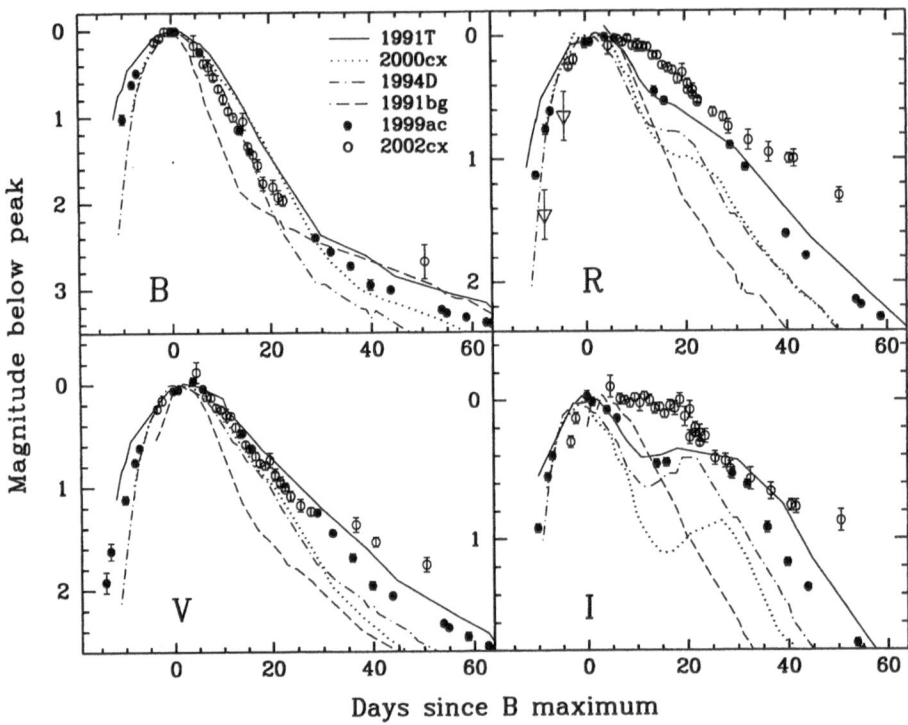

Fig. 3. Comparison (Li *et al.* 2003) between the B, V, R, I light curves of SN 2002cx and those of SN 1991T (Lira *et al.* 1998), SN 2000cx (Li *et al.* 2001b), SN 1994D (Richmond *et al.* 1995), SN 1991bg (Filippenko *et al.* 1992; Leibundgut *et al.* 1993), and SN 1999ac (Li *et al.*, in preparation). All light curves are shifted in time and peak magnitude to match those of SN 2002cx

The premaximum spectrum of SN 2002cx resembles those of SN 1991T-like objects, but with extremely low expansion velocities, the lowest ever measured for a SN Ia. The spectral evolution is dominated by Fe-group element lines, with very weak intermediate-mass element features. The nebular phase was reached unprecedently soon after maximum, despite the low velocity of the ejecta, implying that the ejected mass is low. The nebular-phase spectrum is also quite different from those of other SNe Ia (Fig. 4); there are mysterious emission lines near 7000 Å around 3 weeks after maximum brightness, and other differences as well. At late times, the spectrum is dominated by very narrow Fe II and Co II lines, and the object is very red.

SN 2002cx is inconsistent with the observed SN Ia decline rate vs. luminosity relation, or the spectral vs. photometric sequence. No existing theoretical model successfully explains all observed aspects of SN 2002cx, though the pulsating delayed detonation of a Chandrasekhar-mass white dwarf or the He detonation of a sub-Chandra white dwarf have some promising characteristics and should be pursued further.

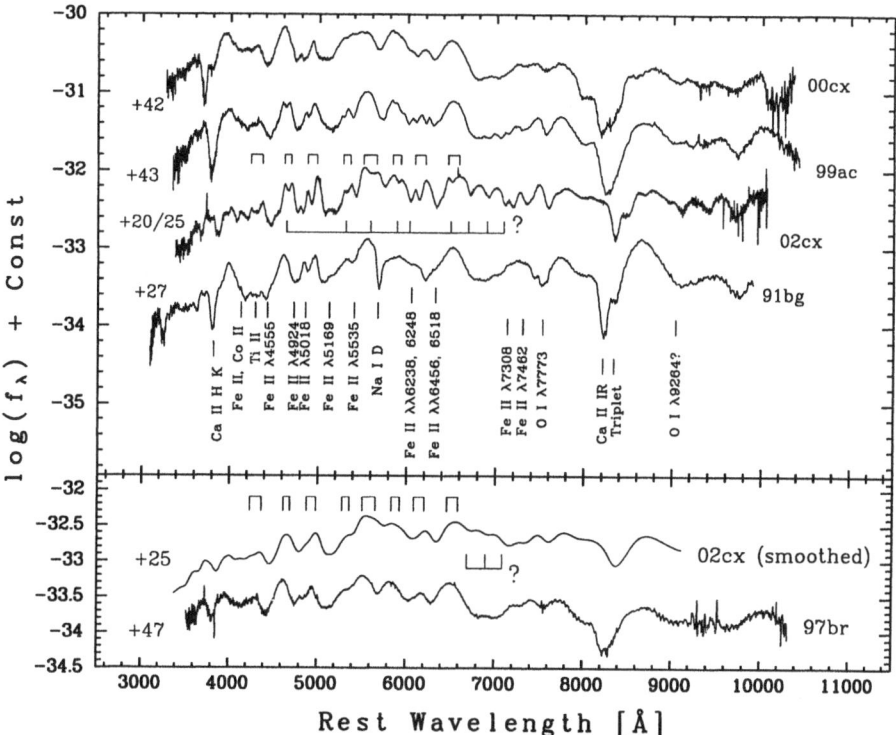

Fig. 4. The spectrum of SN 2002cx at $t = +20/25$ d, shown with spectra of other SNe Ia at older ages (Li *et al.* 2003). The *upper panel* shows the line identifications and the comparison of the spectra. The pairs of short vertical lines above the SN 2002cx spectrum mark possible "double peaks", while these below the SN 2002cx spectrum mark possible additional resolved lines (compared with other SNe Ia). The lower panel shows the comparison between the $t = +25$ d spectrum of SN 2002cx after convolving with a Gaussian function with $\sigma = 2{,}500$ km s^{-1}, and the day +47 spectrum of SN 1997br. Note that although the "double peaks" are gone, additional features seem to be present around 7000 Å in the spectrum of SN 2002cx

4   Studies of Type II Supernovae

We have also used KAIT to obtain excellent light curves of SNe II, with complementary spectra obtained at Lick Observatory and elsewhere. These are being used to study the physical properties of SNe II, and also to derive distances through the expanding photosphere method (EPM), a variant of the Baade (1926) method used to measure distances to variable stars.

In two detailed studies, we derived EPM distances to SN 1999em ($D = 8.2 \pm 0.6$ Mpc; Leonard et al. 2002a) and SN 1999gi ($D = 11.1^{+2.0}_{-1.8}$; Leonard et al. 2002b). In addition to its cosmological use, knowing the EPM distance to SN 1999gi allowed us to set constraints on the upper mass limit of its progenitor star of 15^{+5}_{-3} $M_\odot$, through the analysis of prediscovery images. This is substantially less restrictive than the upper mass limit (9^{+3}_{-2} $M_\odot$) recently found in the same manner by Smartt et al. (2001, 2002). The increased upper limit results mainly from the larger distance derived through EPM than was assumed by the Smartt et al. (2001, 2002) analyses, which relied on less precise (and less recent) distance measurements to NGC 3184.

We have also obtained high signal-to-noise ratio spectropolarimetry of some SNe II-P with the Keck 10-m and Lick 3-m telescopes (Leonard et al. 2001). At early times, SNe II-P appear to be polarized very little, suggesting that any departures from spherical symmetry are small. This is encouraging news for those who attempt to derive EPM distances for SNe II-P: Unlike the empirically based method used to measure distances to SNe Ia, distances derived to SNe II-P rely on the assumption of a spherically symmetric flux distribution during the early stages of development (i.e. the plateau). We plan to obtain spectropolarimetry of additional SNe II-P, in order to much more thoroughly test the fundamental assumption of spherical symmetry in EPM.

However, multi-epoch spectropolarimetry shows that the polarization increased with time (Fig. 5a), implying a substantially spherical geometry at early times that becomes more aspherical at late times when the deepest layers of the ejecta are revealed. In addition, our data on other core-collapse SNe indicates large polarizations for objects that have lost much of their envelope prior to exploding (e.g. SN Ic 2002ap, Fig. 5b; see below and Leonard et al. 2002c). For core-collapse events, then, it seems that the closer we probe to the heart of the explosion, the greater the polarization and, hence, the asymmetry. The current speculation is that the presence of a thick hydrogen envelope dampens the observed asymmetry.

5   The Peculiar SN Ic 2002ap

Although core-collapse SNe present a wide range of spectral and photometric properties, there is growing consensus that much of this variety is due to the state of the progenitor star's hydrogen and helium envelopes at the time of explosion. Those stars with massive, intact envelopes produce Type II-plateau SNe, those that have lost their entire hydrogen envelope (perhaps through stellar winds or mass transfer to a companion) result in SNe Ib, and those that have been stripped of both hydrogen and most (or all) of their helium produce SNe Ic; see Filippenko (1997) for a general review.

Recently, a new subclass of objects has emerged whose members generically resemble SNe Ic (no hydrogen or obvious helium spectral features), but, unlike

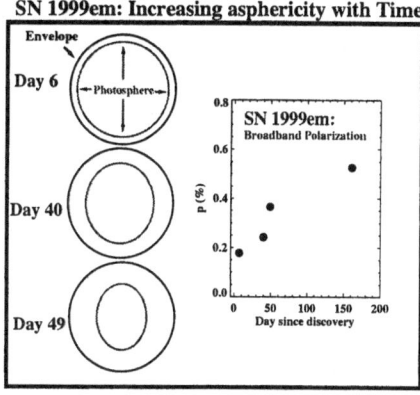

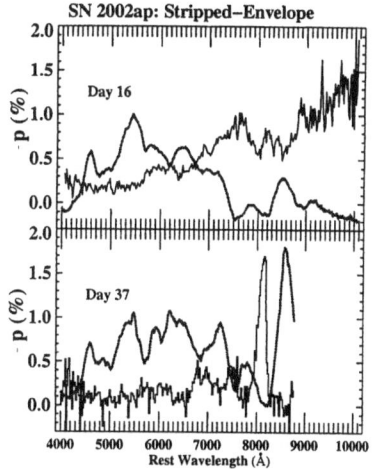

Fig. 5. (a) *(left)* The temporal increase in the polarization of the Type II-P SN 1999em suggests greater asphericity deeper into the ejecta. (b) *(right)* Polarization level (*thin noisy lines*) of the peculiar SN Ic 2002ap, with relative flux (*thick smooth lines*) over-plotted for comparison of features (Leonard *et al.* 2002c)

traditional SNe Ic, have spectra characterized by unusually broad features at early times, indicating velocities in excess of $\sim 30,000$ km s^{-1}. A few also possess inferred kinetic energies exceeding that of "normal" core-collapse SNe by more than a factor of 10 (see, e.g. Nomoto *et al.* 2001). These objects are colloquially referred to as "hypernovae", although not all of them are clearly more luminous or energetic than normal SNe Ic.

Intense interest in hypernovae has been sparked not only by their peculiar spectral features, but also by the strong spatial and temporal association between the brightest and most energetic of these events, SN 1998bw, and the γ-ray burst (GRB) 980425 (e.g. Galama *et al.* 1998). There are only a few generally accepted members of this rare class (e.g. SN 1997dq, SN 1997ef). A related subclass of SNe exhibits many of the characteristics of these objects, but with hydrogen present in the spectra; the clearest examples are SN 1997cy and SN 1999E (Germany *et al.* 2000; Turatto *et al.* 2000; Filippenko 2000, and references therein), and they, too, are sometimes called hypernovae. The hydrogen emission probably comes from the interaction of relatively hydrogen-poor ejecta with circumstellar gas previously expelled by the progenitor star (and richer in hydrogen than the remaining parts of the progenitor).

SN 2002ap in M74 (Mazzali *et al.* 2002) is one of the most recent examples of a peculiar SN Ic of this kind. So far, it has been the brightest supernova of the year 2002. Though not discovered by LOSS (it went off between two KAIT observations, but was discovered during that interval by Yoji Hirose [IAUC 7810]), KAIT's observations set a useful limit on the explosion date.

Figure 6 shows the spectrum of SN 2002ap obtained after the SN reappeared following solar conjunction, about 5 months after maximum brightness. It is characterized by strong emission lines of intermediate-mass elements superimposed on a weak continuum, implying that the SN has entered the nebular phase. Unusual narrow lines are visible on top of some of the broad-line profiles, including especially those of [O I] $\lambda\lambda6300$, 6364 and Mg I] $\lambda4571$.

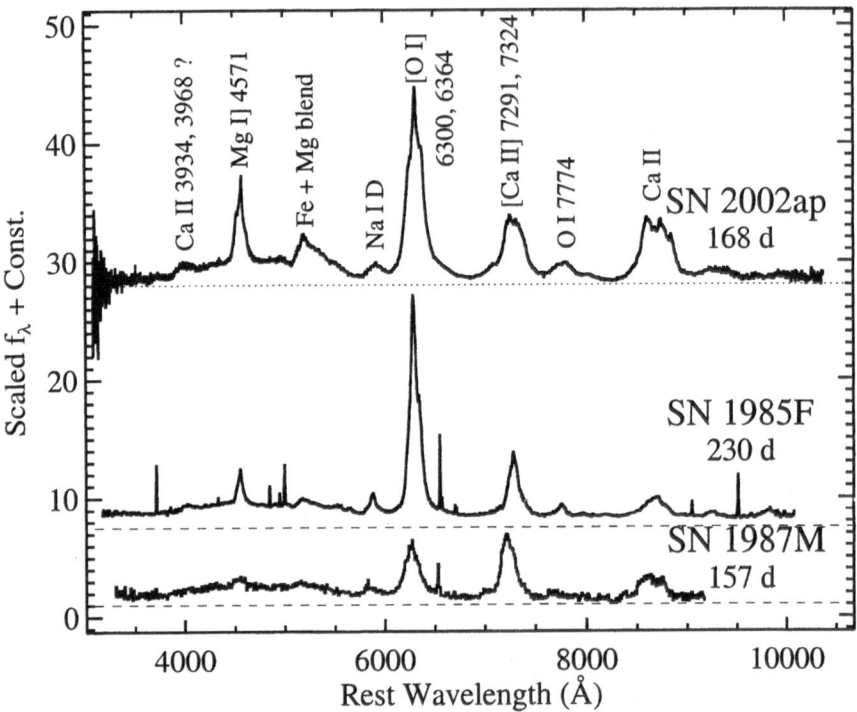

Fig. 6. SN 2002ap (*top*) in the nebular phase, compared with the SN IIb/Ib/Ic 1985F (*middle*) and the "ordinary" Type Ic SN 1987M (*bottom*) at similar epochs; the estimated day since explosion is indicated. The exact spectral classification of SN 1985F is unknown, since it was discovered long after maximum brightness (see Filippenko & Sargent 1986; Filippenko 1997), but it provides the closest match we could find to the spectrum of SN 2002ap. The spectra are scaled so that the height of the [Ca II] $\lambda\lambda7319$, 7324 blend is approximately the same in all three cases

Although there is no guarantee that the three objects were at the same stage of their development (they may evolve at different physical rates, even if they are approximately the same calendar age), one can see in Fig. 6 that relative to SN 1987M, a typical SN Ic, SN 2002ap has *much* stronger [O I] and Mg I] emission. The only SN Ib/Ic we have found comparable to SN 2002ap is SN 1985F, as shown. However, SN 2002ap exhibits a larger Mg I] $\lambda4571$ to [O I] $\lambda\lambda6300$, 6364 ratio than that of SN 1985F, suggesting that we are seeing even closer to the O-Ne-Mg layer in SN 2002ap. Qualitatively, this supports the hypothesis that SN 2002ap (and perhaps other peculiar examples of the SN Ic subclass) have progenitors that are even more highly stripped than normal SNe Ic. There might be other factors to consider as well, but the sequence II-P $\rightarrow$ IIb $\rightarrow$ Ib $\rightarrow$ Ib/c $\rightarrow$ Ic $\rightarrow$ Ic-pec may fundamentally be one dominated by the degree to which the envelope of the progenitor has been stripped.

6   The Progenitors of Core-Collapse SNe

Identifying the massive progenitor stars that give rise to core-collapse SNe is one of the main pursuits of supernova and stellar evolution studies. Using ground-based images of recent, nearby SNe obtained primarily with KAIT, astrometry from the Two Micron All Sky Survey, and archival images from the *Hubble Space Telescope*, we have attempted the direct identification of the progenitors of 16 Type II and Type Ib/c SNe (Van Dyk *et al.* 2002).

We may have identified the progenitors of the Type II SN 1999br in NGC 4900, SN 1999ev in NGC 4274, and SN 2001du in NGC 1365 as supergiant stars with $M_V^0 \approx -6$ mag in all three cases. We may have also identified the progenitors of the Type Ib SN 2001B in IC 391 and SN 2001is in NGC 1961 as very luminous supergiants with $M_V^0 \approx -8$ to -9 mag, and possibly the progenitor of the Type Ic SN 1999bu in NGC 3786 as a supergiant with $M_V^0 \approx -7.5$ mag.

Additionally, we have recovered at late times SN 1999dn in NGC 7714, SN 2000C in NGC 2415, and SN 2000ew in NGC 3810, although none of these had detectable progenitors on pre-supernova images. In fact, for the remaining SNe only limits can be placed on the absolute magnitude and color (when available) of the progenitor. The detected Type II progenitors and limits are consistent with red supergiants as progenitor stars, although possibly not as red as we had expected. Our results for the SNe Ib/c do not strongly constrain either Wolf-Rayet stars or massive interacting binary systems as progenitors.

Acknowledgements

I am grateful to the Committee on Research (U.C. Berkeley) and the conference organizers for providing partial travel support to attend this meeting. My recent research on SNe has been financed by the US National Science Foundation, most recently through grant AST-9987438, as well as by NASA grants AR-8754, GO-9114, GO-9428, and AR-9529 from the Space Telescope Science Institute, which is operated by AURA, Inc., under NASA Contract NAS5-26555. KAIT and its associated science have been made possible with funding or donations from NSF, NASA, the Sylvia and Jim Katzman Foundation, Sun Microsystems Inc., Lick Observatory, the Hewlett-Packard Company, Photometrics Ltd., AutoScope Corporation, and the University of California. I thank R. Foley, D.C. Leonard, and W.D. Li for assistance with the figures.

References

1. W. Baade: Astr. Nachr. **228**, 359 (1926)
2. A. V. Filippenko: ARAA **35**, 309 (1997)
3. A. V. Filippenko: In *Cosmic Explosions*, ed. S. S. Holt, W. W. Zhang (AIP, New York 2000) p. 123
4. A. V. Filippenko, *et al.*: In *Small-Telescope Astronomy on Global Scales*, ed. W. P. Chen, *et al.* (ASP, SF 2001) p. 121
5. A. V. Filippenko, W. L. W. Sargent: AJ **91**, 691 (1986)
6. A. V. Filippenko, *et al.*: AJ **104**, 1543 (1992)
7. T. J. Galama, *et al.*: Nature **395**, 670 (1998)
8. L. M. Germany, *et al.*: ApJ **533**, 320 (2000)
9. M. Hamuy, *et al.*: AJ **120**, 1479 (2000)
10. B. Leibundgut, *et al.*: AJ **105**, 301 (1993)
11. D. C. Leonard, *et al.*: ApJ **553**, 861 (2001)
12. D. C. Leonard, *et al.*: PASP **114**, 35 (2002a)
13. D. C. Leonard, *et al.*: AJ **124**, 2490 (2002b)
14. D. C. Leonard, *et al.*: PASP **114**, 1333 (2002c)
15. W. D. Li: IAUC 7075 (1998)
16. W. D. Li, *et al.*: In *Cosmic Explosions*, ed. S. S. Holt, W. W. Zhang (AIP, New York 2000) p. 103
17. W. D. Li, *et al.*: ApJ **546**, 734 (2001a)
18. W. D. Li, *et al.*: PASP **113**, 1178 (2001b)
19. W. D. Li, *et al.*: PASP **114**, 403 (2002)
20. W. D. Li, *et al.*: PASP, submitted (2003)
21. P. Lira: Masters thesis, Univ. Chile (1995)
22. P. Lira, *et al.*: AJ **115**, 234 (1998)
23. T. Matheson, *et al.*: AJ **121**, 1648 (2001)
24. T. Matheson, *et al.*: IAUC 7903 (2002)
25. P. Mazzali, *et al.*: ApJ **572**, L61 (2002)
26. K. Nomoto, *et al.*: In *Supernovae and Gamma Ray Bursts*, ed. M. Livio, *et al.* (CUP, Cambridge 2001), p. 144
27. S. Perlmutter, *et al.*: ApJ **517**, 565 (1999)
28. M. M. Phillips, *et al.*: AJ **118**, 1766 (1999)
29. P. A. Pinto, R. G. Eastman: New Astronomy **6**, 307 (2001)
30. M. W. Richmond, R. R. Treffers, A. V. Filippenko: PASP **105**, 1164 (1993)
31. M. W. Richmond, *et al.*: AJ **109**, 2121 (1995)
32. A. G. Riess, *et al.*: AJ **116**, 1009 (1998)
33. A. G. Riess, *et al.*: ApJ **560**, 40 (2001)
34. M. Schwartz, W. D. Li: IAUC 7766 (2001)
35. M. Schwartz, W. D. Li: IAUC 7823 (2002)
36. M. Schwartz, *et al.*: IAUC 7514 (2000)
37. S. J. Smartt, *et al.*: ApJ **556**, L29 (2001)
38. S. J. Smartt, *et al.*: ApJ **565**, 1089 (2002)
39. M. Turatto, *et al.*: ApJ **534**, L57 (2000)
40. S. D. Van Dyk, *et al.*: PASP **112**, 1532 (2000)
41. S. D. Van Dyk, *et al.*: PASP in press (astro-ph/0210347) (2002)
42. S. van den Bergh, W. Li, A. V. Filippenko: PASP **114**, 820 (2002)
43. W. M. Wood-Vasey, *et al.*: IAUC 7902 (2002)
44. S. E. Woosley, T. A. Weaver: In *Supernovae, Session IV, Les Houches*, ed. S. Bludman, *et al.* (Elsevier, Amsterdam 1994) p. 63

Optical, Infrared, and Bolometric Light Curves of Type Ia Supernovae

Nicholas B. Suntzeff

Cerro Tololo Inter-American Observatory, National Optical Astronomy Observatories, Casilla 603, La Serena, Chile

Abstract. The thermalized energy from the radioactive decays of ^{56}Ni and ^{57}Ni and their daughter nuclides power the light curves of supernovae near maximum light. The bolometric light curve gives us a fundamental understanding of the energy evolution of a supernova explosion and the amount of radioactive nuclides produced. In this review, I will discuss the bolometric evolution of the Type IIp supernovae SN1987A, and the general class of bolometric light curves of Type Ia thermonuclear explosions.

1   SN1987A

We have been monitoring the photometric properties of the Type IIp SN1987A in the LMC for more than 15 years. Over the first 5 years the optical photometry could be measured from the ground in the typical $1''$ seeing at our observatories. In the near infrared bands of JHK where the seeing is better and the crowding stars (which are typically bluer than the SN) are less of a problem, we could monitor the SN for up to 10 years. Once the supernova faded below the integrated magnitudes of the inner ring which lies at about $1''$ from the supernova debris, HST or ground-based AO imagery is needed to isolate the debris and inner ring evolution. In Fig. 1, I show the structure of the inner ring region.

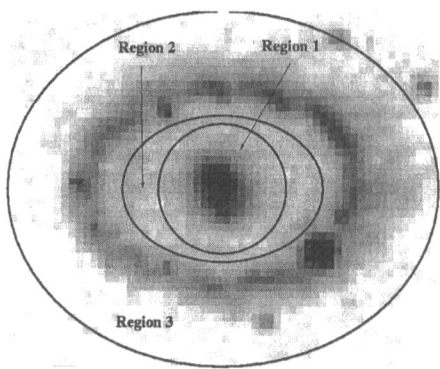

Fig. 1. An HST image of region near the inner ring of SN1987A. Region 1 isolates the expanding debris. Region 2 represents the shocked region between the debris and the inner ring. Region 3 represents the inner ring

From our *HST* images, we have measured the evolution of the brightness of the three regions shown in Fig. 1 with respect to an averaged background region outside of Region 3. For the earlier ground-based data where we can only measure the light from all three regions, we have subtracted off the extrapolated flux from Regions 2 and 3 to derive estimated fluxes for just the debris region. In Fig. 2 we plot the optical light curve for 15 years of evolution. With these corrections, the optical and *HK* photometry from the ground and *HST* flight system equivalents merge well together. The near-infrared photometry in $J/F110W$ shows a large discontinuity presumably due to the radical differences in the two filter systems. The photometry of non-stellar SEDs can show large systematic errors due to the differences in the sensitivity functions of the atmosphere/telescope/filter/detector system. In early photometry of SN1987A, we noted difference of up to 0.4 mag in I photometry due to the differences in the facility I filters [1]. In Table 1, I list the latest photometry for SN1987A

Table 1. Optical Photometry of SN1987A as of May 2002

	U	B	V	R	I
debris	20.91(32)	20.88(12)	21.28(07)	20.96(22)	20.46(04)
decay rate (mag y^{-1})	0.23	0.20	0.24	0.35	0.22

Table 2. Near-Infrared Photometry of SN1987A

days since explosion	J	H	K
4151	19.81(09)	18.29(06)	18.40(02)
5428	20.46(15)	...	...

The leveling off of the light curves after year 5 is due to two causes: the longer decay times for the remaining radioactive nuclides and the freeze-out of the cooling. By year five, the main radioactive energy sources of ^{56}Ni and its daughter nucleus ^{56}Co with e-folding times of 8.8d and 111.3d have decayed away. The nuclides ^{57}Co (the daughter nuclide of ^{57}Ni) and ^{44}Ti with e-folding times of 390d and 87y are left as the main energy input, along with the energy input from a possible (but as yet unseen) pulsar. However, the light curves at this phase no longer represent the prompt thermalization of the input radioactive energy. As shown in [2,3], the recombination and cooling time scales become larger than the expansion time scale, and the debris nebula is not able to cool at the rate of radioactive energy input.

The ultraviolet, optical, and infrared photometry can be integrated to the thermalized "uvoir" energy flux to derive the masses of the synthesized radioactive nuclides. The early time data (less than 1000d) with prompt thermalization and cooling can be simply fit to radioactive decay models as in [4–6]. The late-time bolometric light curves, when fit with models including the freeze-out, can

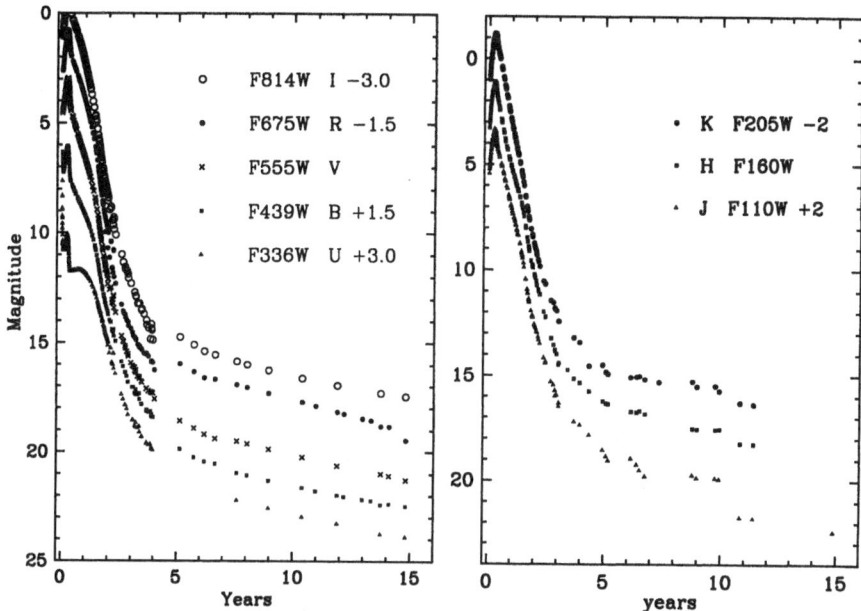

Fig. 2. Optical light curves (left panel) and near-infrared light curves (right panel) for the debris (Region 1) in SN1987A. The data have been shifted by the magnitudes listed in the legend. The data are a combination of ground-based $UBVRIJHK$ and the equivalent HST flight system magnitudes

be used to measure the amount of ^{57}Ni and ^{44}Ti. Fransson and Kozma [3] present the latest values for the amounts of these nuclides where they found the following masses: $M(^{57}Ni, ^{57}Ni, ^{44}Ti) = (0.069, 0.003, 0.0001)M_\odot$. Lundqvist et al. [7] find upper limits on the mass of ^{44}Ti consistent with these numbers based on the non-detection of the 25mμ lines of [FeI] and [FeII] from ISO/SWS. They caution, however, that if dust cooling is important, the limits on the mass of ^{44}Ti could be significantly higher. Only direct detection of the 1.157 MeV line of ^{44}Ti will resolve the ambiguity in the models.

In Fig. 3 I show the late-time photometric data for the ring and the inter-ring region. The ring region suddenly began to brighten at a rate of ~ -0.24 mag y^{-1} in $UBVI$ and ~ -0.12 mag y^{-1} in R since year 13. The brightening of the ring was predicted in [8–10]. The ring was expected to brighten by 7.5mag to about $V = 11$ as the blast wave from the debris struck the inner ring starting around year 16-20. However, in year 10, Pun et al. [11] discovered a single spot brightening, evidently due to an inward protrusion of the inner ring. Our observations here show that the general rapid brightening of the ring began later, around year 13.

The inter-ring region has accelerated its brightening since year 13 and is now brightening at a rate of ~ -0.12 mag y^{-1} in $UBVI$. The inter-ring light is presumably due to the the emission from the reverse shock which is located

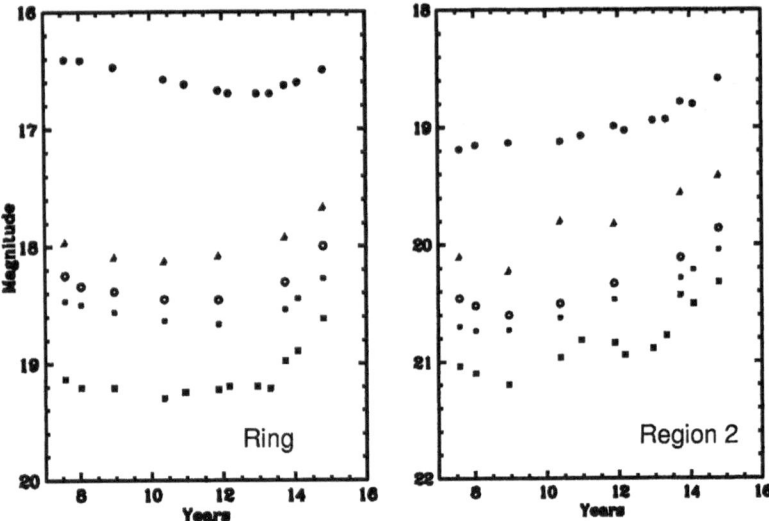

Fig. 3. $UBVRI$ optical light curves for the ring Region 3 (left panel) and the inter-ring Region 2 (right panel). The magnitudes from bottom to top are $BVIUR$

at $\sim 75\%$ of the radius of the inner boundary of the inner ring. This emission was predicted by [10] and discovered by [12] with STIS data from HST. Refined models are presented in [13,14]. The reverse shock formed in the equatorial plane as neutral hydrogen atoms streaming from the debris at 15000 km s^{-1} hit the ionized region at the inner surface of the ring. There is also diffuse light present in the STIS data which may be due to excitation by non-thermal particles accelerated by the shock.

Significant dust formed in SN1987A starting around day 500, and by day 1000, the bolometric flux of the nebula was dominated by dust radiation with a temperature of $T_{dust} \sim 200$K and mid-infrared line emission [6,15]. In Fig. 4 I show the latest data on the mid-infrared evolution of SN1987A with the ISOCAM detection [16] and a detection from OSCIR on the CTIO 4m telescope [17] at 10μm. Without 20μm detections however, we can't estimate the dust temperature and cannot measure the dust emission. If we assume the same bolometric corrections from day 1800, we find the bolometric flux of SN1987A is $\log_{10}(L) \sim 36.0$ erg s^{-1}.

2   The Bolometric Behavior of Type Ia Supernovae

Because there is much less envelope mass above the radioactive energy sources in Type Ia SNe, the bolometric properties of these supernovae are more difficult to interpret because much less of the radioactive energy is thermalized. The physics of the thermalization is discussed in [18,19]. The fraction of the radiation which is thermalized must be transported to the surface and radiated away. If

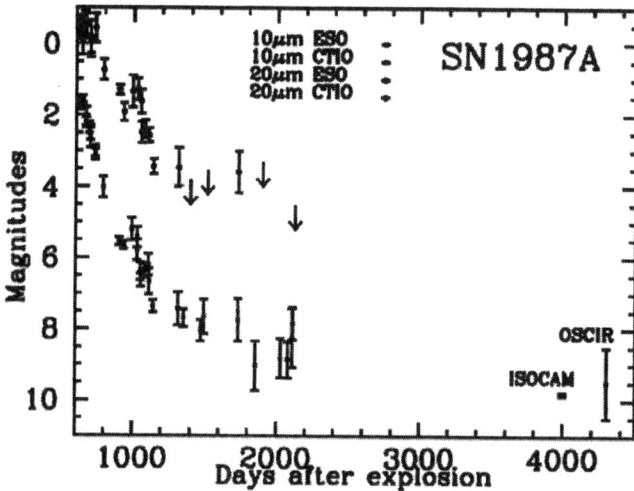

Fig. 4. Mid-infrared light curves of SN1987A

the diffusion time for the energy is short compared to the dynamical time, the energy appears as thermalized "uvoir" radiation. If the diffusion time is long, the trapped radiation will be converted partly into kinetic energy and not seen in the bolometric light curve. With a longer diffusion time, the peak of the bolometric light curve is shifted to later times when the input energy from the radioactive decays is less. This leads to "Arnett's Law" [23] which states that the bolometric luminosity at maximum light is equal to the instantaneous energy released by the radioactive decay. This law was (and is) an important tool to measure the amount of radioactive nickel produced in the explosion.

Continuing the discussion in [18], for the first 60 days since explosion the thermalization time is short compared to the dynamical time, and little radioactive energy is converted into kinetic energy. Thus the integrated uvoir bolometric luminosity represents the fraction of radioactive energy input which has been thermalized. By day 40 more than one half of gamma-ray radiation is leaving the explosion unthermalized. Even at B maximum light some 20 days after explosion, some 15% of the gamma rays are unthermalized. The observed uvoir bolometric light curve is then a function of the rapidly decreasing optical depth to the gamma-rays over the first 100 days and the decreasing energy input from the radioactive nickel nuclides. The combination of these two processes accelerates the light curve decay at a rate faster than the radioactive decay of ^{56}Co after maximum light.

The uvoir bolometric light curve interpretation is also complicated by the efficiency of the trapping of the gamma rays and positrons in the ^{56}Co decay [20–22]. Milne et al. [21] show that the gamma-rays from the ^{56}Co decay carry 30 times more energy than the positrons, but also are more penetrating and can escape more easily. As the nebula expands from days 50 to 200, the energy deposition switches from the ^{56}Co gamma-ray Comptonization to the energy input from the positron annihilations. Some of the positrons, however, can be transported out of the nebula, depending on the magnetic field structure. The bolometric light curve at this phase then becomes dependent on the details of the mixing of the ^{56}Co in the nebula and the properties of the magnetic fields.

While the interpretation of the bolometric light curve requires theoretical knowledge of the optical depths to gamma-rays and the details about the mixing of ^{56}Ni, the construction of the uvoir bolometric light curve is rather simple due to the following coincidence - - *most of the thermalized flux appears at optical wavelengths.* In Fig. 5 I show the cumulative flux of SN1992A at various phases during the first 100 days of evolution [24]. Also shown in the figure is a comparison of a Type Ia SED with that of a Type Ic and II [25,26]. It can be clearly seen that the broad peak of the flux distribution for Type Ia SNe appears in the optical region. In fact, typically 80% or more of the uvoir flux appears in the optical from day –6 onwards.

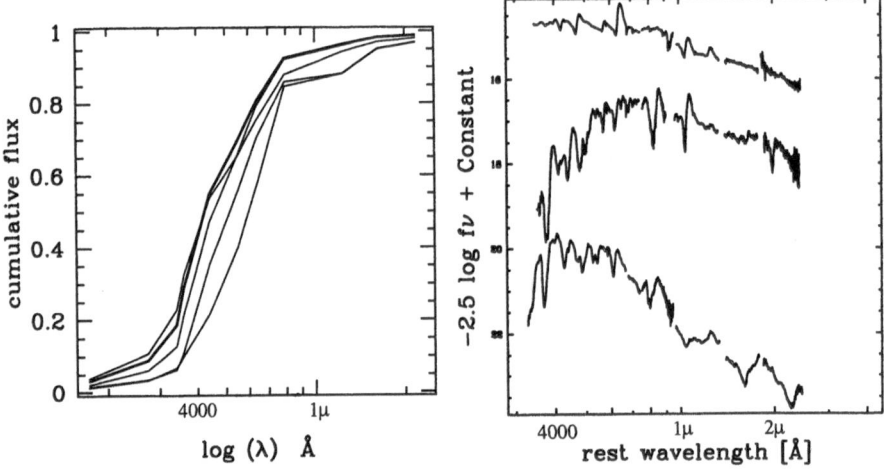

Fig. 5. Flux distributions of Type Ia SNe. The left panel shows the cumulative flux distribution for SN1992A for days –6,0,5,20,80 (from top to bottom at $\lambda = 300$nm), with the distribution for day 0 shown in the darker line. 80% or more of the total uvoir flux appears in the optical window of 300-1000nm. The right panel shows the SEDs (top to bottom) for SN1999em (Type II), SN 1999ex (Ibc) and SN 1999ee (Type Ia) near maximum light (reference [25]). Note that the broad flux peak of the SED appears in the optical

Only a few papers have been published trying to estimate the bolometric light curves (see [22] for a summary). In [27,28], the optical broad-band magnitudes were integrated to provide a magnitude that should be a close surrogate to the uvoir bolometric magnitude. From the estimated bolometric light curve, they found a range of more than a factor in 10 in the ^{56}Ni masses for a group of nearby Type Ia SNe. In [29], a V magnitude was used with a bolometric correction to study the gamma-ray trapping in the late-time light curves, which also showed a significant range in ^{56}Ni masses.

In 1996, I used the small amount of data on supernovae which had space ultraviolet spectra, optical $UBVRI$, and near-infrared JHK data to estimate accurate bolometric fluxes [24]. I found that there was a range in peak bolometric magnitudes (implying a range in nickel masses) and that the bolometric light curves appear to have a small secondary hump in the light curve around days 20-40 corresponding to the secondary maximum in the I band. Such a flux redistribution was unexpected, and points to a rather sudden change in the opacity and cooling in the nebula.

The construction of the bolometric light curve requires the sum of the space ultraviolet, optical, and near-infrared data. The falloff of the flux beyond H is such that the assumption of a R-J law adds only a few percent to the total flux and a R-J extrapolation from H or K is entirely adequate. The extrapolation to the space ultraviolet, however, is larger. Using SN 1992A data, I found that extrapolating the U flux point at 360nm to zero flux at 300nm was a reasonable representation of the true space ultraviolet flux from day –6 onwards, yielding agreement to 5% in the uvoir flux between the full integration and the integration using the ultraviolet extrapolation. It is important to continue to observe nearby SNe with HST in the ultraviolet to understand the diversity of the ultraviolet flux, which may be an indicator of the metallicity of the progenitor.

Our groups at CTIO and LCO have been following nearby Type Ia and Type II supernovae in $UBVRIYJHK$ in part to study the bolometric properties of these supernovae. The new filter Y at 1μm [30] has been designed to fit in a wavelength band which is remarkably free of telluric absorption and provides a flux point between the widely separated I and J filters. Figure 6 [31] shows the SED evolution of the nearby SN 2001el. This figure shows that the SED is well fit by a black-body with $T_{eff} = 14,000$K before maximum light steepening to a R-J law by maximum light in $BVRIJHK$. After maximum light, a flux deficit appears in J and the bands UBV drop away from a thermal distribution.

In Fig. 7 I plot the uvoir bolometric light curves calculated from the integration of the broad-band magnitudes, including an extrapolation to the ultraviolet to account for the unobserved space ultraviolet, and to the infrared using a R-J law extending from I or K. The absolute flux scale has been set by using either SBF, Cepheid, TRGB, or PN distances for the nearby SNe, or a Hubble law of $H_0 = 74$ km s^{-1} Mpc^{-1} for the more distant SNe. The comparison of SN1999ee and SN2001el shows that the location and size of the secondary bump is significantly different despite the fact that the peak luminosities are very similar.

Also plotted in Fig. 7 are the uvoir bolometric light curves for 16 Type Ia SNe. All have $UBVRI$ photometry and 10 have JHK photometry. The sense of this figure is that most SNe have roughly the same peak bolometric luminosity, but the size and placement of the secondary hump is quite variable. For instance, the peculiar SN2000cx has a rather normal bolometric light curve near

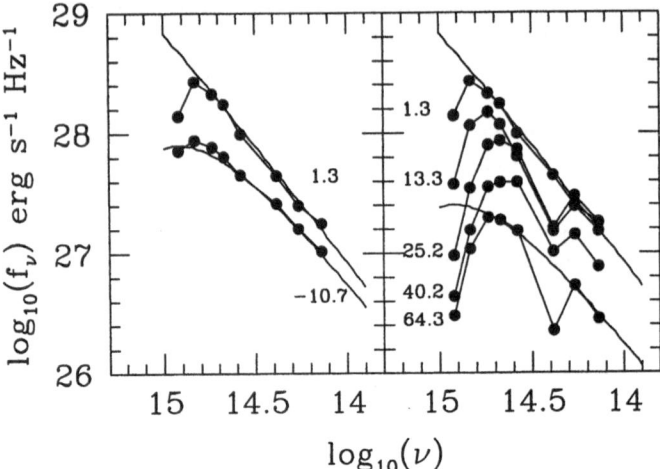

Fig. 6. SED evolution for the Type Ia SN2001el. The left panel shows the SEDs for days (since B_{max}) –10.7 and +1.3. The right panel shows the SEDs for days 1.3, 13.3, 25.2, 40.2, and 64.3. A $T_{eff} = 14,000$K blackbody is fit to the –10.7 and 64.3 data, and a Rayleigh-Jeans law is fit to the data 1.3 data. The SED steepens to the blue as the SN approaches maximum light and then reddens after maximum. The flux deficit in J appears right after maximum light. Note that the IHK data roughly follow the R-J law

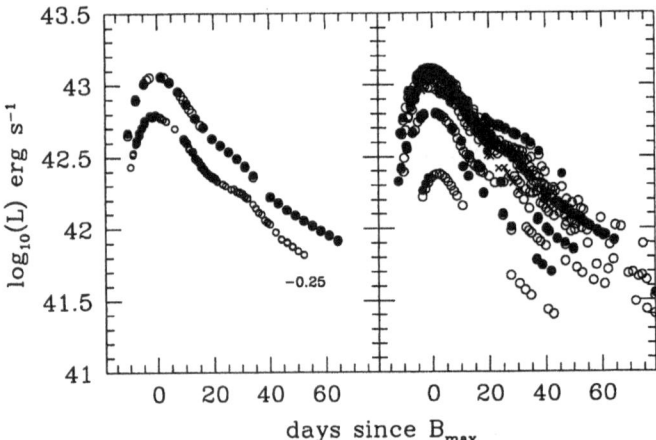

Fig. 7. "UVOIR" bolometric light curves for Type Ia SNe. These light curves include extrapolations to the space ultraviolet and the extrapolation to the red using a R-J law. Closed circles indicate an extrapolation from K and open circles indicate an extrapolation from I. The left panel shows the bolometric light curves for SN2001el (upper curve, $\Delta m_{15} = 1.13$) and SN1999ee (lower curve, $\Delta m_{15} = 0.94$). The SN1999ee data have been shifted by 0.25dex for presentation purposes. The right panel shows the bolometric light curves for 16 Type Ia SNe

maximum light, but almost completely lacks a secondary hump at 30 days. This bolometric behavior reflects the fact that SN2000cx had an anomalously weak I-band secondary maximum. Evidently the peak bolometric magnitude (and thus the ^{56}Ni mass) is not strongly coupled to the opacity and flux redistribution causing the secondary hump.

Finally, in Fig. 8 I show the peak bolometric luminosity versus the light curve shape parameter Δm_{15} for the sample plotted in Fig. 7. Rather than the roughly linear relationship between the intrinsic luminosity in B or V and Δm_{15}, we find that for much of the range of Δm_{15} there is little or no relationship between $\log_{10}(L)$ at peak and Δm_{15}, at least for $\Delta m_{15} < 1.3$.

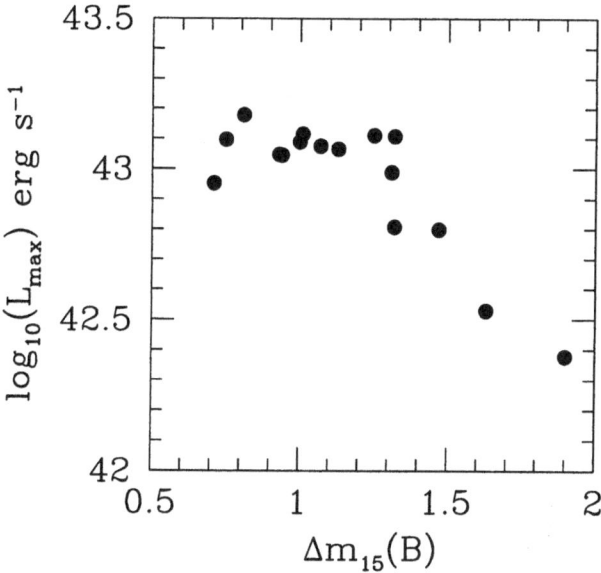

Fig. 8. The peak uvoir bolometric luminosity plotted against the light curve shape parameter of [32] for the SNe shown in Fig. 7

Acknowledgements

I would like to thank my collaborators: M. Phillips and M. Hamuy (LCO); P. Bouchet, P. Candia, K. Krisciunas, R. Schommer (deceased 12 December 2001), and C. Smith, (CTIO); B. Leibundgut (ESO); R. Kirshner, P. Challis and the SInS collaboration; and B. Schmidt (MSSSO) and the High-Z Supernova Team for their help in the collection of these data. This research was supported in part by HST grants GO-07505.02A, GO-08177.6, and GO08641.07A.

References

1. N. Suntzeff, et al.: Astron. J., **117**, 1175 (1988)
2. C. Fransson, C. Kozma Astroph. J. Lett., **408**, L25 (1993)
3. C. Fransson, C. Kozma: New Astronomy Review, **46**, 487 (2002)
4. N. Suntzeff, P. Bouchet: Astron. J., **99**, 650 (1990)
5. P. Bouchet et al.: Astron. Astrop., **245**, 490 (1991)
6. N. Suntzeff, et al.: Astroph. J. Lett., **384**, L33 (1992)
7. P. Lundqvist et al.: Astron. Astrop., **374**, 629 (2001)
8. D. Lou, R. McCray: Astroph. J., **379**, 659 (1991)
9. R. Chevalier, V. Dwarkadas: Astroph. J. Lett., **452**, L45 (1995)
10. K. Borkowski, J. Blondin, R. McCray: Astroph. J. Lett., **476**, L31 (1997)
11. C. Pun, et al.: IAU Circ. 6665 (1997)
12. G. Sonneborn, et al.: Astroph. J. Lett., **492**, L139 (1998)
13. E. Michael, et al.: Astroph. J. Lett., **492**, L143 (1998)
14. E. Michael, et al.: Astroph. J. Lett., **509**, L117 (1998)
15. D. Wooden, et al.: Astroph. J. .Sup., **88**, 477 (1993)
16. J. Fischera, R. Tuffs, H. Völk: Astron. Astrop., **386**, 517 (2002)
17. P. Bouchet: private communication (2002)
18. B. Leibundgut, P. Pinto 1992: Astroph. J., **401**, 49 (1992)
19. B. Leibundgut: Astron. Astrop. Rev., **10**, 179 (2000)
20. P. Ruiz-Lapuente, H. Spruit: Astroph. J., **500**, 360 (1998)
21. P. Milne, et al.: Astroph. J., **559**, 1019 (2001)
22. B. Leibundgut, N. Suntzeff: "Light Curves of supernovae" In *Supernovae & GRBs*, ed. K. Weiler, (Springer-Verlag, 2003) in press
23. W. Arnett: Astroph. J., **253**, 785 (1982)
24. N. Suntzeff: "Observations of Type IA Supernovae". In: *IAU Colloq. 145: Supernovae and Supernova Remnants*, ed. R. McCray, Z. Wang (Cambridge University Press, Cambridge 1996) pp41-51
25. M. Hamuy, et al.: Astron. J., **124**, 417 (2002)
26. M. Hamuy, et al.: Astron. J., **124**, 2339 (2002)
27. G. Contardo, B. Leibundgut, W. Vacca: Astron. Astrop., **359**, 876 (2000)
28. W. Vacca, B. Leibundgut: Astroph. J. Lett., **471**, L37 (1996)
29. E. Cappellaro, et al.: Astron. Astrop., **328**, 203 (1997)
30. L. Hillenbrand, et al.: PASP, **114**, 708 (2002)
31. K. Krisciunas, et al.: Astron. J., in press (astro-ph/0210327) (2003)
32. M. Phillips: Astroph. J. Lett., **413**, L105 (1993)

Infrared Light Curves of Type Ia Supernovae

M.M. Phillips[1], K. Krisciunas[2], N.B. Suntzeff[2], M. Roth[1], L. Germany[3],
P. Candia[2], S. Gonzalez[1], M. Hamuy[4], W.L. Freedman[4], S.E. Persson[4],
P.E. Nugent[5], G. Aldering[5], and A. Conley[5]

[1] Las Campanas Observatory, Carnegie Observatories, Casilla 601, La Serena, Chile
[2] Cerro Tololo Inter-American Observatory, Casilla 603, La Serena, Chile
[3] European Southern Observatory, Casilla 19001, Santiago 19, Chile
[4] Observatories of the Carnegie Institution of Washington, 813 Santa Barbara Street, Pasadena, CA 91101, USA
[5] Lawrence Berkeley National Laboratory, Mail Stop 50-232, 1 Cyclotron Road, Berkeley, CA 94720, USA

Abstract. This paper provides a progress report on a collaborative program at the Las Campanas and Cerro Tololo Observatories to observe the near-IR light curves of Type Ia supernovae. We discuss how the morphologies of the JHK light curves change as a function of the decline rate parameter $\Delta m_{15}(B)$. Evidence is presented which indicates that the absolute magnitudes in the H band have little or no dependence on the decline rate, suggesting that SNe Ia may be nearly perfect cosmological standard candles in the near-IR. A preliminary Hubble diagram in the H band is presented and compared with a similar diagram in V for the same objects. Finally, observations of two peculiar supernovae, 1999ac and 2001ay, are briefly discussed.

1 Introduction

Beginning in March 1999, we have carried out four major observing campaigns with the Swope 1.0 m and du Pont 2.5 m telescopes at the Las Campanas Observatory (LCO) to obtain near-IR light curves of supernovae (SNe) of all types. Data have been obtained for a total of 25 events, with approximately half of these corresponding to Type Ia supernovae (SNe Ia). These observations are part of a long-term collaborative program at LCO and Cerro Tololo Inter-American Observatory to investigate in detail the photometric properties of SNe Ia in the near-IR. These data will be used to study the physics of the explosions and to refine the usage of these objects as cosmological standard candles.

2 *JHK* Light Curve Morphology

Since the pioneering work of Elias and collaborators[1,2], the JHK light curves of SNe Ia have been known to be double-peaked. The secondary maximum occurs ~30 days after the first maximum and is most prominent in the J band. This double-peaked morphology is also observed in the I band[3], and is a function of the decline rate parameter $\Delta m_{15}(B)$[15] in the sense that the secondary maximum occurs later and also is generally stronger in the slowest-declining

SNe Ia[6,18,10]. Except for the very fastest-declining events, the primary maximum in I occurs a few days *before* B maximum[6].

Figure 1 shows plots of the JHK light curves of six SNe Ia covering a range of decline rates. Data for three of these SNe (SN 1999aw, SN 2000bk and SN 2001ba) were obtained as part of our observing program; photometry for the other three (SN 1981B, SN 1986G and SN 1998bu) are taken from the literature[1,5,12,7,9]. As in the I band, we see that 1) the secondary maximum occurs later for the slower-declining events, and 2) the primary maximum occurs a few days before B maximum. The dip between the primary and secondary maxima is much less pronounced in the H and K bands than in J, and the secondary maxima reach nearly the same magnitude as the primary maxima in these two bands. The net effect is that the H and K light curves are fairly flat from a few days before B maximum to 20-35 days after.

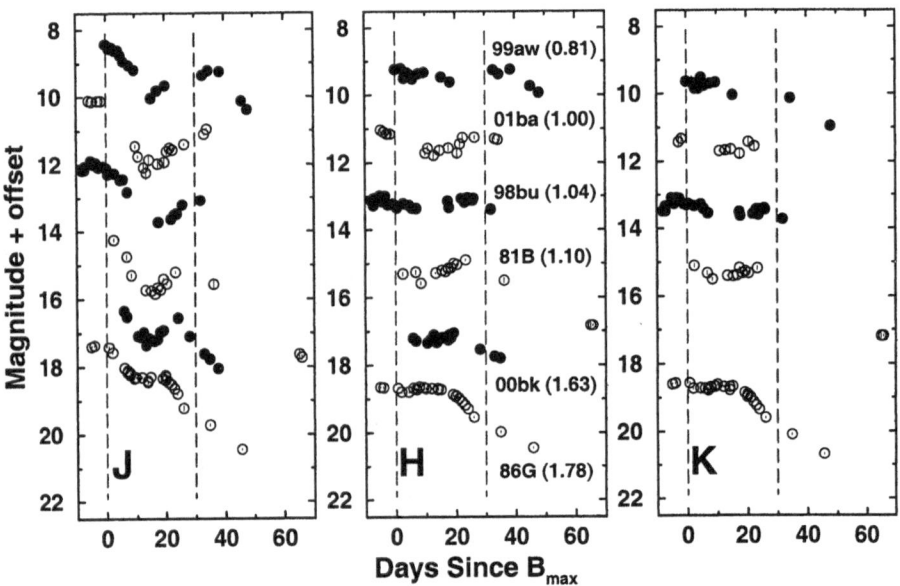

Fig. 1. JHK light curves of SNe Ia representing a range of decline rates. The curves have been arbitrarily shifted in magnitude with respect to each other for the purposes of comparison. The value of $\Delta m_{15}(B)$ is given in parentheses after the name of the SN. Dashed lines are drawn at 0 and 30 days past B maximum to serve as points of reference for the timing of the primary and secondary maxima

We have also observed a few SNe Ia in the Y band at 1.035 μm[8]. At this wavelength – between the I and J bands – the secondary maximum is extremely prominent and, at least in some cases, brighter than the primary maximum. (The Z-band light curve of SN 1999ee shows a similar behavior[19].) We intend to continue collecting data in the Y band for future SNe Ia.

3 Absolute Magnitudes

In the top three panels of Fig. 2 we plot the absolute magnitudes in BVI versus the decline rate parameter $\Delta m_{15}(B)$ for two samples of SNe Ia. The first consists of 16 well-observed nearby SNe Ia which have occurred in host galaxies for which distances have been derived via Cepheids or the surface brightness fluctuations, planetary nebula luminosity function or "tip of the red giant branch" methods. The calibration for all four methods is based on final results of the HST Key Project to measure the Hubble constant[4]. The second sample consists of 50 SNe Ia in the Hubble flow (i.e. with redshifts greater than 3000 km s^{-1}). Distances for these objects were derived from the host galaxy recession velocity assuming $H_0 = 74$ km s^{-1} Mpc^{-1} which gives the best agreement between the two samples. The bottom panel shows the absolute H magnitudes at 10 days after B maximum for the few SNe in the sample with H-band light curves. This epoch was selected rather than maximum light which occurs several days before B maximum and therefore has been observed for very few SNe Ia. Note that all of the data plotted in Fig. 2 have been corrected for both Galactic and host galaxy reddening[17].

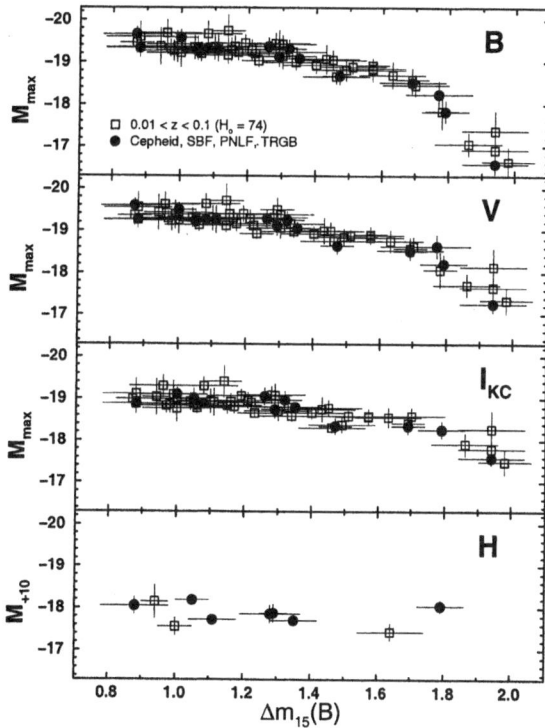

Fig. 2. Absolute magnitudes in $BVIH$ versus $\Delta m_{15}(B)$ for 66 well-observed SNe Ia

In H, the slope of the luminosity versus decline rate relation appears to be essentially flat for $0.8 < \Delta m_{15}(B) < 1.4$. For the eight SNe in this range, we find a weighted mean H-band absolute magnitude at $t = +10$ days of $M(H_{+10})$ $= -17.91 \pm 0.05$. Interestingly, the absolute magnitudes of the two SNe Ia with $\Delta m_{15}(B) > 1.4$ are consistent with this value suggesting that the near-IR luminosities of SNe Ia may have little or no dependence on the decline rate (see also [13]). Further observations are obviously required to confirm this result.

4   Hubble Diagram

One of the great advantages of working in the near-IR is the minimal effect of dust reddening ($A_H \sim 0.1\ A_B$). If the absolute magnitudes of SNe Ia in the near-IR are essentially independent of the decline rate, these objects would then be nearly perfect standard candles at these wavelengths. Hubble diagrams in V and H for seven SNe Ia in the Hubble flow ($z > 0.01$) are shown in Fig. 3. For the V band we have corrected for Galactic and host galaxy extinction, and also for the absolute magnitude versus decline rate relation[17]. The HST Key Project Cepheid distances[4] are assumed for the calibration of the SNe Ia absolute magnitudes. A Hubble constant of 73 ± 4 km s^{-1} Mpc^{-1} is obtained with a small dispersion (0.08 mag), consistent with previous results[17].

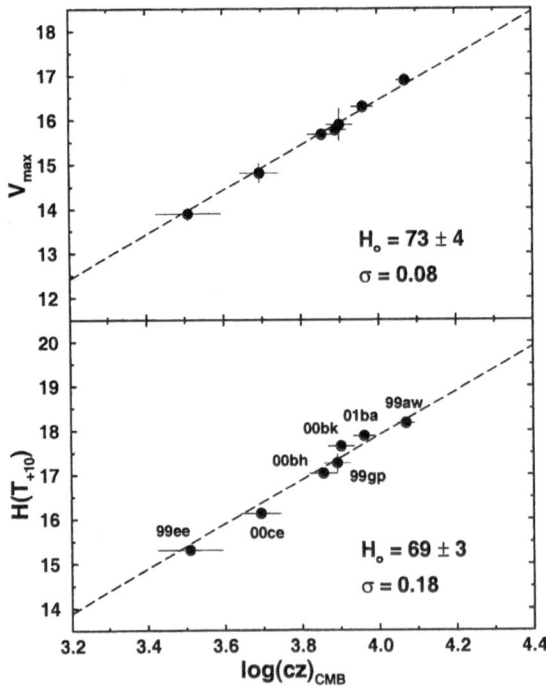

Fig. 3. V and H band Hubble diagrams for seven SNe I with $z > 0.01$

For the H-band points in Fig. 3 we again use the magnitudes at 10 days after B maximum. These have been corrected for dust extinction using the same reddenings assumed for the V data, but no correction has been made for any possible dependence of absolute magnitude on decline rate. A Hubble constant of 69 ± 3 km s^{-1} Mpc^{-1} is implied assuming the value of $M(H_{+10})$ derived in Sect. 3. This agrees to within the errors with the value obtained from the V diagram, although the dispersion of 0.18 mag is a factor of two greater. There are at least two (possibly related) reasons why this might be the case: 1) The H band measurements were made at 10 days after B maximum and the scatter at this epoch may be greater due to the dependence of the light curve shape on the decline rate, and 2) the assumption of a constant absolute magnitude as a function of the decline rate may be incorrect. These hypotheses can only be tested by obtaining more near-IR light curves of SNe Ia which include the maximum-light epoch. Nevertheless, the consistency between the Hubble constants derived independently from the optical and IR data in Fig. 3 already suggests that the methods developed for correcting the optical data for host galaxy reddening and decline rate effects are basically sound.

5 Interesting Objects

5.1 SN 1999ac

SN 1999ac was one of the first SN Ia observed in our program. As shown in Fig. 4, our optical and near-IR light curves begin ~2 weeks before B maximum, making them among the most detailed ever obtained of a SN Ia. Spectra taken before maximum indicated that this SN was a "peculiar" event similar to SN 1991T and SN 1999aa[16]. The optical light curves of SN 1999ac were also peculiar – in particular, the B light curve displayed a slow rise to maximum similar to the light curve of SN 1991T, but then declined much more rapidly than SN 1991T. These photometric peculiarities make it difficult to estimate the host galaxy reddening via standard techniques.

5.2 SN 2001ay

SN 2001ay appeared in the early-type spiral galaxy IC 4423. A spectrum obtained near maximum indicated that the ejecta had unusually high expansion velocities[11]. Figure 5 shows our $UBVRIJH$ light photometry compared with template curves for a "typical" SN Ia. The decline rate of SN 2001ay ($\Delta m_{15}(B) \sim$ 0.6-0.7) is the slowest that we have ever observed. Note the plateau character of the I and J light curves with no strong dip between the two maxima. It is not clear if this photometric peculiarity is related to the high ejecta velocities or is simply a characteristic of all SNe Ia with such slow decline rates. As in the case of SN 1999ac, we cannot use standard methods to derive the host galaxy extinction of SN 2001ay, but this is likely to have been fairly small ($E(B-V) \leq 0.1$) since an echelle spectrum[14] showed Na I D lines at the redshift of IC 4423 which are

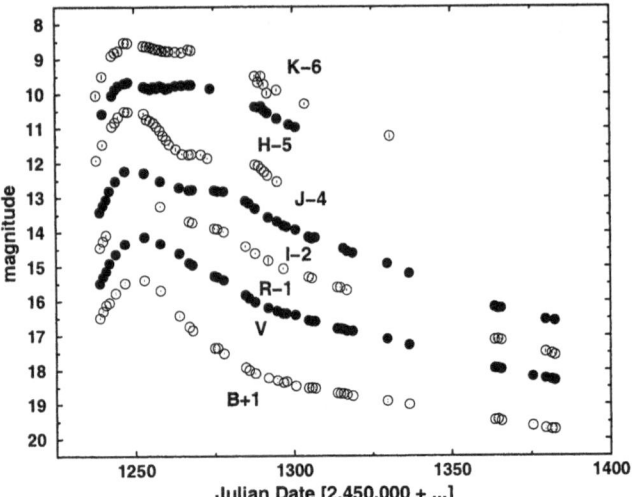

Fig. 4. *BV RIJHK* light curves of SN 1999ac

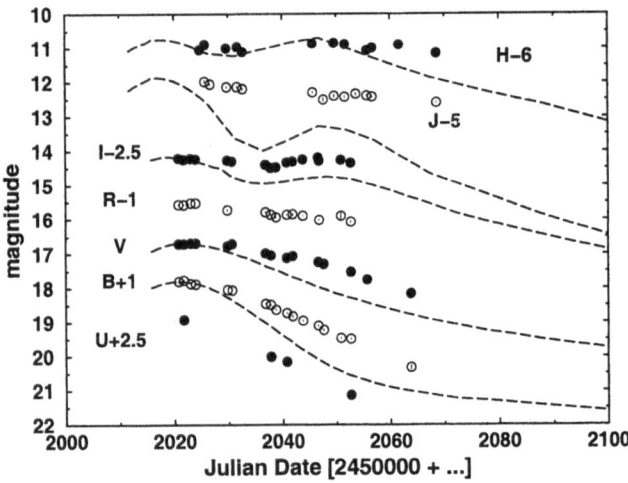

Fig. 5. *UBV RIJH* light curves of SN 2001ay. Dashed lines show template light curves of a "typical" SN Ia

no stronger than those due to the Milky Way. This implies $M(V_{max}) \sim -19.2$, which is surprisingly low for such a slow-declining SN Ia. On the other hand, in the near-IR this object was significantly more luminous $(M(H_{+10}) \sim -18.6)$ than the other SNe Ia we have observed to date (see Sect. 3).

References

1. J.H. Elias, et al.: ApJ **251**, L13 (1981)
2. J.H. Elias, et al.: ApJ **296**, 379 (1985)
3. C.H. Ford, et al.: AJ **106**, 1101 (1993)
4. W.L. Freedman, et al.: Astrophys. J. **553**, 47 (2001)
5. J.A. Frogel, et al.: ApJ **315**, L129 (1987)
6. M. Hamuy, et al.: AJ **112**, 2438 (1996)
7. M. Hernandez, et al.: MNRAS **319**, 223 (2000)
8. L.A. Hillenbrand, et al.: PASP **114**, 708 (2002)
9. S. Jha, et al.: ApJS **125**, 73 (1999)
10. K. Krisciunas et al.: AJ **122**, 1616 (2001)
11. T. Matheson, et al.: IAU Circ. 7612 (2001)
12. Y.D. Mayya, I. Puerari, O. Kuhn: IAU Circ. 6905 (1998)
13. W.P.S. Meikle: MNRAS **314**, 782 (2000)
14. P. Nugent, et al.: IAU Circ. 7612 (2001)
15. M.M. Phillips: Astrophys. J. **413**, L105 (1993)
16. M.M. Phillips, A.V. Filippenko: IAU Circ. 7122 (1999)
17. M.M. Phillips, et al.: AJ **118**, 1766 (1999)
18. A.G. Riess, W.H. Press, R.P. Kirshner: ApJ **473**, 88 (1996)
19. M. Stritzinger, et al.: AJ **124**, 2100 (2002)

Variety in Supernovae

Massimo Turatto[1], Stefano Benetti[1], and Enrico Cappellaro[2]

[1] INAF, Osservatorio Astronomico di Padova,
   Vicolo dell'Osservatorio 5, 35122 Padova, Italia
[2] INAF, Osservatorio Astronomico di Capodimonte,
   Via Moiariello 16, 80131 Napoli, Italia

Abstract. Detailed observations of a growing number of supernovae have determined a bloom of new peculiar events. In this paper we take a short tour through the SN diversity and discuss some important, physical issues related to it. Because of the role of SN Ia in determining the cosmological parameters, it is crucial to understand the physical origin of even subtle, observed differences. An important issue is also the reddening correction. We believe that the measure of interstellar lines on medium resolution spectra of SNe can be used to derive lower limits on the interstellar extinction. A few physical parameters of the progenitor, namely radius, mass, density structure and angular momentum, may explain most of the diversity of core-collapse events. In addition, if the ejecta expand into a dense circumstellar medium the ejecta-CSM interaction may dominate the observed outcome and provide a mean to probe the mass loss history of the SN progenitor in the last stages of its evolution.

1   Introduction

Despite the natural inclination of the human brain to simplify and group phenomena in a numerable amount of classes, there is no doubt that nature is complex, and its complexity increases the more we understand of it. This is valid also in astrophysics and, in particular, in the field of Supernovae (SNe). After the early decades in which the researchers were discovering the universe of cosmic explosions and phenomenologically grouped them in a handful of types, we are at a stage in which the detailed analysis shows the individuality of each event. This does not necessarily contradict previous findings, just reflects our ability to detect, and partially understand, subtler details. Indeed, if up to the 1980s it was possible to isolate only the two major SN types, the utilization of linear detectors has led to a florilegium of types and subtypes. The present classification scheme is complex [37] and accounts, in addition to the intrinsic nature of the explosion mechanisms and progenitors, also for phenomena occurring after the explosion which can dramatically affect the SN display, e.g. the interaction of the ejecta with the circumstellar matter. In the following we discuss some of the most significant issues related to the variety of the SNe.

2   The Diversity of Type Ia SNe

The thermonuclear explosions of accreting white dwarfs produce type Ia SNe, which owing to their high luminosity and accurate calibration are successfully used for determining the geometry of the Universe [18]. This is not to say that SN Ia are standard candles. In fact, even after excluding a number of extreme events or outliers (e.g. SN 1991bg or SN 1991T), spectroscopic and photometric differences among SN Ia do remain. However, empirical relations between the light curve shape and the luminosity at maximum [30,33,29,31] allow to recover type Ia SNe as the best distance indicators up to cosmological distances.

Where the variance of SN Ia comes from is not yet know even if it probably involves differences in the structure and composition of the precursor WD. Analysis of the very early spectra of SN Ia could give fundamental insights in this respect. Unfortunately, because of the very steep rise of luminosity, observing SNIa before maximum is a very demanding task. To address this and other issues related to the physics of type Ia supernovae a new European collaboration has just started (RTN2-2001-00037) which had a cold start with the extensive observations of the early stages of three events, SN 2002bo, SN 2002dj and SN 2002er.

The first data to be analyzed are those of SN 2002bo. The B–V color curve of SN 2002bo is compared in Fig. 1 with those of three other well studied SN Ia after correction for reddening. Indeed, following [31] a total reddening E(B-V)=0.48 mag was estimated for SN 2002bo. This is not unexpected since this SN exploded close to the dust lane of NGC 3190, which obscured SN 2002cv by

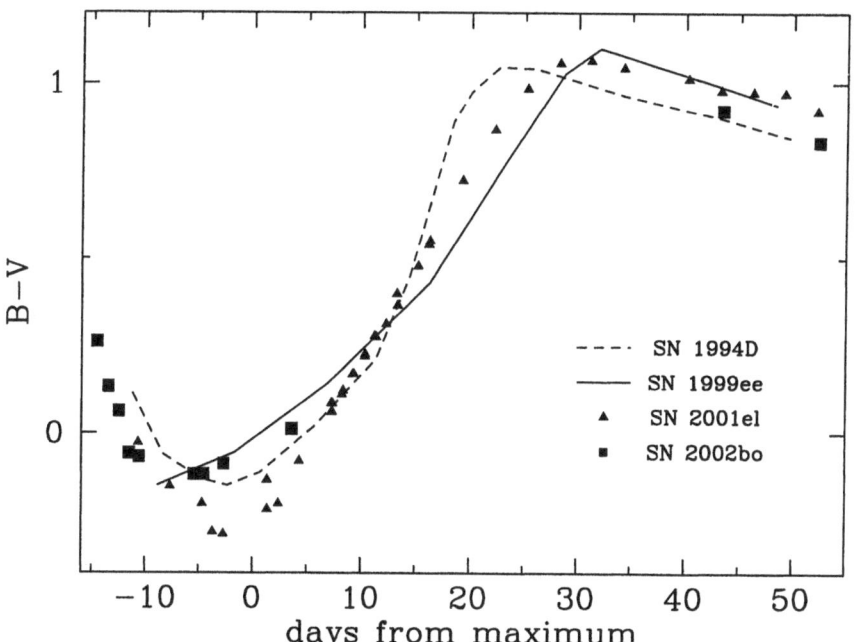

Fig. 1. Comparison of the preliminary B–V light curve of SN 2002bo with those of other normal SNIa, SN 1994D [27], SN 1999ee [36] and SN 2001el [17]. The color curves have been dereddened according to [31]

more than 6 magnitudes [22]. All objects in the figure show the color evolution typical of SN Ia. However, a closer examination of Fig. 1 shows that significant differences in the individual behaviors do exist. For instance, note that SN 1994D reaches its reddest color about 20-25 days after maximum light, i.e. 10 days before SN 1999ee. Other differences are found in the premaximum evolution: while SN 1999ee shows a monotonic reddening starting at least 10 days before maximum, SN 1994D and more clearly SN 2002bo, are very red in the earliest epochs and reach a minimum (blue) color 3-5 days before maximum.

Actually red colors in the early phases of SNIa are predicted by theoretical models [15,16]. This is because the decrease of temperature due to adiabatic cooling occurs before the heating, due to the γ-rays from ^{56}Ni to ^{56}Co radioactive decay, reaches the photosphere. Since at such epochs the observations sample the outermost layers of the exploding stars, the differences in the observed color evolution might reflect differences in the progenitors structure and composition.

Similar considerations can be drawn from the analysis of the spectra. In Fig. 2 four of the earliest available spectra of SNIa (ranging between 14 to 10 days before maximum, that is less than a week from explosion) are shown. Although even small age differences may explain part of the observed diversity,

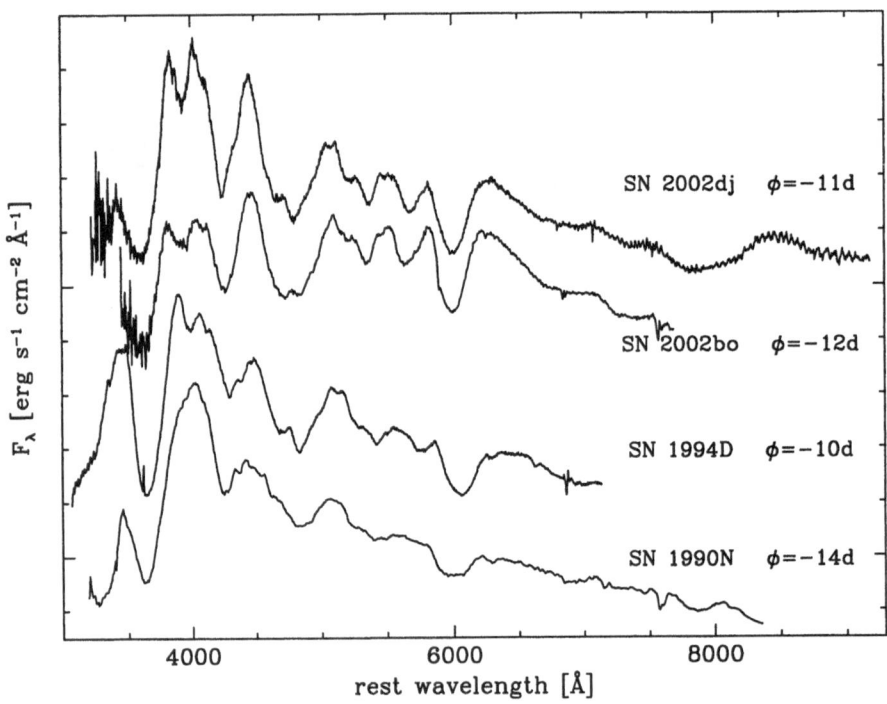

Fig. 2. Comparison of early–time spectra of SN 2002bo and SN 2002dj with those of SN 1990N [19] and SN 1994D (Wheeler, private communication). The flatter slope of the continuum of SN 2002bo is due to reddening

there is no doubt that intrinsic differences exist. In particular, around 6000 Å we note the different profiles of the SiII absorption which in SN 1990N has a flat bottom attributed to the contribution of high velocity CII [13]. Even more striking is the diversity in the blue side where entire absorption bands which are visible in SN 2002dj, SN 2002bo and SN 1994D, e.g. that due to SiII 4128, 4131, CoII 4161, are absent in SN 1990N. A detailed analysis of the sequence of early spectra of SN 2002bi and SN 2002dj is in progress.

It should be stressed that the claims of an accelerated expansion of the Universe relies on the comparison of high–z SNe observed in the optical window with local templates at blue and UV wavelengths, hence any uncertainty on the behavior in these bands reflects on the robustness of the conclusions.

2.1 The Issue of Reddening

In general, the light from SNe is absorbed and reddened by interstellar dust both in the Galaxy and in the host system. While the Galactic component is easily removed using maps of the galactic dust distribution [35] and a standard extinction law, evaluating the extinction in the host galaxy is more difficult.

A widely adopted method is to estimate the color excess by comparing the SN color at selected epochs with that of template SNe. In particular, it has been suggested that the B−V color 2-3 months after the explosion is independent of the SN photometric class [31]. In addition to requiring a fairly good coverage of the SN photometric evolution, this approach relies on the assumption of a uniform behavior for all SN Ia which should be checked by some independent measurement.

In principle, high resolution spectroscopy of the interstellar NaID lines by means of the doublet ratio method can give the gas column density which, in turn, can be converted into reddening assuming an average dust-to-gas ratio. However, the method has the drawback that, because of the need to reach a good S/N and a high spectral resolution it can been applied only to few objects, typically nearby SNe observed in proximity of maximum.

An empirical approach has been applied in the past, relating the EW of interstellar absorption lines, measured on medium resolution SN spectra, to the color excess E(B–V) estimated from color curve comparison [2,32]. The first attempts seemed to suggest the existence of a simple linear relation which would imply a constant dust-to-gas ratio, a unique extinction law and a negligible effect of saturation. The latter might be understood considering that because of the galaxy rotation the various absorbing clouds have different radial velocity components along the line of sight. This spreads the interstellar absorptions at different wavelengths and prevents heavy line saturation. However, with the growth of the event statistics the scatter around the relation appeared to increase significantly and the existence of a relation was questioned.

To review these issues, we have made use of the homogeneous set of E(B–V) estimates for SNIa provided by Phillips et al. [31], supplemented by a few more recent objects measured with a similar prescription. For the SNe of this sample

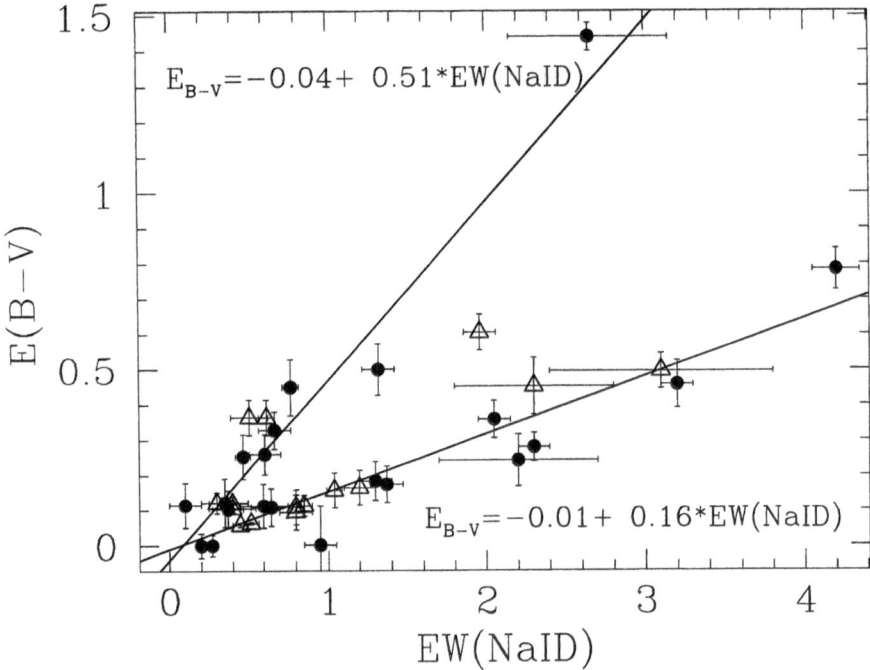

Fig. 3. Color excess of SNIa inside the host galaxy, E(B–V), determined from the tail of the color curves [31], versus EW(NaID) measured on low dispersion spectra (filled circles). The values of SN 1986G (the rightmost point) include also the Galactic component. Open triangles are estimates of Galactic color excess from for SNe of all types [35]

we have searched both in the literature and in our archive for high signal–to–noise, medium–resolution spectra and have measured the EW(NaID) of the host galaxy component. With these data we have redrawn the E(B–V) vs. EW(NaID) plot (Fig. 3).

An accurate examination of the figure shows that the points, although apparently dispersed, do not fill the plane but rather seem to cluster around two lines with significant different slopes, and only one or two objects in between. Most objects lie on the line with smaller slope $(E(B-V) = 0.16 \times EW(NaID))$ which roughly corresponds to the previous linear relation [2]. However, there are other SNIa which for similar EW(NaID) are much more heavily reddened $(E(B-V) = -0.04 + 0.51 \times EW(NaID))$. Interestingly, a similar bivariate behavior seems to occur for a sample of SNe of all types also for the Galactic component, where this time the extinction is derived from standard dust maps (open symbols) [35].

The interpretation of this finding is beyond the aim of the present paper and it is more likely related to different conditions of the ISM. Nevertheless it is worth noting that Fig. 3 tells that strong reddening is present each time large values of EW(NaID) are measured in the spectra. In other words, by entering in the graph with a measurement of the EW(NaID) we get lower limits to the reddening of SNIa.

3   Core-Collapse Supernovae

Several SN types (II, IIb, Ib, Ic, IIn as well as several peculiar objects) are thought to explode via the gravitational collapse of the core. The great observational diversity has not been fully understood even if it clearly involves the progenitor masses and configurations at the time of explosion. Whereas SN IIP are thought to originate from isolated massive stars, a generalized scenario has been proposed in which common envelope evolution in massive binary systems with varying mass ratios and separations of the components can lead to various degrees of stripping of the envelope [23]. According to this scenario the sequence of types IIL–IIb–Ib–Ic is ordered according to a decreasing mass of the envelope.

3.1   Energetics

In the last few years it has become evident that core–collapse SNe can release different amounts of energy in the explosions. A number of objects with low luminosity and kinetic energy has been discovered (e.g. [39,42]). Although these events are extensively discussed in other contributions to this workshop [26,43], we recall here that they have probably progenitors with $M \geq 20 M_\odot$. In fact, if the progenitor mass is large enough core collapse may leave a black hole remnant and late time accretion onto the compact remnant can be a significant source of radiation. Unfortunately the optical signature of this event has not been detected yet [6].

On the other end there are super–energetic SNe, often called 'hypernovae'. The first and most interesting example was SN 1998bw which was discovered while searching for the optical counterpart of GRB980425. SN 1998bw was of type Ic, hence believed to originate from the core collapse of a massive star, stripped of its H and He envelope. It was as bright as a SNIa and the high expansion velocities ($> 3 \times 10^4$ km s^{-1}) indicate that it was unusually energetic ($> 10^{52}$ ergs) [28]. Its very powerful radio emission has been attributed to a mildly relativistic blast wave interacting with a clumpy, stratified CSM deriving from a turbulent mass–loss history [41]. Other SNe (e.g. SN 1997ef, SN 1998ey and SN 2002ap) bear some spectroscopic resemblance to SN 1998bw but are slightly less energetic.

In all these cases the masses of the progenitors estimated by fitting light curves and spectra are larger than in normal core collapse SNe [24]. In a qualitative scenario which may explain this finding, the outcome of the core collapse of stars with $M \geq 20 - 25 M_\odot$ results in under- or hyper-energetic explosions, depending on the angular momentum of the collapsing cores. Faint, slowly expanding SNe like SN 1997D occur because the progenitor envelopes have a large binding energy and relatively little energy remains available for heating up and accelerating the ejecta. Instead if the core of the progenitor is in rapid rotation, owing possibly to the spiraling–in of a companion star, a high energetic, asymmetric explosions may be obtained [21].

An intriguing possibility is that SN 1997cy, observationally classified as SNIIn and possibly the brightest SN ever observed [14,40], and its twin SN 1999E [34], are associated to GRBs. As in the case of other SNIIn, these events show strong ejecta-CSM interaction with explosion energies as high as 3×10^{52} ergs.

3.2   Interaction with the CSM

Very important in determining the outcome of core–collapse SNe is the possible presence of a dense CSM around the progenitor star. Indeed the interaction of the fast ejecta with the slowly expanding CSM generates a forward shock in the CSM and a reverse shock in the ejecta. The shocked material emits radiation in the optical, radio and X–rays with characteristics which depends on the density of both the CSM and the ejecta, and on the properties of the shock [8]. Studying the ejecta–CSM interaction the mass loss history in the late stages of the stellar evolution can be probed.

In some cases it turns out that the interaction begins immediately after the burst indicating that strong wind persisted up to the very last stages of progenitor life. Often the radiation from the shocked region shadows the thermal emission from the ejecta and the spectrum is dominated by strong emission lines with composite profiles, which reflects the different kinematics of the emitting layers. Observationally these SNe are called type IIn.

In most cases, the SN initially expands in an empty space and no interaction occurs. However, in a number of objects the fast expanding ejecta eventually catches the material ejected during remote strong wind events and the SN emission is revived by the interaction. With the noticeable exception SN 1957D, an unclassified, poorly studied object, the optical spectra of late–time revived SNe are dominated by broad, boxy H emission [12,7].

The different behaviors of the flux evolution of Hα for a number of CSM interacting SNII is displayed in Fig. 4. In general, type IIn SNe show a slow evolution from the earliest phases and the Hα can remain almost constant for years, like in the cases SN 1988Z and SN 1995G [1,25].

The onset of the interaction at late times is evident in the well studied SN 1979C, SN 1980K and SN 1986E [12,7]. After a rather normal evolution in which the Hα flux matches the radioactive decay input energy [9] (dotted curve in the left panel Fig. 4), months or years after the explosion the radiation from interaction becomes dominant, the line flux almost halts its decline and the temporal evolution is well reproduced by the circumstellar interaction model (dotted curve in the right panel) [8]. These three SNe belong to the subclass with linear light curves (SNIIL) which are thought to have lost most of their hydrogen envelope before the explosion. Therefore, the delayed onset of ejecta–CSM interaction is not surprising. In addition, all of them have been detected in the radio, supporting the claim that late time optical and radio emission are correlated [7]. Similar behaviors, i.e. linear light curves, flattening of the Hα flux evolution and

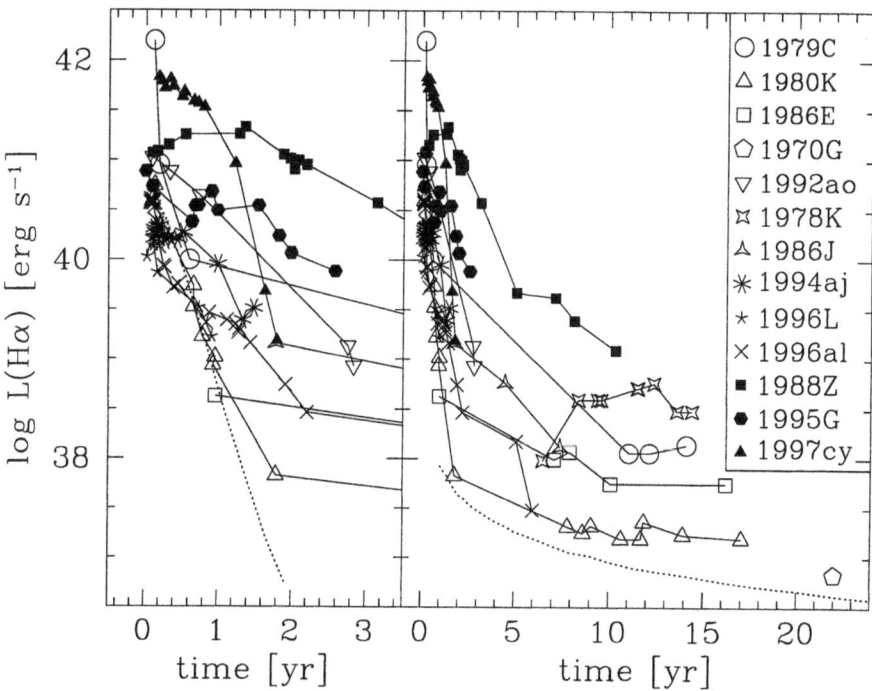

Fig. 4. The long–term Hα evolution of interacting SNII. On the left panel is shown as a dotted curve the radioactive model [9] while on the right-hand panel is the interaction model for SN 1980K [8]. SNe not following the radioactive model require an additional source of energy, which is provided by the interaction with the CSM. SNIIn (filled symbols) decline slowly starting soon after explosion [38,1,40,25]. Some SNII with linear light curves [12,7,4,5,11] (open symbols) show evidence of interaction at late stages. Also the optical observations of the bright radio SNe 1986J and 1978K are reported [10,20] Unpublished data of SN 1992ao, 1996al (ESO 3.6m and NTT) and 1986E (ESO-VLT) come from the Padova archive

boxy line profiles at late time have been exhibited also by SN 1994aj, SN 1996L and SN 1996al, which showed signatures of slowly expanding shells above the photosphere also in the earliest times [4,5].

Although differences between the observations and the models do exist and improved interaction models are needed, we emphasize that the present observational facilities make possible to study in detail the interaction of the ejecta with the CSM (kinematics, density, temperature of the emitting layers) for decades after the explosion which corresponds to probe tens of thousands of years of the progenitor mass loss history.

4   Conclusions

In this paper we have addressed the SN variety. SNIa, which are used as distance indicators up to cosmological distances, can differ as to absolute magnitudes, intrinsic colors at maximum, light and color curve shapes. These differences might be related to the composition and structure of the WD and to variations in the explosion mechanism.

All SNe can be heavily reddened. In the plane EW(NaID) vs. E(B-V), type Ia SNe seem to cluster on two different linear relations, possibly due to different dust-to-gas ratio in the host galaxies. If, on the one hand this prevents that accurate reddening can be obtained only by means of medium dispersion spectroscopy, on the other there is no doubt that significant reddening is always present when interstellar absorption lines are observed (see Fig. 3).

Core–collapse SNe show a wider variety. Although all ignited by the same event, i.e. the core-collapse, the explosion of progenitors with radii, masses, density structures and angular momenta different up to one order of magnitude can release different amounts of energy and variable amounts of heavy and intermediate mass elements. These, in turn, result in significantly different observables (absolute magnitudes, colors, spectral and luminosity evolution, etc.).

Despite the total number of SNe discovered is well over 2300 [3], there is no doubt that we are still scratching the surface of the SN diversity.

References

1. I. Aretxaga, S. Benetti, R.J. Terlevich, A.C. Fabian, E. Cappellaro, M. Turatto, M. Della Valle: MNRAS **309**, 343 (1999)
2. R. Barbon, S. Benetti, E. Cappellaro, L. Rosino, M. Turatto: A&A **237**, 79 (1990)
3. R. Barbon, V. Buondì, E. Cappellaro, M. Turatto: A&AS **139**, 531, (1999) (http://www.pd.astro.it/supern)
4. S. Benetti, E. Cappellaro, I.J. Danziger, M. Turatto, F. Patat, M. Della Valle: MNRAS **294**, 448 (1998)
5. S. Benetti, M. Turatto, P.A. Mazzali: MNRAS **305**, 811 (1999)
6. S. Benetti, et al.: MNRAS **322**, 361 (2001)
7. E. Cappellaro, I.J. Danziger, M. Turatto: MNRAS **227**, 106 (1995)
8. R.A. Chevalier, C. Fransson: ApJ **420**, 268 (1994)
9. N.N. Chugai: MNRAS **250**, 513 (1991)
10. N.N. Chugai, I.J. Danziger, M. Della Valle: MNRAS **276**, 530 (1995)
11. R.A. Fesen: ApJ **413**, L109 (1993)
12. R.A. Fesen, et al.: AJ **117**, 725 (1999)
13. A. Fisher, D. Branch, P. Nugent, E. & Baron: ApJL **481**, L89 (1997)
14. L.M. Germany, D.J. Reiss, B.P. Schmidt, C.W. Stubbs, E.M. Sadler: ApJ **533**, 320 (2000)
15. P. Höflich, C.L. Gerardy, R.A. Fesen, S. Sakai, Shoko: ApJ**568**, 791 (2002)
16. P. Höflich: private communication (2002)
17. K. Krisciunas, et al.: AJ submitted, astro-ph/0210327 (2002)
18. B. Leibundgut: AAR **10**, 179 (2000)
19. B. Leibundgut, et al.: ApJ **371**, L23 (1991)

20. B. Leibundgut, R.P. Kirshner, P.A. Pinto, M.P. Rupen, R.C. Smith, J.E. Gunn, D.P. Schneider: ApJ **372**, 531 (1991)
21. A.I. MacFayden, S.E. Woosley: ApJ **524**, 262 (1999)
22. P. Meikle, S. Mattila, A. Glasse, J. Buckle, A. Adamson: IAUC **7911** (2002)
23. K. Nomoto, K. Iwamoto, T. Suzuki: Phys. Reports **256**, 173 (1995)
24. K. Nomoto, K. Maeda, H. Umeda, T. Ohkubo, J. Deng, P.A. Mazzali: 'Hypernovae and their Nucleosynthesis'. In: 'A massive Star Odyssey, from Main Sequence to Supernova', ed. V.D.Hucht, A.Herrero, C.Esteban (San Francisco; ASP) in press (2002)
25. A. Pastorello, et al.: MNRAS **333**, 27 (2002)
26. A. Pastorello, et al.: 'Faint Core–Collapse Supernovae'. This volume
27. F. Patat, S. Benetti, E. Cappellaro, I.J. Danziger, M. Della Valle, P.A. Mazzali, M. Turatto: MNRAS **278**, 111 (1996)
28. F. Patat, et al.: ApJ **555**, 900 (2001)
29. S. Perlmutter, et al.: ApJ **483**, 565 (1997)
30. M.M. Phillips: ApJ, **413**, L105 (1993)
31. M.M. Phillips, P. Lira, N.B. Suntzeff, R.A. Schommer, M. Hamuy, J. Maza: AJ, **118**, 1766 (1999)
32. M.W. Richmond, R. Treffers, A.V. Filippenko, Y. Paik, Young, B. Leibundgut, E. Schulman, C.V. Cox: AJ **107**, 1022 (1994)
33. A.G. Riess, et al.: AJ, **116**, 1009 (1998)
34. L. Rigon, et al.: MNRAS, in press (2003)
35. D.J. Schlegel, D.P. Finkbeiner, M. Davis: ApJ **500**, 525 (1998)
36. M. Stritzinger, et al.: AJ **124**, 2100 (2002)
37. M. Turatto: 'Classification of Supernovae'. In: 'Supernovae and Gamma–Ray Bursts', ed. K.W. Weiler (Springer, Berlin Heidelberg 2003) in press
38. M. Turatto, E. Cappellaro, I.J. Danziger, S. Benetti: MNRAS **265**, 471 (1993)
39. M. Turatto, et al.: ApJ **498**, L129 (1998)
40. M. Turatto, et al.: ApJ **534**, L57 (2000)
41. K.W. Weiler, N. Panagia, M.J. Montes: ApJ **562**, 670 (2001)
42. L. Zampieri, A. Pastorello, M. Turatto, E. Cappellaro, S. Benetti, G. Altavilla, P. Mazzali, M. Hamuy: MNRAS, in press (2002) (astro-ph/0210171)
43. L. Zampieri, et al.: 'Peculiar, Low Luminosity Type II Supernovae: Site Of Black Hole Formation?'. This volume

Faint Core-Collapse Supernovae*

Andrea Pastorello[1,2], Eddie Baron[3], Stefano Benetti[2], David Branch[3], Enrico Cappellaro[4], Ferdinando Patat[5], Massimo Turatto[2], Luca Zampieri[2], Mario Hamuy[6], Mark Armstrong[7], and Peter Meikle[8]

[1] Department of Astronomy, University of Padova,
   vicolo dell'Osservatorio 2, I-35122 Padova, Italy
[2] INAF - Astronomical Observatory of Padova,
   vicolo dell'Osservatorio 5, I-35122 Padova, Italy
[3] Department of Physics and Astronomy, University of Oklahoma,
   Norman, OK 43019, USA
[4] INAF - Astronomical Observatory of Capodimonte,
   via Moiariello 16, I-80131 Napoli, Italy
[5] European Southern Observatory, Karl Schwarzschild-Str. 2,
   D-85748 Garching bei München, Germany
[6] The Observatories of the Carnegie Institution of Washington,
   813 Santa Barbara St., Pasadena, CA 91101, USA
[7] Rolvenden, Kent, UK
[8] Astrophysics Group, Blackett Laboratory, Imperial College,
   Prince Consort Road, London SW7 2BZ, UK

Abstract. We present the spectroscopic and photometric data of a number of Core-Collapse Supernovae having faint luminosities. These objects cover the whole range of properties from the faint SN 1997D to the normal SN 1998A. Their absolute luminosities at maximum cover a wide interval between about $M_V = $ -13 and -17; the spectral lines at the end of the photospheric phase indicate expansion velocities between 1000 and 4000 km s^{-1}. The analysis of the absolute luminosity of the radioactive tails suggests that there is a wide distribution of Ni masses powering the late-time light curves.

1 Introduction

Core–Collapse Supernovae (CC–SNe), the explosive events terminating the life of massive stars, cover a wide range of properties from the extremely faint 1997D–like events ([20,1,23,11]) to the exceptionally overluminous objects called *Hypernovae* [8], like SN 1998bw (e.g. [18,15]) or SN 1997cy [21]. In the past years a continuity of observational properties among the CC–SNe was noted ([13,14]), and recently a correlation between the intrinsic luminosity, the kinetic energy, the ejected mass and the ^{56}Ni yield has been suggested [6]. Until a few years ago there was a general agreement that the luminosity of a CC–SN was powered by 0.07–0.1$M_\odot$ of ^{56}Ni, but now we know that of ^{56}Ni mass has a wide distribution [11].

---

* Based on observations collected at ESO - La Silla and Paranal (Chile), CTIO (Chile), TNG, WHT and JKT (La Palma, Canary Islands, Spain) and Asiago (Italy)

Among the CC–SNe, also the plateau SNe show heterogeneous behaviours. Here we focus on some peculiar SN II-P events: SN 1998A, a CC–SN underluminous only at early phases (resembling SN 1987A), and some low Ni mass SNe (with SN 1997D–like properties).

2   SN 1998A in IC 2627

SN 1998A, discovered by the PARG SN search [25], is probably the closest known replica of SN 1987A [26]. We contribute to the available photometry with our observations obtained with ESO–La Silla telescopes [12]. The data were reduced applying a template subtraction technique.

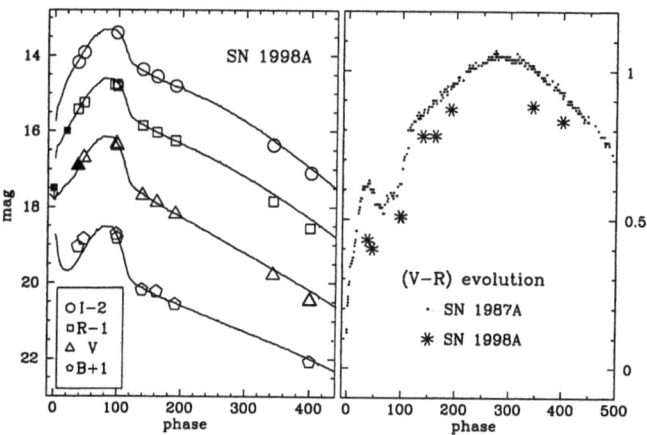

Fig. 1. Left: photometric evolution on SN 1998A (BVRI bands) compared with SN 1987A. The data of SN 1987A (solid curves, [24]) are scaled in magnitude to match the points of SN 1998A. Right: (V-R) colour evolution of SN 1998A and SN 1987A

The results, compared with the available photometry of SN 1987A ([24] and references therein) and shifted in magnitude to match the points of SN 1998A, are shown in Fig. 1. Light curves and colour evolution are similar in both SNe, even if SN 1998A is slightly bluer, in particular at early epochs. A comparison between the absolute light curves shows that SN 1998A is brighter (about 0.6 magnitudes in R, [12]) than SN 1987A.

Also the spectroscopic evolution is reminiscent of SN 1987A. A comparison between the spectra of SN 1998A and SN 1987A at two reference epochs (40 days and about 1 year after explosion) is shown in Fig. 2. The resemblance is surprising, even if the Ba II lines (e.g. Ba II λ 6142 Å) are stronger in SN 1987A. An important difference is the higher expansion velocities of the SN 1998A ejecta: at the end of recombination, the position of the minima of the Fe II lines (multiplet 42) indicates a $v_{exp} \sim 3500$ km s^{-1} [12], against $v_{exp} \sim 2100$ km s^{-1} in SN 1987A.

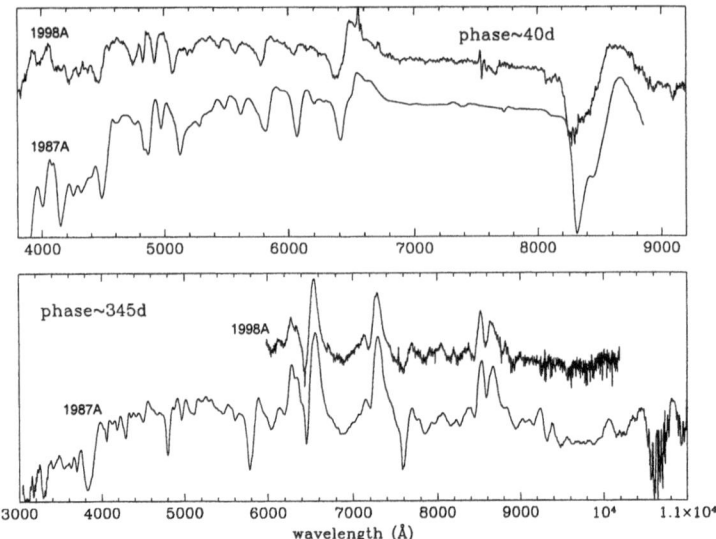

Fig. 2. Comparison between the spectra of SN 1998A and SN 1987A obtained ∼ 40 days (up) and about 1 year (bottom) after explosion

In summary, in spite of general spectro–photometric analogies, there are several observational differences between SN 1998A and SN 1987A: SN 1998A is brighter, bluer and shows higher kinetic energies. Moreover the fainter Ba II lines are in accordance with the lower observed temperatures. This implies that the explosion of SN 1998A was more energetic than that of SN 1987A. The characteristic shape of the light curve, with a secondary peak about 3 months after explosion suggests, in analogy with SN 1987A, that also the progenitor of SN 1998A was a compact blue supergiant. But the physical parameters of these two events were probably different. In order to estimate the parameters of the SN 1998A ejecta, we use the semi–analytical model recently presented by [23]. The model takes in account all the main sources powering the light curve from the photospheric phase (with the exception of shock breakout) up to the radioactive stage. The simultaneous fit of the evolution of the luminosity, the continuum temperature and the photospheric velocity indicates an explosion energy of 6–7 $\times 10^{51}$ erg, an ejected envelope $M_{ej} \approx 25$–$30 M_\odot$ and a Ni mass $M_{Ni} \approx 0.12 M_\odot$.

3   Ni Poor CC–SNe

The explosion of SN 1997D highlighted the existence of a group of SNe II that have low luminosity at all epochs. Assuming a plateau duration of about 60 days, a Ni mass of ∼ $0.002 M_\odot$ was estimated for SN 1997D. All the observational properties (low luminosity, continuum temperatures and expansion velocities) are consistent with two different progenitor models. Turatto et al. [20] proposed a model in which a massive star ($M_{MS} \geq 25 M_\odot$ and R$\leq 300 R_\odot$) undergoes

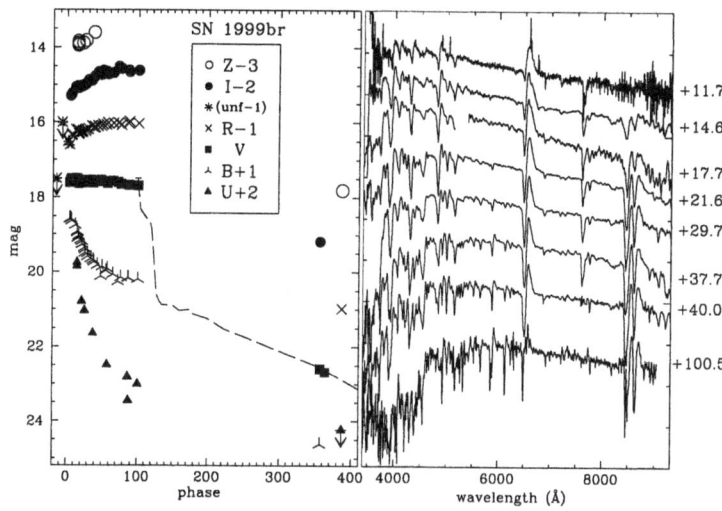

Fig. 3. Left: UBVRIZ light curves of SN 1999br. The dashed line represents the V band light curve of SN 1997D in an arbitrary magnitude scale. Right: spectroscopic evolution of SN 1999br from ∼ 3 weeks to about 100 days after the explosion

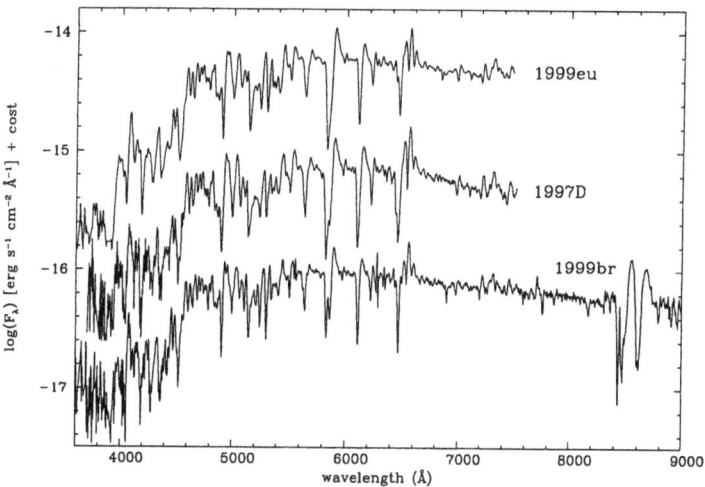

Fig. 4. Comparison among spectra of different Ni poor SNe obtained at the same evolutionary phase (∼ 100 days)

a low explosion energy (∼ 4×10^{50} erg). In this case a considerable fraction of the stellar envelope may remain gravitationally bound to the compact remnant. This material falls back onto the neutron star and eventually turns it into a black hole (BH) ([20,22]). In an alternative model, proposed by [3], a low metallicity star with $M_{MS} \approx 10 M_{\odot}$ and $R \leq 85 R_{\odot}$ produces an underenergetic explosion

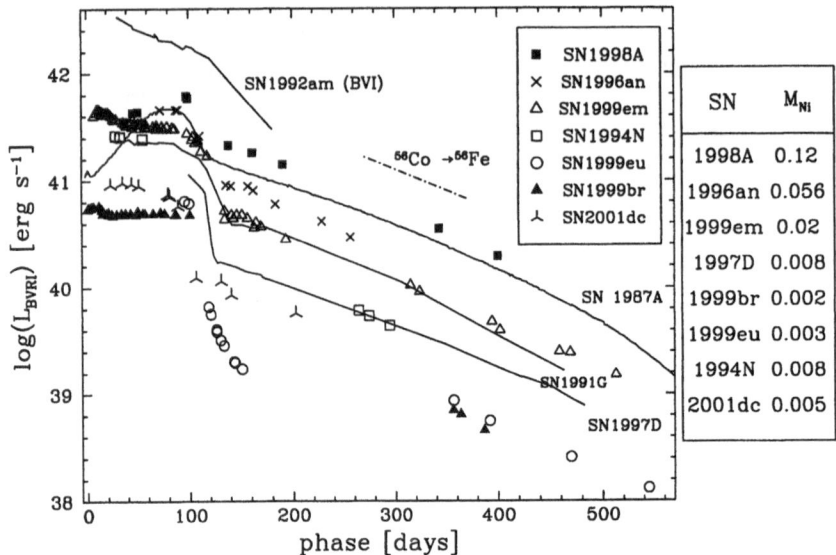

Fig. 5. (a) Comparison among BVRI pseudo–bolometric light curves of type IIP SNe compared with some famous reference objects. (b) Estimates of Ni mass ejected by our sample of SNe II. Sources of the data: SN 1996an from Padova–Asiago SN Archive; SN 1999em from [5,9,4]; SN 1998A from [12]; the SN 1997D–like SNe from [11]

($E_{exp} \approx 10^{51}$ erg). In this scenario no BH is expected to form. Both models were affected by uncertain observational parameters: the luminosity estimate at maximum, the explosion epoch and, consequently, the duration of the plateau.

The recent discovery of other SN 1997D–like events [11], gave tighter constraints for the theoretical estimates [23]. In particular SN 1999br, discovered soon after the explosion, provides the opportunity to follow the complete spectro–photometric evolution of a ^{56}Ni poor event. In Fig. 3 the photometric (left) and spectroscopic (right) evolutions of SN 1999br are shown. All data come from [7] and [11]. The early–time photometry is characterized by a widely extended plateau (lasting about 100 days, much more than ~ 60 days assumed for SN 1997D). The right panel of Fig. 3 shows for the first time the spectroscopic evolution from an almost normal type II appearance to a spectrum showing typically narrow (≤ 1000 km s^{-1}) SN 1997D–like P–Cygni features.

On the basis of spectroscopic analogies with SN 1999br (see Fig. 4), we calibrated in phase the available photometry and spectroscopy of SN 1997D and other similar events (SN 1994N, SN 1999eu, SN 2001dc) [11]. These SNe show great homogeneity in the photometric evolution and the luminosity of the radioactive tails suggests that the light curves are powered by $2-8 \times 10^{-3} M_\odot$ of ^{56}Ni.

The colour curves, the continuum temperature and the expansion velocities evolutions are all similar [11]. Therefore we expect similar progenitor and explosion parameters. On the basis of these new observational data, we recently performed an accurate re-analysis of SN 1997D-like SNe and found that their progenitors are intermediate mass main sequence stars ($M_{MS} \geq 20M_\odot$), 14-17$M_\odot$ of which are ejected [23]. The explosion energy is $6-9\times10^{50}$ erg. These underenergetic SNe are powered by the smallest ^{56}Ni mass among all CC events ([10,11]).

4   The Distribution of Ni Mass Ejected by CC-SNe

The statistics about the Ni masses ejected by the CC-SNe is discussed in several papers (e.g. [19] or [6]), claiming that the typical range of Ni yields peaks in the narrow interval $0.07-0.1M_\odot$. Only a few well studied SNe (low Ni mass events as SN 1991G [2], SN 1994W [17] and SN 1997D [20] or high Ni mass objects as *Hypernovae* or the bright type IIP SN 1992am [16]) deviate from this range. We performed an analysis of the bolometric light curves of a sample of type IIP SNe (Fig. 5) and found that probably a wide range of Ni masses is ejected (see table in Fig. 5).

References

1. S. Benetti et al.: MNRAS, **322**, 361 (2001)
2. E.L. Blanton et al.: AJ, **110**, 2868 (1995)
3. N.N. Chugai & V.P. Utrobin: A&A, **354**, 122 (2000)
4. A. Elmhamdi et al.: accepted by MNRAS, (astro-ph/0209623) (2002)
5. M. Hamuy et al.: ApJ, **553**, 886 (2001)
6. M. Hamuy & P.A. Pinto: ApJ, **566**, 63 (2002)
7. M. Hamuy: accepted by ApJ, (astro-ph/0209174) (2002)
8. K. Iwamoto et al.: Nature, **295**, 672 (1998)
9. D.C. Leonard ApJ, **553**, 861 (2001)
10. K. Nomoto et al.: *Hypernovae and their Nucleosynthesis*. In: *A Massive Star Odyssey, from Main Sequence to Supernova - IAU Symposium N. 212*, 2002, ed. by K.A. van der Hutch, A. Herrero & C. Esteban
11. A. Pastorello et al.: in preparation (2002a)
12. A. Pastorello et al.: in preparation (2002b)
13. F. Patat, R. Barbon, E. Cappellaro, M. Turatto: A&AS, **98**, 443 (1993)
14. F. Patat, R. Barbon, E. Cappellaro, M. Turatto: A&A, **282**, 731 (1993)
15. F. Patat et al.: ApJ, **555**, 900 (2001)
16. B.P. Schmidt et al.: AJ, **107**, 1444 (1994)
17. J. Sollerman, R.J. Cumming, P. Lundqvist: ApJ, **493**, 933 (1998)
18. J. Sollerman et al.: ApJ, **537**, 127 (2000)
19. J. Sollerman: NewAR, **46**, 493 (2002)
20. M. Turatto et al.: ApJ, **498**, 129 (1998)
21. M. Turatto et al.: ApJ, **534**, 57 (2000)
22. L. Zampieri, S.L. Shapiro, M. Colpi: ApJ, **502**, 149 (1998)
23. L. Zampieri et al.: accepted by MNRAS, (astro-ph/0210171) (2002)
24. P.A. Whitelock et al.: MNRAS, **230**, 7 (1989)
25. A. Williams, S. Woodings, R. Martin, A. Verveer, J. Biggs: IAU Circ. 6805 (1998)
26. S.J. Woodings et al.: MNRAS, **301**, 5 (1998)

Peculiar, Low Luminosity Type II Supernovae: Site of Black Hole Formation? *

Luca Zampieri[1], Andrea Pastorello[1,2,3], Massimo Turatto[1], Enrico Cappellaro[4], Stefano Benetti[1], Giuseppe Altavilla[1,2], Paolo Mazzali[5], and Mario Hamuy[6]

[1] INAF – Astronomical Observatory of Padova, Padova, Italy
[2] Department of Astronomy, University of Padova, Padova, Italy
[3] Department of Physics and Astronomy, University of Oklahoma, Norman OK, USA
[4] INAF – Astronomical Observatory of Capodimonte, Napoli, Italy
[5] INAF – Astronomical Observatory of Trieste, Trieste, Italy
[6] The Observatories of the Carnegie Institution of Washington, Pasadena CA, USA

Abstract. A number of supernovae classified as Type II show remarkably peculiar properties such as an extremely low expansion velocity and an extraordinary small amount of ^{56}Ni in the ejecta. We have modelled the available observations of these peculiar Type II supernovae by means of a new semi-analytic light curve code. We find that these events are under-energetic with respect to a typical Type II supernova and that the inferred mass of the ejecta is large. These supernovae are likely to originate from the explosion of a massive progenitor in which the rate of early infall of stellar material on the collapsed core is large. Events of this type could form a black hole remnant, giving rise to significant fallback and late-time accretion.

1 SN 1997D and SN 1999br

Recently, a number of supernovae (SNe) have been discovered that, according to their spectral properties, are classified as Type II but that, at the same time, show remarkably peculiar properties. The first clearly identified example was the exceptionally faint SN 1997D in NGC 1536 [10,4]. Modelling the spectra and light curve Turatto et al. [10] concluded that a low energy explosion in a 26 $M_\odot$ progenitor star could fit the early observations. The scenario of the low energy explosion of a high mass progenitor is also consistent with the late time ($\sim$ 400 days) spectral and photometric data [4]. It was early suggested that SN 1997D could host a black hole remnant formed during the explosion [10] and that the luminosity powered by fallback of envelope material onto the central black hole could emerge at about 1000 days after the explosion [13]. On the other hand, Chugai & Utrobin [5] presented an alternative analysis in which the progenitor was a star at the low end of the mass range of core-collapse SNe (8–12 $M_\odot$).

After SN 1997D, a number of SNe with similar observational properties have been identified [8] (see also Pastorello et al., this Proceedings). As for SN 1997D, these SNe provide a unique opportunity to probe the physics of the explosion and reach a better understanding of both the explosion mechanism and the

---

* Based on observations collected at the European Southern Observatory and Cerro Tololo Inter-American Observatory, Chile (Prop. ESO 63.H-0141 and 66.D-0683).

conditions for the formation of black hole remnants. Here we present the results of a joint analysis of the observations of two peculiar Type II SNe with very low luminosity and expansion velocity, SN 1997D and SN 1999br, which are representative of the properties of the whole group.

Supernova 1997D, serendipitously discovered on 14 January 1997 during an observation of the parent galaxy NGC 1536, is the first, clearly identified example of a peculiar Type II SN with a very low luminosity and expansion velocity [10]. It was detected when it was already decaying from the plateau and was at least 2 mag fainter than a typical Type II SN. The decline rate of the last segment of the bolometric light curve is consistent with complete thermalization of the gamma rays from the radioactive decay of ^{56}Co into ^{56}Fe. Assuming for SN 1997D the same deposition as in SN 1987A, the ejected ^{56}Ni mass is $\sim 10^{-3} - 10^{-2} M_\odot$ [4]. The spectra are dominated by a red continuum and P-Cygni profiles of H I, Ba II, Ca II, Na I and Sc II [8]. The most striking property of these spectra is the very low expansion velocity inferred from the spectral lines (1100-1200 km s^{-1} for H; [4]).

Recently, other SNe with properties similar to those of SN 1997D have been identified [8]. We focus on one particularly representative object, SN 1999br in NGC 4900, discovered on 12 April 1999. It was early recognized that this event has exceptionally low luminosity and that it shows other properties similar to those of SN 1997D, such as spectra with narrow P-Cygni lines, prominent Ba II features and a low continuum temperature. It had a long plateau lasting at least 110 days (with a mean luminosity $\leq 10^{41}$ erg s^{-1}), extensively monitored with the CTIO and ESO telescopes [7]. The spectrum of SN 1999br at ~ 100 days shows narrow metal lines. As for SN 1997D, the red continuum and the strength of Ba II lines are caused by the low temperature of the ejecta. The U-B bands are strongly affected by line blanketing as the temperature decreases.

The spectrum of SN 1997D obtained soon after discovery and that of SN 1999br at $\sim$100 days are strikingly similar in both the continuum and the line components [8,14]. The close resemblance of the two spectra, the low inferred expansion velocity ($\sim$1000 km s^{-1}) and the fact that the two SNe have comparable luminosities strongly suggest that SN 1997D and SN 1999br may be similar events. SN 1997D was discovered at the end of the plateau stage when the light curve was plummeting. The duration and luminosity of the plateau were inferred by comparison between models and observations and the consequent estimate of the explosion epoch (~ 60 days before discovery) was uncertain [10]. On the other hand, the date of the explosion of SN 1999br has an uncertainty of only a few days (thanks to a stringent pre-discovery limit). Then, from the similarity of the observational properties (luminosities and spectra) of SN 1997D and SN 1999br, we tentatively assume that the phase of SN 1997D at discovery is $\sim 90 - 100$ days. This assumption will be checked by means of a detailed comparison of spectral-synthesis models with observations, presently under way.

2   Modelling Low Luminosity Supernovae

In order to determine the physical properties of a SN, the observed light curve
and spectra are compared to model calculations. Full hydrodynamical calculations using realistic pre-supernova models show that the dynamical evolution
of the envelope during shock passage is quite complex. The propagation of the
shock determines how the explosion energy is distributed in the envelope of the
progenitor star. In the present analysis we do not consider the evolution of the
star during this complex phase but rather assume idealized initial conditions
that provide an approximate description of the ejected material after shock (and
possible reverse shock) passage, as derived from hydrodynamical calculations. At
the onset of the evolution, the post-shock, ejected envelope is assumed to have
spatially constant density ρ and total mass $M = 4\pi\rho R^3/3$. Mass conservation
gives $\rho = \rho_0(R_0/R)^3$, where R_0 and ρ_0 are the initial radius and density of the
envelope. In reality, the outer part of the star develops a steep power-law density structure that affects the light curve during the first few days after shock
breakout (which is not included). However, this is only $\sim 1\%$ by mass of the
star, while most of the stellar material resides in the inner part with roughly
constant density. The latter region dominates the light curve after 10–20 days.
The expansion velocity of the envelope as a function of interior (Lagrangian)
mass $m(r) = 4\pi\rho r^3/3$ is $v(m) = v_0(m/M)^{1/3}$ (homologous expansion), where
v_0 is the initial velocity of the outermost shell. The initial thermal+kinetic energy of the envelope is $E = (3/10)Mv_0^2/f_0$, where f_0 is the initial fraction of
kinetic energy (assumed fixed and equal to $1/2$). Elements are assumed to be
completely mixed throughout the envelope. In our simple spherically symmetric
model, their distribution depends only on r (or m). In particular, Hydrogen, Helium and Oxygen are assumed to be uniformly distributed, whereas ^{56}Ni is more
centrally peaked [14]. The total ^{56}Ni mass is denoted by M_{Ni}. Our assumptions
about the velocity and elemental distributions should be regarded as a potential
source of systematic uncertainty in the present model.

   The evolution of the post-shock SN envelope is computed using a spherically
symmetric, semi-analytic model that can provide a robust estimate of the parameters of the ejecta of Type II SNe. The novelty of the present approach lies
mainly in the fact that the physical properties of the envelope are derived by
performing a simultaneous comparison of the observed and simulated light curve,
evolution of the line velocity and continuum temperature. The model follows the
approach originally introduced by [1] and later developed by [2]. The treatment
of the motion of the recombination front follows in part [9]. The evolution is
computed solving the energy balance equation for the envelope, including all the
relevant energy sources powering the SN, and is schematically divided in 3 phases
from the photospheric up to the late nebular stages. The model has been tested
against numerical radiation-hydrodynamic computations in spherical symmetry
under the same assumptions, giving good agreement (for details see [14], [15]).

   Following the discussion in §1, we assume that the phase of SN 1997D at
discovery is $\sim 90-100$ days and that the early light curve and the evolution of the
line velocity resemble closely those of SN 1999br. Figure 1 shows a comparison

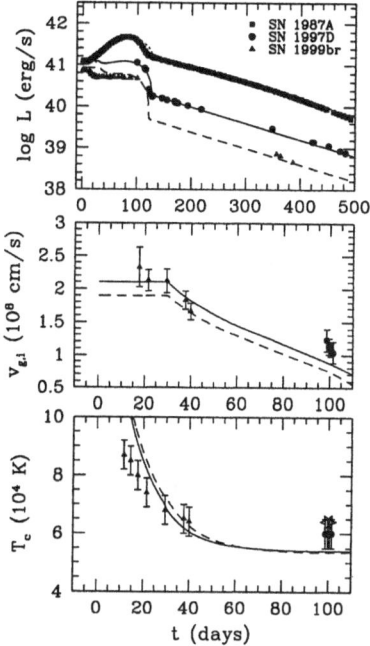

Fig. 1. UBVRI luminosity, L, velocity of metal (Sc II) lines, $v_{g,i}$, and continuum temperature, T_c, as a function of time for SN 1997D and SN 1999br. The light curve of SN 1987A is also shown for comparison. The adopted distance moduli and the estimated total reddenings are from [8]. The lines represent the model curves. The velocities in the mid panel are those of the gas at the position of the recombination wavefront. The line velocity and continuum temperature inferred from the model refer to the photospheric epoch. The asterisk is the continuum temperature inferred from the spectral synthesis model of SN 1997D at discovery [10] moved to a phase ~ 100 d

of the observations with the results of our semi-analytic model calculations. Considering the approximations adopted in the model, the general agreement with observations is satisfactory. The main parameters of the ejected envelope required to reproduce the observations are listed in Table 1.

The estimated value of E indicates that both events were rather under-energetic compared to a typical Type II SN. The inferred ^{56}Ni mass of SN 1999br is extremely small, testifying that the energy available to produce and eject nucleosynthetic elements was very low. Furthermore, the post-explosion envelope is rather compact. The ejected envelope masses are quite large, comparable to those required to reproduce the plateau of typical Type II SNe. In particular, the ejected mass of SN 1997D is almost three times larger than that estimated by Chugai & Utrobin [5] and only 30% smaller than that inferred by Turatto et al. [10]. It is worth noting that the gross properties of the light curve and line velocity of SN 1997D and SN 1999br can be roughly accounted for by SN 1987A-like parameters, but simply decreasing v_0 (and hence E) and M_{Ni}.

Table 1. Parameters (defined in the text) of the semi-analytic model

	R_0 $(10^{12}$ cm$)$	M $(M_\odot)$	M_{Ni} $(M_\odot)$	v_0 $(10^8$ cm s$^{-1})$	E $(10^{51}$ erg$)$
SN 1987A	5	18	7.5×10^{-2}	2.7	1.6
SN 1997D	9	17	8×10^{-3}	2.1	0.9
SN 1999br	7.5	14	2×10^{-3}	1.9	0.6

Although it is not straightforward to determine the error in the parameters, we estimate that the intrinsic uncertainty of the "fit" is not larger than $\sim 30\%$ (see also [15]). Additional sources of systematic errors are related to the approximations introduced in the present analysis, in particular to the choice of the initial conditions. Although it is known that significant mixing occurred in SN 1987A, this effect may be less pronounced in other SNe. The light curve is very sensitive to the prescription for mixing and to the actual velocity distribution as a function of mass $v(m)$. If the innermost Helium and heavier elements layers did not mix appreciably with the Hydrogen envelope, they would not produce any observable effect (having very low energy). In this case, the estimated value of M would simply refer to the Hydrogen envelope mass. Then, in general, M represents a lower limit to the total mass of the ejecta. Furthermore, the estimate of the envelope mass is sensitive to uncertainties in the actual value of the gas opacity κ and the details of the recombination physics. Here we adopted $\kappa = 0.2$ cm^2 g^{-1}. Only larger opacities and/or a delayed onset of recombination could in principle allow for smaller envelope masses, but this is not in agreement with what was found for SN 1987A and other Type II SNe [15]. The fact that the line velocities at ~ 100 days are slightly larger than what is predicted by the model (see Fig. 1) seems to indicate that the assumption of spatially constant density is only approximately correct and that more mass is concentrated in the innermost, low velocity part of the envelope. This effect may reduce the estimated energy E because the bulk of the kinetic energy is carried by the outer, high velocity layers.

We emphasize that the results for SN 1997D are based on the assumption that it is an event similar to SN 1999br. However, although the inferred parameters of the ejecta of SN 1997D rely on this hypothesis, the estimates for SN 1999br are certainly valid because they are not affected by uncertainties on the early light curve, the duration of the plateau and the evolution of the line velocity.

Determining the mass of the progenitor is quite a difficult task and, unless a pre-discovery identification is available, the inferred value is usually rather uncertain. Adding to the mass of the ejecta reported in Table 1 the mass of the collapsed core ($\approx 2M_\odot$), the progenitors of SN 1997D and SN 1999br have at least 19 and 16 $M_\odot$, respectively. We stress that these are lower bounds and the actual values of the progenitor masses are likely to be larger. Therefore, the present estimates situate SN 1997D and SN 1999br in the intermediate mass range and rule then out an origin from a low mass progenitor.

Understanding why the energy of these SNe is low is of paramount importance in connection with the hydrodynamics of the explosion and the formation of the central compact object. Because the mass of the iron core before the explosion does not vary significantly with progenitor mass M_* [11], the gravitational potential energy liberated during the collapse of the core is roughly independent of M_*. To unbind a spherically symmetric envelope, the SN shock must overcome the ram pressure of the gas that started to accrete after the collapse of the core (see e.g. [6]). With increasing progenitor mass both the ram pressure and the binding energy of the envelope become larger. Therefore, comparatively less energy remains available to heat up and accelerate the ejecta. The final luminosity and expansion velocity are then small and the resulting SN is under-energetic. Because of the low kinetic energy acquired by the envelope, a large fraction of the stellar material (in particular the innermost layers) is likely to remain gravitationally bound to the core after shock passage and fall back onto it. This is confirmed also by the small amount of ^{56}Co present in the ejecta. The fallback of stellar material may also turn the newly formed neutron star into a black hole. If this happens, the late time fallback onto the central black hole may give rise to detectable emission of radiation with a characteristic power-law decay [12,3]. The detection of this emission in the late time light curve would provide the first direct evidence for the presence of a black hole in the site of its formation.

We acknowledge support from MIUR through grant Cofin MM02905817 and ASI under grant ASI I/R/70/00. M. H. acknowledges support by NASA through Hubble Fellowship grant HST-HF-01139.01-A awarded by the Space Telescope Science Institute, which is operated by the Association of Universities for Research in Astronomy, Inc., for NASA, under contract NAS 5-26555.

References

1. Arnett, W. D., 1980, ApJ, 237, 541
2. Arnett, W. D., Fu, A., ApJ, 1989, 340, 396
3. Balberg, S., Zampieri, L., Shapiro, S. L., 2000, ApJ, 541, 860
4. Benetti, S. et al., 2001, MNRAS, 322, 361
5. Chugai, N. N., Utrobin, V. P., 2000, A&A, 354, 557
6. Fryer, C. L., 1999, ApJ, 522, 413
7. Hamuy, M. et al., 2003, in preparation
8. Pastorello, A. et al., 2002, MNRAS, submitted
9. Popov, D. V., 1992, ApJ, 414, 712
10. Turatto, M. et al., 1998, ApJ, 498, L129
11. Woosley, S. E., Weaver, T. A., 1995, ApJS, 101, 181
12. Zampieri, L., Colpi, M., Shapiro, S. L., Wasserman, I., 1998, ApJ, 505, 876
13. Zampieri, L., Shapiro, S. L., Colpi, M., 1998, ApJ, 502, L149
14. Zampieri, L. et al., 2002a, MNRAS, submitted
15. Zampieri, L. et al., 2002b, in preparation

Evidence for Core Collapse in the Type Ib/c SN 1999ex

Mario Hamuy[1], Maximilian Stritzinger[2], M.M. Phillips[1], Nicholas B. Suntzeff[3], José Maza[4], and Philip A. Pinto[5]

[1] Carnegie Observatories, 813 Santa Barbara Street, Pasadena, CA 91101, USA
[2] Department of Physics, The University of Arizona, Tucson, AZ 85721, USA
[3] Cerro Tololo Inter-American Observatory, Casilla 603, La Serena, Chile
[4] Departamento de Astronomía, Universidad de Chile, Casilla 36-D, Santiago, Chile
[5] Steward Observatory, The University of Arizona, Tucson, AZ 85721, USA

Abstract. We present optical and infrared spectra of SN 1999ex, which are characterized by the lack of strong hydrogen lines, weak optical He I lines, and strong He I $\lambda10830,20581$. SN 1999ex provides a clear example of an intermediate case between pure Ib and Ic supernovae, which suggests a continuous spectroscopic sequence between SNe Ic to SNe Ib. Our $UBVRIz$ photometric observations of SN 1999ex started only one day after explosion, which permitted us to witness an elusive transient cooling phase that lasted 4 days. The initial cooling and subsequent heating due to $^{56}Ni \rightarrow {}^{56}Co \rightarrow {}^{56}Fe$ produced a dip in the light curve which is consistent with explosion models involving core collapse of evolved massive helium stars, and not consistent with light curves resulting from the thermonuclear runaway of compact white dwarfs.

1 Introduction

In a rare occurrence the spiral galaxy IC 5179 ($cz=3498\ km\ s^{-1}$) produced two supernovae (SNe) in an interval of only three weeks. The first object (SN 1999ee) was a Type Ia event discovered by us 10 days before maximum [16]. The early discovery motivated us to use the YALO and 0.9-m telescopes at the Cerro Tololo Inter-American Observatory in order to secure nightly $UBVRIz$ photometric observations of this event, and the YALO and Las Campanas 1-m and 2.5-m telescopes to obtain JHK photometry. Although the second object (SN 1999ex) exploded three weeks later and was promptly present in our CCD images we did not notice its presence. Its discovery had to await independent observations obtained at Perth Observatory [14]. Once SN 1999ex was reported to the IAU Circulars we initiated an optical and infrared (IR) spectroscopic followup using the European Southern Observatory NTT and Danish 1.5-m telescope at La Silla and the VLT at Cerro Paranal. Our spectroscopic and photometric observations of SN 1999ex constitute an unprecedented dataset, which provides support to our understanding of the nature of core collapse SNe. In this paper we show some of our observations and their interpretation. For a detailed report of our observations the reader is referred to [8,23,13].

2 Spectroscopic Observations

Figure 1 compares the near-maximum optical spectra of SN 1999ex to those of the prototype of the Ib class SN 1984L [9], the Type Ic SNe 1994I [5] and 1987M [4]. The first spectrum of SN 1999ex was characterized by a reddish continuum and several broad absorption/emission features due to Ca II H&K $\lambda\lambda$3934,3968, Na I D $\lambda\lambda$5890,5896 and the Ca triplet with a clear P-Cygni profile. This spectrum bore quite some resemblance to that of the Type Ic SN 1994I [5]. However, SN 1999ex showed evidence for He I absorptions (shown with tick marks) of moderate strength in the optical region, thus suggesting the existence of an intermediate Ib/c case. Our observations of SN 1999ex provide a clear link between the Ib and Ic classes and suggest that there is a continuous sequence of SNe Ib and Ic objects.

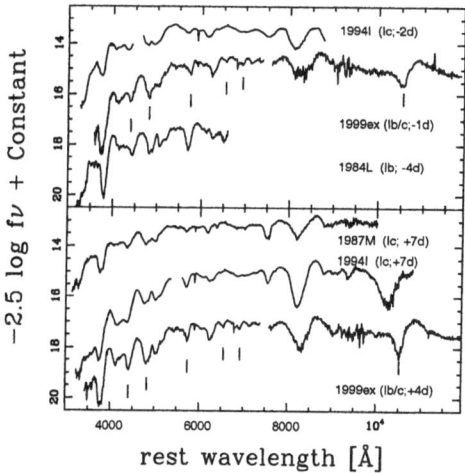

Fig. 1. Comparison of spectra of the Type Ib/c SN 1999ex with the prototype of the Ib class SN 1984L [9], the Type Ic SNe 1994I [5] and 1987M [4]. *Tick marks* indicate the He I lines in the SN 1999ex spectra. The strengths of the helium lines gradually increase from the Type Ic to the Ib SN, and SN 1999ex provides a link between these two subclasses. Time (in days) since B maximum is indicated for each spectrum

The presence of helium in the atmosphere of SN 1999ex can be further examined in our IR observations shown in Fig. 2. The strong feature near 1.05 μm probably had a significant contribution from He I λ10830, although it could be blended with lines of C I and Si I [17,1]. If this feature was He I λ10830 it would imply an expansion velocity of 6,000-8,000 km s^{-1}, which matches very well the velocities derived from the Fe and Na lines. This identification is supported by the presence of He I λ20581 with the same expansion velocity as deduced from He I λ10830. Our IR spectra of SN 1999ex provide unambiguous proof that helium was present in the atmosphere of this intermediate Ib/c object. A detailed atmosphere model could be very useful at placing specific limits on the helium mass in the ejecta of SN 1999ex and constraining the nature of its progenitor. Evidently, K-band spectroscopy is probably the best tool to explore the presence or absence of helium in the atmospheres of SNe Ib and Ic.

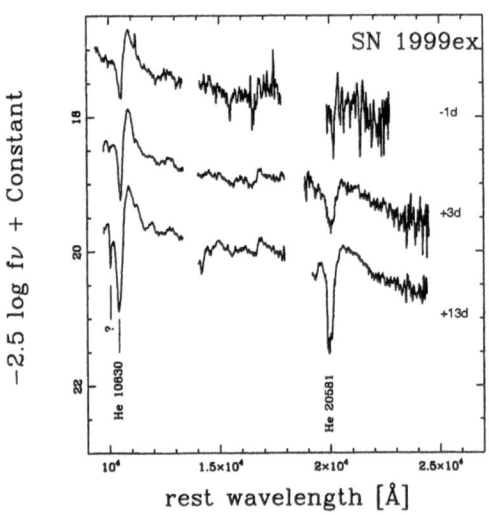

Fig. 2. IR spectroscopic evolution of SN 1999ex. The two most prominent features are attributed to He I. Time (in days) since B maximum is indicated for each spectrum

The question of whether or not SNe Ic have helium is still controversial. A detailed inspection of the spectra of SN 1994I led Filippenko et al. [5] to conclude that weak He I lines were probably present in the optical region and that He I $\lambda10830$ was very prominent, although its Doppler shift implied an unusually high expansion velocity $\sim16,600$ km s^{-1} as can be seen in the bottom panel of Fig. 1. Clocchiatti et al. [3] confirmed these observations and found evidence that He I $\lambda5876$ was also present in SN 1994I with a blueshift of $\sim17,800$ km s^{-1}. They also reported high velocity He I $\lambda5876$ in the spectra of the three best-observed Type Ic SNe (1983V, 1987M, and 1988L), which led them to conclude that most, and probably all, SNe Ic have optical He I lines. This conclusion, however, was recently questioned by Millard et al. [17] and Baron et al. [1] by means of spectral synthesis models, which showed that the 1.05 μm feature could be accounted with lines of C I and Si I. Moreover, Baron et al. [1] argued that the feature attributed to He I $\lambda5876$ in the spectrum of SN 1994I could be a blend of other species. Recently, Matheson et al. [15] compiled and analyzed a large collection of spectra of SNe Ib and Ic and did not find compelling evidence for helium in the spectra of SNe Ic. Evidently, there is no consensus yet about this issue. The existence of an intermediate Ib/c object such as SN 1999ex suggests a continuous transition between SNe Ib and SNe Ic so it is likely that some SNe Ic have some helium in their atmospheres.

3   Photometric Observations

Figure 3 shows the $UBVRIz$ light curves of SN 1999ex. Clearly the observations began well before maximum light thanks to our continuous followup of IC 5179 owing to the prior discovery of SN 1999ee. The first detection occurred on JD 2451481.6 in all filters. Excellent seeing images obtained on the previous night allowed us to place reliable upper limits to the SN brightness, which permitted us to conclude that the explosion took place on JD 2451480.5 (± 0.5). Along with the Type Ic hypernova 1998bw [6], these are the earliest observations of a SN Ib/c. The most remarkable feature in this figure is the early dip in the U and B light curves – covering the first 4 days of evolution – after which the SN steadily rose to maximum light. Similar initial upturns have been observed before in SN Ic 1998bw [6], SN II 1987A [7], and SN IIb 1993J [20,19]. For these SNe II it is thought that the initial dip corresponded to a phase of adiabatic cooling that ensued the initial UV flash caused by shock emergence, which super-heated and accelerated the photosphere. The following brightening is attributed to the energy deposited behind the photosphere by the radioactive decay of $^{56}Ni \rightarrow ^{56}Co \rightarrow ^{56}Fe$. This similarity suggests that the progenitor of SN 1999ex

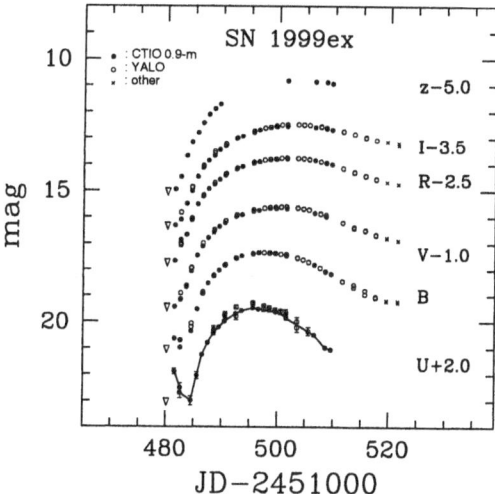

Fig. 3. $UBVRIz$ light curves of SN 1999ex measured with the YALO (*open points*) and CTIO 0.91-m telescopes (*closed points*). Upper limits derived from images taken on JD 2451480.5 are also included (*open triangles*). The *solid line* through the U data is drawn to help the eye to see the initial upturn

was a massive star that underwent core collapse. Woosley et al. [24] computed Type Ib SN models consisting of the explosion of an evolved 6.2 M$_\odot$ helium star. Their Figure 7 shows the bolometric luminosities of three models with different explosion energies and ^{56}Ni nucleosynthesis, all of which show an initial peak followed by a dip a few days later and the subsequent brightening caused by ^{56}Ni $\rightarrow$ ^{56}Co $\rightarrow$ ^{56}Fe making them good models for SN 1999ex.

In order to compare the observations with the models we computed a bolometric light curve for SN 1999ex by performing blackbody (BB) fits to our BVI photometry (Fig. 4). Among the three SN Ib models of Woosley et al. the one with kinetic energy of 2.7×10^{51} ergs and 0.16 M$_\odot$ of ^{56}Ni provides the best match to SN 1999ex. The agreement is remarkable considering that we are not attempting to adjust the parameters. The initial peak and subsequent dip have approximately the right luminosities although the evolution of SN 1999ex was somewhat faster. The following evolution is well described by the model.

The observation of the tail of the shock wave breakout in SN 1999ex and the initial dip in the light curve provides us with an insight on the type of progenitor system for SNe Ib/c. Several different models have been proposed as progenitors for this type of SNe. One possibility is an accreting white dwarf which may explode via thermal detonation upon reaching the Chandrasekhar mass [22,2]. These models are expected to produce light curves with an initial peak that corresponds to the emergence of the burning front, a fast luminosity drop due to adiabatic expansion, and a subsequent rise caused by radioactive

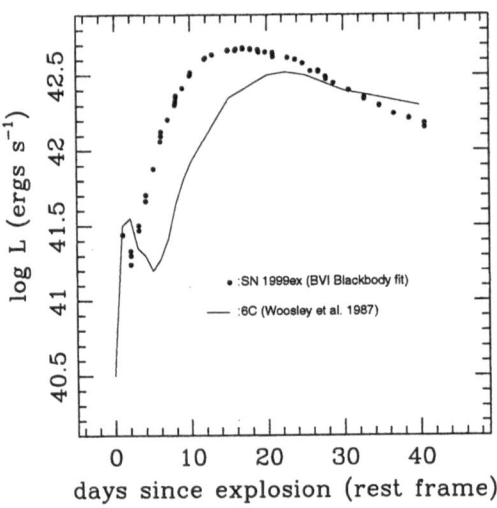

Fig. 4. Bolometric light curve of SN 1999ex derived from blackbody fits to the BVI magnitudes, assuming $E(B - V)_{host}$=0.28 and a distance of 51.2 Mpc (*closed circles*). The *solid line* is the 6C hydrogenless core bounce SN model of Woosley et al. [24]

heating. Given the compact nature of the white dwarf ($\sim$1,800 km) the cooling time scale by adiabatic expansion is only a few minutes [11] and the light curve is entirely governed by radioactive heating [10]. Hence, these models are not expected to show an early dip at a few days past explosion as is observed in SN 1999ex. The second and more favored model for SNe Ib/c consists of core collapse of massive stars ($M_{ZAMS} > 8$ M$_\odot$) which lose their outer H envelope before explosion. Within the core collapse models, there are two basic types of progenitor systems: 1) a massive ($M_{ZAMS} > 35$ M$_\odot$) star which undergoes strong stellar winds and becomes a Wolf-Rayet star at the time of explosion [25], and 2) an exchanging binary system [21,18,12] for less massive stars. The resulting SNe have light curves containing an initial spike followed by a dip. Since the initial radii of these progenitors are $\sim$100 times greater than that of white dwarfs, the dip occurs several days after explosion [24,21,25], very much like SN 1999ex.

Although a detailed modeling of SN 1999ex is beyond the scope of this paper, our fortuitous observations lend support to the idea that SNe Ib/c are due to core collapse of massive progenitors rather than thermonuclear disruption of white dwarfs.

M.H. acknowledges support provided by NASA through Hubble Fellowship grant HST-HF-01139.01-A awarded by the Space Telescope Science Institute.

References

1. E. Baron, D. Branch, P.H. Hauschildt, A.V. Filippenko, R.P. Kirshner: ApJ **527**, 739 (1999)
2. D. Branch, K. Nomoto: A&A **164**, L13 (1986)
3. A. Clocchiatti, J.C. Wheeler, M.S. Brotherton, A.L. Cochran, D. Wills, E.S. Barker, M. Turatto: ApJ **462**, 462 (1996)
4. A.V. Filippenko, A.C. Porter, W.L.W. Sargent: AJ **100**, 1575 (1990)
5. A.V. Filippenko et al.: ApJ **450**, L11 (1995)
6. T.J. Galama et al.: Nature **395**, 670 (1998)
7. M. Hamuy, N.B. Suntzeff, R. González, G. Martin: AJ **95**, 63 (1988)
8. M. Hamuy et al.: AJ **124**, 417 (2002)
9. R.P. Harkness et al: ApJ **317**, 355 (1987)
10. P. Höflich, A. Khokhlov: ApJ **457**, 500 (1996)
11. P. Höflich: private communication (2002)
12. K. Iwamoto, K. Nomoto, P. Höflich, H. Yamaoka, S. Kumagai, T. Shigeyama: ApJ **437**, L115 (1994)
13. K. Krisciunas et al.: AJ, in preparation (2002)
14. R. Martin, A. Williams, S. Woodings: IAU Circ. 7310 (1999)
15. T. Matheson, A.V Filippenko, W. Li, D.C. Leonard, J.C. Schields: AJ, **121**, 1648 (2001)
16. J. Maza, M. Hamuy, M. Wischnjewsky, L. González, P. Candia, C. Lidman: IAU Circ. 7272 (1999)
17. J. Millard et al.: ApJ **527**, 746 (1999)
18. K. Nomoto, H. Yamaoka, O.R. Pols, E.P.J. van den Heuvel, K. Iwamoto, S. Kumakai, T. Shigeyama: Nature **371**, 227 (1994)

19. M.W. Richmond, R.R. Treffers, A.V. Filippenko, Y. Paik, B. Leibundgut, E. Schulman, C.V. Cox: AJ **107**, 1022 (1994)
20. B.P. Schmidt et al.: Nature **364**, 600 (1993)
21. T. Shigeyama, K. Nomoto, T. Tsujimoto, M. Hashimoto: ApJ **361**, L23 (1990)
22. R.A. Sramek, N. Panagia, K.W. Weiler: ApJ **285**, L59 (1984)
23. M. Stritzinger et al.: AJ **124**, 2100 (2002)
24. S.E. Woosley, P.A. Pinto, P.G. Martin, T.A. Weaver: ApJ **318**, 664 (1987)
25. S.E. Woosley, N. Langer, T.A. Weaver: ApJ **411**, 823 (1993)

The Dusty Type IIn Supernova 1998S

Peter Meikle[1], Alexandra Fassia[1], Thomas R. Geballe[2], Peter Lundqvist[3], Nikolai Chugai[4], Duncan Farrah[1], and Jesper Sollerman[3]

[1] Blackett Laboratory, Imperial College, Prince Consort Road, London SW7 2BW, UK
[2] Gemini Observatory, 670 N. A'ohoku Place, Hilo, HI 96720, USA
[3] Stockholm Observatory, AlbaNova, Dept. of Astronomy, Stockholm SE 106 91, Sweden
[4] Institute of Astronomy, RAS, Pyatnitskaya 48, 109017, Russia

Abstract. The type IIn SN 1998S is one of the most remarkable core-collapse supernovae ever observed. It underwent a complex interaction with a substantial circumstellar medium, resulting in radiation at wavelengths from radio to X-rays. IR and optical observations have revealed a wide variety of broad and narrow emission lines. Examination of the SN/CSM interaction and of the ejecta spectra has allowed us to deduce that the supernova probably arose from a massive, RSG progenitor having a large (>3200 AU radius), dusty circumstellar disk. SN 1998S also developed one of the strongest, most persistent infrared excesses ever seen in a supernova. IR/optical monitoring of SN 1998S has been carried out to nearly 1200 days post-explosion. This includes coverage to wavelengths as long as 4.7 μm, making SN 1998S only the second supernova (after SN 1987A) to be observed in this spectral region. Fading of the central and redshifted components of the late-time H I and He I line profiles suggests strongly that dust condensed in the ejecta. However, it is less clear whether the strong late-time IR emission arose from this dust, or from an IR echo in the dusty CSM. One interesting possibility is that dust condensed in the cool dense shell between the outer and reverse shocks, thus simultaneously producing both the line obscuration and the IR emission.

1 Introduction

One of the challenges of supernova research is to obtain evidence about the nature and environment of the progenitor. For type IIn supernovae, the progenitors must have undergone one or more mass-loss phases before explosion. By using the 'illumination' of the resulting circumstellar medium (CSM) by the supernova explosion we can get clues about the progenitor, even at distances of tens of Mpc.

More than 30 years ago, it was suggested [1–3] that supernovae could be important sources of interstellar dust. More recent work [4–6] still invokes core-collapse SNe as significant dust contributors. However, the number of supernovae in which dust has been detected is relatively small and, prior to SN 1998S, only for SN 1987A had dust condensation been convincingly demonstrated [7–14].

An opportunity to address both the progenitor problem and the question of supernovae as dust sources has been provided by the occurrence of the type IIn SN 1998S. This has become the most intensively-studied type IIn event [15–30].

2   Early-Time Behaviour

2.1   Early-Time Optical Spectral Evolution

The earliest optical spectra of SN 1998S show a blue continuum with emission features superimposed. A rough blackbody fit yields T∼25,000 K, but with a blue excess [18, 24]. The emission lines are identified with H I (Balmer series), He I, He II, C III and N III. The high-ionisation carbon and nitrogen lines are also commonly observed in Wolf-Rayet stars [18]. The emission lines have a broad base (*e.g.* Hα FWZI∼20,000 km/s), but a narrow 'peaked' unresolved centre. The lines are symmetrical about the local standard of rest. This is quite surprising since at such an early phase most of the receding part of the supernova should be occulted by the photosphere. In fact, Chugai [23] has shown that this constitutes some of the earliest evidence of a strong ejecta-CSM wind interaction. The broadening results from Thomson scattering in a radiatively-accelerated CSM wind lying immediately above an opaque, relatively cool dense shell (CDS) at the ejecta-wind interface [31,32]. The blue excess can also be attributed to the CDS, since the significant optical depth can yield an increase in continuum absorptive opacity with wavelength, due to both bound-free (*i.e.* the Paschen Continuum) and free-free processes [18].

By about 2 weeks after the explosion, the emission lines had essentially disappeared. (Following Fassia et al. [24], we adopt JD 2450875.2 as zero epoch. This was probably a few days post-explosion.) This disappearance is attributed to the dense inner-CSM being overrun by the ejecta. Nevertheless, the CDS remained optically-thick in the Paschen continuum until around 40-50 days, and this accounts for the lack of strong broad lines from the ejecta during the ∼2–6 weeks era. However, during this time we can see weak unresolved lines superimposed on the continuum. This is due to the flash-ionisation of the undisturbed wind (see below). After ∼40–50 days, broad, square-shaped lines in Hα and the Ca II triplet formed. Such line profiles are characteristic of emission from a shocked ejecta/CSM shell.

2.2   Early-Time Infrared Spectral Evolution

SN 1998S is unique in that it allowed the first-ever good IR spectroscopic coverage of a type IIn event [17,24]. In the *J*-band, we see Paschen β, Paschen γ and He I 1.083 μm lines. Their evolution was similar to that seen in the optical. At the earliest times broad-based, peaked profiles were present. These faded by day 17, being replaced by broad, square-shaped profiles by day 44. Between days ∼10 and ∼60 a strong, unresolved He I 1.083 μm CSM line was superimposed on the ejecta/CSM broad lines.

By day 44, the *HK*-band was dominated by Paschen α. By day 108, first-overtone CO emission was clearly present in the *K*-band. The presence of CO in core-collapse supernovae is increasingly regarded as ubiquitous. In all cases where *K*-band observations have been carried out in the period 3-6 months post-explosion, CO has been detected. There are now seven known cases: 87A (IIpec)

[13,33-35], 95ad (IIP) [36], 98S (IIn) [17,24], 99dl (IIP) [37], 99em (IIP) [37,38], 99gi (IIP) [38], 00ew (Ic) [38]. Modelling of the SN 1998S spectra suggests a CO velocity of ~2000 km/s [17,24]. From this, Fassia et al. [24] deduced a core mass of 4 $M_\odot$ implying a massive progenitor. The actual mass of CO derived was 10^{-3} $M_\odot$. The low-excitation rotation-vibration states of CO mean that it is a powerful coolant. Its presence is suspected to be a necessary condition for dust condensation to occur in the ejecta.

On day 130, Fassia et al [24] succeeded in measuring the IR flux out to a wavelength of 3.8 μm (L'-band). This revealed a remarkable IR excess of $K - L' = +2.5$. The most plausible interpretation of this is emission from warm dust. But where was the dust located? For dust condensing in the ejecta to produce such a large, early L' flux the lowest possible velocity of the dust-forming region would be 11,000 km/s, and that includes the assumption that the temperature is close to the dust evaporation temperature of ~1500 K. Such high velocities were seen only in the extreme outer zones of the H/He envelope. No metals were seen at such high velocities. Fassia et al. [24] concluded that the IR excess at this epoch cannot, therefore, have been due to grain condensation in the ejecta. It must instead have been produced by an IR echo of the maximum-light luminosity from pre-existing dust in the CSM.

2.3 Narrow Lines

Of particular interest are the high-resolution echelle spectra of SN 1998S obtained at the WHT by Bowen et al. [15] and Fassia et al. [24] on days 17 and 36. These observations succeeded in resolving the narrowest CSM lines. From forbidden lines such as [OIII] 5007 Å, an undisturbed CSM velocity of about 40 km/s is obtained, which is characteristic of an RSG wind. Fassia et al. [24] also deduced a centre of mass redshift velocity of +847 km/s. Between day 17 and day 36 the [OIII] 5007 Å profile changed from having a red deficit, to being quite symmetrical about the SN centre of mass. This is attributed to the effect of the finite light travel time across the CSM [24]. As the initial ionising flash from the supernova propagated across the CSM, it took longer for the resulting nebular emission to reach us from the far side. This allowed confirmation that the CSM really was expanding. Making some simple assumptions, the echo geometry indicates that the CSM extended to at least ~2100 AU.

The narrow [OIII] 5007 Å line persisted for at least a year [28]. Assuming a maximum ejecta velocity of 10,000 km/s [18,24], it can be deduced that the unshocked CSM must have extended to at least 2000 AU, which is consistent with the lower limit derived from the echo interpretation. In fact, later observations have shown that the CSM extended to at least 3200 AU (see below). From the intensity ratio of [OIII] (4959 Å + 5007 Å) to [OIII] 4363 Å, Fassia et al. [24] infer a wind density of at least 1.5×10^6 cm^{-3}, implying a CSM mass exceeding 0.005 $M_\odot$, and a mass-loss rate exceeding around 2×10^{-5} $M_\odot$/yr. This is consistent with the radio/X-ray estimate of around 10^{-4} $M_\odot$/yr [30].

The behaviour of the allowed H I, He I CSM lines was more complex. Not only did they exhibit asymmetric P Cygni profiles, but there were clearly two velocity components. The slower component is attributed to the same origin as the forbidden lines *viz.* the photo-ionised, unaccelerated CSM. The profile of this component was probably a combination of emission from the recombination cascade together with a classical P Cygni line due to scattering from the populated excited levels (resulting from the recombination cascade). The broad absorption component is more difficult to explain. It extends to a velocity of around 350 km/s which is too fast for a red supergiant wind. It may be that, as in the case of SN 1987A, the SN 1998S progenitor went through a fast-wind phase prior to explosion [24]. An alternative explanation is that CSM close to the supernova was accelerated by photospheric photons, or by relativistic particles from the ejecta/CSM shock [26]. Another possibility is that the faster component arose in shocked clumps within the CSM wind [26].

2.4   Bolometric Light Curve

SN 1998S was exceptionally luminous, reaching a de-reddened $M_B = -19.6$ [16]. This is around 10 times the typical luminosity of a type II SN. The excellent coverage achieved in the optical and IR allowed Fassia et al. [16] to examine the bolometric light curve. Both blackbody and UVOIR fits indicate that the total energy radiated in the first 40 days exceeded 10^{50} ergs, which is again 10 times the typical value for type II SNe. Between days 90 and 130, the bolometric light curve is well-reproduced by the radioactive decay luminosity of 0.15 $M_\odot$ ^{56}Ni. However, by this era the ejecta/CSM shock energy must also have contributed a minor contribution.

2.5   Polarisation

Spectropolarimetry by Leonard et al. [18] and Wang et al. [25] indicate asymmetry in the material responsible for the observed radiation. Leonard et al. favour a highly flattened CSM, with possibly some asymmetry in the ejecta. In contrast, Wang et al. favour ejecta asymmetry as the main cause of the polarisation.

3   SN 1998S at Late Times

Most of the observed features described in the previous section can be attributed to the interaction of the supernova with a pre-existing, dusty, possibly flattened CSM. SN 1998S remained observable from X-rays to radio for over 3 years [30,39]. This persistence was due to the ongoing conversion of the SN kinetic energy to radiation via the ejecta/shock interaction. To obtain further insights into this phenomenon, regular observations continued during this phase. In this section, I shall consider two aspects of this *viz.* the IR emission and the nature and evolution of the line profiles.

3.1   The Infrared Spectral Energy Distribution at Late Times

Infrared monitoring of SN 1998S continued at UKIRT up to day 1191 [39]. Observations were extended as far as the M-band (4.7 μm). Other than SN 1987A, this is the only time that such longwave IR radiation has been detected from a supernova. The IR excess persisted throughout this period. Between days 326 and 819, plausible blackbody fits to the de-reddened $HKL'M$ photometry (1.6–4.7 μm) are obtained. (We exclude the J-band to avoid contaminating the analysis with the very strong He I 1.083 μm emission.) The derived temperature and velocity declined from around 1400 K and 4000 km/s on day 326, to 930 K and 2000 km/s on day 819. However, for the latest photometry (days 1042 and 1191) it was not possible to achieve a single-temperature fit. On day 328, IR spectroscopy to 2.5 μm was acquired at UKIRT. This revealed that, while there was a small contribution due to Paschen α, the IR excess was due primarily to a smooth continuum rising to longer wavelengths. We conclude that the late-time IR excess was due to thermal emission from warm dust.

We can now pose two, possibly connected, questions: what powered the IR emission from the dust, and where was the dust located? The total energy emitted by the dust in the 1.6–4.7 μm region between days 300 and 1200 was about 10^{49} ergs. This is a factor of 10 more than could be supplied by the decay of the daughter products of 0.15 M$_\odot$ ^{56}Ni over the same period. We can therefore immediately rule out radioactivity as the source of the IR energy. There are two other possible energy sources. We know that the energy of the early light curve amounted to $\sim 10^{50}$ ergs. Thus, one possibility is that 10% of this was channeled into the IR emission via an IR echo from CSM dust. On the other hand, it is likely that of order 10^{51} ergs was stored in the kinetic energy of the ejecta. It would take only 1% of this to account for the IR emission, through the heating of either pre-existing (CSM) or newly-condensed (ejecta) dust. At 130 days, the huge L'-band flux argues strongly in favour of emission from pre-existing dust *i.e.* an IR echo. However, such an argument is less convincing at the later times being considered here. The blackbody fits produce velocities of 4000–2000 km/s which could, just conceivably, have arisen from dust condensation in the ejecta. To try to distinguish between the IR echo and dust condensation scenarios, we now examine the line profiles at late times.

3.2   Hα and He I 1.083 μm Profiles

The Hα profile of SN 1998S changed quite dramatically between day 97 and day 1086. On day 97 the profile had the form of a broad, steep-sided, fairly symmetrical line spanning ± 7000 km/s across the base [24]. This appearance persisted to at least day 140 [18]. However, by the time the supernova was recovered in the second season, the shape was remarkably different. The day 240 (relative to our adopted zero epoch) spectrum of Gerardy et al. [17] shows that the profile had developed a triple-peak structure, comprising a central peak close to the rest-frame velocity, and two outlying peaks at, respectively, ± 4500 km/s. Gerardy et al. suggest that the outermost peaks could have been produced by an emission zone having a ring or disk structure seen nearly edge-on, and resulting from the SN shockwave collision with the disk/ring. Following Chugai & Danziger [40] they also suggest that the central peak might have been due to shocked wind clouds. Spectra obtained on days 276 [17] and 288 [39] show a remarkable fading of the central and redshifted peaks with respect to the blueshifted peak. Subsequent spectra to day 640 show a continuation of this trend [ref. 39, R. Fesen private communication, A. Filippenko private communication]. In addition, the strong blue peak shifted in velocity to -3500 km/s, presumably due to a slowing of the shock as it encountered an increasing mass of CSM. Finally, when the supernova was recovered in the 4th season on day 1086, it was found that the profile had undergone another dramatic change [39]. While the blue-shifted peak at about -3500 km/s persisted, the central peak had grown in relative strength to about twice the height of the blue peak. An explanation for this latest behaviour is still being investigated. IR (J-band) spectra were obtained at UKIRT between days 225 and 1185 [39]. IR spectra were also acquired by Gerardy et al. on days 276 and 370 [17]. The form and evolution of the strong He I 1.083 μm profile was very similar to that of Hα. The Hα and He I 1.083 μm lines persisted to >1100 days by which time their extreme blue limbs were still at a velocity of $\sim$5000 km/s. This indicates that the CSM must have extended to >3200 AU.

3.3 Source of the IR Emission and Location of the Dust

The relatively sudden fading of the central and redshifted components of the H I, He I lines immediately suggests dust condensation in the ejecta. This could have caused obscuration of the central and receding regions. A similar effect was observed in the ejecta line profiles of SN 1987A [12,41] and, more recently, in the type IIP SN 1999em [42]. The presence of CO emission from SN 1998S as early as 130 days lends credence to the dust condensation scenario. Moreover, the effect is comparably strong at 0.66 μm and 1.08 μm, suggesting that the dust quickly became optically thick. It is difficult to see how pre-existing dust in the CSM could have produced such an effect. Indeed, pre-existing dust would probably have been evaporated by the initial flash out to a radius of at least several thousand AU [43]. On the other hand, there may actually have been two dust zones present. In this more complicated scenario, the IR emission would be due to an IR echo of the SN peak luminosity from pre-existing dust in the CSM. The line-profile obscuration, however, would be due to possibly cooler dust condensing in the ejecta.

We note an interesting coincidence. Throughout the second year, the magnitude and evolution of the velocities of the blue-shifted H I, He I peaks were similar to those derived from the blackbody fits to the IR fluxes. It has been recognised for many years (e.g. ref. 31) that the interaction of the supernova ejecta with a dense CSM will produce outer and reverse shocks. When radiative cooling is important at the reverse shock front, the gas undergoes a thermal instability, cooling to $\sim$10,000 K, thus forming a dense, relatively cool zone - the 'cool dense shell' or CDS. Line emission from low-ionisation species in the CDS will be produced [32]. We believe that this emission was responsible for the blueshifted and redshifted peaks of the Hα and He I line profiles. An exciting possibility, which still requires further study, is that dust may have formed in the CDS at the ejecta/wind interface. If cooling in the outer layer of the CDS, shielded from the reverse shock X-ray/UV radiation, brought the temperature to below the condensation temperature, dust could have formed and survived there. In a similar process, suggested by Usov [44], dust may form in the colliding winds of Wolf-Rayet stars. Rayleigh-Taylor or convective instabilities [32] might have produced opaque clumps of dust, totally obscuring the central and receding parts of the supernova, while at the same time allowing some of the line radiation to escape from the approaching component of the CDS. Thus, this scenario can simultaneously account for the strong IR flux, the obscuration effect and the velocity coincidence with the line profiles.

4 Summary

The detailed study of the type IIn SN 1998S indicates that it probably arose from a massive, RSG progenitor having a large (>3000 AU), dusty circumstellar disk. The excess IR emission at early times was due to an IR echo from this disk. At late times the origin of the strong IR emission is less clear. It may be that a 'double-dust' scenario applies where the line obscuration was due to dust condensation in the ejecta, while the IR emission arose from an IR echo from the dusty CSM. Alternatively, dust condensation in the cool dense shell may account for both the line obscuration and the IR emission. More detailed modelling of the data will be required in order to test this 'single-dust' scenario.

References

1. F. Cernuschi, F.R. Marsicano, I. Kimel: Annales d'Astrophysique **28**, 860 (1965)
2. F. Cernuschi, F. Marsicano, S. Codina: Annales d'Astrophysique **30**, 1039 (1967)
3. F. Hoyle, N.C. Wickramasinghe: Nature **226**, 62 (1970)
4. R.D. Gehrz: 'Sources of Stardust in the Galaxy'. In: *Interstellar Dust, IAU Symposium 135*, ed. by L.J. Allamandola, A.G.G.M. Tielens (Kluwer, Dordrecht) (1989) p. 445
5. A.G.G.M. Tielens: 'Carbon Stardust: From Soot to Diamonds'. In: *Carbon in the Galaxy, NASA Conference Publication 3061*, ed. by J.C. Tarter, S. Chang, D. De Frees (NASA) (1990) p. 59
6. E. Dwek: ApJ **501**, 643 (1998)
7. L.B. Lucy, I.J. Danziger, C. Gouiffes, P. Bouchet: 'Dust Condensation in the Ejecta of SN 1987A'. In: *Structure and Dynamics of the Interstellar Medium, Proceedings of IAU Colloq. 120, Granada, Spain, April 17–21, 1989* ed. by G. Tenorio-Tagle, M. Moles, J. Melnick (Springer-Verlag, Berlin) (1989) p. 164
8. S.H. Moseley et al.: Nature (**340**, 697 (1989)
9. P.A. Whitelock et al.: MNRAS **240**, 7P (1989)
10. D.H. Wooden: Observations and interpretation of the infrared emission from supernova 1987A. PhD thesis, University of California at Santa Cruz (1989)
11. N.B. Suntzeff, P. Bouchet: AJ **99**, 650 (1990)

12. I.J. Danziger, L.B. Lucy, P. Bouchet, C. Gouiffes: 'Molecules Dust and Ionic Abundances in Supernova 1987A'. In: *Supernovae. The Tenth Santa Cruz Workshop in Astronomy and Astrophysics, July 9-21, 1989, Lick Observatory.* ed. by S.E. Woosley (Springer-Verlag, New York) (1991) p. 69
13. W.P.S. Meikle et al.: MNRAS **261**, 535 (1993)
14. P.F Roche, D.K. Aitken, C.H. Smith: MNRAS **261**, 522 (1993)
15. D.V. Bowen, K.C. Roth, D.M. Meyer, J.C. Blades: ApJ **536**, 225 (2000)
16. A. Fassia et al.: MNRAS **318**, 1093 (2000)
17. C.L. Gerardy, R.A. Fesen, P. Höflich, J.C. Wheeler: AJ **119** 2968 (2000)
18. D.C. Leonard, A.V. Filippenko, A.J. Barth, T. Matheson: ApJ **536**, 239 (2000)
19. E.J. Lentz et al.: ApJ **547**, 406 (2000)
20. Q.-Z. Liu et al.: A&A Suppl. Ser. **144**, 219 (2000).
21. B. Roscherr, B. E. Schaefer: ApJ **532**, 415 (2000)
22. G.C. Anupama, T. Sivarani, G. Pandey: A&A **367**, 506 (2001)
23. N. Chugai: MNRAS **326**, 1448 (2001)
24. A. Fassia, et al.: MNRAS **325**, 907 (2001)
25. L. Wang, A.D. Howell, P. Höflich, J.C. Wheeler: ApJ **550**, 1030 (2001)
26. N.N. Chugai et al.: MNRAS **330**, 473 (2002)
27. C.L. Gerardy et al.: ApJ **575**, 1007 (2002)
28. R.A. Gruendl, Y-H. Chu, S.D. Van Dyk, C.J. Stockdale: AJ **123**, 2847 (2002)
29. W. Li et al.: PASP **114**, 403 (2002)
30. D. Pooley et al.: ApJ **572**, 932 (2002)
31. R.A. Chevalier, C. Fransson: 'Supernova Interaction with a Circumstellar Wind and the Distance to SN 1979C'. In: *Supernovae as Distance Indicators. Harvard-Smithsonian Center for Astrophysics, Cambridge, MA, USA, September 27–28, 1984.* ed. N. Bartel (Springer-Verlag, New York) (1985) p. 123
32. R.A. Chevalier, C. Fransson: ApJ **420**, 268 (1994)
33. R.M. Catchpole, I.S. Glass: IAU Circ. **4457** (1987)
34. P. McGregor, A.R. Hyland: IAU Circ. **4468** (1987)
35. J. Spyromilio, W.P.S. Meikle, R.C.M. Learner, D.A. Allen: Nature **334**, 327 (1988)
36. J. Spyromilio, B. Leibundgut: MNRAS **283**, L89 (1996)
37. J. Spyromilio, B. Leibundgut, R. Gilmozzi: A&A **376**, 188 (2001)
38. C.L. Gerardy et al.: PASJ, submitted (2002). Also astro-ph/0207480
39. W.P.S. Meikle, A. Fassia, P. Lundqvist, T. Geballe, N. Chugai, J. Sollerman, D. Farrah: MNRAS, in preparation.
40. N.N. Chugai, I.J. Danziger: MNRAS **268**, 173 (1994)
41. J. Spyromilio, W.P.S. Meikle, D.A. Allen: MNRAS **242**, 669 (1990).
42. A. Elmhamdi et al.: these proceedings.
43. E. Dwek: ApJ **297**, 719 (1985)
44. V.V. Usov: MNRAS **252**, 49 (1991)

SN IIP 1999em:
Observations Until Dust Formation

A. Elmhamdi[1], I.J. Danziger[2], N. Chugai[3], and A. Pastorello[4]

[1] SISSA / ISAS, via Beirut 4, 34014 Trieste, Italy
[2] Osservatorio Astronomico di Trieste, via G.B. Tiepolo 11, 34131 Trieste, Italy
[3] Institute of Astronomy, Russian Academy of Sciences, Pyatnitskaya 48, 109017 Moscow, Russia
[4] Osservatorio Astronomico di Padova, vicolo dell'Osservatorio 5, 35122 Padova, Italy

Abstract. SN 1999em in NGC 1637 has attracted much attention among type II P supernovae. The widely spaced and critical observations of this object (photometry and spectroscopy) provide a unique opportunity to test our knowledge of the physics of core collapse SNe and help in understanding the different parameters that characterize the pre-SN evolution as well as some behaviour that such events may show (Bochum event, asymmetry, dust formation).

Discussion

We present photometry and spectra of SN 1999em until ~ 640 days after explosion. For more details about observational data and their analysis see Elmhamdi et al. (2002). We construct the "bolometric" light curve by integrating the flux in the UBVRI bands (Fig. 1). Comparing then the tail luminosities with SN 1987A we obtain an amount of ejected ^{56}Ni $\approx 0.022\ M_\odot$. This is a smaller value than that obtained for the prototype SN IIP 1969L ($\sim 0.07\ M_\odot$; Turatto et al. 1993) and similar to SN 1991G ($\sim 0.024\ M_\odot$; Blanton et al. 1995). Together with

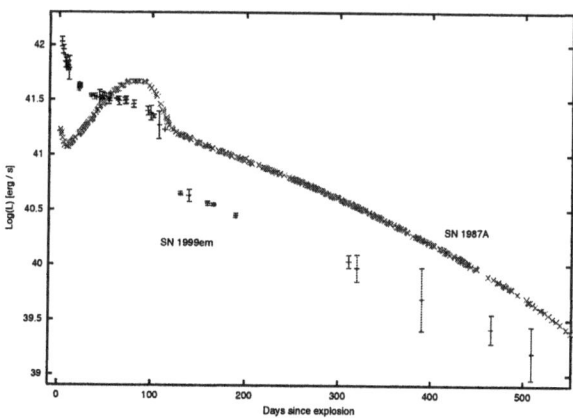

Fig. 1. The UBVRI bolometric light curve of SN 1999em with that of SN 1987A for comparison

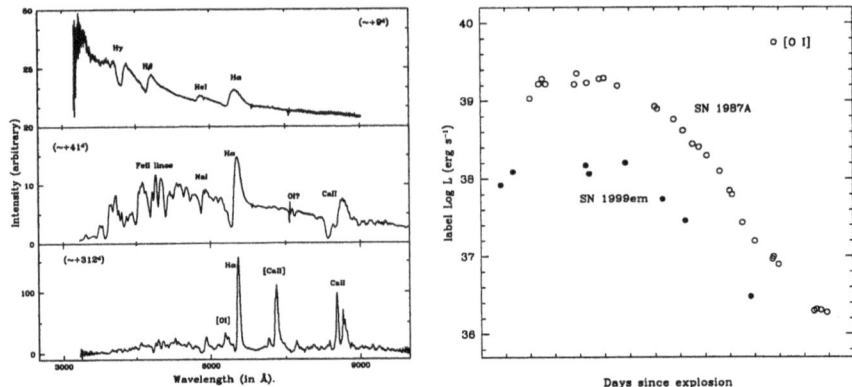

Fig. 2. Left: Sample of spectra during different phases of evolution with some line identifications. Right: The luminosity evolution of the [O I] 6300,64 Å doublet

observationally determined quantities, use is made of the type IIP SN light curve models proposed by Litvinova & Nadyozhin (1985) and the analytical model developed by Popov (1993) to recover three important parameters of the progenitor star, namely, the explosion energy (E), ejecta mass (M) and pre-supernova radius (R). The input parameters of the models are: the plateau duration (t_p), absolute V-magnitude at the plateau (M_V) and the velocity at the photospheric phase measured from weak absorption lines. We obtain results, consistent with both models: E $\sim (0.5 - 1) \times 10^{51}$ erg, R $\sim 120 - 150$ $R_\odot$ and a possible range for the main sequence mass of $12 - 14$ $M_\odot$. A supergiant $G_0 - G_5$ progenitor is therefore indicated. These results with their associated uncertainties serve as a starting point for more detailed computations and modelling.

A sample of spectra is shown in Fig. 2 (left panel) covering the different phases of evolution. The right panel reports the temporal evolution of the luminosity of the [O I]6300,64Å doublet compared to that of SN 1987A. One notes that the luminosity of [O I] doublet at the epoch of ~ 1 year is powered by the γ-ray deposition and by ultraviolet emission arising from the deposition of γ-rays in oxygen-poor material and one can write the [O I] doublet luminosity as:

$$L(6300) = \eta \frac{M_O}{M_{ex}} L(^{56}\text{Co}), \qquad (1)$$

where η is the efficiency of transformation of the energy deposited in oxygen into the [O I] doublet radiation, M_O is the mass of oxygen, M_{ex} is the "excited" mass in which the bulk of radioactive energy is deposited. We can then estimate the oxygen mass assuming similar η and M_{ex} for both SN 1999em and SN 1987A and a factor 15 lower luminosity of [O I] doublet and a factor 3.4 lower ^{56}Ni mass for SN 1999em. An oxygen mass range for SN 1987A of $1.5 - 2M_\odot$ (Fransson et al. 1993; Chugai 1994) indicates a mass range of $0.3 - 0.4M_\odot$ for SN 1999em. This corresponds to a main sequence mass of $13 - 14M_\odot$ according to the nucleosynthesis computations (Woosley & Weaver 1995), in good agreement with results derived using observational data and analytical models.

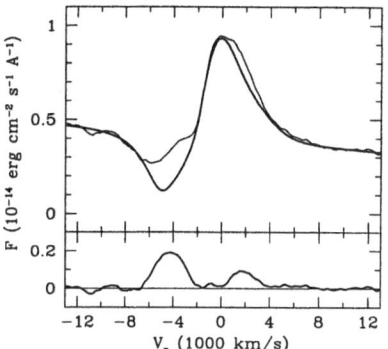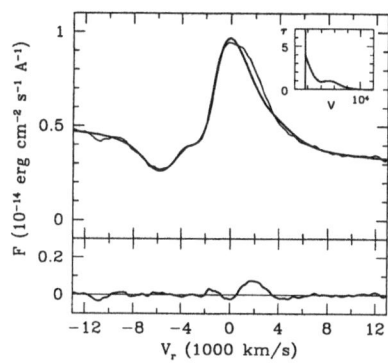

Fig. 3. The Hα on day 97. The overplotted (thick line) is a profile for a spherical model. The residual (bottom panel) shows two peaks, which may be interpreted as an evidence for the presence of bipolar excitation regions in the atmosphere. Right panel: The overplotted (thick line) is the profile for the spherically symmetric model with the non-monotonic behaviour of the optical depth shown in window (up right corner). The residual shows the excess in the red, which may indicate overexcitation in the far hemisphere

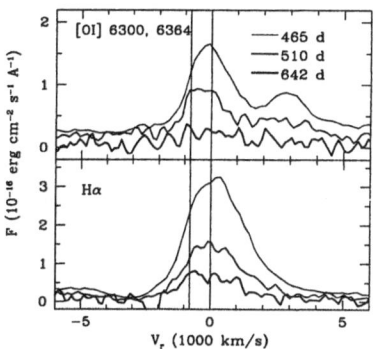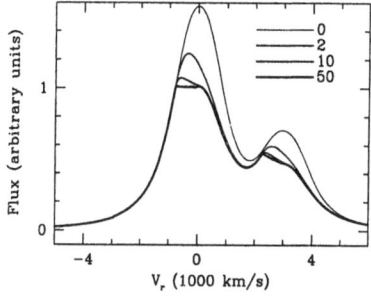

Fig. 4. Left: The oxygen [O I] doublet and Hα in the latest three epochs. Right: The effect of the opaque core on the [O I] doublet. Lines of different thickness correspond to different optical depths of the core. Note, for $\tau \gg 10$ the profile is flat-topped

In addition, as did SN 1987A at early phases, SN 1999em shows some interesting behaviour in their spectra especially for Hα. This latter shows a P-Cygni profile with blue shifted emission peaks at early times which can be understood in the context of the model developed by Chugai (1988) in the case of SN 1987A as being due to the reflection of photons by the photosphere. At later photospheric phases, the development of fine structures resembling the "Bochum event" in SN 1987A becomes evident. Two explanations are possible, a bi-polar jet model proposed by Lucy (1988) and underexcitation of hydrogen combined with ^{56}Ni asymmetry (Fig. 3).

We also point out a pronounced evolution in the [O I]6300,6364Å profile between days 465 and 510, which is also seen in the Hα profile (Fig. 4, left). We interpret this rapid change (time scale of the order 0.1) as being due to dust condensation as was the case for SN 1987A (dust formation around day 526; Danziger et al. 1989, Phillips & Williams 1991). Some clear differences between the two events in manifesting the dust formation are noted, namely, the flattening in the peak of [O I]6300Å profile for SN 1999em, while the line preserves a round-topped shape in the case of SN 1987A. We suggest this difference is due to the high optical depth in the case of SN 1999em with a velocity of the dusty sphere of about 800 km s^{-1} (Fig. 4, right), while SN 1987A had an order of unity dust-optical-depth with velocity about 1800 km s^{-1} (Lucy et al. 1989). On the other hand the earlier emergence of the dust in SN 1999em is probably related to the lower amount of ^{56}Ni which in turn implies that the condensation temperature was reached earlier in SN 1999em.

References

1. Blanton E.L. et al. 1995, AJ, 110.2868
2. Chugai N.N. 1988, SvAL, 14, 334
3. Chugai N.N. 1994, ApJ, 428, 17
4. Danziger I.J., Bouchet P., Gouiffes C., Lucy L.B. 1989, IAUC 4746
5. Elmhamdi A. et al., 2002, submitted to MNRAS.
6. Fransson C. et al. 1993, in Supernova and Supernova remnants, MCCRAY and WANG, Xian, China, Colloquium 145 p.211
7. Litvinova I.Y. & Nadozhin D.K. 1985, SVAL, 11, 145
8. Lucy L.B. 1988, in SN 1987A in the Large Magellanic Cloud, eds. Kaplan M. & Michlitsianos, A.G., Cambridge University Press, p. 323
9. Lucy L.B., Danziger,I.J., Gouiffes, G., Bouchet, P. 1989, In Structure and dynamics of the Interstellar Medium, ed. G. Tenorio-Tagle et al., IAU Colloquium No. 120 (Springer-Verlag), p. 164
10. Phillips M.M., Williams R.E., 1991, in Supernovae, ed. S.E. Woosley (New York: Springer), p. 36
11. Popov D.V. 1993, ApJ, 414, 712
12. Turatto M., Cappellaro E., Benetti S and Danziger I.J. 1993, MNRAS 265, 471
13. Woosley S.E., Weaver T.A., 1995, ApJ, 101, 181

Physical Properties of SNe IIP Derived from a Comparison of Theoretical Models with Observations

Dmitrij K. Nadyozhin[1,2]

[1] Institute of Theoretical and Experimental Physics, Moscow, Russia
[2] Max-Planck-Institut für Astrophysik, Garching, Germany

Abstract. The hydrodynamical modelling of the Type II plateau supernova light curves predicts a correlation between three observable parameters (the plateau duration Δt, the absolute magnitude M_V and photospheric velocity u_{ph} at the middle of the plateau) and three physical ones (the explosion energy E, the mass of the envelope expelled M, and presupernova radius R). The correlation is used to estimate E, M, and R for a dozen of well-observed SNe IIP. The resulting value of E varies within a factor of 5 $\left(0.5 \lesssim E/10^{51} \, \mathrm{erg} \lesssim 2.7\right)$, whereas the envelope mass turns out to be within the limits $10 \lesssim M/M_\odot \lesssim 30$. The presupernova radius is typically about $(200 - 600) \, R_\odot$.

A new method of determining SNe IIP distances is proposed. It is based on an additional assumption of a possible correlation between E and the mass of ^{56}Ni produced to power the post-plateau light curve tail by means of the ^{56}Co decay. The method is thought to work for the SNe IIP with well-observed bolometric light curves both during the plateau and radioactive tail phases.

The formulae approximating the correlation between the intrinsic supernova parameters $(E, \, M, \, R)$ and the observational ones $(M_V, \, \Delta t, \, u_{ph})$ are given by [11]:

$$\lg E = 0.135 \, M_V + 2.34 \lg \Delta t + 3.13 \lg u_{ph} - 4.205 \,, \tag{1}$$

$$\lg M = 0.234 \, M_V + 2.91 \lg \Delta t + 1.96 \lg u_{ph} - 1.829 \,, \tag{2}$$

$$\lg R = -0.572 \, M_V - 1.07 \lg \Delta t - 2.74 \lg u_{ph} - 3.350 \,, \tag{3}$$

where E is expressed in units of 10^{51} erg, M and R are in solar units, Δt in days, and u_{ph} in $1000 \, \mathrm{km \, s^{-1}}$, whereas M_V can be expressed through the apparent magnitude V by the relation:

$$M_V = V - A_{V \, \mathrm{tot}} - 5 \lg(D/1 \mathrm{Mpc}) - 25 \,, \tag{4}$$

where D is the distance to a supernova and $A_{V \, \mathrm{tot}}$ is the total absorption towards the supernova.

We have chosen 13 supernovae whose observational data are collected in Table 1. The last column gives the references which served as the main sources for the data listed in columns 2–7. The heliocentric recession velocities v_{rc} of host galaxies (column 3) were used to determine the supernova distances D. First, v_{rc} was converted into v_{220} – the recession velocity corrected for a Virgocentric

Table 1. Input data

SN	Host galaxy	v_{rc} km s^{-1}	A_{Vtot} mag	V mag	Δt days	u_{ph} km s^{-1}	Ref.
1968L	NGC 5236	516	0.219	12.0	80	4100	[12,15,22]
1969L	NGC 1058	518	0.203	13.4	100	4000	[12,15]
1986L	NGC 1559	1292	0.099	14.7	110	4000	[5]
1988A	NGC 4579	1519	0.136	15.0	110	3000	[12,15,5,13,21]
1989L	NGC 7339	1313	0.423	16.5	120	2700	[5,17]
1990E	NGC 1035	1241	1.083	16.0	120	4000	[15,5,16,2]
1991al	LEDA 140858	4572	0.318	17.0	90	6000	[5]
1992af	ESO 340-G038	6000	0.171	17.3	90	6000	[5,17]
1992am	anon 0122-04	14600	0.464	19.0	110	4800	[5,18]
1992ba	NGC 2082	1104	0.193	15.43	100	2900	[5,17]
1999cr	ESO 576-G034	6069	0.324	18.6	100	3600	[5]
1999em	NGC 1637	717	0.314	14.0	110	3000	[5-7,1,3]
1999gi	NGC 3184	592	0.65	15.0	110	2900	[14,19,10,9]

Table 2. The supernova physical properties

SN	v_{220} km s^{-1}	D Mpc	M_V mag	E 10^{51} erg	M $M_\odot$	R $R_\odot$	M_{Nio} $M_\odot$
1968L	291	4.85	−16.65	0.83	10.3	286	
1969L	766	12.77	−17.33	1.05	13.0	595	
1986L	1121	18.68	−16.76	1.56	23.5	251	0.027
1988A	1179	19.65	−16.60	0.67	14.5	452	0.080
1989L	1556	25.93	−16.69	0.57	14.5	618	
1990E	1238	20.63	−16.66	1.98	31.9	200	0.047
1991al	4476	74.60	−17.68	2.61	17.6	347	0.11
1992af	6000	100.00	−17.87	2.46	15.9	445	0.22
1992am	14600	243.33	−18.40	1.66	13.9	1321	0.36
1992ba	1096	18.27	−16.07	0.57	13.7	272	0.027
1999cr	6069	101.15	−16.75	0.90	14.5	368	0.080
1999em	743	12.38	−16.78	0.63	13.2	569	0.051
1999gi	707	11.78	−16.01	0.72	18.7	226	0.025*

*Derived for $V = 17.^m86$ at $t = 174.3$ d from [9] and Hamuy's [5] recipe (section 5.3 of his Thesis) to convert V into luminosity L.

infall of 220 km s^{-1} as provided by [8], and then the distance D was calculated as $D = v_{220}/H_0$ with $H_0 = 60$ km s^{-1} Mpc^{-1}. The resulting values of v_{220} and D are given in Table 2 together with M_V, E, M, and R calculated according to Eqs. (4), (1)–(3), respectively. We see that the expelled mass, explosion energy, and presupernova radius remain within reasonable limits: $(10 - 30)\, M_\odot$, $(0.6 - 2.6) \times 10^{51}$ erg, and $(200 - 1300)\, R_\odot$, respectively. We believe that these results can serve as a good guiding line in the elucidation of the intrinsic supernova properties by means of a tight interaction between the observational and theoretical projects.

The last column of Table 2 gives the mass of ^{56}Ni for some supernovae estimated by reducing the radioactive-tail luminosities from [5] to the distances given in column 3. The nickel mass M_{Ni0} can be found from the relation:

$$M_{Ni0} = \frac{D^2}{145} Q, \quad Q \equiv F_{41}(t) \exp\left(\frac{t}{111.3}\right), \tag{5}$$

where M_{Ni0} is in $M_\odot$, t in days and D in Mpc. The quantity $F_{41}(t)$ is the tail luminosity measured at time t in units 10^{41} erg s^{-1} under an assumption that the supernova is at the distance $D = 1$ Mpc. Equation (5) contains a single observational parameter Q that depends neither on t nor on D. Thus, it is absolutely unimportant at which t the tail luminosity is actually measured – one has only to be sure that the supernova really entered its tail phase. If the value of M_{Ni0} were known, one would easily find the distance D from Eq. (5) for the Q-value known from observations. So, we have to look for a way to estimate M_{Ni0} independently.

Let us assume now that there exists a statistically admissible correlation between the supernova explosion energy E and M_{Ni0}. Using this correlation in combination with Eqs. (1) and (4), one can find D and then E, M_{Ni0}, M, and R. To demonstrate this idea, we assume that E is proportional to the energy $E_{np \to Ni}$ released owing to the recombination of free neutrons and protons into ^{56}Ni:

$$E = \xi \, E_{np \to Ni} = 16.6 \, \xi \, M_{Ni0} = 0.1145 \, \xi \, D^2 \, Q, \tag{6}$$

where, as usual, E is in 10^{51} erg, M_{Ni0} in $M_\odot$ and D in Mpc and ξ is an

Table 3. The tail-calibrated supernova physical properties ($\xi = 1$)

SN	D Mpc	M_V mag	E 10^{51} erg	M $M_\odot$	R $R_\odot$	M_{Ni0} $M_\odot$	Q
1986L	29.67	−17.76	1.14	13.7	944	0.067	0.0113
1988A	15.21	−16.05	0.79	19.6	217	0.048	0.0299
1990E	29.16	−17.41	1.57	21.3	539	0.094	0.0161
1991al	85.31	−17.97	2.38	15.0	509	0.14	0.00286
1992af	86.45	−17.55	2.71	18.8	293	0.16	0.00317
1992ba	19.85	−16.25	0.53	12.4	346	0.032	0.0119
1999em	11.08	−16.54	0.68	15.0	414	0.041	0.0485
1999gi	14.53	−16.46	0.63	14.5	411	0.038	0.0259

adjustable parameter. Inserting E from Eq. (6) and M_V from Eq. (4) into Eq. (1) and solving for D, we obtain:

$$\lg D = -0.374 \lg(\xi \, Q) + 0.0505 \, (V - A_{V\,tot}) + 0.875 \lg \Delta t + 1.17 \lg u_{ph} - 2.482, \tag{7}$$

where D is in Mpc, Δt in days, and u_{ph} in 1000 km s^{-1}. The resulting distances D, M_{Ni0}, and all others values in question are given in Table 3 for 8 supernovae selected from Table 2 with their Q-values (the last column of Table 3) being derived from observed tail-luminosities presented in [5], excluding SN 1999gi for which Q was estimated for the observational data published in [9]. The differences in M_{Ni0} between Tables 2 and 3 are due to the differences in corresponding distances D.

Comparing the distances D from Table 3 (designated as D_{tail} hereafter) with D from Table 2 we can see a wonderful coincidence. For 6 SNe from the list, D_{tail} coincide with D within the limits $\pm 23\%$ and only for two SNe (1986L and 1990E) D_{tail} are by a factor of 1.5 larger than D. The two SNe differ from others in Table 2 by having the largest masses M. In Table 3, their masses already are not so prominent. The coincidence seems to be really remarkable because D_{tail} comes from the intrinsic supernova properties, and have nothing to do with the recession velocities and Hubble constant.

Recently, an important project has been started. Its ultimate aim is to discover the supernova progenitors (presupernovae) or at least to impose conclusive constraints on their properties by inspecting the prediscovery images of stellar environment nearby the supernovae. The first (distance-dependent!) upper mass limits were obtained for the progenitors of the SNe 1999em and 1999gi [19,20,9]. Our masses M for these SNe (Tables 2, 3) are not in a serious confrontation with these observations *as long as the corresponding distances are about* 11 Mpc *or larger*.

We have to note that the results, obtained here with the aid of the correlations demonstrated by our hydrodynamical models and approximated by Eqs. (1)–(3), should be considered carefully as far as it concerns individual supernovae: the corresponding rms deviations for this approximation are about $\pm 30\%$ for E and M. The errors in the observational values (especially in u_{ph} and Δt) could modify E and M by another factor of 1.3. So, it seems reasonable to assume an uncertainty factor of ~ 1.5 for the *individual* values of E and M in Tables 2 and 3. As to the presupernova radius R, it can be overestimated by a factor of 2–3 for the SNe with large values of M_{Ni0} [4], such as 1991al, 1992af, and perhaps 1992am; this can happen because Eqs. (1)–(3) do not take into account the radioactive heating consistently.

Inspecting Tables 2 and 3, one can observe that the explosion energy E does not correlate with the presupernova mass M (the difference of $1.5 - 2\, M_\odot$ between a presupernova mass and expelled envelope mass M is unimportant here). For nearly the same mass $M \approx 15\, M_\odot$, the explosion energy E can differ by a factor of 4 – compare, for instance, SN 1991al with SN 1999em. It is hard to attribute this result only to the statistical errors mentioned above. So, it is reasonable to expect another parameter to be involved in the supernova mechanism. It could be the rotation (and magnetic fields) inherited by the collapsing stellar core or nonspherical jet-like perturbations of a random nature arising from the macroscopic neutrino-driven advection below the accretion shock, which could launch the outgoing blast wave earlier when the recombination nuclear energy stored in a hot neutron-proton gas is not yet as large as it should be in the case of spherical symmetry. If this is correct one may expect that the asphericity of the explosion *anticorrelates* with the explosion energy.

In conclusion we emphasize the necessity of constructing a new grid of the SN IIP hydrodynamic models based on current evolutionary presupernova models and taking into account ^{56}Ni as an additional parameter in a consistent way. Such a grid would allow to create more precise analytic approximations for a number of correlations between supernova physical parameters and observable properties of the SN IIP.

Acknowledgements

It is a pleasure to express my deep gratitude to the Max-Planck-Institut für Astrophysik for hospitality and financial support. The work was supported also by the Swiss National Science Foundation and the Russian Foundation for Basic Research (Project 00-02-17230).

I am grateful to G.A. Tammann and W. Hillebrandt for fruitful and encouraging discussions and B. Parodi for acquainting me with the FORTRAN code converting v_{rc} into v_{220}.

I thank M. Hamuy for sending a copy of his Thesis where a good portion of the observational data used in this work was taken from.

References

1. E. Baron, D. Branch, P.H. Hauschildt et al.: ApJ **545**, 444 (2000)
2. S. Benetti, E. Cappellaro, M. Turatto et al.: A&A **285**, 147 (1994)
3. A. Elmhamdi, I.J. Danziger, N. Chugai et al.: MNRAS (2002) in press
4. E.K. Grassberg, D.K. Nadyozhin: AZh **63**, 1137 (1986) [Sov. Astron. **30**, 670 (1986)]
5. M. Hamuy: Type II Supernovae as Distance Indicators. Ph.D. Thesis, University of Arizona (2001)
6. M. Hamuy, P.A. Pinto, J. Maza et al.: ApJ **558**, 615 (2001)
7. M.P. Haynes, L. van Zee, D.E. Hogg et al.: AJ **115**, 62 (1998)
8. R.C. Kraan-Korteweg: A&AS **66**, 255 (1986)
9. D.C. Leonard, A.V. Fillipenko, W. Li et al.: AJ **124**, 2490 (2002)
10. W. Li, A.V. Fillipenko, S.D. Van Dyk et al.: PASP **114**, 403 (2002)
11. I.Yu. Litvinova, D.K. Nadyozhin: Sov. Astron. Lett. **11**, 145 (1985)
12. E. Patat, R. Barbon, E. Cappellaro, M. Turato: A&AS **98**, 443 (1993)
13. P. Ruiz-Lapuente, M. Kidger, R. Lopez, R. Canal: AJ **100**, 782 (1990)
14. E.M. Schlegel: ApJ **556**, L25 (2001)
15. B.P. Schmidt, R.P. Kirshner, R.G. Eastman: ApJ **395**, 366 (1992)
16. B.P. Schmidt, R.P. Kirshner, R. Schild et al.: AJ **105**, 2236 (1993)
17. B.P. Schmidt, R.P. Kirshner, R.G. Eastman et al.: ApJ **432**, 42 (1994)
18. B.P. Schmidt, R.P. Kirshner, R.G. Eastman et al.: AJ **107**, 1444 (1994)
19. S.J. Smartt, G.F. Gilmore, N. Trentham et al.: ApJ **556**, L29 (2001)
20. S.J. Smartt, G.F. Gilmore, N. Trentham et al.: ApJ **565**, 1089 (2002)
21. M. Turatto, E. Cappellaro, S. Benetti, I.J. Danziger: MNRAS **265**, 471 (1993)
22. R. Wood, P.J. Andrews: MNRAS **167**, 13 (1974)

The Properties of SN 2002ap
and Other Type Ic Hypernovae

Paolo A. Mazzali[1], Ken'ichi Nomoto[2], Jinsong Deng[2], Keiichi Maeda[2], and Yulei Qiu[3]

[1] INAF – Osservatorio Astronomico di Trieste, Trieste, Italy
[2] Astronomy Dept., University of Tokyo, Tokyo, Japan
[3] National Astronomical Observatory, Chinese Academy of Science, Beijing, China

Abstract. The recent Type Ic SN 2002ap is spectroscopically a member of the group of Type Ic 'hypernovae' including SN 1998bw/GRB980425, SN 1997ef, and SN1997dq, the properties of which are reviewed. Analysis of the spectra and of the light curve of SN 2002ap suggest that it ejected less mass and exploded with a smaller kinetic energy than the other hypernovae, but yet the kinetic energy was significantly larger (4-10 times) than in normal core-collapse SNe confirming the hypernova classification. The progenitor of SN 2002ap may have been less massive than those of other hypernovae, suggesting a correlation between progenitor mass and SN properties.

1 Introduction

The discovery of a supernova (SN 1998bw) in the error box of a Gamma-Ray Burst (GRB980425, [5]) was the first evidence of a possible relation between the enigmatic GRBs and a well-studied astrophysical phenomenon.

SN 1998bw was of Type Ic and so it originated from the explosion of the CO core of a massive star, stripped of its H and He envelopes. However, SN 1998bw was different from previously known SNe Ic: it was a much more energetic event (explosion kinetic energy $E_K \sim 3 \cdot 10^{52}$ erg), as indicated by the very broad absorption lines, which reach velocities of ~ 30000 km s^{-1} [10]. Because of the broad line features and the high E_K, SN 1998bw was called a 'hypernova'. Also, SN 1998bw ejected a mass of $\sim 10 M_\odot$, much more than typical SNe Ic, indicating a progenitor mass of $\sim 40 M_\odot$. Finally, SN 1998bw synthesised $\sim 0.5 M_\odot$ of radioactive ^{56}Ni, giving rise to a bright, extended light curve.

Further studies revealed that the ejecta of SN 1998bw are probably rather asymmetric [19], although the degree of this asymmetry is controversial [9,14]. Asymmetric models have been able to reproduce some of the observed features, such as the light curve or the late-time spectra. The latter are peculiar, in that the [Fe II] lines are broader than the [O I] 6300Å line [19]. The estimate of E_K is smaller in the asymmetric models ($2 - 10 \cdot 10^{51}$ erg), because the SN was observed near the direction of the jet which probably also caused the GRB.

Although SN 1998bw was certainly an exceptional case, other peculiar SNe Ic were later discovered or recognised. They have in common very broad absorption features indicative of a high ejecta velocity and hence of a large E_K. Analysis has shown that their properties are not constant indicating a range of progenitors.

One particularly well studied object is SN 1997ef. Its properties were similar to those of SN 1998bw although not as extreme. SN 1997ef produced $\sim 0.15\,M_\odot$ of ^{56}Ni and ejected $\sim 8\,M_\odot$ of material with $E_K \sim 8 \cdot 10^{51}$ erg [11]. The progenitor probably had a mass of $\sim 35\,M_\odot$. Further spectroscopic analysis [17] revealed that the observed line widths can only be reproduced if a small amount of matter is moving at $v \sim 30000\,\mathrm{km\,s^{-1}}$, which is not as fast as in SN 1998bw but still makes the E_K estimate increase to $\sim 2 \cdot 10^{52}$ erg. A similar result was obtained also for SN 1998bw [3].

In the case of SN 1997ef, there is some evidence that 1D explosion models may not reproduce the observations perfectly. This is derived from the presence of significant line absorption at low velocity ($\sim 2000\,\mathrm{km\,s^{-1}}$). This is well below the supposed cutoff in the ejecta distribution in velocity which results from the formation of a compact remnant (probably a black hole for both SN 1998bw and SN 1997ef). However, the inconsistent relative width of the Fe and O lines is not seen in the late-time spectra of SN 1997ef, which may suggest that either this event was not very asymmetric or, if it was, that it was not observed near the jet direction. This is consistent with the apparent lack of an associated GRB (a possible correlation is only weak [31]).

Other SNe Ic have been observed that are very similar to SN 1997ef. One example is SN 1997dq [15], whose earliest available spectrum looks very similar to those of SN 1997ef at an epoch of 3–4 months. Unfortunately, early observations for SN 1997dq are not available. However, nebular spectra of SN 1997dq are available at epochs when spectra of SN 1997ef are not. This allowed us to estimate the ^{56}Ni mass of both SNe to be $\sim 0.17 M_\odot$ [21]. Single spectra of SN 1999ey [7] and SN 2002bl [1] suggest that these objects are also similar to SN 1997ef.

It appears, as we have discussed, that the properties of these energetic, broad-lined SNe Ic are varied, as confirmed by the latest example of a SN Ic hypernova.

2 SN 2002ap

The Type Ic SN 2002ap was discovered in M74 on 2002 January 30 [8]. It was immediately recognised as a hypernova from its broad spectral features [6 and references therein]. Luckily, the SN was discovered very soon after it exploded: the discovery date was Jan 29, while the SN was not detected on Jan 25 [22].

Figure 1a shows the maximum-light spectra of SN 2002ap, of the hypernovae SN 1998bw and SN 1997ef and of the normal SN Ic 1994I, and Fig. 1b shows their V-band light curves. If line width is the distinguishing feature of a hypernova, then clearly SN 2002ap is a hypernova, as its spectrum resembles that of SN 1997ef much more than that of SN 1994I. Line blending in SN 2002ap and SN 1997ef is comparable. However, some individual features that are clearly visible in SN 1994I but completely blended in SN 1997ef can at least be discerned in SN 2002ap (e.g. the Na I–Si II blend near 6000 Å and the Fe II lines near 5000 Å). Therefore, spectroscopically SN 2002ap appears to be located just below SN 1997ef in a 'velocity scale', but far above SN 1994I, as seems to be con-

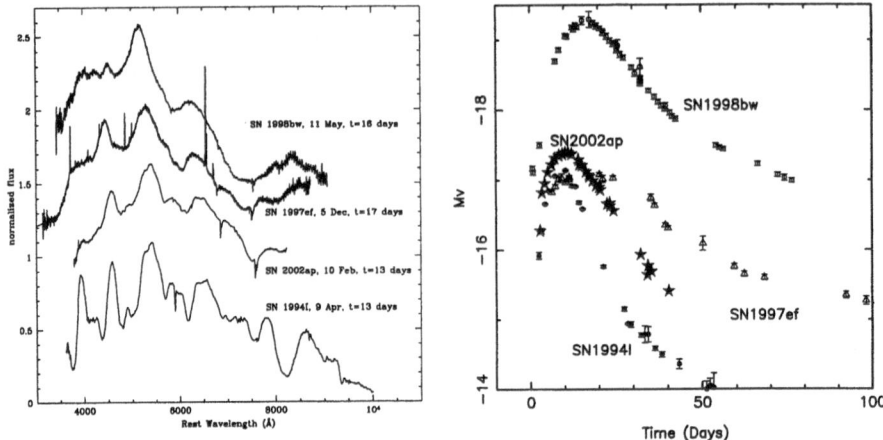

Fig. 1. a *(left)* The near-maximum spectra and **b** *(right)* the absolute V-band light curves of Type Ic SNe and hypernovae: SN 1998bw, SN 1997ef, SN 2002ap and SN 1994I [20]

firmed by the properties of the light curve. The light curve of SN 2002ap peaks earlier than both hypernovae 1998bw and 1997ef, but later than the normal SN 1994I, suggesting an intermediate value of the ejecta mass M_{ej}.

Using a distance to M74 of 8 Mpc ($\mu = 29.5$ mag; [26]), and a combined Galaxy/M74 reddening of $E(B - V) = 0.09$ mag [29], the absolute magnitude of SN 2002ap is $M_V = -17.4$. This is comparable to SN 1997ef and fainter than SN 1998bw by almost 2 mag. Since peak brightness depends on the ejected ^{56}Ni mass, SN 2002ap, SN 1997ef and SN 1994I appear to have synthesized similar amounts of ^{56}Ni. Estimates were $\sim 0.07\,M_\odot$ for SN 1994I [23] and $0.13\,M_\odot$ for SN 1997ef [17]. The ^{56}Ni mass for SN 2002ap is estimated to be $\sim 0.07\,M_\odot$, which is similar to that of normal core-collapse SNe such as SN 1987A and SN 1994I.

The spectral evolution of SN 2002ap follows closely that of SN 1997ef at a rate about 1.5 times faster. The model which best reproduces the spectra and the light curve of SN 2002ap has $M_{ej} = 2.5$–$5\,M_\odot$ and $E_K = 4 - 10 \cdot 10^{51}$ erg (see Fig. 2a). Both M_{ej} and E_K are much smaller than in SN 1998bw and SN 1997ef (but they could be larger if a significant amount of He is present).

Although SN 2002ap appears to lie between classical core-collapse SNe and hypernovae, it should be regarded as a hypernova because its kinetic energy is distinctly higher than for classical core-collapse SNe. In other words, the broad spectral features that characterize hypernovae are the results of a high kinetic energy. Also, SN 2002ap was not more luminous than normal core-collapse SNe. Therefore brightness alone should not be used to discriminate hypernovae from normal SNe, while the criterion should be a high kinetic energy accompanied by broad spectral features. Further examples of hypernovae are necessary in order to establish whether a firm boundary between the two groups exists.

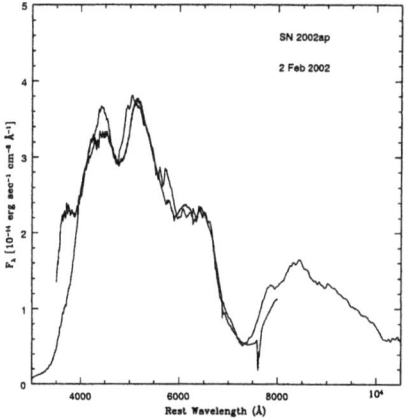 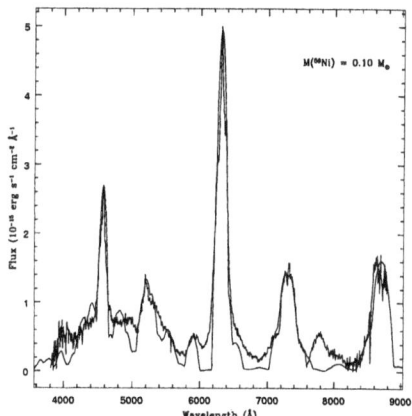

Fig. 2. a *(left)* A spectrum of SN 2002ap observed on Feb. 2, 2002, compared to a synthetic spectrum obtained with our Monte Carlo code [18,16].
b *(right)* A spectrum of SN 2002ap observed at Beijing Astronomical Observatory on July 11, 2002, compared to a synthetic NLTE spectrum computed for a ^{56}Ni mass of 0.1 $M_\odot$ and a total mass of ≈ 1.3 $M_\odot$below 6000 km s^{-1}

For the derived values of E_K, $M_{\rm ej}$, and $M(^{56}{\rm Ni})$ we can constrain the progenitor's main-sequence mass $M_{\rm ms}$ and the remnant mass $M_{\rm rem}$. The model which is most consistent with our estimates is one with a C+O core mass $M_{\rm CO} \approx 5\,M_\odot$, $M_{\rm rem} \approx 2.5\,M_\odot$and $E_K = 4 \cdot 10^{51}$ erg. Such a core forms in a He core of mass $M_\alpha \approx 7\,M_\odot$, corresponding to $M_{\rm ms} \approx 20 - 25\,M_\odot$. The non-detection of the progenitor in pre-discovery images of M74 [27], suggests that the progenitor lost mass through binary interaction. The spiral-in of a companion star is probably required for the progenitor to lose its hydrogen and some (or most) of its helium envelope before collapse [24]. This would suggest that the progenitor was in a state of high rotation. A high rotation rate and/or envelope ejection may also be necessary conditions for the birth of a hypernova.

SN 2002ap was not apparently associated with a GRB. This is actually not so surprising, since E_K of SN 2002ap is about a factor of 5-10 smaller than that of SN 1998bw, as also indicated by the weak radio signature [2]. The available data show no clear signature of asymmetry except perhaps for some polarisation [12,13,32], which is smaller than that of SN 1998bw. This suggests that the degree of asphericity is smaller in SN 2002ap and that the possible 'jet' may have been weaker making GRB generation more difficult. It is also possible that a GRB did occur, but that it was not observed because of an unfavourable orientation with respect to the line-of-sight. Asymmetric models indeed show more 'normal' properties away from the jet axis.

Nebular spectra, which developed at an epoch of about 4 months, do not show the peculiar narrow [O I] 6300Å line as in SN 1998bw, but they do show a narrow core in both [O I] and Mg II] 4571Å, which may be interpreted as a signature of asymmetry. The mass of ^{56}Ni estimated from modelling these spectra with a 1D NLTE code (Fig. 2b) is $\approx 0.1\,M_\odot$.

3   Properties of Hypernovae

The estimated progenitor mass and explosion energy of SN 2002ap are both smaller than those of previous 'hypernovae' such as SN 1998bw and SN 1997ef, but larger than those of normal core-collapse SNe. There appears to be a correlation between M_{ms} and E_K. Figure 3a shows the relation as obtained from fitting the optical light curves and spectra for various SNe/hypernovae [25]. Hypernovae appear to lie at the massive end of SN progenitors.

Our analysis suggests that the E_K may also be related to M_{ms}. Figure 3b shows $M(^{56}\text{Ni})$ against M_{ms} [25]. The amount of ^{56}Ni appears to increase with increasing M_{ms} of the progenitor except for SN II 1997D [30].

Further observational examples are needed to confirm these relations. In particular, it is unclear what fraction of massive stars with $M \gtrsim 20\,M_\odot$ explode energetically. Massive core-collapse SNe with either a normal explosion energy (e.g., SN 1984L [28]) or a very small one (SN 1997D [30]) also appear to exist.

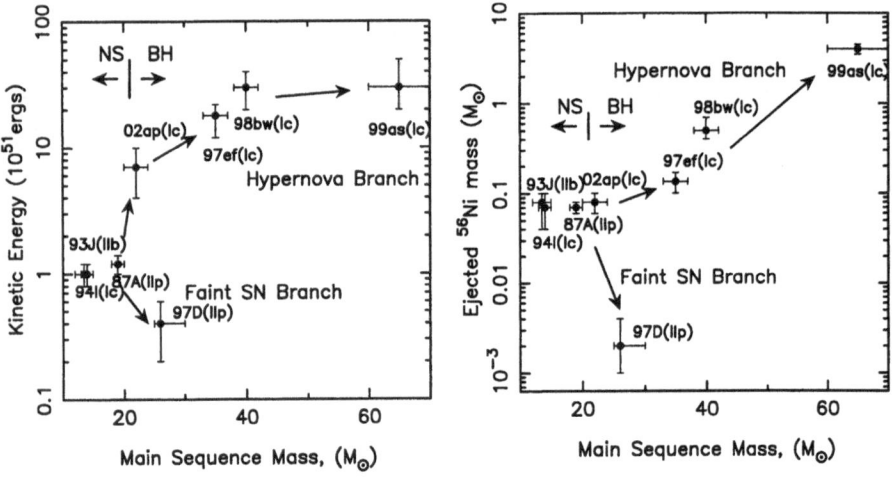

Fig. 3. a *(left)* Explosion energies and **b** *(right)* ejected ^{56}Ni mass against main sequence mass of the progenitors for several core collapse supernovae/hypernovae [25].

This trend might be explained as follows. Stars with $M_{ms} \lesssim 25\,M_\odot$ form a neutron star, producing $\sim 0.08 \pm 0.03\,M_\odot$ of ^{56}Ni (e.g. SN IIb 1993J, SN Ic 1994I and SN 1987A, although SN 1987A may be a borderline case between neutron star and black hole formation). Stars with $M_{ms} \gtrsim 25\,M_\odot$ form a black hole (e.g. [4]); whether they become hypernovae or faint SNe may depend on the angular momentum in the collapsing core. For SN 1997D, because of the large gravitational potential, the explosion energy was so small that most ^{56}Ni fell back onto a compact remnant; such fall-back might cause the collapse of the neutron star into a black hole. The core of SN II 1997D might not have had a large angular momentum: if the progenitor had a massive H-rich envelope, core angular momentum may have been transported to the envelope possibly via a magnetic field. Hypernovae such as SN 1998bw, SN 1997ef and SN 2002ap may have had rapidly rotating cores owing possibly to the spiraling-in of a companion star in a binary system. Mass-loss rate and binarity certainly influence the outcome.

The result that the mass of ^{56}Ni synthesized in core-collapse supernovae may be a function of the main-sequence mass of the progenitor star is important for the chemical evolution of galaxies.

References

1. M. Armstrong, A.V. Filippenko et al.: *IAU Circ.* **7845** (2002)
2. E. Berger, S.R. Kulkarni, D.A. Frail: *IAU Circ.* **7817** (2002)
3. D. Branch: Supernovae and Gamma-Ray Bursts (Proceedings of the STScI Symposium, Baltimore, USA), eds. M. Livio, N. Panagia, K. Sahu (Cambridge), 96 (2001)
4. E. Ergma, E.P.J. van den Heuvel: A&A, **331**, L29 (1998)
5. T. Galama et al.: Nature **395** 670 (1998)
6. A. Gal-Yam, E.O. Ofek, O. Shemmer: MNRAS, **332**, L73 (2002)
7. P. Garnavich, S. Jha, R. Kirshner: *IAU Circ.* **7066** (1998)
8. Y. Hirose: *IAU Circ.* **7810** (2002)
9. P. Höflich, J.C. Wheeler, L. Wang: ApJ **521** 179 (1999)
10. K. Iwamoto, P.A. Mazzali et al.: Nature **395**, 672 (1998)
11. K. Iwamoto et al.: ApJ **534**, 660 (2000)
12. K.S. Kawabata, D. Jeffery, G. Kosugi, T. Sasaki, et al.: ApJ **580**, L39 (2002)
13. D. Leonard, A.V. Filippenko, R. Chornock, R. Foley: PASP **114**, 1333
14. K. Maeda, T. Nakamura, K. Nomoto, P.A. Mazzali, F. Patat, I. Hachisu: ApJ **565**, 405 (2002)
15. T. Matheson, A.V. Filippenko, W. Li, D.C. Leonard, & J.C. Shields: AJ, **121**, 1648 (2001)
16. P.A. Mazzali: A&A **363**, 705 (2000)
17. P.A. Mazzali, K. Iwamoto, K. Nomoto: ApJ **545**, 407 (2000)
18. P.A. Mazzali, L.B. Lucy: A&A **279**, 447 (1993)
19. P.A. Mazzali, K. Nomoto, F. Patat, K. Maeda: ApJ **559**, 1047 (2001)
20. P.A. Mazzali et al.: ApJ, **572**, L61, (2002)
21. P.A. Mazzali et al., in preparation. (2002)
22. S. Nakano, R. Kushida, W. Li: *IAU Circ.* **7810** (2002)
23. K. Nomoto et al.: Nature **371**, 227 (1994)
24. K. Nomoto et al.: Supernovae and Gamma-Ray Bursts (Baltimore, USA), eds. M. Livio, N. Panagia, K. Sahu (Cambridge), 144 (2001)
25. K. Nomoto et al.: IAU Symposium 212, A massive Star Odyssey, from Main Sequence to Supernova, eds. V.D. Hucht, A. Herrero, & C. Esteban (San Francisco; ASP), in press (astro-ph/0209064)
26. M.E. Sharina, I.D. Karachentsev, N.A. Tikhonov: A&AS, **119**, 499 (1996)
27. S.J. Smartt, P. Vreeswijk, E. Ramirez-Ruiz, et al.: ApJ **572**, L147 (2002)

28. D.A. Swartz, J.C. Wheeler: ApJ, **379**, L13 (1991)
29. M. Takada-Hidai, W. Aoki, G. Zhao: PASJ, submitted (2002)
30. M. Turatto, et al.: ApJ, **498**, L129 (1998)
31. L. Wang, J.C. Wheeler: ApJ Letters **504**, 87 (1998)
32. L. Wang, D. Baade, P. Höflich, J.C. Wheeler, et al.: ApJ, submitted (astro-ph/0206386) (2002)

Freeze-Out Effects in Hydrogen and Helium Lines of SN 1987A at the Early Photospheric Epoch

Victor P. Utrobin[1,2] and Nikolai N. Chugai[3]

[1] Institute of Theoretical and Experimental Physics, 117259 Moscow, Russia
[2] European Southern Observatory, D-85748 Garching, Germany
[3] Institute of Astronomy of Russian Academy of Sciences, 109017 Moscow, Russia

Abstract. We have developed a time-dependent model of ionization, excitation and energy balance of the atmosphere of SN 1987A at the photospheric epoch to study early behavior of hydrogen and helium lines. The ionization freeze-out effects play a key role in producing both the strong Hα during nearly the first month and the He I 5876 Å scattering line on day 1.76. Using an extended reaction network between hydrogen molecules and their ions demonstrates that ion-molecular processes are likely responsible for the blue peak in the Hα profile at the Bochum event epoch.

1 Introduction

We understand spectrum formation of SNe II not better than we do the hydrogen and helium spectra of SN 1987A at the early photospheric epoch. Despite many attempts to explain Hα and He I 5876 Å lines we are still far from satisfactory results [1–4]. The stumbling block is that any model produces too weak Hα and He I 5876 Å at the early photospheric epoch.

Yet, it has already become clear that a residual ionization (freeze-out effects) may result in the enhanced excitation of hydrogen compared to steady-state regime [5,6]. However, in our previous study [6] we assumed simple laws for the behavior of electron temperature in the atmosphere. Here we will confirm a vital role of freeze-out effects in the hydrogen excitation at the photospheric epoch using the upgraded model in which time-dependent chemical kinetics is solved along with time-dependent energy balance. Moreover, here we consider neutral helium, not only hydrogen, and will show a key role of the time-dependent kinetics for He I 5876 Å line too.

We also recapitulate here our previous results on modeling the blue peak of Hα at the Bochum event phase upon the basis of the time-dependent kinetics with hydrogen-composed species [6]. This will demonstrate that not only first-order effects, like freeze-out, but subtle molecular processes may be also important for hydrogen spectrum formation in SN II.

2 Model and Input Physics

The atmosphere model is based on a hydrodynamic model [7] with a $15M_\odot$ ejecta and a kinetic energy of 1.9×10^{51} erg. The radiation field is treated in the

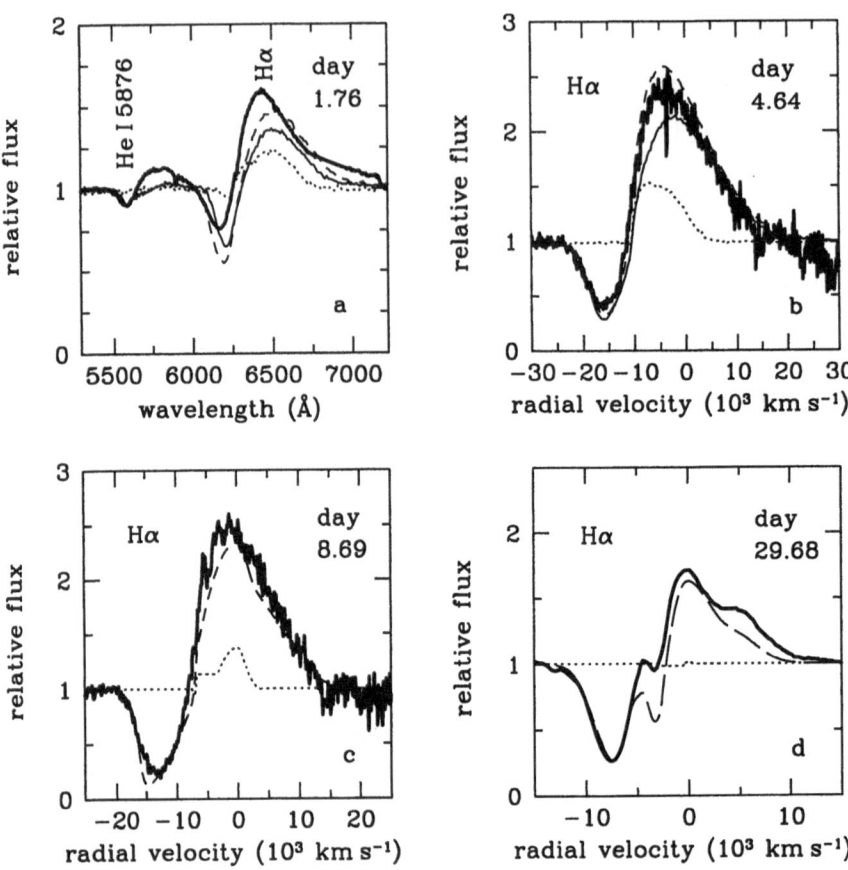

Fig. 1. Time evolution of Hα from day 1.76 to 29.68 (**a-d**) and He I 5876 Å on day 1.76 (**a**) in SN 1987A. The observed spectra from day 1.76 to 8.69 [12] and on day 29.68 [13] (*thick solid line*) are compared to those computed with the full time-dependent energy balance (*thin solid line*), with adiabatic evolution of the kinetic temperature (*short dashed line*), with the electron temperature equal to the local radiative temperature (*long dashed line*), and for the steady state (*dotted line*)

approximation of sharp photosphere. For t<1.8 days the radiation field in continuum is described by the photospheric radius and effective temperature, taken from the hydrodynamic model, and at later epoch by the empirical values [8] and by the UV and optical observations [9]. The line radiation transfer is treated in the Sobolev approximation [10,11]. Electron temperature in the atmosphere is determined from solving the time-dependent energy equation.

The following elements and molecules are calculated in non-LTE chemical kinetics: H, He, C, N, O, Ne, Na, Mg, Si, S, Ar, Ca, Fe, H^-, H_2, H_2^+, and H_3^+. All elements but H are treated with three ionization stages. The level populations of H and He I are calculated for 15 and 29 levels, respectively, while other

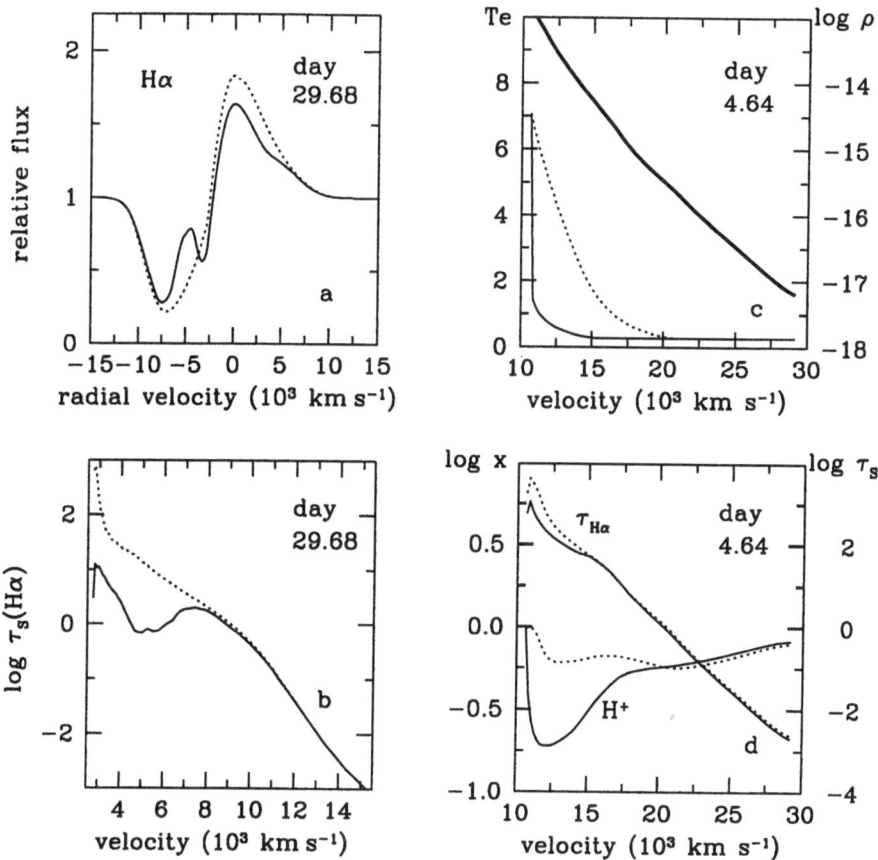

Fig. 2. *Left panel:* The role of molecular reactions for the Bochum event. (a) Hα profiles on day 29.68 calculated in time-dependent approach with (*thin solid line*) and without (*dotted line*) molecular reactions. (b) The behavior of the Sobolev optical depth of Hα in both models. *Right panel:* Physical conditions in the supernova atmosphere ($v >$ 10700 km/s) on day 4.64. (**c**) The density (*thick solid line*) and electron temperature (in units of 1000 K). The temperature is given for the model with the full energy balance (*thin solid line*) and for that with the adiabatic approximation (*dotted line*). (**d**) The behavior of the Sobolev optical depth for Hα and fractional abundance of H^+ for both models

atoms and ions are assumed to consist of the ground state and continuum. The reaction network involves all bound-bound and bound-free, radiative and collisional processes for atoms and ions, and 7 radiative and 37 collisional processes for molecules.

3   Results and Discussion

We have developed the time-dependent chemistry model coupled with the time-dependent energy balance at photospheric phase and investigated spectra of SN 1987A. The fit between the observed and calculated Hα profiles on days 1.76, 4.64, 8.69, and 29.68 and those of He I 5876 Å line on day 1.76 is fairly good in Fig. 1 indicating that our model reflects the reality reasonably well. However, we admit that on day 1.76 our assumption on the sharp photosphere may be responsible for the deficit of the calculated Hα and He I 5876 Å net radiation.

The time-dependent effects in the Hα and He I 5876 Å lines are remarkable in producing the strong absorption both in Hα and He I 5876 Å lines on day 1.76 (Fig. 1a). Moreover, the relative contribution of time-dependent effects to Hα increases in time (Fig. 1). No doubts, the ionization freeze-out plays a key role in producing the ionization and excitation of hydrogen during at least the first month. At later epochs the non-thermal ionization and excitation from radioactive ^{56}Ni decays become important [7].

After about day 20 molecular processes neutralizing ionized hydrogen (mainly $H^- + H^+ \rightarrow 2H$) dominate over recombination close to the photosphere. This results in the non-monotonic radial dependence of the Hα Sobolev optical depth that is blamed for the blue peak in Hα at Bochum event phase (Figs. 2a,b). Earlier we have used the adiabatic approximation for electron temperature [6]. While it is a good approximation for the outer layers, we demonstrate that the full time-dependent energy balance should be solved to produce more confident results in the time-dependent models (Figs. 2c,d).

4   Conclusions

Our present study of the time-dependent effects and the influence of molecules and radioactivity in the photospheric epoch of SN 1987A may be summarized as follows:

- the time-dependent effects are a key prerequisite for the strong Hα line during nearly the first month and the He I 5876 Å line on day 1.76;
- the additional excitation from radioactive ^{56}Ni decays is not required to account for the Hα line during about the first month;
- the molecular processes among hydrogen-composed species turn out to be an important factor of hydrogen neutralization which is manifested by the emergence of the blue peak of Hα at the Bochum event phase.

We believe that these effects are important for normal SNe II-P at the plateau epoch as well.

Acknowledgements

V.P.U. is grateful to F. Patat and B. Leibundgut for hospitality at the ESO and the financial support. This work is partially supported by the RFBR (project 01-02-16295).

References

1. R.G. Eastman, R.P. Kirshner: Astrophys. J. **347**, 771 (1989)
2. W. Schmutz et al.: Astrophys. J. **355**, 255 (1990)
3. R.C. Mitchell et al.: Astrophys. J. **556**, 979 (2001)
4. R.C. Mitchell et al.: Astrophys. J. **574**, 293 (2002)
5. N.N. Chugai: 'Pre-Discovery Hard X- and Gamma-Ray Luminosity of SN 1987A from Optical Spectra'. In: *Supernovae*. ed. by S.E. Woosley (Springer-Verlag, New York 1991) pp. 286–290
6. V.P. Utrobin, N.N. Chugai: Astron. Lett. **28**, 386 (2002)
7. V.P. Utrobin: Astron. and Astrophys. **270**, 249 (1993)
8. R.M. Catchpole et al.: Mon. Not. Roy. Astron. Soc. **229**, 15P (1987)
9. C.S.J. Pun et al.: Astrophys. J. Suppl. Ser. **99**, 223 (1995)
10. V.V. Sobolev: *Moving envelopes of stars* (Harvard University Press, Cambridge 1960)
11. J.I. Castor: Mon. Not. Roy. Astron. Soc. **149**, 111 (1970)
12. M.M. Phillips et al.: Astron. J. **95**, 1087 (1988)
13. R.W. Hanuschik, J. Dachs: Astron. and Astrophys. **205**, 135 (1988)

The Collisionless Shock in SN 1006 – Resolving the Narrow Component with UVES*

Jesper Sollerman[1], Parviz Ghavamian[2], and Peter Lundqvist[1]

[1] Stockholm Observatory, AlbaNova, SE-106 91 Stockholm, Sweden
[2] Department of Physics and Astronomy, Rutgers University, 136 Frelinghuysen Rd., Piscataway, NJ 08854-8019, USA

Abstract. Using the VLT/UVES we have resolved the narrow Hα and Hβ components in the fast collisionless shock of SN 1006. The FWHM of those components is merely 21 km s^{-1}. Comparing the narrow component measurement of SN 1006 with those of other Balmer-dominated SNRs shows that the width of the narrow component does not correlate in a simple way with the shock velocity. The implications for the pre-heating mechanism responsible for the width of the narrow components in Balmer dominated supernova remnants is discussed.

1 Introduction

1.1 Physics of Balmer-Dominated Collisionless Shocks

Supernova (SN) explosions produce some of the strongest shocks in nature. Such shocks are often non-radiative, losing only a negligible fraction of their internal energy to radiative cooling. The low density of the pre-shock gas makes the temperature equilibration between electrons and ions behind the shock very inefficient in terms of normal Coulomb heating.

Unlike a radiative shock, the optical emission behind a non-radiative shock is produced entirely by collisional excitation. If the preshock medium is significantly neutral, the emission from a non-radiative shock is dominated by hydrogen line emission (i.e. it is Balmer-dominated). The Balmer line profiles are quite remarkable. When a non-radiative shock hits partially neutral gas, cold hydrogen atoms overrun by the shock are collisionally excited. The radiative decay of these excited neutrals produces narrow-component Balmer emission with a line width determined by the pre-shock temperature. On the other hand, charge exchange between cold neutrals and protons produces fast hydrogen atoms with the velocity distribution of the postshock protons. Collisional excitation of these fast neutrals produces broad Balmer line emission [3,4].

Supernova shocks are also believed to be the sites of cosmic ray acceleration and SN 1006 has been suggested as a prime candidate [1]. One way to obtain observational constraints on the cosmic ray energetics is from the Balmer dominated supernova remnants (e.g. [10]).

During the last ten years, high-resolution spectra of Balmer dominated shocks have revealed narrow line widths of $\sim$30 − 50 km s^{-1}. This is clearly higher

---

* Based on observations collected at ESO, Paranal, Chile.

than expected from the thermal width of the undisturbed interstellar matter and suggests that the gas is somehow being pre-heated by a precursor before it enters the shock [12]. The precursor could be due to ionization by energetic particles leaking out of the shock and into the upstream gas. The particles may be cosmic rays, and such precursors are predicted by diffusive shock acceleration models [2]. Alternatively, the particles heating the pre-cursor could be escaping high speed neutrals [12].

The correlation between the shock velocity and the width of the narrow component provides a potentially important discriminant between the cosmic ray and neutral precursors. Heating of the pre-shock gas should be rather insensitive to the shock velocity for a cosmic ray precursor, while the opposite should be true for a fast neutral precursor (e.g. [8,12,10]). We have therefore obtained high-resolution observations of a fast Balmer dominated supernova shock, the one in SN 1006.

1.2 SN 1006

The optical remnant of SN 1006 appears as a faint network of filaments dominated by the hydrogen Balmer lines [14,9].

Only recently was a very deep spectrum of SN 1006 obtained [7]. Summing up emission along $51''$ of the slit, also the $\lambda 6678$ line of neutral helium was detected. These observations, together with a detailed non-radiative shock model, confirmed a very low degree of electron-proton equilibration at the shock front, $T_e/T_p \leq 0.07$, and the shock speed was determined to be 2890 ± 100 km s^{-1}. These results agree well with earlier analysis of ultraviolet lines from SN 1006, and make SN 1006 the fastest Balmer-dominated shock known in an evolved supernova remnant (but see also SN 1987A [13]). However, the observations by Ghavamian et al. [7] were not of high enough spectral resolution to resolve the narrow component of the Balmer lines. Here, we present high resolution spectral observations of one position of the collisionless shock in SN 1006 that, for the first time, clearly resolve the narrow component.

2 Observations and Reductions

The observations presented in this paper were obtained with the Ultraviolet and Visual Echelle Spectrograph (UVES) on the Very Large Telescope (VLT) on Paranal, Chile. The VLT/UVES observations were conducted in visitor mode on June 13, 2001.

The set-up covered the wavelength region $4760 - 6840$ Å. UVES has a short slit with a length of $12''$. We used a slit width of $1.8''$ positioned as shown in Fig. 1. The red arm is equipped with a mosaic of two 2k×4k CCDs, where the spatial scale is $0.18''$ pixel^{-1} and a scale of $0.022 - 0.026$ Å pixel^{-1}. The position shown in Fig. 1 was observed in 7 exposures of 15 minutes each, in total 6300 seconds.

The spectra were interactively reduced using the UVES-pipeline. The reduction separates the two CCDs and performs bias subtraction and flat-fielding of the data. Wavelength calibration was done by comparison to ThAr arc lamps. More than 1000 spectral lines were fitted for both of the CCDs with an RMS of better than 0.008 Å. The resolution for arc-lines with the 1.8″ slit is about 13 km s^{-1}, or 0.29 Å at the position of Hα.

The pipeline finally extracts the spectrum of the reduced frames. We averaged the signal over 5.5″ along the slit, and subtracted the skylines from the region at the very end of the slit.

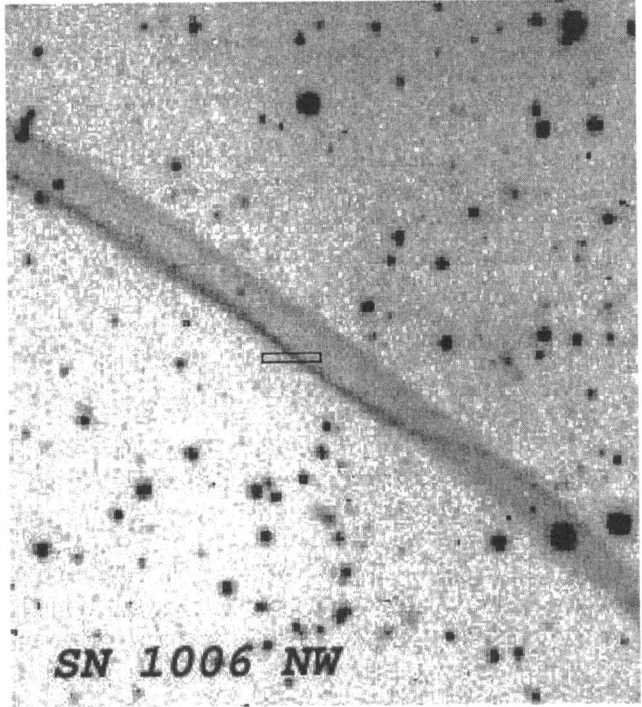

Fig. 1. The position of the UVES slit on SN 1006. The image is from Winkler & Long [15]

3   Results

In Fig. 2 we show the fully reduced Hα line from the part of SN 1006 shown in Fig. 1. The main result is that the narrow Balmer line is indeed resolved. We measure the FWHM by a Gaussian fit to be 24 km s^{-1}. This is clearly broader than the surrounding skylines, which show a FWHM of about 12 km s^{-1}. Thus, correcting the width of the Hα line for the instrumental resolution, we arrive at an intrinsic width of 21 km s⁻¹.

Apart from Hα, the only other line detected is Hβ. While this line is more noisy, the width can nevertheless be measured by a Gaussian fit. The width corrected for instrumental resolution is 21 km s⁻¹, consistent with the line width of Hα. The nearby skylines have widths of slightly more than 12 km s⁻¹, consistent with the resolution obtained in the arc lamp spectral lines.

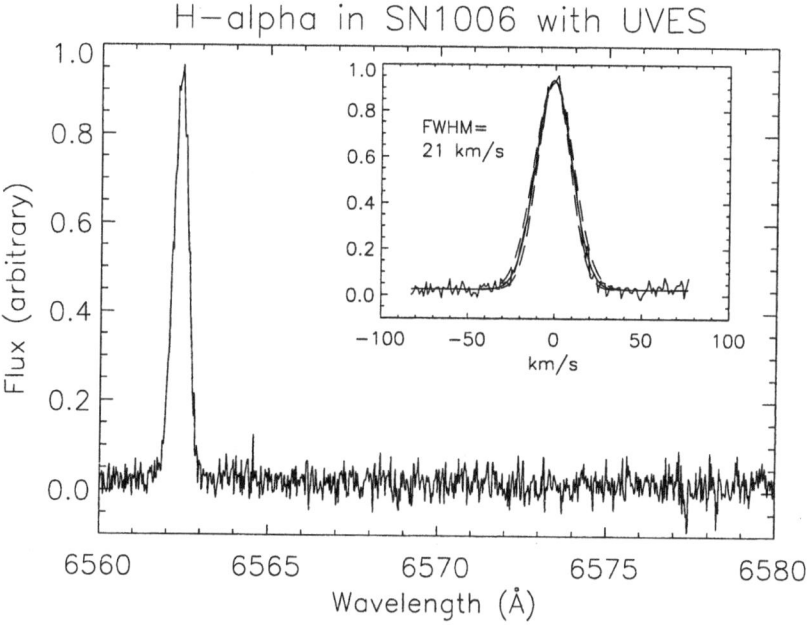

Fig. 2. The narrow Hα line detected in SN 1006. The insert shows the line in velocity-space, with the best Gaussian fit of FWHM=21 km s⁻¹ (full line). The broken lines indicate Gaussian fits with FWHM which are 10% wider and narrower

4   Discussion

4.1   Comparison to Other Balmer Dominated Supernova Remnants

The narrow component in several other Balmer dominated shocks have been spectroscopically resolved. Smith et al. [12] studied four remnants in the LMC and found narrow component widths of $30 - 50$ km s⁻¹. The same year, Hester et al. [8] investigated the Balmer dominated parts of the Cygnus loop and found similar values for the narrow components. More data and more sophisticated analysis have been obtained since and Table 1 gives an up-to date summary.

The shocks in Table 1 of RCW 86, Cygnus and Tycho were modeled in detail by [6], and SN 1006 by [7]. The shock velocities for the three LMC remnants in Table 1 were estimated by Raymond [10] based on the observations by Smith et al. [11] and extrapolating the relation between shock velocity and equilibration. The main conclusion from Table 1 is the similar widths of the narrow components for a very large range of shock velocities.

Table 1. Balmer dominated supernova remnants

SNR	Shock velocity km/s	Narrow component FWHM km/s	References
Cygnus[a]	300-400	28-35	[6,8]
RCW 86	580-660	– –	[6]
0505-67.9	440-880	32-43	[11,10,12]
0548-70.4	700-950	32-58	[11,10,12]
0519-69.0	1100-1500	39-42	[11,10,12]
0509-67.5	– –	25-31	[12]
Tycho	1940-2300	44 ± 4	[6,5]
SN 1006	2290 ± 100	21 ± 3	[7]; This work

[a] Not exactly same region of SNR measured for broad and narrow components.

4.2   Implications for the Cosmic Ray Production in Supernova Shocks

It has been noted that the widths of the narrow components correspond to a rather high kinetic temperature [12,8]. A FWHM of $30 - 50$ km s^{-1} corresponds, if purely thermal, to a kinetic temperature of $20\,000 - 50\,000$ K. This is clearly higher than expected in the undisturbed ISM where neutral hydrogen exists.

One process, which may heat the upstream gas to suprathermal velocities, is the diffusion of energetic particles from the postshock region. Smith et al. [12] concluded that the only viable explanation for the broad narrow components in the LMC remnants was the heating of the preshock medium by some sort of shock precursor, due to either fast neutrals or cosmic rays streaming ahead of the shock.

Although some of the broad component neutrals undoubtedly cross into the preshock region, their energy distribution and number density are very sensitive to a number of parameters, most importantly the preshock velocity [12]. Such a neutral precursor model thus has problems in explaining the relatively small range of narrow component widths in Balmer-dominated remnants summarized in Table 1.

The new data of SN 1006 show that the resolved narrow component of the Balmer lines has a width of merely 21 km s^{-1}. This corresponds to a kinetic temperature of ~ 9500 K. Our data thus indicate that the fastest of the known Balmer dominated shocks displays the most narrow of all pre-cursor lines (Table 1). Whatever mechanism broadens the narrow component in slower shocks is likely to operate also at higher Mach numbers. This clearly does not favor a model where the pre-heating is simply correlated with the shock velocity, as suggested for fast neutrals [12,8].

In fact, the process heating the pre-shock gas in SN 1006 appears to be less efficient than in the other SNRs listed in Table 1. A kinetic temperature of less than 10,000 K can hardly be called suprathermal.

5 Summary

With a shock velocity of 2900 km s^{-1} SN 1006 is the fastest of the Balmer dominated shocks in our sample. We have resolved the narrow Hα and Hβ lines and measured their FWHM to be 21 km s^{-1}. This is lower than measured in any other Balmer dominated supernova remnant, and disfavors any precursor mechanism which simply correlates the shock velocity with the degree of pre-heating. The corresponding low kinetic temperature of the pre-shock medium in SN 1006 may not even require a particle precursor to heat the upstream gas.

References

1. G. E. Allen, R. Petre, E. V. Gotthelf: ApJ **558**, 739 (2001)
2. R. Blandford, D. Eichler: Physics Report **154**, 1 (1987)
3. R. A. Chevalier, J. C. Raymond: ApJ **225**, L27 (1978)
4. R. A. Chevalier, R. P. Kirshner, J. C. Raymond: ApJ **235**, 186 (1980)
5. P. Ghavamian, J. Raymond, P. Hartigan, W. P. Blair: ApJ **535**, 266 (2000)
6. P. Ghavamian, J. Raymond, R. C. Smith, P. Hartigan: ApJ **547**, 995 (2001)
7. P. Ghavamian, P. F. Winkler, J. Raymond, K. S. Long: ApJ **572**, 888 (2002)
8. J. J. Hester, J. Raymond, W. P. Blair: ApJ **420**, 721 (1994)
9. R. P. Kirshner, P. F. Winkler, R. A. Chevalier: ApJ **315**, 589 (1987).
10. J. C. Raymond: Space Science Reviews **99**, 209 (2001)
11. R. C. Smith, R. P. Kirshner, W. P. Blair, P. F. Winkler: ApJ **375**, 652 (1991)
12. R. C. Smith, J. C. Raymond, J. M. Laming: ApJ **420**, 286 (1994)
13. G. Sonneborn, C. S. J. Pun, R. A. Kimble, et al.: ApJ **492**, 139 (1998)
14. S. van den Bergh: ApJ **208**, L17 (1976)
15. P. F. Winkler, K. S. Long: ApJ **491**, 829 (1997)

Prospects for SNIa Explosion Mechanism Identification Through SNRs

Carles Badenes[1,2] and Eduardo Bravo[1,2]

[1] Dpt. Física i Eng. Nuclear, Universitat Politècnica de Catalunya, Diagonal 647, 08028 Barcelona, Spain
[2] Institut d'Estudis Espacials de Catalunya, Gran Capità 2-4, 08034 Barcelona, Spain

Abstract. We present the first results from an ongoing work aimed to use supernovae remnants to discriminate among different type Ia supernovae explosion models. We have computed the hydrodynamic interaction of supernova ejecta with the interstellar medium obtaining the evolution of the density, temperature and ionization structure of the remnant. We have used ejecta profiles obtained from 1D hydrodynamic calculations of the different explosion mechanisms that are currently under debate. We have analyzed the best indicators that allow to discriminate among the different explosion mechanisms, taking into account the diversity of scenarios proposed for the presupernova evolution of the binary system, and the uncertain amount of electron heating in collisionless shocks.

1 Thermonuclear Supernovae Remnants

Modern X-ray observatories, like *Chandra* and *XMM-Newton* give the opportunity to collect a huge amount of high quality, high resolution, data on supernova remnants (SNRs). Whereas the atomic codes that allow to compute and fit the X-ray spectra have experienced a considerable improvement in the last years, this is not the case of our knowledge of the evolution of SNRs. One step towards a more complete understanding of these objects is to couple one-dimensional hydrodynamical calculations to an accurate ionization code.

The explosion mechanism(s) responsible for thermonuclear supernovae (SNIa) is still a matter of strong debate. In spite of the advances made in recent years in the understanding of the physics of the flame, there is still no definitive answer to the question of what is the mode of propagation of a thermonuclear flame through a massive C+O white dwarf. To answer this, we need to exploit every piece of evidence that can hold a signature of the flame propagation history. Here, we present the first results from an ongoing work aimed to use SNRs to discriminate among different SNIa explosion models. We have made a systematic exploration of the parameter space of explosion mechanisms including pure detonations, pure deflagrations, delayed detonations, pulsating delayed detonations, and sub-Chandrasekhar models, and we have obtained a set of explosion ejecta profiles with the aid of a 1D SN hydrodynamics code. Then, for each one of the profiles, we have computed the hydrodynamic interaction of the supernova ejecta with a uniform interstellar medium (ISM), obtaining the evolution of the density, temperature and ionization structure of the remnant. In order to be sure that SNR properties allow to discriminate between explosion models it is necessary to look for the effects of other unknowns of the problem, mainly the presupernova evolution, which determines the ambient medium structure at the time of SN explosion, and the physics of collisionless shocks.

1.1 Presupernova Evolution

We have tested four simple models of presupernova mass loss due to binary wind (for details see [1] and [2]), with parameters chosen in order to reproduce the gross features of different evolutionary scenarios (references can be found in [1]). The structure of the ambient medium at the time of SN explosion has been computed with a SNR hydrocode which included radiative losses. Presupernova winds interact with the ISM producing the typical forward–shock/contact–discontinuity/reverse–shock structure, and providing:

- A cold dense shell appropriated for the formation of dust. This shell could also play a role in the formation of a light echo.
- A close circumstellar shell which could interact early with the SN ejecta (but only in our wind model D: low velocity wind active till the SN explodes).

We have computed the interaction of SN ejecta with the ambient medium sculpted by the four wind models, together with the interaction of the same ejecta profile with a constant density ISM. In Fig. 1 we show the expansion parameter of the forward shock (defined as $\eta = d(\ln R_{shock})/d(\ln t)$). As can be seen, the dynamical properties of SNR are mostly sensitive to the presupernova history at ages larger than that of Tycho's SNR (1.357×10^{10} s).

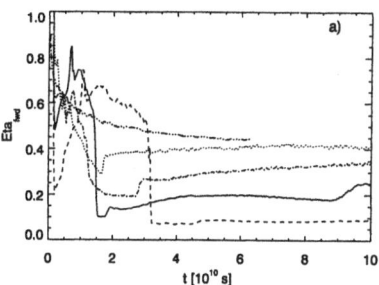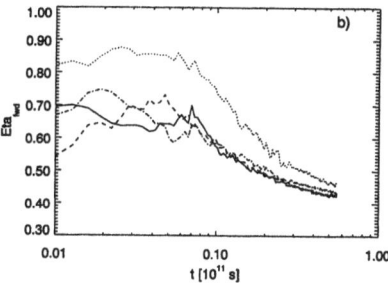

Fig. 1. Expansion parameter of the forward shock as a function of time. (a) Wind models: A (*solid line*), B (*dotted line*), C (*dashed line*), and D (*dash-dotted line*), and uniform ISM (*triple-dot-dashed line*). (b) Explosion models: Delayed detonation (*solid line*), deflagration (*dotted line*), pulsating delayed detonation (*dashed line*), and sub-Chandrasekhar (*dash-dotted line*)

1.2   Explosion Mechanism

We have explored different explosion mechanisms, whose ejecta are characterized by their density and chemical profiles, obtained with a SNIa 1D hydrocode (see [2] for details). We have computed the interaction of each SN ejecta profile with a uniform density ISM ($\rho_{ISM} = 10^{-24}$ g·cm^{-3}) with the same SNR hydrocode described previously, and we have obtained the ionization structure of the SNR as a function of time. In principle, there are two ways to discriminate between the different ejecta profiles: one is the dynamic evolution of the SNR, the other is its X-ray emission.

Dynamical Evolution of the Supernova Remnant The dynamics of young SNRs is sensitive to the explosion mechanism up to a few hundred years after the explosion. As can be seen in Fig. 1, our main result is that the expansion parameter of the deflagration model is the largest at early times due to the sharp density contrast at the point of flame quenching.

Collisionless Shocks Physics The physical processes that are at work in collisionless shocks physics are not well understood, the main uncertainty being the amount of electron heating in the shocks. The heavy plasma ejected by SNIa poses additional problems, because the electron number density depends on the local chemical composition and the degree of ionization of each element (which is out of ionization equilibrium, [3]).

We have computed carefully the ionization evolution of the plasma, and have checked the sensitivity of the results to the amount of electron heating at the shock. In Fig. 2 we show the results from two calculations in which we have assumed either no electron heating at the shock or a moderate electron heating (1% of the ion temperature). Coulomb interactions between electrons and ions

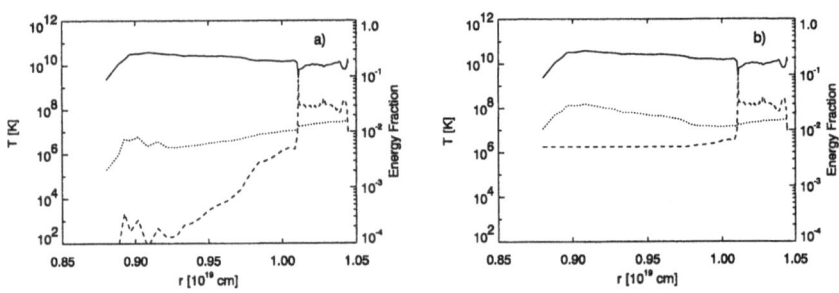

Fig. 2. Ion (*solid line*) and electron (*dotted line*) temperatures in the region between the reverse shock (left) and the contact discontinuity (right), together with the fraction of internal energy in the electron gas (*dashed line*) for the delayed detonation model at the age of Tycho's SNR. (**a**) Assuming no electron heating at the shock front. (**b**) Assuming a moderate amount of electron heating at the shock front (see text for details)

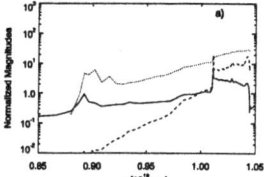 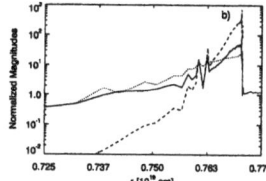 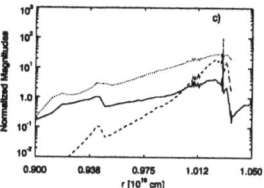

Fig. 3. Normalized density ($\rho/10^{-24}$ g·cm^{-3}, *solid line*), electron temperature ($T_e/10^6$ K, *dotted line*), and ionization timescale ($n_e t/10^9$s·cm^{-3}, *dashed line*), in the region between the reverse shock (left) and the contact discontinuity (right). Note the different r scales. (**a**) Delayed detonation model. (**b**) Deflagration model. (**c**) Pulsating delayed detonation model

were included in both calculations. Our main results concerning the influence of electron heating at the shock are:

- The ionization timescale does not depend on it at all.
- The electron temperature near the reverse shock (and, thus, the high energy X-ray continuum) depends strongly on it.

X-Ray Emissivity Finally, we address the problem of the X-ray emissivity associated to each explosion model. Up to now, we have not concluded this part of the project, but some preliminary results are already available. In Fig. 3 we show the density, temperature and ionization timescale profiles of the deflagration, pulsating delayed detonation and delayed detonation models at the age of Tycho's SNR. Our results show that the most model sensitive characteristics are:

- The ionization timescale close to the contact discontinuity and, thus, the centroid of the Fe/S Heα complex of X-ray lines.
- The density profile and, thus, the emission measure and the X-ray luminosity.
- The high energy continuum is mainly dependent on the assumed physics of the collisionless shocks.

This work has been supported by the MCYT grants EPS98–1348 and AYA2000–1785 and by the DGES grant PB98-1183–C03-02. C. B. is very indebted to CIRIT for a grant.

References

1. C. Badenes, E. Bravo: ApJ **556**, L41 (2001)
2. C. Badenes, E. Bravo: 'The Imprint of Presupernova Winds on Supernova Remnant Evolution: Towards More Realistic Models for Type Ia Supernova Remnants and their Spectra'. In: *New Visions of the X-ray Universe in the XMM-Newton and Chandra Era, at ESTEC, The Netherlands, November 26–30, 2001.* ed. F. Jansen (ESA, 2002), in press
3. K. Borkowski, W. Lyerly, S. Reynolds: ApJ **548**, 820 (2001)

Light Curves of Type Ia Supernovae as a Probe for an Explosion Model

Elena Sorokina[1] and Sergey Blinnikov[2]

[1] Sternberg Astronomical Institute, 119992 Moscow, Russia
[2] ITEP, 117218, Moscow, Russia

Abstract. We present theoretical UBVI- and bolometric light curves of SNe Ia for several explosion models, computed with our multi-group radiation hydro code. We employ our new corrected treatment for line opacity in the expanding medium. The results are compared with observed light curves. Our goal is to find the most viable thermonuclear SN model that gives good fits not only to a typical SN Ia light curve, but also to X-ray observations of young SNIa remnants. It appears that classical 1D SNIa models, such as deflagration W7 and detonation DD4, fit the light curves not so well as a new 3D deflagration model by Reinecke et al (which is averaged over angles for our LC modelling). This model seems good also in reproducing X-ray observations of Tycho SNR. We believe that the main feature of this model which allows us to get correct radiation during the first month, as well as after a few hundred years, when an SNR forms, is a strong mixing that pushes material enriched in iron and nickel to the outermost layers of SN ejecta.

1 Introduction

At the moment, there are many models of the thermonuclear explosion of a star, that lead to the event we know as a Type Ia Supernova (SN Ia). Some of them were discussed in the talk by J.Niemeyer [9]. Only a few parameters, such as kinetic energy and total ^{56}Ni production, can be derived directly from the explosion modelling and compared with the observational values. The subsequent evolution of the exploded star gives us much more possibilities to compare models and to decide which one fits observations best by reproducing more details in SN Ia light curves and spectra. We will focus here on the broad-band UBVI and bolometric light curve computations for SN Ia models.

There are several effects in SN physics which lead to difficulties in the light curve modelling of any type of supernova. For instance, an account should be taken correctly for deposition of γ photons produced in decays of radioactive isotopes, mostly ^{56}Ni and ^{56}Co. After being emitted, γ photons travel through the ejecta and can finish up in either thermalization or in non-coherent scattering processes. To find this, one has to solve the transfer equation for γ photons together with hydrodynamical equations. The full system of equations should involve also radiative transfer equations in the range from soft X-rays to infrared for the expanding medium. There are millions of spectral lines that form SN spectra, and it is not a trivial problem to find a convenient way to treat them even in the static case. The expansion makes the problem much more difficult to solve: hundreds or even thousands of lines give their input into emission and absorption at each frequency.

On the first glance, modelling of SNe Ia seems easier than of other types of supernovae, since the hydro part is very simple: coasting stage starts very early, there are no shocks and no additional heating from them.

On the other hand, much more difficulties arise in the radiation part. SNe Ia become almost transparent in the continuum at the age of a few weeks. This means that NLTE effects are stronger than for other types of supernovae. Radiation is decoupled from matter within the entire SN Ia ejecta even before maximum light (which occurs around the 20th day after explosion), see e.g. [4] or Fig. 2 in [15]. In this case, one cannot ascribe to radiation the gas temperature, or any other temperature, since the SN Ia spectrum differs strongly from a blackbody one. One has to solve a system of time-dependent transfer equations in many energy groups instead, with an accurate prescription for treatment of a huge number of spectral lines, which are the main source of opacity in this type of SN [2,11].

2   Method

We compute broad-band UBVI and bolometric light curves of SNe Ia with the multi-energy radiation hydro code STELLA. Time-dependent equations for the angular moments of intensity in fixed frequency bins are coupled to Lagrangean hydro equations and solved implicitly. Thus, we have no need to ascribe any temperature to the radiation: the photon energy distribution may be quite arbitrary.

While radiation is nonequilibrium in our approximation, ionization and atomic level populations are assumed to be in LTE. The effect of line opacity is treated as an expansion opacity according to Eastman & Pinto [5] and to our new recipes [15]. We have compared the results and found that infrared band-passes (in which the ejecta is the most transparent) are especially sensitive to the treatment of opacity (see Fig. 4 in [15] for the comparison of SN Ia light curves calculated with these two approaches), so one should be very careful on this point. To simulate NLTE effects we used the approximation of the absorptive opacity in spectral lines. NLTE results [2] and ETLA approach [11] demonstrate that fully absorptive lines give us an acceptable approximate description of NLTE effects.

We treat γ-ray opacity as a pure absorptive one, and solve the γ-ray transfer equation in a one-group approximation following [17]. The heating by decays $^{56}Ni \rightarrow {}^{56}Co \rightarrow {}^{56}Fe$ is taken into account.

In the calculations of SN Ia light curves we use up to 200 frequency bins and up to ~ 400 zones as a Lagrangean coordinate.

3   Light Curves of SNe Ia

In our previous work [16] we studied the behaviour of the broad-band optical and ultraviolet light curves for classical 1D models of SNe Ia: Chandrasekhar-mass models W7 (deflagration) [10] and DD4 (delayed detonation) [19], and sub-Chandrasekhar-mass models LA4 (helium detonation; actually, an averaged 2D model) [8] and WD065 (detonation with low ^{56}Ni production) [14]. Here we concentrate on a new 3D deflagration model by M.Reinecke et al. [12] (MR, hereafter). Working in more dimensions they have to involve less free parameters than was needed in 1D, so they get their model almost from the "first principles".

Table 1. Parameters of SN Ia models

Model	Mass ($M_\odot$)	E_{kin} (10^{51} erg)	M_{56Ni} ($M_\odot$)
W7	1.38	1.2	0.6
DD4	1.38	1.23	0.6
MR	1.4	0.46	0.42

The main features of the 3D model are compared with the ones of classical 1D models in the Table. From the first glance it seems that the light curves for the models which differ so much could not be similar.

Nevertheless, Fig. 1 demonstrates that they are similar in many details. The possible reasons for this can be understood if one has a look at the element distribution over the ejecta. The compositions for W7 and MR models are shown in Fig. 2.

At the moment of our light curve computation the full nuclide yields for MR were not yet obtained. Therefore, the model consisted of the elements, which were chosen as representative examples for the energy release calculation, namely "Fe" for iron group elements that were divided onto 80% of ^{56}Ni and 20% of ^{56}Fe, "Mg" for intermediate mass elements and unburned C and O in equal proportion. The instabilities that have developed in the 3D model were not supposed to be so huge in approximate 1D models of explosion. This has led to the differences in the nickel distribution over the ejecta: it is mixed to the outermost layers in the 3D model. These layers became much more opaque than in the 1D models, and, despite having less than half of kinetic energy, the 3D model has a photospheric velocity comparable to that in the 1D models. It is probably still a bit too low, so the light curve is wider, since the ejecta expand a bit slower and photons are locked inside them for a bit longer time. The broad-band light curves for the MR model fit the observations of one of the typical SN Ia SN 1994D in U and B bands surprisingly well, while classical 1D models, such as W7 and DD4, show faster declines in the optical than observed. Unfortunately, the bolometric light curve for the MR model (Fig. 3) is somewhat too slow. The ejecta must expand with higher speed to let photons diffuse out faster.

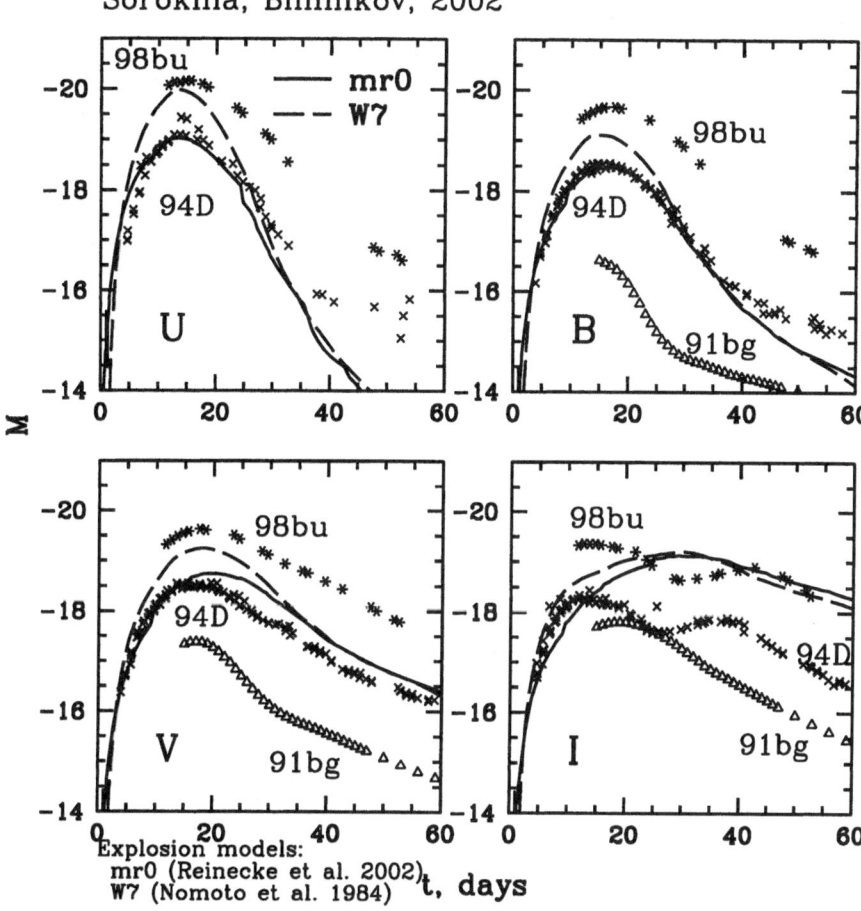

Fig. 1.

Since our light curve code is 1D, we cannot model 3D effects directly. We tried to estimate them by averaging the model over different solid angles. The main question we wished to answer was: how different can the SN Ia look like for an observer from different viewing angles. To start with, we just compared the models averaged over cones with opening angles of 14° with the one averaged over the whole 4π [13]. We tried to choose two limiting cases: the angles with minimal and maximal nickel abundances, i.e. one with nickel bubble extended to the surface, and another one between such bubbles with nickel only in the center. Since total ^{56}Ni mass differs in these models, the light curves are also very different (Fig. 4). They just simulate different models, but not a model observed from different angles.

In reality, the total energetics one observes is defined by total ^{56}Ni mass. So we have to preserve it when averaging the model over different directions. In the next experiment we have taken the same solid angle with low ^{56}Ni abundance

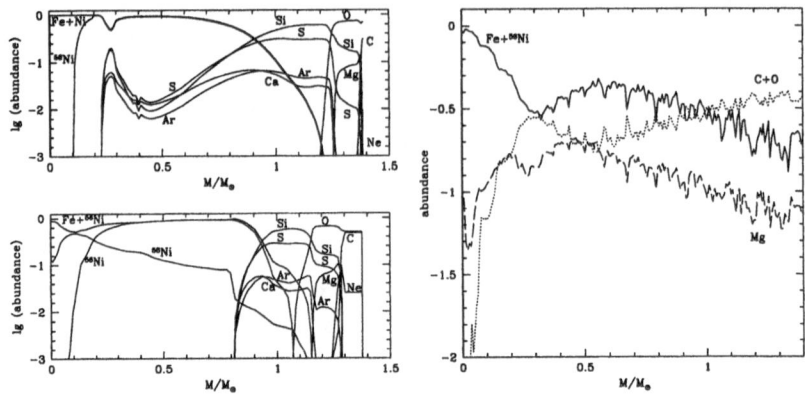

Fig. 2.

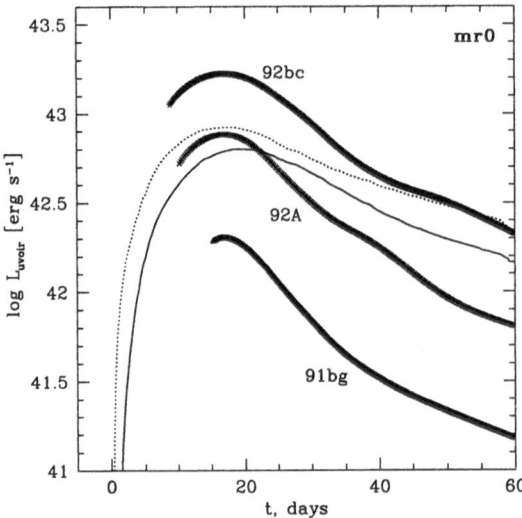

Fig. 3. Total (dotted) and *UVOIR* (solid) bolometric luminosity of MR model compared with observations [3]

and enhanced the mass of ^{56}Ni in every mass zone by the same ratio, so that the total ^{56}Ni mass became equal to the one of the original 3D model. The resulting light curve is compared with the fully averaged case in Fig. 5. Since ^{56}Ni mass fixes the brightness of SNe Ia [1], the light curves become equally bright, but the maximum is shifted in time due to different energetics and mixing. We believe that differences between 1D and 3D light curve calculations should not be much stronger than the difference between these two light curves, when other physical assumptions are fixed.

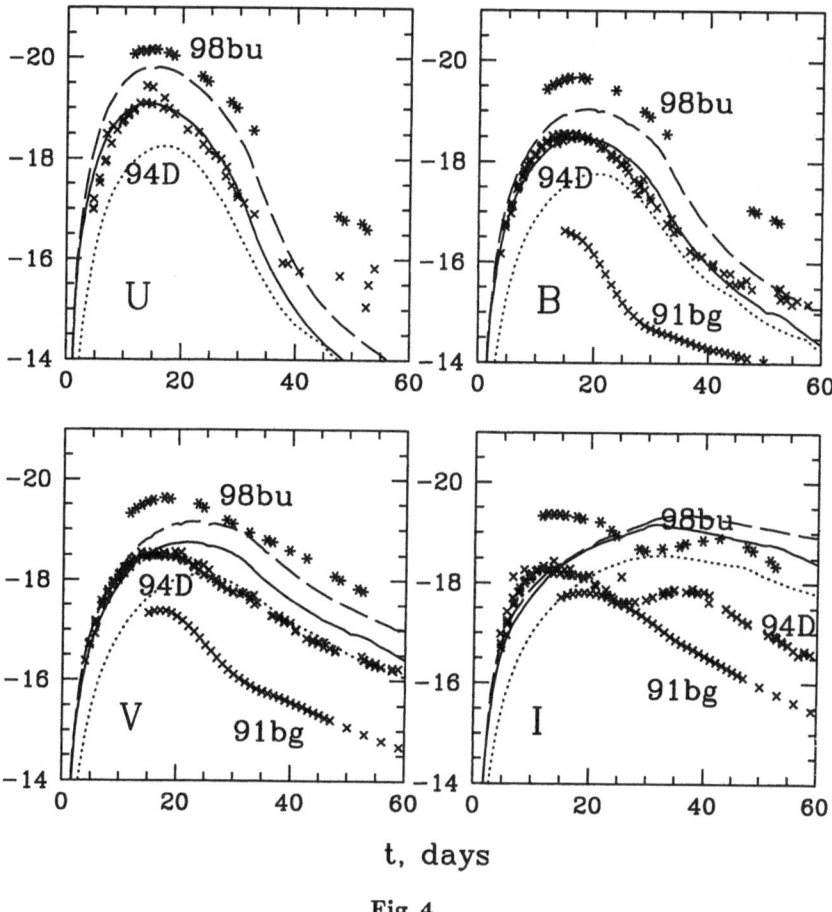

Fig. 4.

There is also another reason which allows us to believe that the MR model is better than the classical 1D models. The Tycho SNR is believed to be the remnant of SN Ia. We calculated in detail the X-ray emission of the Tycho SNR, allowing for time-dependent ionization and recombination and compared the computed X-ray spectra and images in narrow filter bands with XMM-Newton observations. Our preliminary results [7] show that all Chandrasekhar-mass models produced similar X-ray spectra at the age of Tycho, but they differ strongly in the predictions of how the remnant should look like in the lines of different ions due to the very different distribution of elements in the ejecta for 1D and 3D models. We found that W7 and DD4 models produce rather a wide ring in Fe lines, while it is narrow for the MR model. The image for the latter model is very similar to what is observed.

Sorokina, Blinnikov, Reinecke 2002

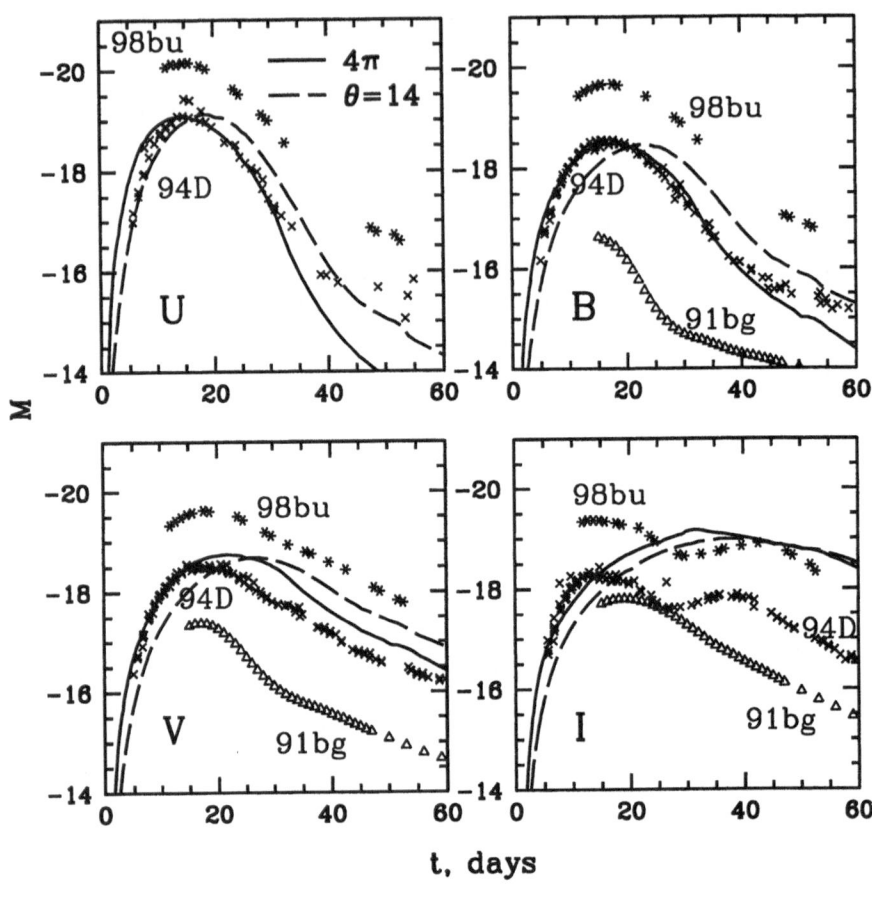

Fig. 5.

4 Conclusions

The new 3D SN Ia model MR [12] is very appealing. Yet it is not a final one: a detailed post-processing of nucleosynthesis changes of the composition has been done very recently [18], and it is not yet checked in the light curve calculation. Our light curve computations are also preliminary, since more work is needed on the expansion opacity. Hopefully, none of the required improvements will spoil the light curve of this model and its X-ray spectra of the SNR stage, since the specific qualities of the model can be primarily explained by the enrichment of the outermost layers of SN ejecta by Fe and Ni.

The SN light curve modelling still has a lot of physics to be added, such as a 3D time-dependent radiative transfer, including as much as possible of NLTE effects, which are especially essential for SNe Ia [6]. All this will improve our understanding of thermonuclear supernovae.

Acknowledgements

The authors are grateful to Wolfgang Hillebrandt, to the organizers of the conference and to the MPA staff for all their help, to Stan Woosley for his hospitality at UCSC where a major part of the work on expansion opacity was done, and to Bruno Leibundgut for providing us with the data [3] in electronic form.

The work is supported in Russia by RFBR grant 02-02-16500, in the US, by NASA grant NAG5-8128.

References

1. W.D. Arnett: ApJ **253**, 785 (1982)
2. E. Baron, P.H. Hauschildt, A. Mezzacappa: MNRAS **278** 763 (1996)
3. G. Contardo, B. Leibundgut, W. D. Vacca: A&A **359**, 876 (2000)
4. R.G. Eastman, In: *Thermonuclear Supernovae*. ed. by P. Ruiz-Lapuente et al. (Kluwer Academic Pub., Dordrecht, 1997) p. 571.
5. R.G. Eastman, P.A. Pinto: ApJ **412**, 731 (1993)
6. P. Höflich: Workshop on Stellar Atmosphere Modeling, 8-12 April 2002 Tuebingen. Eds: I. Hubeny, D. Mihalas, K. Werner, astro-ph/0207103 (2002).
7. D.I. Kosenko, E.I. Sorokina, S.I. Blinnikov, P. Lundqvist, K.A. Postnov: Submitted to 34th COSPAR Sci. Assembly, Houston, 10-19 October 2002.
8. E. Livne, D. Arnett: ApJ **452**, 62 (1995)
9. J. Niemeyer: 'Explosion Models of Type Ia Supernovae'. In this volume.
10. K. Nomoto, F.-K.Thielemann, K. Yokoi: ApJ **286**, 644 (1984)
11. P.A. Pinto, R.G. Eastman: ApJ **530**, 757 (2000)
12. M. Reinecke, W. Hillebrandt, J. C. Niemeyer: A&A **386**, 936 (2002)
13. M. Reinecke: private communication (2002)
14. P. Ruiz-Lapuente et al.: Nature **365**,728 (1993)
15. E.I. Sorokina, S.I. Blinnikov: *Nuclear Astrophysics, 11th Workshop at Ringberg Castle, Tegernsee, Germany, February 11-16, 2002*, ed. by E. Müller, W. Hillebrandt, pp. 57-62
16. E.I. Sorokina, S.I. Blinnikov, O.S. Bartunov: Astron. Letters **26**, 67 (2000) S.I. Blinnikov, E.I. Sorokina: A&A **356**, L30-L32 (2000)
17. D.A. Swartz, P.G. Sutherland, R.P. Harkness: ApJ **446**, 766 (1995)
18. C. Travaglio: private communication (2002)
19. S.E. Woosley, T.A. Weaver. In: *Supernovae*, ed. by J. Audouze et al. (Elsevier Science Publishers, Amsterdam 1994), p.63

Inflection Points on the Light Curves of Type Ia Supernovae

N.N. Pavliouk and D.Yu. Tsvetkov

Sternberg Astronomical Institute, Universitetski pr.13, 119992, Moscow, Russia

Abstract. While the distribution of supernova absolute magnitudes at peak is often investigated, there was no systematic study of absolute light curve behavior near the inflection point. A type Ia supernova has the maximum of color $B - V$ just near this point. And the maximum of this color corresponds to the minimum of photospheric temperature. So we could expect the light from type Ia supernova at inflection point to be a better standard candle than at the peak. We determined positions of inflection points on the light curves of type Ia supernovae and studied the distributions of absolute magnitudes M_B and M_V for them for the first time. The standard deviation of M_B at the inflection point is $0^m.47$, which is approximately a factor of 1.5 smaller than at the peak of the light curve.

1 Introduction

Some authors put forward the hypothesis that all type Ia supernovae (SNe), or some of them, are good standard candles at maximum (e.g., [1,2]). In other studies, this hypothesis was invariably questioned (e.g., [3,4]).

In addition, the idea of Pskovskii [5] that there is a relation between the absolute magnitude of SNe Ia at maximum and the rate of decline (β) reappeared in [6], where the light curve parameter Δm_{15} was proposed instead of β. It was supposed that type Ia supernovae were better standard candles after correcting the luminosity at maximum light for the rate of decline [7] or for the difference in the shapes of the light curves in different bands [8].

It is now clear there are several sequences among SNe Ia and they are not as homogeneous a class as they were previously thought to be.

Much less attention was given to another important point on the light curve. This is the so-called inflection point, noted by Pskovskii [5,9]. At this point, a change from the rapid brightness decline after maximum to a slower exponential decline occurs. The $B - V$ color index of SNe, which characterizes the temperature, reaches its maximum near the inflection point.

2 Estimations of the Observed and Absolute Magnitudes of Supernovae at the Inflection Points

Pskovskii [5,9] determined the position of the inflection point by drawing two straight lines by the least-squares method, one of which corresponds to the segment of rapid decline immediately after the maximum, and the other corresponds to the segment of slow exponential decline that follows.

With the help of the Sternberg Astronomical Institute catalog of SNe light curves [10] we fit the observed light curve more accurately by a set of individual linear segments rather than by two straight lines. The position of the inflection points for a light curve is determined taking into account the contribution of all linear segments of the smooth curve in their vicinity.

The methods of calculating the absolute magnitudes of SNe depend on the methods used for determining the distances to SNe and allowing for the interstellar extinction. Whereas for nearby galaxies there are several means of determining their distances, for distant galaxies the only possibility is to calculate the distances from the Hubble law. It is also necessary to correct the observed radial velocities of galaxies for known large-scale motions.

Let us denote the distance to a galaxy determined from the Hubble law with $H_0 = 75$ km s^{-1} Mpc^{-1} by D_{75}. Tully [11] used the same value of H_0 to calculate the distances to nearby galaxies.

The distance $D_{75} = 55$ Mpc (which is the distance to SN 1991ag) was taken as the boundary. For $D_{75} > 55$ Mpc we reduced the velocities of galaxies to the system of cosmic microwave background (CMB) radiation. Otherwise, we used the velocities relative to the Galaxy's standard of rest that were corrected for the motion toward the center of the Virgo cluster of galaxies by the method described in [12].

For the absolute magnitude to be calculated, the interstellar extinction, which is normally described by the color excess $E(B - V)$ and by the total-to-selective extinction ratio R, must be known in addition to the distance. Note that, even in the Galaxy, R_B can vary in a fairly wide range, approximately from 3 to 5. In our computations, we took the standard values which are commonly used to estimate the interstellar extinction in the Galaxy: $R_B = 4$ and $R_V = 3$.

The method of calculating the color excess from the equivalent widths of the interstellar Na I D$_1$ and D$_2$ lines (e.g., [13–15]) is considered reliable for the Galaxy; however, it requires spectra with a sufficiently high resolution which are available only for few SNe. The accuracy of this method is not very high, especially for low extinction.

The most popular method of estimating $E(B - V)$ involves determining the displacement of the color curves of the SNe under study from the color curve of SN Ia for which the interstellar extinction inside the parent galaxy can be assumed to be zero. As the standard color curve, we used the color curve of SN 1972E; the most accurate photoelectric color determinations covering a long interval are available for this star. Since it exploded far from the galaxy's center, the extinction inside the parent galaxy and the influence of the galactic background on the photometric accuracy must be negligible. Unfortunately, this method does not allow us to distinguish the effect of extinction from an intrinsic difference in the color curves, which is definitely observed for SNe Ia.

3   Results of the Calculations of the Absolute Magnitudes for Type Ia Supernovae at Inflection Points

We found the mean B absolute magnitude (M_B) of SNe Ia at the inflection point corrected for the extinction by shifting the color curves to be $-16^m.64 \pm 0^m.47$ for our sample of 52 SNe, without the K correction being applied or $-16^m.71 \pm 0^m.47$ with the K correction. The standard deviation for the magnitudes at maximum light is $0^m.7$, which is approximately a factor of 1.5 larger. Thus, SNe Ia exhibit a smaller scatter in luminosity at the inflection point than at the point of maximum, provided that none of SNe Ia is excluded from the sample. In the case with extinction determined from equivalent width of Na I lines M_B is $-16^m.75 \pm 0^m.61$ for the sample of 12 SNe. In the V band, we found M_V to be $-17^m.49 \pm 0^m.44$, $-17^m.55 \pm 0^m.44$ (the sample of 48 SNe), and $-17^m.54 \pm 0^m.55$ (the sample of 11 SNe), respectively.

Figure 1 shows the distribution of all SNe from our sample on the absolute magnitude M_B. It differs markedly from normal distribution: the number of SNe with luminosities below the mean decreases only slightly as the luminosity decreases to $-15^m.7$, and then it is abruptly cut off.

There are no marked differences in the means that were obtained by the two methods of correction for the extinction. This implies that for most SNe Ia the intrinsic scatter of $(B - V)$ at the inflection point is not large and the extinction determined from the shift of the color curves is fairly reliable.

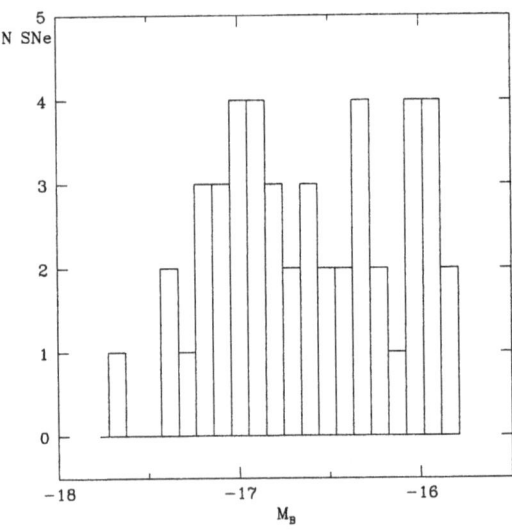

Fig. 1. The distribution of SNe Ia on absolute magnitude at the inflection point

4   Conclusions

Our main conclusion is that, when studying the light curves of SNe Ia, their behavior near the inflection point must be analyzed. The standard deviation of the absolute magnitudes for SNe at this point is approximately a factor 1.5 smaller than that at the point of maximum ($0^m.47$ and $0^m.7$, respectively).

The photospheric temperature of SNe Ia near the inflection point reaches a minimum and shows a smaller scatter than in the other segments of the light curve. Consequently, at the inflection point, the color excess should be a better indicator of the interstellar extinction towards SNe.

We showed that the difference between the mean absolute magnitudes that were calculated by determining the extinction by two different methods – by comparing the color curves of SNe with the one for SN 1972E and from the intensities of interstellar Na I absorption lines – are not statistically significant.

SNe Ia are not perfect standard candles at the inflection point because the value of deviation $0^m.47$ is anyway too high. But the absolute magnitude of SNe Ia at this point can also be used for cosmological studies, supplementing data on absolute magnitudes at maximum.

References

1. D.L. Miller, D. Branch: Astron. J. **100**, 530 (1990)
2. B. Leibundgut, G.A. Tammann: Astron. Astrophys. **230**, 81 (1990)
3. D.Yu. Tsvetkov: Peremennye Zvezdy **22**, 279 (1986)
4. S. van den Bergh: Publ. Astron. Soc. Pacific **108**, 1091 (1996)
5. Yu.P. Pskovskii: Astron. Zhurn. **54**, 1188 (1977)
6. M.M. Phillips: Astrophys. J. **413**, 105 (1993)
7. M. Hamuy, M.M. Phillips, J. Maza, et al.: Astron. J. **109**, 1 (1995)
8. A.G. Riess, W.H. Press, R.P. Kirshner: Astrophys. J. **473**, 88 (1996)
9. Yu.P. Pskovskii: Astron. Zhurn. **61**, 1125 (1984)
10. D.Yu. Tsvetkov, N.N. Pavlyuk, O.S. Bartunov: *Catalog of Supernova Light Curves* (http://mira.sai.msu.su/sn/snlight 2002)
11. R.B. Tully: *Nearby Galaxies Catalog* (Cambridge University Press, Cambridge 1988)
12. D.L. Miller, D. Branch: Astron. J. **103**, 379 (1992)
13. L.M. Hobbs: Astrophys. J. **191**, 381 (1974)
14. M.V. Penston, J.C. Blades: MNRAS **190**, 51 (1980)
15. R. Barbon, S. Benetti, L. Rosino, et al.: Astron. Astrophys. **237**, 79 (1990)

Gamma-Rays from Supernovae

Roland Diehl

Max Planck Institut für extraterrestrische Physik, D-85741 Garching, Germany

Abstract. Gamma-ray line astronomy is capable to probe recent and current supernova nucleosynthesis in the nearby universe. Breakthroughs have not occurred yet, but the ^{56}Ni measurements of thermonuclear supernovae are already very interesting: SN1991T has shown a gamma-ray signal from ^{56}Co decay consistent with expectation for this superluminous event; on the other hand, SN1998bu did not show any hints for such gamma-rays despite being closer and observed much longer. – The ^{44}Ti detection from the Cas A supernova remnant stimulated search for more possibly otherwise occulted young SNRs in ^{44}Ti gamma-rays. The apparent paucity of ^{44}Ti sources in the inner Galaxy seems to suggest that ^{44}Ti ejection is from rather non-typical core-collapse supernovae. – Interstellar ^{26}Al ($\tau = 1.04 \cdot 10^6$ y) is attributed to massive stars as dominating sources. The relative contributions of supernovae and Wolf-Rayet stars can probably only be answered from gamma-ray measurement of correlated ^{60}Fe. – High spectral-resolution experiments such as INTEGRAL (launched on Oct 17, 2002) will add key diagnostics of SNe through measurements of gamma-ray line shapes of radioactive isotopes.

1 Introduction

Nucleosynthesis in cosmic sources produces also radioactive isotopes. Their decay can be observed with gamma-ray telescopes once the overlying material density falls below a few grams per cm^2. Although just seven isotopes are promising for such studies, they span decay times from weeks to millions of years, and hence address specific production reactions and nucleosynthesis sites [7].

Shortlived radioactive isotopes decay already inside the object, and deposit their energy to emerge as scattered or thermalized radiation. So ^{56}Ni decay with τ=8.8 d powers the supernova light curve. Measuring 'early' ^{56}Ni gamma-rays from these supernovae could probe the efficiency of such energy conversion, and the large-scale distribution and mixing of fresh isotopes. Radioactive isotopes with decay times above the unfolding time of the supernova explosion are easier to observe: ^{56}Co and ^{44}Ti gamma-rays have been seen from supernovae (τ=111 d and 89 y, respectively).

For more long-lived radioactive isotopes, many source events may occur during a decay time. Given the rather coarse spatial resolution of gamma-ray telescope of the order of degrees, apparently diffuse emission may result, as is the case for ^{26}Al and ^{60}Fe (τ=1.04 10^6 y and τ=2.0 10^6 y, respectively). The interpretation of such measurements therefore requires a model for the source population and their spatial distribution. This is even more important for measurements of positron annihilation gamma-rays, because compact accreting objects and pulsars are expected to be also sources of positrons, besides supernova nucleosynthesis.

Supernova aspects addressed by gamma-ray line observations are: (1) core collapse nucleosynthesis in the inner region of the collapse, where explosive Si burning and freeze-out occur near the edge of the mass cut separating the compact remnant star from ejected material (^{44}Ti/^{56}Ni ratio and velocities), (2) flame velocity in thermonuclear supernovae (amount of ^{56}Ni), (3) mixing of nucleosynthesis products within massive stars and their supernovae, driving the return of chemically-enriched elements from massive stars through stellar winds and supernovae (^{26}Al and ^{60}Fe).

2   Thermonuclear Supernovae and ^{56}Ni

In thermonuclear explosions of CO white dwarfs following central carbon ignition the flame speed evolution will determine the nucleosynthesis and in particular the total amount of ^{56}Ni produced. Any of the SN Ia scenarios (sub-Chandrasekhar, deflagration, delayed detonations and pulsating delayed detonation models) has been shown to be capable to produce a wide variety of ^{56}Ni masses ranging from $\simeq 0.1$ to 1 M$_\odot$ [28,16,15]. The sensitivity of present γ-ray instruments limits the observations of type Ia supernovae to events within about 15 Mpc. SN1991T at 13 Mpc was marginally (3-5σ) detected with COMPTEL [26,25]. SN1998bu provided a second and seemingly better opportunity at a distance of 11.6 Mpc [14]. Yet, no gamma-rays from ^{56}Ni decay could be observed despite 14 (rather than 2 for SN1991T) weeks of total exposure [13].

The COMPTEL limit for the 1238 keV line of $2.3 \cdot 10^{-5}$ photons cm^{-2} s^{-1} constrains the visible ^{56}Ni mass to below 0.35 M$_\odot$ if the supernova is simply assumed to have been transparent to gamma-rays at the time of observations. This is probably not the case, however, and much of the γ-ray energy should be deposited in the supernova over this time window. Detailed Monte Carlo energy transport calculations show that with 0.7-0.8 M$_\odot$ [23] of ^{56}Ni derived for SN1998bu from processed supernova radiation at lower frequencies, COMPTEL should have seen ^{56}Co γ-rays at least due to those models which turn more rapidly from deflagration into detonation (W7DT) or partially-produce radioactivity in their outer ejecta (HECD) [13].

3   Core-Collapse Supernovae and ^{44}Ti

1.156 MeV γ-rays following ^{44}Ti decay have been detected from the 330-year old Galactic supernova remnant Cas A [18,38]. SN1987A's late light curve has been interpreted as powered by a similar amount of ^{44}Ti [12]. Cosmic ^{44}Ca arises from ^{44}Ti nucleosynthesis only. Chemical evolution models suggest that an 'average core collapse' supernova should at least produce as much as the supernova-model predicted yield of several 10^{-5} M$_\odot$ [37]. The above two core

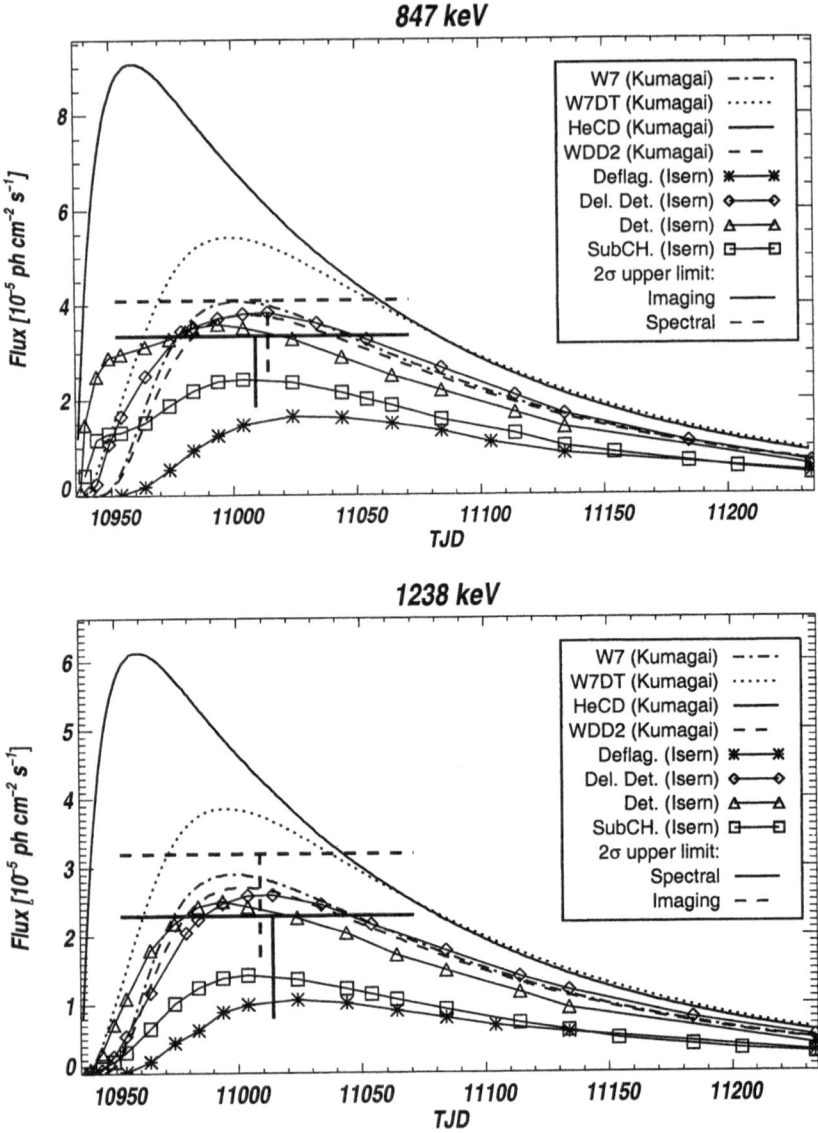

Fig. 1. Gamma-ray light curves from SN1998bu. COMPTEL flux limits favor models with embedded/mixed ^{56}Ni [13]. (Top: 847 keV; bottom: 1238 keV line)

collapse events and their association with ^{44}Ti production appear to be in line with these models and the concept of standard core collapse events. There is no other well-understood candidate supernova where one may test core-collapse supernova ^{44}Ti production.

^{44}Ti decay can be observed in gamma-rays throughout the Galaxy; its decay time scale of 89 years implies that the current rate of supernovae, directly related to the observation of the current population of massive stars, should be represented through a few ^{44}Ti gamma-ray sources. But then locations of ^{44}Ti sources in the Galaxy should match the locations of young massive stars with their short evolution times. Massive stars occur in spiral arms and in the inner Galaxy most frequently [2], and/or probably should be distributed as γ-rays from ^{26}Al decay suggest (as ^{26}Al is understood to originate predominantly from massive stars; see Fig. 3).

The COMPTEL survey of the plane of the Galaxy clearly did not show the expected few sources in the inner Galaxy; candidate sources near the noise level appear in outer Galaxy parts at best [17,11,33]. A Monte Carlo study of supernova frequencies as they would be consistent with gamma-ray data has been compared to an identical analysis constrained by historical supernova records [35]. From Fig. 2 it can be seen that indeed the rate of ^{44}Ti producing core collapse events seems systematically lower than the rate compatible with the historical supernova record (the latter includes state-of-the art Galactic extinction). Production of ^{44}Ti may be related to exceptional events, possibly involving asymmetries induced by rotation or magnetic fields. Is well-studied Cas A a typical supernova, or rather an anomalous case of a high-^{44}Ti yield supernova? And what powers SN1987A now?

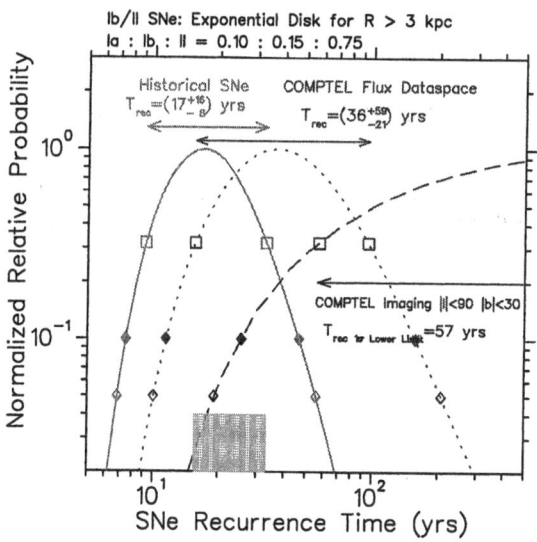

Fig. 2. Supernova rate probability distributions from Monte Carlo simulations of standard events distributed in the Galaxy. Constraints are from historical supernova observations (solid), and alternatively from ^{44}Ti gamma-rays (dashed, dotted). The latter suggest a lower event rate [35]

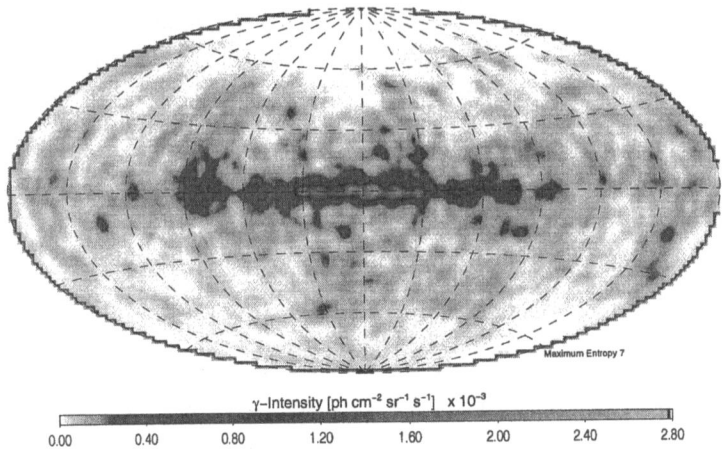

γ–Intensity [ph cm^{-2} sr^{-1} s^{-1}]   x 10^{-3}

0.00   0.40   0.80   1.20   1.60   2.00   2.40   2.80

Fig. 3. ^{26}Al emission image from 9 COMPTEL mission years [30] (Maximum-Entropy deconvolution). Massive stars dominate on the Galactic scale; but how much is from SNe, how much from WR stars? A ^{60}Fe image would show the SN part

4   Sources of ^{26}Al

The all-sky image in ^{26}Al gamma-rays has been updated from the 9-year survey of COMPTEL [30] (see Fig. 3). From the significant spatial structures, and from comparisons of different types of tracers for the candidate sources (supernovae, Wolf-Rayet stars, AGB stars and novae), it had been concluded that massive stars constitute the dominating sources throughout the Galaxy, and novae as well as less massive AGB stars play a minor role [32]. The irregularity of ^{26}Al emission along the plane of the Galaxy with prominent emission from regions of recent massive star formation activity such as the spiral-arm tangents and the Cygnus and Vela regions is more plausibly related to young stellar populations (supernovae and WR stars) with their individually higher yields than to older populations which would be more smoothly distributed from their formation history and lower individual yields. Comparison of ^{26}Al with tracers such as free electrons, dust, and gas [9,5,22,21] is consistent with this interpretation.

But an open issue remains whether explosive nucleosynthesis and ejection of pre-explosive stellar ^{26}Al through core-collapse supernovae dominates or Wolf-Rayet stars with their strong mass loss release the bulk of ^{26}Al. Metallicity trends are opposite for supernovae and WR stars, but the WR population in the more metal-rich inner Galaxy is poorly known from direct observations. The detection of ^{60}Fe and its spatial distribution would help: in supernovae, ^{60}Fe and ^{26}Al should both be produced, even from the same mass zones [36], while WR stars only produce ^{26}Al and negligible amounts of ^{60}Fe. If supernovae therefore dominate the Galactic ^{26}Al budget, supernova nucleosynthesis theory predicts the ^{60}Fe gamma-ray intensity to be about 15% of the ^{26}Al gamma-ray intensity. This value is at the edge of current instrumental sensitivities, although upper limits at this flux level indicate that either supernova nucleosynthesis does not predict the ^{26}Al/^{60}Fe yield ratio correctly or that WR stars do play a significant role [27,8]. The discovery of ^{60}Fe in seafloor sediments of the southern pacific [19] confirms the basic prediction of nucleosynthesis, and even suggests that the solar vicinity saw a supernova at only several 10 pc distance some 3–4 million years ago with ejecta material sweeping over the solar system. The INTEGRAL survey should achieve sufficient sensitivity to detect ^{60}Fe gamma-rays at the 10% level of the ^{26}Al gamma-ray flux and add spectroscopy to assess the issues about high ^{26}Al velocities in the ISM and thus the environment of ^{26}Al-ejecting events [34].

Detection of ^{26}Al from the Vela region at intensity levels comparable to the inner Galaxy [10] was a big surprise, and opened the prospect to detect ^{26}Al from a nearby ($\leq$ 500 pc) source, the Vela SNR. The observed emission however shows a pronounced peak at l=267^{o}($\pm1^{o}$), although it is clearly extended beyond what would be expected from a single point source.

A multi-source interpretation has been suggested, with the known sources plus emission from otherwise occulted sources within the Vela Molecular Ridge [6]. COMPTEL does not detect any one of the three prominent individual nearby sources, once one of the candidate source tracers which has been identified from large-scale Galactic model fitting is adopted as a scalable spatial model for the undiscriminated ^{26}Al content of the region. The Vela SNR 1.809 MeV surface brightness is too low (consistent with nucleosynthesis models), despite its new distance value of 250 pc [4]. The 1.809 MeV emission peak lies at the edge of the $\simeq$ 7°diameter Vela SNR and probably is not related to it. The RXJ0852.0-4622 SNR with its 2°diameter overlaps the emission peak attributed to the Vela Molecular Ridge, hence could contribute [1]; this cannot be resolved by COMPTEL. Too little is known about this object to allow conclusions for nucleosynthesis; the upper limit on a contribution above the galaxy-wide dust emission tracer model excludes distances below 120 pc [6]. The WR source γ^2 *Velorum* is located further south in a low-^{26}Al emission region and should not confuse results from these supernovae [29].

In the Cygnus Region, prominent ^{26}Al emission is mainly attributed to the Cyg OB2 association [31], one of the richest clusters of young stars in the Galaxy, and possibly the driver of the Cygnus region star formation [24]. The Cygnus Loop supernova remnant at $\simeq$440 pc distance [3] is probably too distant for being prominent in ^{26}Al emission [31].

References

1. Aschenbach B., Iyudin A.F., Schönfelder V., A&A **350**, 997 (1999)
2. Bartunov O.S., Makarova I.N., Tsvetkov D.Y., A&A **264**, 428 (1992)
3. Blair W.P. *et al.*, AJ **118**, 942 (1999)
4. Cha A.N. *et al.*, ApJ **515**, L25 (1999)
5. Chen W., Gehrels N. and Diehl R., ApJ **444**, L57 (1995)
6. Diehl R. *et al.*, Astroph. Letters and Communications **38**, 357 (1999)
7. Diehl R. and Timmes F.X., PASP **110**, 748, 637 (1998)

8. Diehl R. *et al.*, AIP **410**, 1109 (1997)
9. Diehl R. *et al.*, A&AS, 120, 4, 321 (1996)
10. Diehl R. *et al.*, A&A **298**, L25 (1995)
11. Dupraz C. *et al.*, A&A **324**, 683 (1997)
12. Fransson C., and Kozma C., NewAstRev. 46, 8-10 (2002)
13. Georgii R., *et al.*, A&A **394**, 517 (2002)
14. Hjorth, J. and Tanvir, N.R., ApJ **482**, 68 (1997)
15. Höflich P. *et al.*, ApJ **568**, 791 (2002)
16. Höflich, P., and Khokhlov, A., ApJ **457**, 500 (1996)
17. Iyudin A.F.*et al.* (1998), Astroph. Lett. Comm., **38**, 383
18. Iyudin, A. F.*et al.*, A&A **284**, L1 (1994)
19. Knie K. *et al.*, Phys.Rev.Lett. **83**, 18 (1999)
20. Knödlseder J.*et al.*, A&A **345**, 813 (1999)
21. Knödlseder J., ApJ **510**, 915 (1999)
22. Knödlseder *et al.*, A&A **344**, 68 (1999)
23. Leibundgut B., Astr.Astroph.Rev. **10**, 179 (2000)
24. Massey P. and Thomson A.B., AJ **101**, 1408 (1991)
25. Morris, D. J.*et al.*, in preparation for ApJ (2002)
26. Morris, D. J.*et al.*, AIP, **410**, 1084 (1997)
27. Naya J.E. *et al.*, ApJ **499**, L169 (1998)
28. Nomoto, K., Thielemann F.-K., Yokoi K., ApJ **286**, 644 (1984)
29. Oberlack U.G. *et al.*, A&A **353**, 715 (2000)
30. Plüschke S. *et al.*, ESA-SP 459, 55 (2001)
31. Plüschke S. *et al.* AIP **510**, 44 (2000)
32. Prantzos N. & Diehl R., Phys. Rep. **267**, 1 (1996)
33. Schönfelder V. *et al.*, AIP **510**, 54 (2000)
34. Sturner S. & Naya J.E., *ApJ* **526**, 200 (1999)
35. The L.-S. *et al.*, *AIP* **510**, 64 (2000)
36. Timmes F.X. *et al.*, ApJ **449**, 204 (1996)
37. Timmes F.X. *et al.*, ApJ **464**, 332 (1996)
38. Vink J. *et al.*, ApJ **560**, L79 (2001)
39. Winkler C. *et al.*, A&AS **120**, 637 (1996)

X-Ray and Radio Bright Type Ic SN2002ap – A Hypernova Without an Associated GRB

F.K. Sutaria[1], A. Ray[2], and P. Chandra[2,3]

[1] The Open University, Milton Keynes, MK7 6AA, U.K.
[2] Tata Institute of Fundamental Research, Bombay 400 005, India
[3] Joint Astronomy Programme, Indian Institute of Science, Bangalore 560 012, India

Abstract. Combined X-ray (0.3 – 10 keV) and Radio (0.61 and 1.42 GHz) observations of the type Ic SN 2002ap are used here, to determine the origins of the prompt X-ray and radio emission from this source. Our analysis of the XMM-Newton observations suggests that the prompt X-ray emission originates from inverse Compton scattering of photospheric thermal emission by energetic electrons. In addition, we use the reported multifrequency VLA observations of this supernova. We compare the early radiospheric properties of SN 2002ap with those of SN 1998bw (type Ic) and SN 1993J (type IIb), to contrast the prompt emission from a GRB associated SN and other supernovae without such counterparts.

1 Introduction

The nature and origins of X-ray and radio emission from supernovae is a subject of much interest. This is mainly because early (few days to years) emission in these bands occurs from a shocked, dense CSM around the progenitor, and this can be used to infer the mass loss rate and the nature of the progenitor. Further, early synchrotron radio emission can also occur from relativistic electrons in amplified magnetic fields within the ejecta itself and its detection constrains theoretical models on emission and reabsorption mechanisms in shocked matter.

SN2002ap is an energetic ($E_{explosion} \sim 4 - 10 \times 10^{51}$ ergs [8]), one of the closest (in M 74, distance $d = 7.3$ Mpc [12]), type Ic SN which exploded on Jan. 28 ± 0.5, 2002 [8]. It was also one of the earliest detections in multiple bands, as in SN1987A and SN1993J. Interest in highly energetic ("hypernova") type Ib/Ic SNe has been intense after the temporal and positional near-coincidence of GRB 980425 with the type Ic SN1998bw implied a possible GRB-SN connection for the long duration GRBs.

We report our findings from the combined, near-simultaneous observations of SN2002ap with XMM-Newton, GMRT [14] and VLA [1]. For a detailed analysis of the X-ray data and early 610 and 1420 MHz observations of SN2002ap with GMRT (Giant Meterwave Radio Telescope), we refer the reader to [14].

2 XMM-Newton and GMRT Observations and Analysis

SN2002ap was detected by XMM-Newton EPIC CCDs between 5.025 to 5.42 days after explosion [10] (See also [13], [14]), with an effective exposure time of $\sim$ 25.5 ks. After subtracting the contribution from the nearby "contaminating" X-ray source CXU J013623.4+154459 in the 50″ spectral extraction region on the EPIC-PN CCD, we estimate the 0.3-10 keV flux from SN2002ap as $1.07^{+0.63}_{-0.31} \times 10^{-14}$ ergs cm^{-2} s^{-1}. Because the source is very faint, both thermal bremsstrahlung model ($N_H = 0.49 \times 10^{22}$ cm^{-2}, $kT = 0.8$ eV) and a simple powerlaw ($N_H = 0.42 \times 10^{22}$ cm^{-2}, spectral index $\alpha = 2.5$) fit the sparse spectra well, with $\chi^2/d.o.f. = 1.2/20$. Simultaneous XMM-Newton Optical Monitor (OM) observations of M74 in the UVW1 band yield a flux of $(7.667 \pm 0.002) \times 10^{-15}$ erg cm^{-2} s^{-1} Å^{-1}.

SN 2002ap was observed with the Giant Meterwave Radio Telescope (GMRT) at 0.61 GHz 8.56 days after explosion, and at 1.42 GHz 42 days after explosion. The exposure time for each observation was 4 hours. We did not detect the SN in either observation and the upper limits on the flux have been tabulated in Table 1.

3   Results and Discussion

The earliest radio detection of SN2002ap was $\sim$ 4.5 days after the explosion, in the 8.46 MHz band, and the frequency of the peak radio flux declined gradually from 8.46 GHz to 1.43 GHz over a period of 10 days from the explosion epoch (VLA observation, [1]) – a clear indication of the presence of relativistic electrons. This wavelength dependence of the radio turn-on can be due to either free-free absorption (FFA) in the thermally excited, homologously expanding matter overlying the interaction region, or synchrotron self-absorption (SSA) by the relativistic e^- responsible for radio emission, accelerated in regions close to the expanding interaction region [3]. In the case of SN 2002ap, the combined GMRT upper limits [14] and VLA observations [1] are best fit by the SSA model (Fig. 1) in the optically thin, spherically symmetric limit, consistent with the early optical observations. The corresponding values of spectral index α, magnetic field B, and the radius of the radiosphere R_r are tabulated in Table 2.

Table 1. GMRT observations of SN2002ap

Date of Obs.	ν (MHz)	Resolution (arcsec)	2σ Flux mJy	RMS mJy/ beam
5 Feb'02	610	9.5 x 6	< 0.34	0.17
8 Apr'02	1420	8 x 3	< 0.18	0.09

Table 2. Best Fit parameters of the SSA Model for SN2002ap on day 8.96. F_p is the peak flux emitted at frequency ν_p, B is the ambient magnetic field in the radiosphere at radius R_r.

α	ν_p GHz	F_p μJy	R_r cm.	B G
0.8	2.45	397	3.5×10^{15}	0.29

The early radio and X-ray turn-ons imply that the mass-loss rate of the progenitor was relatively low, at $\dot{M} \leq 6 \times 10^{-5}$ $M_\odot$ yr^{-1} ([14], using VLA detection by [1], and the SSA models for X-ray absorption by [2]).

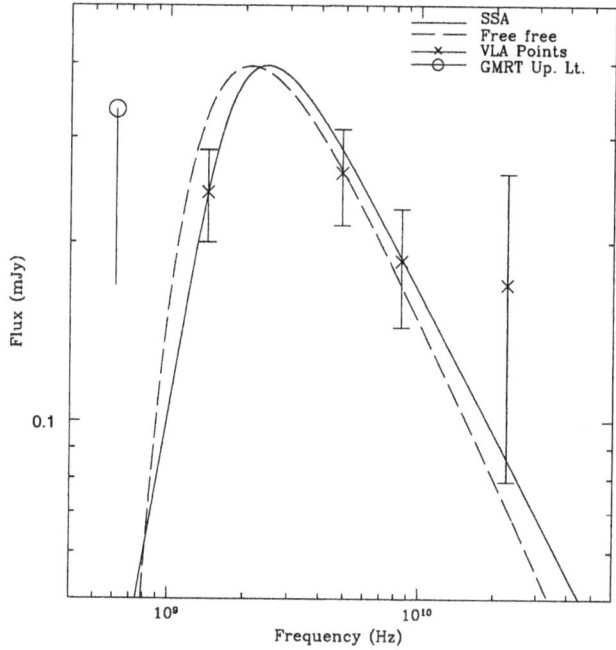

Fig. 1. Best fit synchrotron self-absorption spectrum (solid line) for SN2002ap radio emission on day 8.96 with energy spectral index = -0.8. Also shown is the free-free absorption model (dashed line) with parameters used taken from [1]. The VLA data is marked with a "×", while the GMRT upper limit at 610 MHz is marked by a "○"

Table 3. Comptonising Plasma Properties at t = 5d with x-ray energy index $\gamma = 3$ and $R_{opt} = 3.4 \times 10^{14}$ cm

Scenario	$\dot{M}_{-5}$ $\times 10^{-5}\ M_\odot\ yr^{-1}$	u_{w1} 10 km/s	τ_e $\times 10^{-4}$	T_e $10^9 K$
Wolf-Rayet	1.5	58	4	2
Case-BB Binary	10	58	25	1.5

3.1   Origins of X-Ray Emission

Using the optical observations of SN 2002ap [9] at the same epoch as the XMM-Newton observation, we find that the radius of the optical photosphere $R_{opt.} = 3.4 \times 10^{14}$ cm. We note that the radiosphere in the SN (Table 2) was well outside this optical photosphere at a similar epoch. However, Compton optical depth considerations ([14], [4]) suggest that the X-ray production region is closer to the optical photosphere.

Table 4. Least-squares fitted SSA equipartition parameters for SN1993J, SN1998bw and SN2002ap $\sim$ 11 days after explosion.

SNe	ν_p	F_p	θ_{eq}	U_{eq}	B_0	R_0
	GHz	mJy	μ as	$\times 10^{44}$ ergs	G	$\times 10^{15}$ cm
2002ap	2.45	0.48	39.0	6.9	0.47	4.80
1998bw	5.5	50.4	112.4	35000	0.23	68.4
1993J	30.5	22.3	17.6	5.0	3.54	1.08

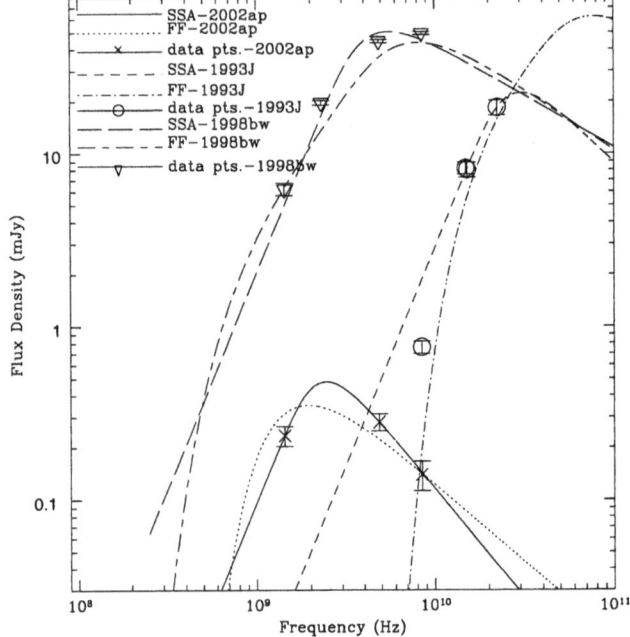

Fig. 2. Comparison of spectra of three SNe (SN1993J [15], SN1998bw [6] and SN2002ap [1]) near day 11 after explosion. Solid lines show the best fit synchrotron self absorption model and dashed lines show the corresponding free free absorption model

Free-free emission mechanisms are likely to be dominated by Compton cooling. The high ejecta velocity ($v \geq 20\,500$ km s^{-1} on day 3.5), coupled with high implied temperature of the shocked circumstellar medium ($T \sim 10^9$ K) and a limited cool absorbing shell suggests that the high energy photons would have a flat tail, up to $\sim$ 100 keV. By contrast, the X-ray spectra is quite soft with thermal bremsstrahlung $T_B = 0.8$ keV. Hence, free-free absorption is unlikely to explain the observed X-ray spectra. However, Compton cooling is the dominant radiative mechanism when the optical photospheric temperature is $T_{\text{eff}} \geq 10^4$ K.

Using the Monte Carlo simulations of Compton scattered spectra for the shocked CSM [11], we find that the observed flux by XMM can be accounted for both in terms of energetics and spectrum, if the electron plasma has $T_e \sim$ few times of 10^9 K (See [14]). Table 3 gives the properties of Comptonised plasma for two likely progenitor systems ([7,5]) for this event.

Finally, we compare the rarely observed early radio spectra of SN2002ap with that of SN1993J (type IIb) and SN1998bw (type Ic, GRB-association), 11 days after explosion. In Fig. 2, we fitted the SSA model to the data, and in Table 4 we estimate the energy U_{eq} in the radiating plasma and the magnetic field B_0, from equipartition arguments, and determine corresponding the angular radius θ_{eq} of the radio-sphere. Comparing θ_{eq} for the 3 SNe, the hydrodynamic shock in SN1998bw causes more rapid expansion leading to larger radio-sphere, while SN2002ap is similar to SN1993J across type classification irrespective of presence or absence of H- and He-envelope in the progenitor.

We thank the XMM-Newton helpdesk and the staff of GMRT, especially S. Bhatnagar, for many useful communications during data processing.

References

1. Berger, E., Kulkarni, S.R., & Chevalier, R.A., 2002, ApJ, 577, L5
2. Chevalier, R.A., & Fransson, C., 1994, ApJ, 420, 268
3. Chevalier, R.A., & Fransson, C., 2003, in "Supernovae and Gamma-Ray Bursts", ed. K. Weiler (Springer-Verlag)
4. Fransson, C., 1982, A&A, 111, 140
5. Habets, G., 1985, Ph.D. thesis, University of Amsterdam
6. Kulkarni, R., Frail, D.A., Wieringa, M.H. et al., 1998 Nature 395, 663
7. Langer, N., & Heger, N., 1999 in Proc. IAU Symp No. 193 ed. K. van der Hucht et al., Astronomical Society of the Pacific
8. Mazzali, P.A., Deng, J., & Maeda, K. et al., 2002, ApJ, 572, L61
9. Meikle, P., Lucy, L., Smartt, S. et al., 2002, IAUC7811
10. Pascual, P., Riestra, R., Garcia, B. et al., 2002, IAUC 7821
11. Pozdnyakov, L.A., Sobol, I.M., & Sunyaev, R.A., 1977, Soviet Astron. 21, 708
12. Sharina, M.E., Karachentsev, I.D., & Tikhonov, N.A., 1996, A&AS, 119, 499
13. Soria, R., Kong, A.K.H., 2002, ApJ, 572, L33
14. Sutaria, F.K., Chandra, P., Bhatnagar, S. and Ray, A., astro-ph/0207137, accepted for publication, A&A
15. van Dyk, S.D., Weiler, K.W., Sramek, R.A. et al., 1994, ApJ 432 L115-L118

Search for Very High Energy γ-Rays from the SNR 1006 with the HEGRA CT1 Air Imaging Cherenkov Telescope

Vincenzo Vitale[1] and the HEGRA Collaboration

[1] Max Planck Institute for Physics, Föhringer Ring 6, 80805 Munich, Germany

Abstract. The observation of γ-rays from SuperNova Remnants (SNRs) allows one to study the acceleration of charged particles ([1,2]) by SuperNova blast waves. The remnant of SN 1006 is a candidate for such studies: its 2 rims (NE and SW) have been detected as X-ray non-thermal sources by ASCA [3] and ROSAT [4] satellites, while TeV γ-ray emission has been observed from NE rim by the CANGAROO collaboration [5]. Both KeV and TeV radiation can be explained by synchrotron and inverse Compton emission of the same electron population (energy $\sim$50 TeV [6]). The HEGRA Collaboration searched for TeV emission from SNR 1006 from 1999 to the 2002. A preliminary analysis of these data is presented. From the experimental location one can observe SN1006 only at large Zenith angle (ZA), for this the data analysis requires substantial modifications. The analysis shows evidence for 5.5 σ excess at NE rim.

1 Cosmic Ray Acceleration

According to many theories, cosmic rays, up to thousands of TeV per unit of charge, should originate mainly from SuperNova Remnants [7]. The main arguments in favour of SNRs are:

1. the SN explosions are frequent enough to power the Galactic Cosmic Rays (GCRs), using around 10 and 30% of the kinetic energy which they release;
2. the diffusive shock acceleration can transfer the energy from shocks to the charged particles of the interstellar medium (ISM) and produce a power law energy spectrum;
3. the difference between the energy spectral index of GCRs at the supposed source (SNRs) and that observed at Earth could be attributed to the GCRs propagation.

The presence of very high energy charged particles in the neighbours of SNRs can be studied from the resulting γ-rays.

2 Detection of V.H.E. γ Using Imaging Air Cherenkov Telescopes

Current imaging air Cherenkov telescopes (IACTs) proved to be able to detect gamma rays in the range energy from 100's of GeV to 10's of TeV. IACTs are sensitive to Cherenkov light produced by the showers of secondary particles developed in the interactions of gamma rays or hadronic cosmic rays with the upper atmosphere. IACTs comprise a large mirror of modest optical quality and a camera of photomultipliers placed in the focal plane. The camera allows for the fast acquisition of shower images ($\sim 10\,$nsec). The shower image gives informations about the type of primary particle (nucleus or γ), its energy and incoming direction; this information can be extracted by using a defined parametrization of the images (Hillas parameters [8]). Applying selection cuts to the image parameters it is possible to strongly suppress the dominant hadronic background. The image features have a strong zenith angle dependence.

3 The Observation of SN1006

The HEGRA collaboration searched for γ emission from SN1006 with the modified prototype CT1 telescope, which is located in La Palma (28.8°, 17.8°,2250m a.s.l.), Canary Islands; then SN1006 can be observed only at large ZA (>70°).

3.1 High Zenith Angle Observations

Data are pre-selected according to optimal atmospheric conditions and performance of the telescope, then a 'cleaning' procedure is applied to minimize the noise

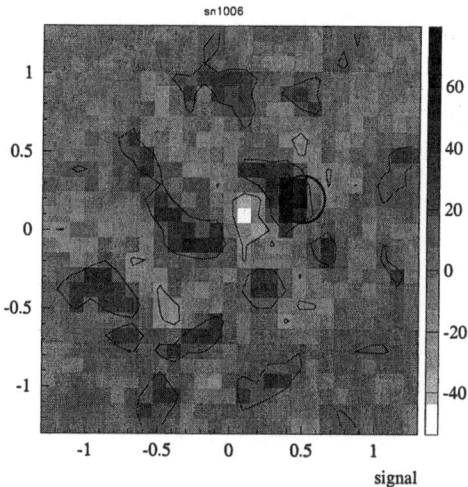

Fig. 1. Sky map of the excess (number of events after selection cuts and estimated background subtraction) over ΔRA (x axis [°])-ΔDEC (y axis [°]), pointing to the center of the SNR1006. Data from 1999-00-01, 219 hours of obs. The black circle points to the North East Rim, from where TeV γ-rays have been observed by the CANGAROO collaboration. The excess found at this position has a significance of 5.5 *sigma*, it is the largest in the FoV. Estimating high statistics background from a circular segment with radius 0.4° and width 0.08° (excluding the NE rim position) the excess has also higher significance

coming from night sky background light. Calculation of the Hillas parameters is performed using the distributions of photoelectrons in the images [9]. For high Zenith Angle (ZA) observations the atmospheric depth and the distance between core of the shower and observer increase ($\sim \frac{1}{cos(ZA)}$). Therefore shower images undergo major changes:

1. reduction of images dimensions compared to the low ZA observations;
2. smaller distance between image c.o.g. and source location (usually at the center of the camera).

These effects have been studied on real hadronic data as function of ZA. This study allows one to transform a well tested set of cuts for γ/h separation at low ZA to cuts for high ZA observations. The γ/h separation cuts for low ZA have been tested on experimental data containing strong signals, which have been measured, obtaining results in agreement with the standard analysis, both for centered and offset sources. In order to study the level of fluctuations of the pure background, both high and low γ/h separation cuts have also been applied to data taken pointing to sky regions without sources (Off-Data). High ZA cuts include also a section for the rejection of cosmic muons directly interacting with photomultipliers.

3.2   Method for the Search of Extended Sources

The gamma-hadron separation for CT1 data relies on the assumption to have a point like source in the center of the Field of View. The search for extended

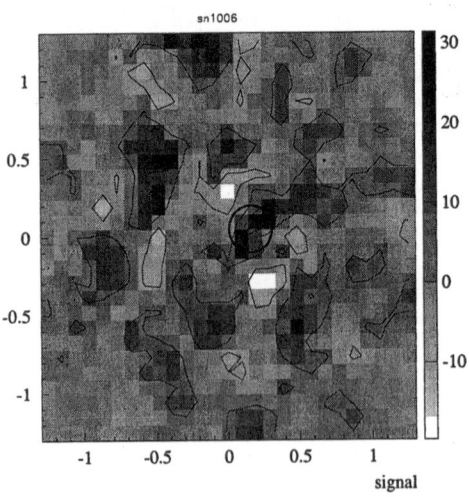

Fig. 2. Sky map of the excess (number of events after selection cuts and estimated background subtraction) over ΔRA (x axis [°])-ΔDEC (y axis [°]), pointing around the NE rim of SNR 1006. Data from 2002, 93 hours of obs. The black circle points to the North East Rim. The excess near the center has a significance of 3 *sigma*

and offset sources requires a modified strategy. The FoV is divided in a grid with spacing given by 1σ angular resolution of the telescope. Then each point is assumed as a possible source of radiation and the excess is calculated. The significance of the found excess has eventually been corrected by number of trials (i.e. number of grid points). For the NE rim of SNR1006 one has to deal basically with a single trial test or with very few trials. The pointing has been checked using known position of the stars in the FoV, looking at the level of anode currents of photomultipliers; the pointing of telescope was found to be correct within $0.1°$ (large ZA). The sky map of the estimated background, the excess sky map and its significance have been obtained for all the data samples taken on the SN1006. Data from 1999, 2000, and 2001 have been merged (Fig. 1), but not those from 2002 (Fig. 2) for which the telescope pointing is different.

4 Results and Conclusions

A preliminary analysis of data taken by HEGRA CT1 telescope on the SNR 1006 has been performed. The largest excess found in the FoV of the telescope has a significance of 5.5 σ (single trial) for the combined 1999, 2000, 2001 data sample (219 hours) and $\sim$3 σ for 2002 sample (93 hours). For both samples the excess is located around the NE Rim of the remnant. During the 2002 observations the telescope pointing has been deliberately shifted of 0.4 deg. Dedicated Monte Carlo simulation for γ-ray air showers at large ZA are now in preparation in order to allow one to cross check the results, and the energy and efficiency calibrations. If the excess is interpreted as signal and assuming $3 \times 10^5\text{m}^2$ for the effective collection area, 20 ± 5 TeV for energy threshold and a pure power law (spectral index = 2.3) for the energy spectrum, then the following integral flux is estimated: $\Phi(\text{E}>25~\text{TeV}) = (6 \pm 5) \times 10^{-14}cm^{-2}s^{-1}$ The observations of SN1006 made by CT1 are an example of the capabilities of IACTs at large ZA.

References

1. Drury L.O'C., Aharonian F., Völk H.J. 1994, A&A 287, 959
2. Naito T., Takahara F.1994, J.Phys.G 20, 477
3. Koyama, K. et al., Nature 378, 255-258, 1995
4. Willingale, R., et al., Mon.Not.R.Astron.Soc., 278, 749, 1996
5. Tanimori, T., et al., Astrophys.J., 497, L25-L28, 1998
6. Tanimori, T., et al., I.C.R.C. 2001, Hamburg
7. Baring M.G., et al., Snowbird Workshop on TeV Gamma-Ray Astronomy, Towards a Major Atmospheric Cherenkov Detector - VI, 173, AIP, New York, ISBN 1-56396-938-6
8. Hillas A.M. Proc. 19th ICRC (La Jolla) 3, 445
9. Petry D., Ph.D. Thesis, 1997

Part VI

Supernovae and Their Environment

Circumstellar Interaction Around Supernovae

Roger A. Chevalier

Department of Astronomy, University of Virginia, P.O. Box 3818,
Charlottesville, VA 22903, USA

Abstract. Circumstellar interaction has been observed around all types of massive star supernovae, especially at radio and X-ray wavelengths. The interaction shells in Type Ib/c supernovae appear to be moving rapidly, although SN 1998bw remains the only case for which the evidence points to relativistic expansion. Type IIP supernovae have relatively low mass loss rates, consistent with moderate mass single stars at the ends of their lives. Type IIb and some IIL and IIn supernovae can have strong mass loss leading up to the explosion, which occurs in a star with little H on it. The mass loss may be driven by the luminosity of the late burning stages.

1   Introduction

During their lifetimes, massive stars undergo mass loss during various evolutionary phases. When they finally end their lives as supernovae, the interaction reveals the mass loss processes leading up to the explosion. and the structure of the exploding star. Recent multiwavelength observations have revealed many facets of the interaction and we now have observations showing circumstellar interaction around all types of massive star supernovae. Circumstellar interaction can be related to late massive star evolution and to the physical processes in the dense gas interaction. Analysis of the radio and X-ray emission from supernovae yields the energy in high velocity ejecta and can be used to set limits on the amount of relativistic ejecta; the relation of supernovae to gamma-ray bursts can thus be explored.

A recent review of circumstellar interaction with an emphasis on the relevant physical processes is in [9]. Here, I briefly review recent developments, emphasizing the relation of circumstellar interaction to the various supernova types. In § 2, I discuss the wide range of multiwavelength observations of circumstellar interaction that is now available. Interaction around the various types of supernovae is discussed in § 3. Conclusions are in § 4.

2   Multiwavelength Observations of Interaction

One of the most sensitive ways of finding or observing supernovae with circumstellar interaction is through their nonthermal, radio synchrotron emission. The radio light curves of extragalactic supernovae show an early absorbed phase, so that there is a rise to a maximum followed by an approximately power law decline [58]. Figure 1 shows the relationship between the observed peak luminosity

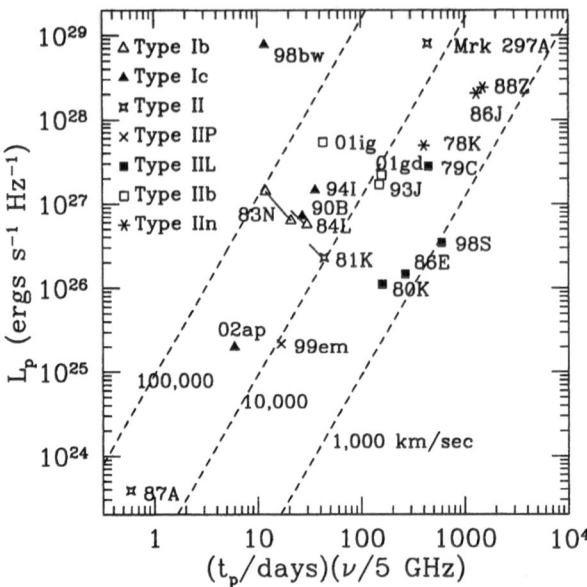

Fig. 1. Peak luminosity and corresponding epoch for the well-observed radio super-novae. The dashed lines give curves of constant expansion velocity, *assuming* synchrotron self-absorption (see [7])

and the age at peak luminosity (from [7], with updates). Recent detections are SN 2001ig [48], SN 2001gd [52], and SN 2002ap [4]. All types of massive star supernovae are represented, including Type Ib/c which are thought to have lost their hydrogen envelopes before the explosion, Type IIb which have lost almost all their hydrogen envelopes before the explosion, Type IIn with narrow line optical emission, Type IIL with linear optical light curves, Type IIP with plateau light curves, and the peculiar SN 1987A. Interestingly, the different types of supernovae populate different parts of this diagram. Although the numbers are still small, some trends are showing up in the interaction properties. The Type Ib/c supernovae peak early and have a range of luminosity, while the Type IIL and Type IIn SNe peak late and also have a range of luminosity. The Type IIb are intermediate between these and the one Type IIP is of low luminosity. The early peak of the Type Ib/c supernovae is an indication that their radio emitting regions expand relatively rapidly [7].

The other wavelength region where young supernova remnants are broadly observed is X-rays. Figure 2 shows the relation between the X-ray and radio luminosity of young remnants. Here, the radio luminosity is estimated from νL_ν at 5 GHz. X-ray emission has been detected from the full range of massive star supernovae, and there is a clear correlation of the X-ray and radio emission. The X-ray luminosities, primarily from [26], are not all in the same X-ray band, but this should not have a large effect on the correlation. X-ray line emission has

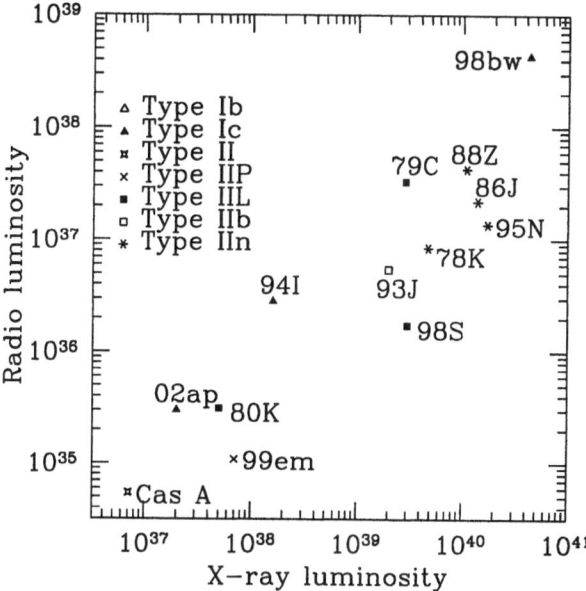

Fig. 2. Radio luminosity vs. X-ray luminosity of supernovae, both in ergs s^{-1}. The X-ray data are primarily from [26]

been detected in a number of cases (SN 1987A [43], SN 1986J [24], SN 1993J [54], SN 1998S [46], Cas A), showing that the emission is thermal, at least in these cases. The correlation of the X-ray and radio emission shows that both of these give a measure of the strength of circumstellar interaction, even though they involve different radiation mechanisms. Radio emission provides a good indicator of whether a particular event is suitable for observation with the space missions *Chandra* and *XMM*.

At UVOIR (ultraviolet/optical/infrared) wavelengths, circumstellar interaction is generally not observed in Type Ib/c and Type IIP supernovae, but is observed in the spectra of Type IIL, IIb, and IIn supernovae. The reason appears to be the relatively low densities and high velocities present in Type Ib/c and Type IIP supernovae, so that radiative shock waves are not present; in addition, the X-ray luminosities are not sufficient to illuminate the expanding ejecta to observable levels. In the cases with stronger circumstellar interaction, UVOIR emission is potentially observable from shocked, cooled gas, from freely expanding ejecta, and from unshocked circumstellar gas. In the case of SN 1995N, all of these components have been observed [21]; ultraviolet *HST* observations are especially useful for observing the ejecta.

Strong infrared emission, presumably from dust, has been observed from the Type IIL supernovae SN 1979C and 1980K [13] and, recently, from a number of Type IIn supernovae [15,12,23]. The preferred explanation has been emission from radiatively heated circumstellar dust. Another possible source of emission is collisionally heated dust in the hot, shocked gas, as is observed in a number of Galactic supernova remnants, including Cas A [14]. The high observed luminosity implies that the source is not dust formed in the freely expanding ejecta.

3   Circumstellar Interaction and Supernova Types

3.1   Type Ib/c Supernovae and "Hypernovae"

The best observed Type Ic supernova was SN 1994I in M51, which had a rapid decline after maximum light and can be interpreted as an explosion with an ejecta mass of only $0.9M_\odot$ [30,62]. It is the only normal Type Ic supernova to have been detected in X-rays [27,29]. Chevalier [7] applied a synchrotron self-absorption model to the available radio data on SN 1994I and found an implied shock radius of about 1.2×10^{16} cm at an age of 36 days, giving a shock velocity of $39,000$ km s^{-1}. Application of the asymptotic power law density profile of [40] ($\rho \propto r^{-10.1}$) yields an outer shock radius $R = 1.7 \times 10^{16}(M/M_\odot)^{-0.32}(E/10^{51}$ ergs$)^{0.44}(\dot{M}_{-5}/v_{w3})^{-0.12}$ cm, where M is the ejecta mass, E is the explosion energy, $\dot{M}_{-5} = \dot{M}/10^{-5}M_\odot$ yr^{-1} is the mass loss rate, and $v_{w3} = v_w/10^3$km s^{-1} is the wind velocity [4]. The approximate agreement of the hydrodynamic model with the radio radius may be improved by using a more accurate density model for the exploded star, because the density profile is not fully in the asymptotic power law regime for such a small mass star. Immler et al. [29] have used *Chandra* observations of SN 1994I to deduce a presupernova mass loss rate of $1 \times 10^{-5}M_\odot$ yr$^{-1}(v_w/10$ km s$^{-1})$. For a Wolf-Rayet star wind velocity $v_w \approx 10^3$ km s^{-1}, the inferred mass loss rate of $1 \times 10^{-3}M_\odot$ yr^{-1} is surprisingly high. However, Immler et al. [29] used a model for the interaction from Chevalier [6], which was developed before the recognition of Type Ib/c supernovae as a separate type and was based on a Type Ia interpretation. If the high mass loss rate holds up in a more detailed model, it may be evidence for mass loss from a companion star, as has previously been suggested based on radio observations of Type Ib/c supernovae [56].

Much attention has recently been focused on Type Ic supernovae because of the occurrence of SN 1998bw, apparently associated with the gamma-ray burst GRB 980425 [22]. The strong radio emission from the source at early times (Fig. 1) implied the presence of relativistic motions, further supporting the GRB connection [33,35]. SN 1998bw also had an unusual optical spectrum, with velocities in lines extending out to $\sim 60,000$ km s^{-1} [22]. The interpretation of the supernova light curve and spectra led to an explosion energy estimate of $(2-4) \times 10^{52}$ ergs, extraordinarily high for a supernova [31,61]. Other Type Ic supernovae have been discovered with broad lines in their spectra, including SN 1997ef, SN 2002ap, and SN 2002bl. Because of the high energies inferred for these events, Mazzali et al. [42] and others have called these objects hypernovae.

Outside of SN 1998bw, broad-lined Type Ic supernovae have not been directly connected with GRBs. In the case of SN 2002ap, there are limits on any corresponding GRB [25]. Despite its broad line spectrum, SN 2002ap was a weaker radio source than the normal SN 1994I (Fig. 1; [4]). Its position in Fig. 1 indicates a higher M_{ej} or lower E than SN 1994I, plus a lower circumstellar density. The model of Mazzali et al. [42] has $M_{ej} = 2.4M_\odot$, which can account for part of the difference, but there is no evidence for the high energy $(4 \times 10^{51}$ ergs) that they infer for the explosion. X-ray observations of SN 2002ap with *XMM* also showed a surprisingly low luminosity [47]. This emission is sensitive to the distribution of high velocity gas and thus sets additional constraints to those obtained from optical observations. Sutaria et al. [53] suggested that inverse Compton is the most likely X-ray emission mechanism because of the relatively soft spectrum of the source. Approximate modeling of the radio emission in [4] showed that the energy in high velocity ejecta is small. The claim of high energies in broad line SN Ic is based on modeling the broad optical absorption lines. However, the radio emission probably gives a more direct estimate of the energy in high velocity ejecta and the possible relation to GRBs. The broad line Type Ic SN 2002bl was also a much weaker radio source than SN 1998bw [3].

The combination of radio and X-ray observations with a hydrodynamic model for the exploded star that is consistent with the observed optical supernova emission should make it possible to estimate the mass loss density around the supernova progenitor. This information will be useful in determining the progenitor evolution leading to an explosion. Type Ib/c supernovae occur at higher rates than expected for single stars that have lost their envelopes and binary evolution is a likely factor (e.g., [59]). The mass loss properties remain uncertain parameters for the presupernova evolution of these objects.

3.2 Type IIb Supernovae

The Type IIb supernova SN 1993J was in M81 at a distance of only about 3.6 Mpc and has become the second best observed case of circumstellar interaction, after SN 1987A. The exploded star in this case is thought to be an initially $12 - 15\ M_\odot$ star that lost most of its envelope mass as a result of binary interaction [45,60].

VLBI imaging shows an approximately symmetric, but variable, shell of radio emission [2]. The ratio of shell thickness to radius is $\Delta R/R \sim 0.3$ [37], which is larger than expected in standard interaction models [5]. A possible explanation is that inhomogeneities in the wind lead to a broader region. Although this specific situation has not been modeled, there is evidence that shock interaction with a clumpy medium gives rise to a broader, more turbulent shocked medium than would otherwise be present [32]. Relatively large scale inhomogeneities may be related to the variability that has been observed in the radio structure [2]. An examination of the optical line profiles of the supernova also gives an indication of inhomogeneity. Without inhomogeneities, the corrugation of the reverse shock wave is relatively small [10], which would produce a more box-like Hα line than was observed [39].

There has been some confusion regarding the ambient density profile for SN 1993J. Assuming a form $\rho_w \propto r^{-s}$, both radio [57,20,38,44] and X-ray [28] studies have led to $s = 1.5 - 1.7$. This is a surprising result because it implies a mass loss rate decreasing as a power law in time leading up to the supernova; Immler et al. [28] deduce a gradual reduction in $\dot{M}$ by > 10 based on the X-ray light curve from *ROSAT* observations. However, the initial radio studies assumed that free-free absorption was the primary absorption process. Fransson & Björnsson [19] showed that other physical processes, especially synchrotron self-absorption, are important for the radio curve and a detailed model yields $s = 2$, which is the expected value close to a massive star. The X-ray result of Immler et al. [28] appears to be due to an interpretation of the slowly decaying X-ray flux in terms of an adiabatic shocked region. In fact, the soft X-ray emission in the *ROSAT* band is dominated by emission from the reverse shock, which we expect to be radiative [20]. The X-ray spectra from *ASCA* give further evidence for a cool shell formed downstream from a radiative shock [55]. In the radiative case with $s = 2$, the X-ray emission is expected to be relatively constant when it is not absorbed [8]. Swartz et al. [54] have obtained a *Chandra* spectrum of SN 1993J at an age of 7 years. It shows cool and hot components that can be identified with the reverse shock wave and the forward shock, respectively. The cool component itself requires two temperature components (0.35 and 1.01 keV) to fit the spectrum. This may be an indication that the emission is from a radiatively cooling shock front.

3.3   Type IIn/IIL Supernovae

The Type IIn and IIL designations are not necessarily distinct; for example, SN 1998S had a light curve that led to a IIL designation [36], but also had narrow lines, leading to a IIn designation [34]. SN 1995N and SN 1998S are the best observed recent supernovae in the IIn category. There have been recent *HST* [21] and X-ray [18,46] observations of these supernovae. They both have X-ray luminosities in the range $10^{40} - 10^{41}$ erg s^{-1}, are strong radio emitters, and show narrow lines in their optical spectra. Both supernovae show evidence for the reverse shock front moving into heavy element rich gas. In the case of SN 1995N, the UV/optical spectra show evidence for emission from photoionized gas just inside the reverse shock front, including the OI $\lambda7774$ recombination line, but there is no H emission associated with this component [21]. *Chandra* spectra for SN 1995N are not yet available, but summed *Chandra* spectra of SN 1998S show evidence for an overabundance of the Si group elements [46]. Thus, although these supernovae were of Type II, with hydrogen in their spectra at early times, they developed evidence for a reverse shock front moving into enriched material within several years of the explosion. In addition, Leonard et al. [34] noted that the photospheric spectrum of SN 1998S at an age of 25 days resembled that of a Type Ic supernova, with absorption lines of OI $\lambda7774$ and SiII $\lambda6355$ and weak Hα emission. The implication is that strong mass loss just before the supernova almost entirely removed the hydrogen envelope.

Additional observations of SN 1998S have shown a number of interesting features. The evolution of the optical spectrum has been interpreted as indicating several regions of mass loss through which the supernova shock front is moving [34,15]. The supernova, which was discovered on 3 March, 1998, showed symmetric broad lines with FWZI (full width at zero intensity) of $\sim 20,000$ km s^{-1} until

12 March, 1998. The sudden disappearance of the lines can be interpreted as due to the shock wave overtaking an inner dense region. Chugai [11] has recently interpreted the broad lines as due to scattering in a dense circumstellar medium out to $\sim 10^{15}$ cm. This requires $\dot{M}_{-5}/v_{w1} \gtrsim 300$ in this region. Leonard et al. [34] find little evidence for subsequent circumstellar interaction until an age of 108 days. The later X-ray and radio observations imply $\dot{M}_{-5}/v_{w1} \sim (10-20)$ [46]. An early infrared excess implies the presence of circumstellar dust [15], which places additional constraints on the circumstellar density and dust content.

Fassia et al. [16] observed narrow emission lines and inferred that the outer circumstellar medium had a velocity of $40-50$ km s^{-1}. From the time evolution and velocities, they inferred that the outer wind started at least 170 years ago, stopped about 20 years ago, and the inner dense wind may have started less than 9 years ago. The variability might be related to the mass loss variability observed in late type stars.

The structure of red supergiant winds is uncertain, because they are generally quite distant. However, red giant AGB (asymptotic giant branch) stars are known to have a semi-periodic shell structure, observed in scattered light (e.g., IRC +10216, [41]) as well as in the *HST* images of several planetary nebulae (e.g., [1]). The characteristic time scale for the shells, several 100 years, is indicative of an instability in the acceleration mechanism for the wind. Similar mass loss processes are likely to occur in red supergiants; the mass loss characteristics of IRC +10216 are $v_w = 14$ km s^{-1} and $\dot{M} = 2 \times 10^{-5} M_\odot$ yr^{-1}, which are comparable to values for red supergiants. The shell properties for IRC +10216 can be determined from observations of dust scattered light [41]. The shells have a density contrast of a factor 3 over the rest of the wind and contain a substantial fraction of the total mass in the wind. If the supernova is expanding at $\sim 10^4$ km s^{-1}, the interaction should lead to structure in the supernova light curves on a timescale of ~ 100 days.

It is also possible that the mass loss properties are determined by the late nuclear burning phases just before the explosion. When there is very little H envelope left on a star, the nuclear processes can play a role in the mass loss properties. Both SN 1995N and SN 1998S appear to have exploded at a time of strong mass loss when the H envelope was almost entirely lost. SN 1993J also exploded with a modest H envelope (few $0.1 M_\odot$); in this case, the small envelope mass has been attributed to binary interaction [45,60]. However, the mass loss from the envelope should then be especially strong when the star first becomes a red supergiant and is not very powerful at the time of the explosion. SN 1993J in fact had strong mass loss at the time of the explosion and in the cases of SN 1995N and SN 1998S, the mass loss was even stronger. These observations suggest the possibility that the heightened luminosity in the late burning stages plays a role in driving the strong mass loss that leads to the loss of the H envelope.

3.4   Type IIP Supernovae

There is little evidence for circumstellar interaction around Type IIP supernovae, but the evidence that we have indicates a low circumstellar density. Figure 1 shows the relatively low radio luminosity of SN 1999em. The X-ray luminosity of SN 1999em is correspondingly low (Fig. 2; [46]) and the X-ray luminosity of the Type IIP SN 1999gi is similarly low [50]. Approximate models for the X-ray emission lead to a mass loss rate of $(1 - 2) \times 10^{-6} M_\odot$ yr$^{-1} (v_w/10$ km s$^{-1})$.

Information on the initial mass of supernovae comes from studies of their stellar environments, especially with *HST*; such studies have been carried out for the Type IIP SNe 1999em and 1999gi [51]. These studies indicate that these Type IIP supernovae may have come from $8 - 12$ $M_\odot$ stars. Type IIP supernovae are compatible with single star evolution and initial masses $\sim 8 - 15 M_\odot$ and relatively little mass loss during the evolution. This is consistent with the evolutionary models of Schaller et al. [49] who find that these stars end their lives as RSGs with $\dot{M} \sim 2 \times 10^{-6} M_\odot$ yr^{-1}, in agreement with the properties noted above.

4   Conclusions

Early circumstellar interaction gives rise to a number of interesting physical processes in dense media, including radiative $\sim 1,000$ km s^{-1} shock waves, particle acceleration in fast shock waves, and the emission of infrared dust echoes. The conditions in circumstellar interaction approach those in quasars; in fact, some supernovae which are likely circumstellar interactors have spectra similar to Seyfert galaxies [17], but the physical situation is better understood for the supernova case. Analysis of the observations shows the mass loss processes leading to the supernova.

The observational support for these studies is promising. The *Chandra* and *XMM-Newton* X-ray missions have made the X-ray emission from nearby supernovae accessible. *SOFIA* and *SIRTF* should contribute to our understanding of infrared dust emission from supernovae. Together with ground-based observations, the detailed picture of supernova interaction with circumstellar matter should give considerable insight into the late evolution of massive stars.

I am grateful to Claes Fransson for collaboration and discussion of these topics. This research was supported in part by NASA grant NAG 5-8232 and grant GO-08648.03-A from the Space Telescope Science Institute, which is operated by the Association of Universities for Research in Astronomy, Inc., under NASA contract NAS 5-26555.

References

1. B. Balick, J. Wilson, A.R. Hajian: AJ **121**, 354 (2001)
2. N. Bartel, et al.: Science **287**, 112 (2000)
3. E. Berger, et al.: GCN No. 1266 (2002)

4. E. Berger, S. R. Kulkarni, R. A. Chevalier: ApJ **577**, L5 (2002)
5. R.A. Chevalier: ApJ **258**, 790 (1982)
6. R.A. Chevalier: ApJ **285**, L63 (1984)
7. R.A. Chevalier: ApJ **499**, 810 (1998)
8. R.A. Chevalier, C. Fransson: ApJ **420**, 268 (1994)
9. R.A. Chevalier, C. Fransson: In *Supernovae and Gamma-Ray Bursts*. ed. by K. W. Weiler, (Springer, Berlin 2003) in press (astro-ph/0110060)
10. R.A. Chevalier, J.M. Blondin, R.T. Emmering: ApJ **392**, 118 (1992)
11. N.N. Chugai: MNRAS **326**, 1448 (2001)
12. E. DiCarlo, et al.: ApJ **573**, 144 (2002)
13. E. Dwek: ApJ **274**, 175 (1983)
14. E. Dwek: ApJ **322**, 812 (1987)
15. A. Fassia, et al.: MNRAS **318**, 1093 (2000)
16. A. Fassia, et al.: MNRAS **325**, 907 (2001)
17. A.V. Filippenko: AJ **97**, 726 (1989)
18. D.W. Fox, et al. MNRAS **319**, 1154 (2000)
19. C. Fransson, C.-I. Björnsson: ApJ **509**, 861 (1998)
20. C. Fransson, P. Lundqvist, R.A. Chevalier: ApJ **461**, 993 (1996)
21. C. Fransson, et al.: ApJ **572**, 350 (2002)
22. T.J. Galama, et al.: Nature **395**, 670 (1998)
23. C.L. Gerardy, et al.: ApJ **575**, 1007 (2002)
24. J.C. Houck, J.N. Bregman, R.A. Chevalier, K. Tomisaka: ApJ **493**, 431 (1998)
25. K. Hurley, et al.: GCN No. 1252 (2002)
26. S. Immler, S., W. H. G. Lewin: In *Supernovae and Gamma-Ray Bursts*. ed. by K. W. Weiler, (Springer, Berlin 2003) in press (astro-ph/0202231)
27. S. Immler, W. Pietsch, B. Aschenbach: A&A **331**, 601 (1998)
28. S. Immler, B. Aschenbach, Q. D. Wang: ApJ **561**, L107 (2001)
29. S. Immler, A. S., Wilson, Y. Terashima: ApJ **573**, L27 (2002)
30. K. Iwamoto, K. Nomoto, P. Höflich, H. Yamaoka, S. Kumagai, T. Shigeyama: ApJ **437**, L115 (1994)
31. K. Iwamoto, et al.: Nature **395**, 672 (1998)
32. B. Jun, T.W. Jones, M.L. Norman: ApJ **468**, L59 (1996)
33. S.R. Kulkarni, et al.: Nature **395**, 663 (1998)
34. D.C. Leonard, A.V. Filippenko, A.J. Barth, T. Matheson: ApJ **536**, 239 (2000)
35. Z.-Y. Li, R. A. Chevalier: ApJ **526**, 716 (1999)
36. Q.-Z. Liu, J.-Y. Hu, H.-R. Hang, Y.-L. Qiu, Z.-X. Zhu, Q.-Y. Qiao: A&A **144**, 219 (2000)
37. J.M. Marcaide, et al.: Science **270**, 1475 (1995)
38. J.M. Marcaide, et al.: ApJ **486**, L31 (1997)
39. T. Matheson, A.V. Filippenko, L.C. Ho, A.J. Barth, D.C. Leonard: AJ **120**, 1499 (2000)
40. C.D. Matzner, C.F. McKee: ApJ **510**, 379 (1999)
41. N. Mauron, P.J. Huggins: A&A **359**, 707 (2000)
42. P.A. Mazzali, et al.: ApJ **572**, L61 (2002)
43. E. Michael, et al.: ApJ **574**, 166 (2002)
44. A.J. Mioduszewski, V.V. Dwarkadas, L. Ball: ApJ **562**, 869 (2001)
45. K. Nomoto, T. Suzuki, T., Shigeyama, S. Kumagai, H. Yamaoka, H. Sato: Nature **364**, 507 (1993)
46. D. Pooley, et al.: ApJ **572**, 932 (2002)
47. P. Rodriguez Pascual, et al.: IAUC No. 7821 (2002)

48. S. Ryder, K. Kranz, E. Sadler, R. Subrahmanyan: IAUC No. 7777 (2001)
49. G. Schaller, D. Schaerer, G. Meynet, A. Maeder: A&AS, **96**, 269 (1992)
50. E.M. Schlegel: ApJ **556**, L25 (2001)
51. S.J. Smartt, G.F. Gilmore, C.A. Tout, S.T. Hodgkin: ApJ, **565**, 1089 (2002)
52. C.J. Stockdale, et al.: IAUC No. 7830 (2002)
53. F.K. Sutaria, P. Chandra, S. Bhatnagar, A. Ray: A&A, accepted (astro-ph/0207137) (2002)
54. D.A. Swartz, K.K. Ghosh, M.L. McCollough, T.G. Pannuti, A.F. Tennant, K.Wu: ApJ, submitted (astro-ph/0206160) (2002)
55. S. Uno, et al.: ApJ **565**, 419 (2002)
56. S.D. van Dyk, R.A. Sramek, K.W. Weiler, N. Panagia: ApJ **409**, 162 (1993)
57. S.D. van Dyk, K.W. Weiler, R.A. Sramek, M.P. Rupen, N. Panagia: ApJ **432**, L115 (1994)
58. K.W. Weiler, N. Panagia, M. J. Marcos, R.A. Sramek: ARA&A **40**, 387 (2002)
59. S. Wellstein, N. Langer: A&A **350**, 148 (1999)
60. S.E. Woosley, R.G. Eastman, T.A. Weaver, P. A. Pinto: ApJ **429**, 300 (1994)
61. S.E. Woosley, R.G. Eastman, B.P. Schmidt: ApJ **516**, 788 (1999)
62. T.R. Young, E. Baron, D. Branch: ApJ **449**, L51 (1995)

Constraining Circumstellar Matter in SNe Ia – High-Resolution Optical Studies with VLT/UVES*

Peter Lundqvist[1], Jesper Sollerman[1], Bruno Leibundgut[2], E. Baron[3], Claes Fransson[1], and Ken'ichi Nomoto[4]

[1] Stockholm Observatory, AlbaNova, Department of Astronomy, SE-106 91 Stockholm, Sweden
[2] European Southern Observatory, Karl-Schwarzschild-Strasse 2, D-85748 Garching bei München, Germany
[3] Department of Physics and Astronomy, University of Oklahoma, Norman OK 73019-0225, USA
[4] Department of Astronomy and the Research Center for the Early Universe, University of Tokyo, Tokyo 113-0033, Japan

Abstract. Since April 2000, we have conducted a program using the ESO/VLT to search for early circumstellar signatures from nearby supernovae.[1] Until now, two Type Ia supernovae have been observed, SNe 2000cx and 2001el. Here we report on results for SNe 2000cx, and we discuss how the observations can be used together with detailed modeling to derive an upper limit on the putative wind from the progenitor system. For a hydrogen-rich wind with velocity 10 km s^{-1}, the mass loss rate for the progenitor system of SN 2000cx is $\dot{M} \lesssim 9 \times 10^{-6}$ M$_\odot$ yr^{-1}. This rivals existing limits from radio and X-rays of previous SNe Ia which, however, all need revision.

1 Introduction

The origin of Type Ia supernovae (SNe Ia) is still unknown. While it is generally believed that they result from the explosion of a white dwarf in a binary system, we are ignorant about the nature of the companion star. Branch et al. [1] list possible types of systems, and argue that the most likely system is a C-O white dwarf which accretes matter from the companion, either through Roche lobe overflow, or as a merger with another C-O white dwarf.

To reveal the nature of SNe Ia, we need methods to discriminate between possible progenitor scenarios. In non-merging scenarios a wind from the companion star is expected. If the wind is ionized and dense enough, it could reveal itself in the form of narrow lines before being overtaken by the supernova blast wave, just as in narrow-line core-collapse supernovae, SNe IIn. If hydrogen dominates the wind Hα would be emitted, and if helium dominates He I λλ5876, 10830 and He II λ4686 may be prominent.

---

* Based on observations collected at the European Southern Observatory, Paranal, Chile.

[1] See http://www.astro.su.se/~peter/sntoo.html

So far, no narrow lines, except interstellar, have been seen in spectra of SNe Ia. A pioneering observational and theoretical study was that of SN 1994D [2]. The spectrum was obtained $\sim$ 6.5 days after explosion with a spectral resolution of $\sim$ 30 km s^{-1} and covered Hα. The analysis of the non-detection involved full time-dependent photoionization calculations to estimate the narrow Hα emission from the tentative wind. The somewhat refined analysis in [3] gave an upper limit on the mass loss from the progenitor system of $\dot{M} \lesssim 2.5 \times 10^{-5}$ M$_\odot$ yr^{-1}. This assumed cosmic abundances for the wind and a wind speed of $v_w = 10$ km s^{-1}.

It was argued in [2] that observations as early as possible are needed because the densest part of the wind is quickly overtaken by the expanding supernova ejecta. This motivated us to start a Target of Opportunity (ToO) program using the Ultraviolet and Visual Echelle Spectrograph (UVES) on ESO's Very Large Telescope (VLT). UVES is a high-resolution two-arm cross-dispersed echelle spectrograph, where both arms can be operated simultaneously using a dichroic beamsplitter [4]. The spectral resolution of UVES is $\approx 50,000$, i.e., 6 km s^{-1} in the region of Na I D and Hα. So far, we have observed two SNe Ia, SN 2000cx and SN 2001el. Here we describe some results for SN 2000cx, as well as results from refined photoionization models. A complete description of the results for SN 2000cx can be found in [5]. Results for SN 2001el will be discussed elsewhere.

2   Observations of SN 2000cx

SN 2000cx was discovered on July 17.5 (UT) [6]. It was located in the outskirts of NGC 524, an S0 galaxy with a recession velocity of 2421 km s^{-1} [7]. The galactic reddening in the direction of NGC 524 is modest, $E(B-V) = 0.083$ [8]. A low dispersion spectrum obtained on July 23 revealed this supernova to be a SN Ia [9], strongly resembling the overluminous SN 1991T a few days before maximum (see also [10]). SN 2000cx continued to rise, in a similar fashion to SN 1994D [10] and reached maximum B brightness ($B \approx 13.4$) on July 27 [10]. However, the post maximum decline in B was much slower than for SN 1994D, whereas V, R and I declined faster than for most other SNe Ia [10]. According to Li et al. [10], the post-maximum spectral evolution of SN 2000cx indicates a high kinetic energy, as well as a high yield of ^{56}Ni, perhaps in excess of 1 M$_\odot$.

Our first VLT/UVES observation of SN 2000cx was made on 2000 July 20.3 UT (epoch 1), $\sim$ 7 days before maximum, and a second spectrum was obtained on 2000 July 25.3 UT (epoch 2). On epoch 1, the supernova was observed using two dichroic settings, enabling simultaneous observations in the red and the blue and effectively covering the full optical spectrum. The total integration time was 2.7 hours. On epoch 2, only one mode (covering Hα), and three integrations of 3600 s each were obtained. Wavelength calibration was done by comparison to ThAr arc lamps, and flux calibration was verified by comparing to low-resolution spectra from [10]. The flux is expected to be accurate to better than 15%. For further details about the observations and reductions see [5].

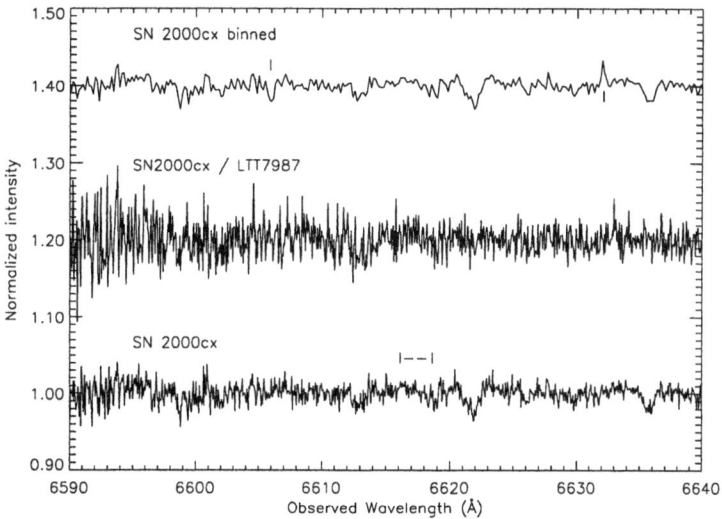

Fig. 1. Normalized spectrum of SN 2000cx on July 25.3 (UT) in the region around Hα (marked by a horizontal dashed line in the bottom spectrum). The bottom shows the spectrum of SN 2000cx with full resolution, whereas the spectrum at the top has been binned. The features at 6622 Å and 6636 Å are erased in the divided spectrum, SN 2000cx over standard star LTT7987 (middle), and are thus not intrinsic to the supernova. In the top spectrum we have added synthetic 3σ features at 6606 Å (absorption) and 6632 Å (emission), assuming a Gaussian profile with FWHM of 18 km s^{-1}. No such features can be seen in the real data at the position of Hα in the divided spectrum, and Hα is therefore absent down to the limit of sensitivity. The middle and top spectra have been shifted upward for clarity

None of our spectra show any sign of narrow Balmer lines at (or near) the velocity of the galaxy, neither in emission, nor in absorption. Figure 1 shows the region of Hα for epoch 2. To estimate a 3σ detection limit, we measured the RMS in the region ± 500 km s^{-1} around the expected position of the Hα line. We used the supernova spectrum to do this, as the noise is actually enhanced in the divided spectrum, due to the lower signal in the standard star spectrum. For epoch 2, and assuming an unresolved line (6 km s^{-1}), we claim a 3σ detection limit on the equivalent width of $\sim$ 2.5 mÅ. For a corresponding limit on an emission line, we used the flux calibrated spectrum to achieve a 3σ limit of 2.6×10^{-17} ergs cm^{-2} s^{-1}. We have also searched for emission lines at He I $\lambda5876$, He II $\lambda4686$ and [O III] $\lambda5007$ in our data, but found no such lines.

3 Model Calculations

To obtain an upper limit on the density of the putative circumstellar gas around SN 2000cx we have modeled the line emission in a way similar to that of [2,3], although with some important refinements (see [5]). We assume that the supernova ejecta have a density profile of $\rho_{ej} \propto r^{-7}$ and that the ejecta interact with the wind of a binary companion, which has a density profile of $\rho_w \propto r^{-2}$. During the first few hours after explosion, the radius of the outermost supernova ejecta is comparable to the presumed binary separation and the ejecta expand into an asymmetric environment. But already after ~ 1 day the interaction between the ejecta and the wind occurs far away from the center of explosion, which justifies a spherically symmetric treatment. The power-law density distributions makes it possible to use similarity solutions for the expansion and structure of the interaction region [11]. Calculations are started at $t_0 = 1.0$ day, and at this epoch we assume that the maximum velocity of the ejecta is $V_{ej} = 4.5 \times 10^4$ km s^{-1}. With the adopted slopes of the ejecta and circumstellar gas, the maximum ejecta velocity evolves as $V_{ej} \propto t^{-0.2}$. At 1 day, the velocities of the circumstellar shock and the reverse shock going into the ejecta are $V_s \sim 4.5 \times 10^4$ km s^{-1} and $V_{rev} \sim 1.1 \times 10^4$ km s^{-1}, respectively.

The ionizing radiation from the interaction region consists of free-free emission from the shocked ejecta and circumstellar gas, and Comptonized photospheric radiation in the shocked gas [5]. The free-free emission from the circumstellar shock can be neglected in the present analysis. With the parameters chosen in our models [5], the temperature of the reverse shock evolves as $T_{rev} \sim 2.4 \times 10^9 \, t_{day}^{-0.4}$ K, where t_{day} is the time in days since explosion.

In [2,3] it was assumed that the reverse shock was the sole source of ionizing radiation (apart from a weak preionization by the white dwarf). However, recent model calculations [12,13] show that there could be a phase of ionizing radiation from the photosphere within the first ~ 10 days after the explosion. We have included the results from the w7jzl155.ph model [13]. This uses the W7 model [14] as input and has a peak in far-UV emission between days $4 - 6$.

Equipped with the evolution of the ionizing radiation, we have calculated the time dependent ionization and temperature structure of the unshocked circumstellar gas. The code we use is an updated version of that used in [2,15]. Models were made for values of $\dot{M}$ in the range $(1 - 300) \times 10^{-7}$ M$_\odot$ yr^{-1}, assuming $v_w = 10$ km s^{-1}. For low mass loss rates, the photospheric radiation dominates the ionization and its soft photons can only heat the wind to temperatures in the range $(5 - 10) \times 10^3$ K, whereas for $\dot{M} \gtrsim 5 \times 10^{-6}$ M$_\odot$ yr^{-1}, the ionizing radiation from the reverse shock dominates, and the ionized wind close to the shock is heated to $\gtrsim 5 \times 10^4$ K. For low $\dot{M}$ values almost the full wind is ionized, whereas for values in the upper range, only the inner part of the wind becomes ionized. This means that line emission from the wind comes from a large volume for low $\dot{M}$ values, whereas for $\dot{M} \gtrsim 1 \times 10^{-6}$ M$_\odot$ yr^{-1}, the emitting volume is small. In particular, Hα is produced at the expanding ionization front where it is boosted by collisional excitation. In Fig. 2 we show the line emission produced in models with different wind densities. Models have been made for different He/H-ratios of the circumstellar gas, as well as for full equipartition between electrons and ions of the shocked ejecta and for a lower electron temperature, $T_e = T_{rev}/2$. Coulomb heating is expected to be sufficient to establish full equipartition for $\dot{M} \gtrsim 5 \times 10^{-6}$ M$_\odot$ yr^{-1} and in Fig. 2 solid

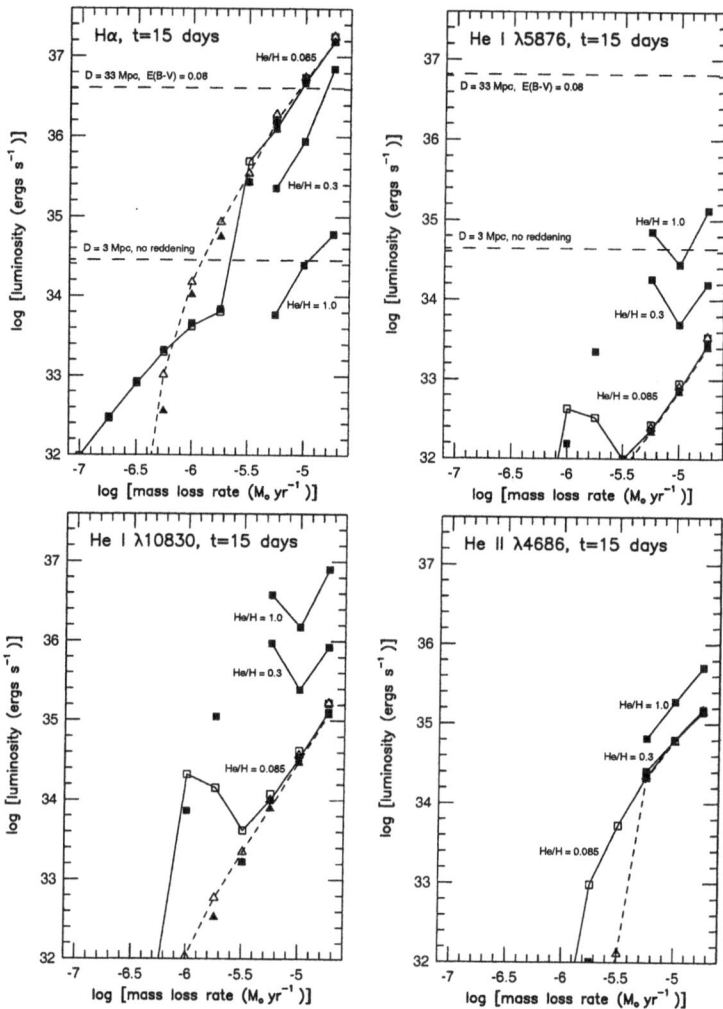

Fig. 2. Line luminosities at 15 days (i.e. roughly our epoch 2) as a function of mass loss rate, assuming $v_w = 10$ km s^{-1}. Squares show models in which ionizing radiation is only produced by the reverse shock, while in models marked by triangles we have also included the photospheric emission from the model w7jzl155.ph [13]. Filled symbols are for temperature equipartition between electrons and ions behind the reverse shock, whereas for open symbols $T_e = T_{rev}/2$. Models are shown for three values of the number density ratio He/H. The observed limit for SN 2000cx ($D = 33$ Mpc) is also shown

and short-dashed lines make a crossover between models around that value. At the flux levels relevant to SN 2000cx, the modeled line fluxes are insensitive to equipartition and the photospheric spectrum; for cosmic abundances we obtain a limit from Hα which is $\dot{M} \lesssim 9 \times 10^{-6}$ M$_\odot$ yr^{-1}. SN 2001el was closer and we have deeper exposures, so its limit is expected to be lower. It is evident, however, that the Hα flux is sensitive to the He/H-ratio. For high He/H the most promising line to observe would be He I $\lambda10830$.

Our Hα limit rivals radio and X-ray limits [16,17]. However, it is emphasized that these need revision as the efficiency of radio generation is poorly known [2], synchrotron self-absorption has not been considered in the radio models and the only existing X-ray limit [17] made erroneous interpretations for SN 1992A. As discussed in [5], a more likely limit for SN 1992A is $\dot{M} \lesssim 1.3 \times 10^{-5}$ M$_\odot$ yr^{-1} (for $v_w = 10$ km s^{-1}). Considerably improved X-ray limits are expected from early observations with Chandra or XMM.

We thank Sergei Blinnikov and Elena Sorokina for making the model results of w7jzl155.ph available to us. This work was supported by the Swedish Research Council, the Swedish Space Board, and the Royal Swedish Academy of Sciences. PL is a Research Fellow at the Royal Swedish Academy supported by a grant from the Wallenberg Foundation.

References

1. D. Branch, M. Livio, L.R. Yungelson, F.R. Boffi, E Baron: PASP **107**, 1019 (1995)
2. R.J. Cumming, P. Lundqvist, L.J. Smith, M. Pettini, D.L. King: MNRAS **283**, 1355 (1996)
3. P. Lundqvist, R.J. Cumming: 'Supernova progenitor constraints from circumstellar interaction: Type Ia'. In: *Advances in Stellar Evolution, Workshop on Stellar Ecology in Marciana Marina, Elba, Italy, June 23–29, 1996*, ed. by R.T. Wood, A. Renzini (CUP, Cambridge 1997) pp. 293–296
4. S. D'Odorico, L. Kaper: *UVES Users Manual* (2000)
5. P. Lundqvist, J. Sollerman, E. Baron, C. Fransson, B. Leibundgut, K. Nomoto, J. Spyromilio: ApJ, to be submitted (2002)
6. C. Yu, M. Modjaz, W.D. Li: IAU Circ., 7458 (2000)
7. G. de Vaucouleurs, A. de Vaucouleurs, J.R. Corwin, R.J. Buta, G. Paturel, P. Fouque: *Third reference catalogue of Bright galaxies*, (Springer, New York 1991)
8. D.J. Schlegel, D.P. Finkbeiner, M. Davis: ApJ **500**, 525 (1998)
9. R. Chornock, D.C. Leonard, A.V. Filippenko, W.D. Li, E.L. Gates, K. Chloros: IAU Circ., 7463 (2000)
10. W.D. Li, A.V. Filippenko, E. Gates, R. Chornock, A. Gal-Yam, E.O. Ofek, D.C. Leonard, M. Modjaz, R.M. Rich, A.G. Reiss, R.D. Treffers: PASP **113**, 1178 (2001)
11. R.A. Chevalier: ApJ **258**, 790 (1984)
12. S.I. Blinnikov, E. Sorokina: A&A **356**, L30 (2000)
13. S.I. Blinnikov, E.I. Sorokina: 'Prospects of investigations of thermonuclear supernovae in far UV' (in Russian). In: *Scientific prospects of the space ultraviolet observatory SPECTRUM-UV, Workshop in Moscow, Russia, November 16–17, 2000*, ed. by B.M. Shustov, D.S. Wiebe (GEOS, Moscow 2001) pp. 84–89
14. K. Nomoto, F.-K. Thielemann, K. Yokoi: ApJ **286**, 644 (1984)
15. P. Lundqvist, C. Fransson: ApJ **464**, 924 (1996)
16. F. Boffi, D. Branch: PASP **107**, 347 (1995)
17. E.M. Schlegel, R. Petre: ApJ **418**, L53 (1993)

X-Ray Emission Caused by Circumstellar Interaction in Supernovae

Tanja K. Nymark and Claes Fransson

Stockholm Observatory, S-106 91 Stockholm, Sweden

Abstract. We present a model for the emission from the cooling region behind the reverse shock created by the interaction between the supernova ejecta and the circumstellar medium. Assuming steady state we use a simple hydrodynamical model to compute temperature and velocity for selected densities behind the shock. This is then used to compute the ionization balance for 13 elements (H, He, C, N, O, Ne, Mg, Si, S, Ar, Ca, Fe and Ni) at each density. Full multilevel calculations are performed for most ions, yielding the population of each level, from which we compute the emitted X-radiation. Applications to X-ray emission from the interaction region between supernova ejecta and pre-supernova wind will be discussed.

1 Circumstellar Interaction

The interaction between supernova ejecta and the circumstellar medium creates a strongly emitting region consisting of the outward going circumstellar shock and the reverse shock, which moves backward into the ejecta (Fig. 1). If the circumstellar medium is dense, as is the case for most type II supernovae, the interaction is strong and the interaction region is set up early. The shocked circumstellar material is hot and the emission is mainly free-free and Compton emission. The shocked ejecta behind the reverse shock is cooler and denser than the region behind the outer shock, and therefore line emission is strong, and the material cools fast. This creates a cool shell behind the shock, which absorbs most of the emission, so at early times this emission cannot be observed. As the cool shell becomes optically thin, the emission can penetrate out and eventually the reverse shock comes to dominate the X-ray emission from the supernova.

2 One Temperature Versus Many

The shocked gas is often modelled with two components, one temperature for the outer shock and one for the reverse shock. This is not sufficient to reproduce the emission from the radiative reverse shock, where the temperature falls rapidly and many zones of different temperature contribute to the emission. Our model combines hydrodynamic modelling with an emission code giving a better fit to the spectrum (Fig. 2).

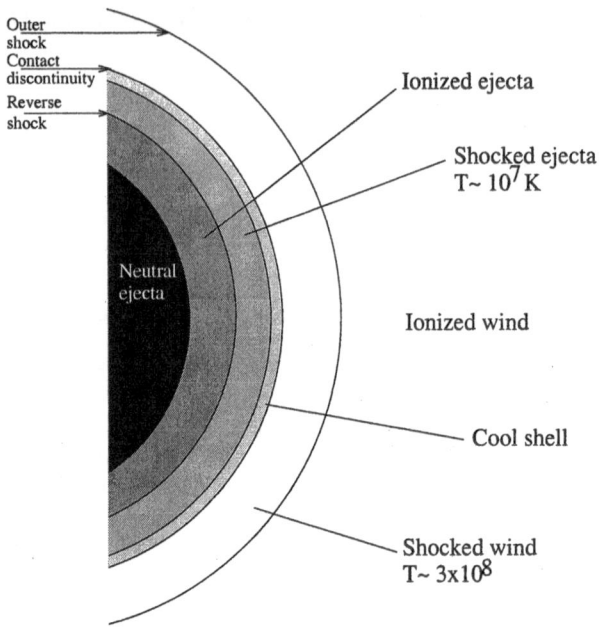

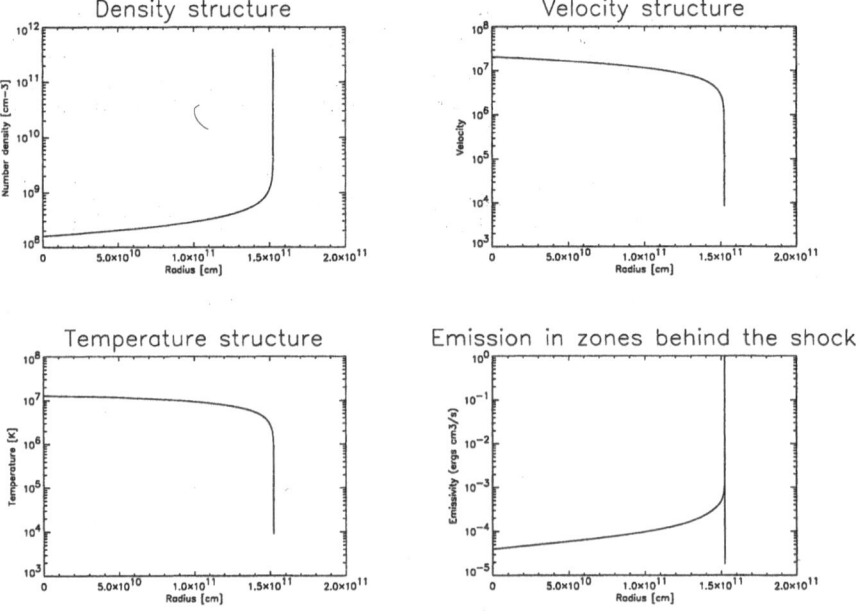

Fig. 1. Interaction of supernova ejecta with the progenitor wind. The four plots show the structure of the shocked ejecta behind the reverse shock

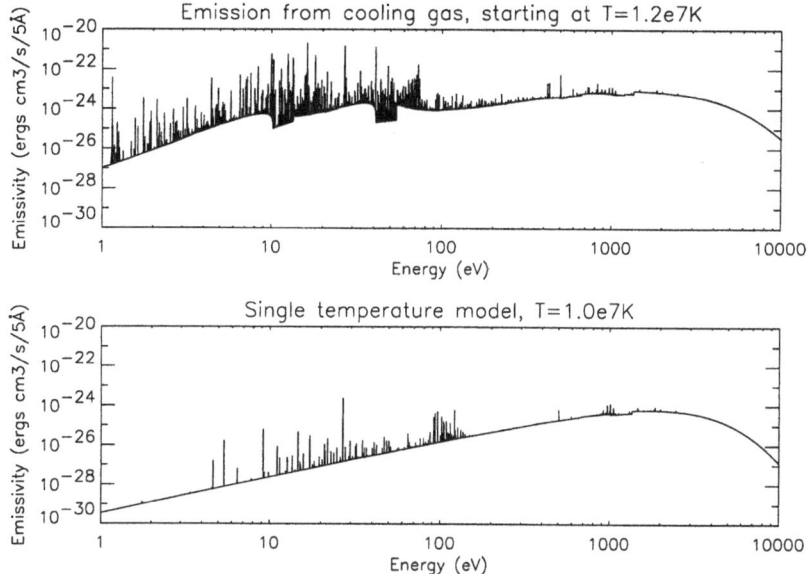

Fig. 2. Spectrum created by our model (upper) and by a single temperature model of T= 10^7 K (lower)

3   Shock Structure

We assume that the interaction region is thin compared to the size of the ejecta. This is a good approximation as long as the cooling time is short compared to the expansion time, which is certainly the case for the radiative shock. In this case it is also a good approximation to describe the region behind the shock as a steady state flow, for which the input parameters are given by a self-similar solution. We solve the steady state equations in a fine grid behind the shock, thus determining the density, temperature and velocity of the shocked ejecta. As Fig. 1 shows, the size of the region is of the order of 10^{12} cm. A typical size of the supernova is 10^{15} cm, so the assumption of a thin shell clearly holds.

4   Spectral Code

Once the temperature and density of each zone is known, the ionization structure can be computed. Assuming steady state also here, we obtain the ionization state of each element in each zone. We then solve the full multi-level equations for most ions, yielding the populations of a number of levels. We obtain the emission in each line from the level populations, according to $P_{ij} = n_i E_{ij} A_{ij}$. The continuum emission is computed by the method of Mewe et al [3]. Figure 3 shows the cooling as a function of temperature in our model, as well as the contribution from each element.

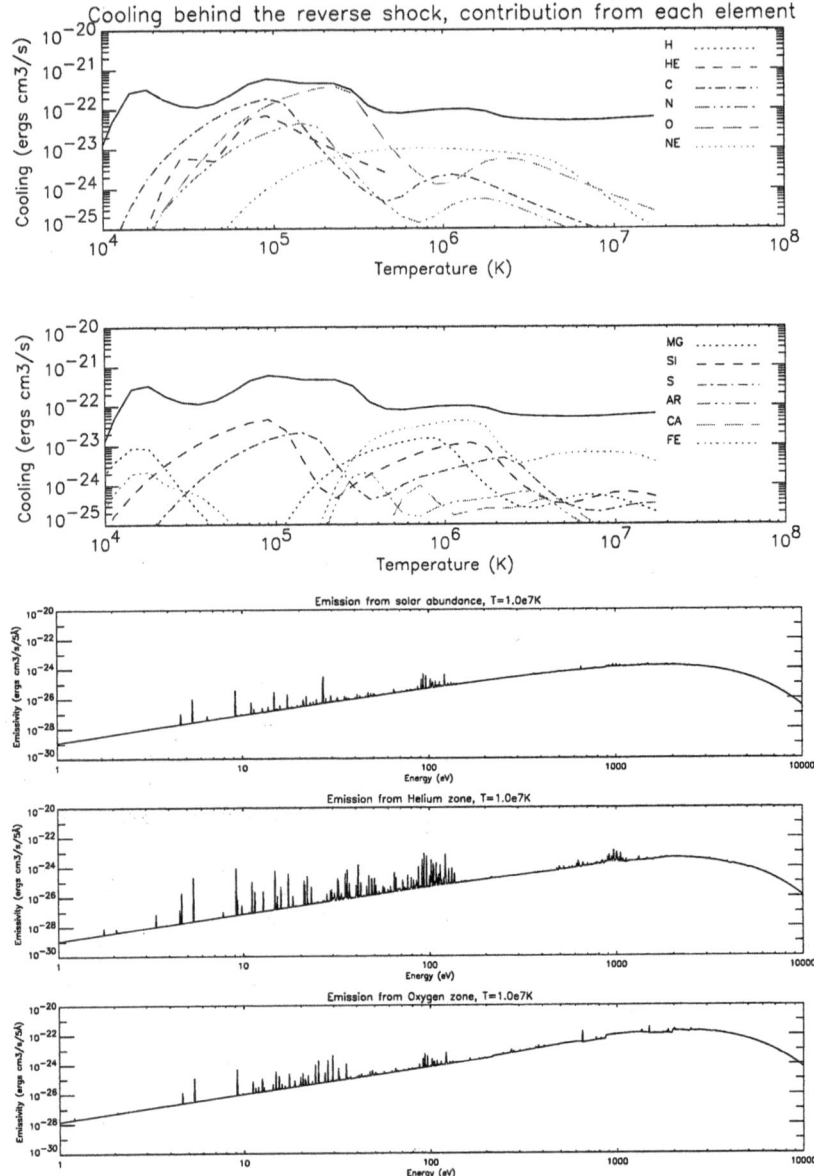

Fig. 3. The two upper panels show the cooling curve, with contribution from each element. The three lower panels illustrate the abundance effects in the spectrum of the reverse shock: emitted spectrum with solar abundance (upper), and with the reverse shock in the helium zone (middle) and in the oxygen-rich zone (lower) of the ejecta

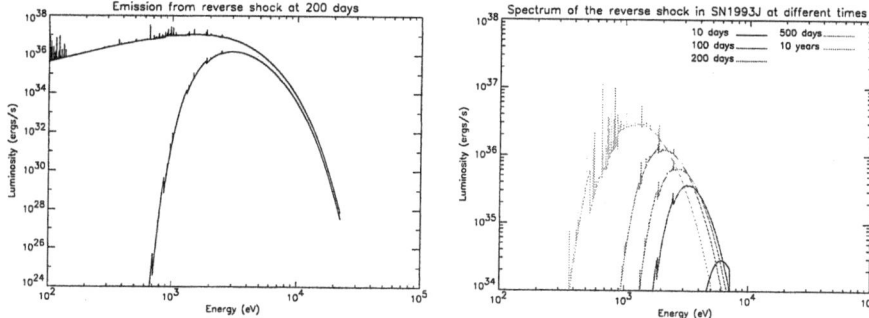

Fig. 4. Emission from the reverse shock in SN1993J. The left hand figure illustrates the importance of absorption in reducing the observed emission from the reverse shock, while the right hand figure shows how the spectrum evolves with time as the cool shell becomes optically thin to ever lower energies

5 Abundance Effects

As the reverse shock travels backwards into the ejecta, it passes through regions of different abundances reflecting the onion-like structure in the atmosphere of the progenitor. This affects the emitted spectra as elements normally rare come to dominate the line emission (Fig. 3).

6 SN 1993J

We have applied our model to the spectrum of SN1993J, assuming an ejecta density profile $\rho_{sn} = \rho_0(\frac{t}{t_0})^{-3}(\frac{v}{v_0})^{-n}$, with n=25, and a maximum velocity of the ejecta at 10 days of $V_{max}(10) = 2 \cdot 10^4$ km/s. For the mass loss rate and wind velocity of the progenitor we assume $\frac{M_{-5}}{u_{w1}} = 4.0$ [1]. Figure 4 shows the time evolution of the spectrum from the reverse shock for these parameters. We clearly see how the emission from the reverse shock, which at first cannot be seen, becomes visible and eventually very strong, as the absorption in the cool shell decreases.

7 Discussion

We have developed a model which by combining a hydrodynamic model with ionization balance and multi-level calculations can successfully reproduce the emission from the cooling gas behind the reverse shock created by circumstellar interaction. This can be used to trace the density structure and abundances in the ejecta and thus give invaluable information on the progenitor.

References

1. C. Fransson, P. Lundqvist, R. Chevalier: Ap.J. **461**, 993 (1996)
2. R. Chevalier, C. Fransson: Ap.J. **420**, 268 (1994)
3. R. Mewe, J.R. Lemen, G.H.J. van den Oord: A & AS **65**, 511 (1986)

Light Echoes in Type Ia Supernovae

Ferdinando Patat

European Southern Observatory, Karl-Schwarzschild-Str. 2, 85748 Garching, Germany

Abstract. Light echoes are a promising tool to probe the environment where SNe Ia explode and an independent source of information on the progenitor's nature. After giving a brief introduction to the phenomenon I review the two known cases, i.e. SN 1991T and SN 1998bu. We then present the results obtained from the test case of SN 1998es in NGC 632. This object was classified as a 1991T-like event and was affected by a strong reddening, a fact which made it a good candidate for a light echo detection.

1 Introduction

In the last few years the study of light echoes in supernovae (SNe) has become rather fashionable, since they provide a potential tool to perform a detailed tomography of the SN environment (Crotts [4], and references therein) and, in turn, can give important insights into the progenitor's nature.

The problem has been addressed by several authors (see for example Chevalier [2], Schaefer [16] and Sparks [18]) and therefore here we will give only a brief introduction to the light echo phenomenon in the single scattering approximation. For this purpose, let us imagine the SN immersed in a dusty medium at a distance d from the observer. If we then consider the SN event as a radiation flash, whose duration δt_{SN} is so small that $c\,\delta t_{SN} \ll d$, at any given time the light echo generated by the SN light scattered into the line of sight is confined in an ellipsoid, whose foci coincide with the observer and the SN itself.

In the SN surroundings, this ellipsoid can be very well approximated by a paraboloid, with the SN lying in its focus. If we introduce a coordinates system centred on the source, with the y axis coinciding with the line of sight (see Fig. 1), then the equation of this paraboloid can be written as

$$y = \frac{1}{2c\Delta t}(x^2 + z^2) - \frac{c\Delta t}{2} \tag{1}$$

where Δt is the time delay between the arrivals of scattered and direct photons, which left the SN at the same instant. It is clear that the paraboloid can be regarded as the locus containing all scattering elements which produce a constant Δt and for this reason we will refer to it as the iso–delay surface.

Let us now consider an infinitesimal volume element dV lying on such an iso-delay surface defined by a given Δt. If $L(t)$ is the number of photons emitted

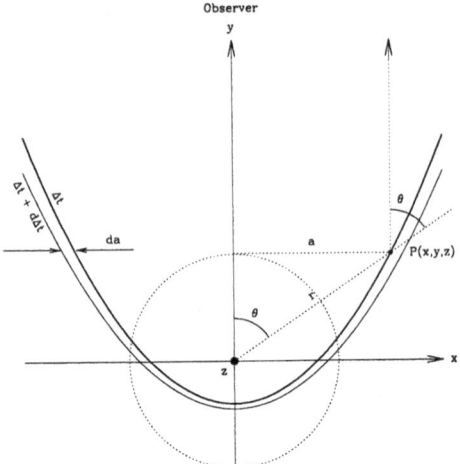

Fig. 1. Geometry of the problem. The z axis is perpendicular to the sheet plane

per unit time by the SN, the flux of scattered photons per unit time and unit area which reach the observer at time t with a delay Δt is

$$df = \frac{L(T - \Delta t)}{4\pi r^2} \times n(r) \times a_s \times \omega_s \frac{\Phi(\theta)}{4\pi d^2} \times dV \tag{2}$$

where r is the distance of the volume element to the SN, $n(r)$ is the number density of scattering particles, a_s is the scattering cross-section, ω_s is the dust albedo, θ is the scattering angle defined by $cos(\theta) = y/r$ and $\Phi(\theta)$ is the scattering phase function, normalized in order to have $\int_{4\pi} \Phi(\theta)d\Omega = 1$. For the sake of simplicity, in the following we will assume that the dust density depends only on $r = y + c\Delta t$ and that the scattering medium is characterized by an axial symmetry around the y axis. With this simplifying hypothesis, the infinitesimal volume element can be expressed as $dV = 2\pi a\, da\, dy$, where a is the impact parameter (see Fig. 1).

Then, if we set

$$f(\Delta t) = \frac{c\, a_s\, \omega_s}{2} \int_{-c\Delta t/2}^{y_m} \frac{n(r)\Phi(\theta)}{r} dy \tag{3}$$

we can finally write the total scattered flux received by the observer at time t as:

$$F(t) = \frac{1}{4\pi d^2} \int_0^t L(t - \Delta t)\, f(\Delta t)\, d\Delta t \tag{4}$$

while the total flux is simply $F_T(t) = L(t)/4\pi d^2 + F(t)$. From Eq. 4 it is clear that the outcoming scattered signal is the result of a convolution, and therefore at any given time t the observer will receive a combined signal, which contains a mixture of photons emitted by the SN in the whole time range $0 \leq \Delta t \leq t$.

As usual in this kind of studies (cf. Chevalier [2]), we have adopted the Henyey & Greenstein scattering phase function $\Phi(\theta)$, which includes the degree of forward scattering through the parameter $g = cos(\theta)$ ($g=0$ corresponds to an isotropic scattering while $g=1$ indicates a complete forward scattering) Both empirical estimates and numerical calculations (White [21]) give $g \approx 0.6$, while for the dust albedo we have adopted $\omega_s \approx 0.6$ (Mathis, Rumpl & Nordsiek [11]).

Eq. 4 can be solved numerically, using the observed light curve of a typical SN Ia to get $L(t)$, but some instructive results can be obtained analytically. In fact, the light curve of a SN Ia can be well approximated by a flash ($L(t) = L_0$ for $t \leq \delta t_{SN}$ and $L(t)=0$ for $t > \delta t_{SN}$), and this makes the solution of Eq. 4 very simple. The flash duration can be computed from the observed light curve as $L_0 \, \delta t_{SN} = \int_0^{+\infty} L(t) \, dt$, and turns out to be of the order of 0.1 years for a SN Ia.

After normalizing to the SN luminosity at maximum, we have $F_T(t)/F_T(0) = \delta t_{SN} \, f(t)$, so that one is left with the solution of Eq. 3 only. In general, this requires a numerical integration, even though a few analytical solutions exist (Chevalier [2]). The problem becomes particularly simple if one assumes an isotropic scattering ($\Phi(\theta) = 1/4\pi$) and a constant density ($n(r) = n_0$). Under these assumptions, the resulting echo luminosity has a very low time dependency and therefore, once the light echo radiation overcomes that of the SN, the light curve is expected to decline very slowly (~ 0.05 mag yr^{-1}) and, depending on the dust cloud extension, to brighten for many years.

What is interesting to note here is that the integrated echo luminosity is mostly governed by the dust density and depends only mildly on the geometry of the scattering material. For typical values of the cross section (6×10^{-22} cm^2) and dust density (10 cm^{-3}), the light echo is expected to take place at about 10 magnitudes below the SN maximum, and thus it would dominate the total luminosity starting at about two years after the explosion. At such epochs, the color of the object is expected to turn blue and the spectrum to show an abrupt change from the nebular appearance to a rather weird aspect, being a mixture of all spectra shown by the SN all the way back to the explosion. This effect, which makes the study of late phases particularly cumbersome in such *unlucky* cases, was actually seen in the two known events, which are described in the next section.

2 Known Light Echoes in Supernovae of Type Ia

The first detection of a light echo in a SN Ia occurred for SN 1991T (Schmidt et al. [17]), which started to deviate from the usual exponential decline at about 1.5 years past maximum, when it settled on a much slower decline tail and showed unprecedented blue spectra. The existence of a light echo was definitely confirmed by the HST images taken between 1996 and 1998, which have indeed shown a resolved structure (2R~0″.4) evolving with time (Sparks et al. [19]). This implied the presence of dust within ~50 pc from the SN.

The only other known case is that of SN 1998bu (Cappellaro et al. [1]), which so far has shown a photometric and spectroscopic behaviour very similar to that of SN 1991T. Due to this similarity, Cappellaro et al. [1] have modelled both light echoes using a thin sheet of dust, perpendicular to the line of sight and located at a distance D in front of the SN. The implied optical depth was kept fixed at the value inferred from the estimated extinction A_V, i.e. 0.46 for SN 1991T (Phillips et al. [14]) and 0.86 for SN 1998bu (Jha et al. [10]), so that the echo and the observed reddening are consistently explained. In this respect, we note that the total equivalent widths EW of NaI D lines for the two objects are 1.6 Å (Patat et al., in preparation) and 0.7 Å (Munari et al. [12]) respectively. Therefore, if a relation between EW and A_V really exists, we do not have understood it yet very well (see also Turatto, these proceedings).

With these simplifying assumptions, the solution of Eq. 3 becomes very simple and contains one single free parameter, namely the distance D between the SN and the dust cloud. The reproduction of the observed data, both spectroscopic and photometric, is fairly good (see Cappellaro et al. [1] and Fig. 2 here) and allows one to estimate this distance. In the case of SN 1991T, this turns out to be ~40pc, while for 1998bu the inferred value is ~80pc. At the distance of NGC 3368, this implies a diameter of about 0″.3, and therefore the light echo was expected to be resolved by HST. This was confirmed by WFPC2 observations performed at 762 days after maximum by Garnavich et al. [5], who reported the presence of two main features: a ring with a diameter of 0″.48 and a barely resolved inner disk (2R~0″.14), which indicates the presence of scattering material within 10pc from the SN.

Now, it is quite intriguing that both SNe exploded in SAB-liner galaxies (NGC 4527 and NGC 3368) which show an enhanced star forming rate. Moreover, both SNe were slow decliners, even though we must say that SN 1998bu was not so extreme as SN 1991T and did not share its spectroscopic peculiarities. These facts go in the same direction of the body of evidence collected in the last ten years concerning the possible relation between the SN characteristics and the host galaxy morphological type (van den Bergh & Pazder, [20], Hamuy et al. [6,7], Howell [8]). These studies have rather convincingly demonstrated that the sub-luminous events are preferentially found in early type galaxies (E/S0), while the super-luminous ones (1991T-like) tend to occur in spirals (Sbc or later). What is important to underline here, is that 1991T-like events seem to be associated with young environments and are, therefore, the most promising candidates for the study of light echoes. Or, in turn, if echoes are detected around such kind of SNe, this would strengthen their association with sites of relatively recent star formation.

The sample of known light echoes is far too small to draw any significant conclusion and therefore we badly need to improve on the statistics.

3   SN 1998es in NGC 632: A Test Case

The most promising case during the last years was represented by SN 1998es in NGC 632. This SN, in fact, was classified as a 1991T-like by Jha et al. [9], who also noticed that the parent galaxy was classified as an S0, hosting a nuclear starburst (Pogge & Eskridge [15]). Moreover, the SN was reported to be projected very close to a star forming region and a high resolution spectrum showed interstellar NaI D features very similar to those of SN 1991T (Patat et al., in preparation), indicating the presence of a significant amount of material in front of the SN (EW=1.46 Å).

A 1991T-like echo at the distance of NGC632 (43 Mpc, H_0= 75 km s^{-1} Mpc^{-1}) would have a magnitude $B \sim 24$ and therefore an 8m-class telescope is required. For this purpose, we have imaged NGC 632 in B and V with VLT-FORS2 on September 25, 2001. A careful analysis of the data did not reveal any stellar object at the expected position. Due to the good image quality achieved in the 1800 sec B stacked image (0.″64) and the smooth galaxy background, we could set an upper limit for the integrated echo luminosity of $B \geq 25.5$.

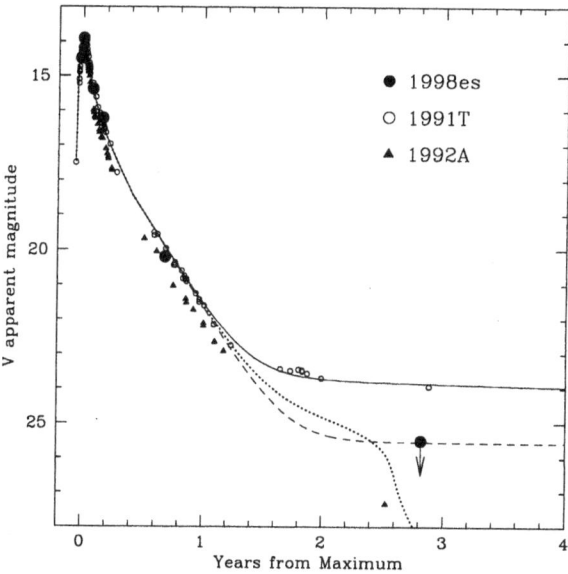

Fig. 2. Light curve of SN 1998es

There are two possible explanations for this non-detection. The first comes from the fact that, in the hypothesis of a thin sheet and $D \gg ct$, the echo luminosity is proportional to τ/D, where τ is the dust optical depth along the line of sight to the SN. Therefore, if we assume that τ is the same as in the case of SN 1991T (as the high resolution spectra seem to tell us), it is sufficient to place the dust sheet ~4 times further away ($D \approx 150$ pc) to produce a light echo 1.5 mag fainter than in 1991T (see Fig. 2, dashed line). Another reasonable explanation, given the poor time coverage of SN 1998es, is the possible confinement of the dust to a small cloud in front of the SN. In fact, if R is the radius of a spherical cloud at a distance D from the SN, the iso-delay surface will stop to intercept the scattering material approximately for $ct \sim R^2/2D$. For example, for $D=120$ lyr and $R=25$ lyr, the echo would last 2.6 years only. In Fig. 2 we have plotted the numerical solution one obtains in this case (dotted line). We note that, unfortunately, it is impossible to put meaningful constraints for D and R, due to the very poor photometric coverage at late phases.

The FORS2 images were very deep, the combined B frames reaching a 5σ peak limiting magnitude of ~27 for unresolved objects. Despite the fact that such deep data did not show the light echo, they led to the serendipitous discovery of a very faint tidal tail (Patat, Carraro & Romaniello [13]), which extends for at least 50 kpc. This discovery strongly supports the suggestion made by Chitre & Joshi [3] that NGC 632 is probably the result of a galaxy merger, which in turn is responsible for the observed nuclear/circum-nuclear starburst.

Light echoes have several applications, which should be exploited in the future. Being generated by reflections, they are strongly polarized, with a maximum in the polarization degree at $\theta = \pi/2$. This, in principle, can give a direct estimate of the distance (Sparks [18]). The HST light echo follow-up also allows one to determine the distance to the SN, provided that the geometry of the dust distribution is to some extent known.

Finally, and in our opinion most important of all, more sophisticated models applied to high quality photometric and spectroscopic data can provide useful information on the dust properties, its distribution around the progenitor and possibly give some hints on its nature.

References

1. E. Cappellaro, F. Patat, P.A. Mazzali et al.: ApJ **549**, L215 (2001)
2. R. Chevalier: ApJ **308**, 225 (1986)
3. A. Chitre & U.C. Joshi: A&AS **139**, 105 (1999)
4. A.P.S. Crotts: ApJ **333**, L51 (1988)
5. P.M. Garnavich, R.P. Kirshner & P. Challis: AAS Meeting 199, #47.01 (2001)
6. M. Hamuy, M.M. Phillips, N.B. Suntzeff et al.: AJ **112**, 2391 (1996)
7. M. Hamuy, S.C. Trager, P.A. Pinto et al.: AJ **120**, 1479 (2000)
8. D.A. Howell: ApJ **554**, 193 (2001)
9. S. Jha, P.M. Garnavich, P. Challis & R.P. Kirshner: IAU Circ. n. 7054 (1998)
10. S. Jha, P. Garnavich, R.P. Kirshner: ApJS **125**, 73 (1999)
11. J.S. Mathis, W. Rumpl & K.H. Nordsiek: ApJ **217**, 425 (1977)
12. U. Munari, R. Barbon, L. Tommasella & M. Rejkuba: IAU Circ. n.6903 (1998)
13. F. Patat, G. Carraro & M. Romaniello: Friulian Journal of Science, (2002) in press
14. M.M. Phillips, P. Lira, N.B. Suntzeff et al.: AJ, **118**, 1766 (1999)
15. R.W. Pogge & P.B. Eskridge: ApJ **106**, 1405 (1993)
16. B. Schaefer: ApJ **323**, L47 (1987)
17. B.P. Schmidt, R.P. Kirshner, B. Leibundgut et al.: ApJ **434**, L19 (1994)

18. W.B. Sparks: ApJ **433**, 29 (1994)
19. W.B. Sparks, F. Macchetto, N. Panagia et al.: ApJ **523**, 585 (1999)
20. S. Van den Bergh & J. Pazder: ApJ **390**, 34 (1992)
21. R.L. White: ApJ **229**, 954 (1979)
22. J. Xu, A.P.S. Krotts & W.E. Kunkel: ApJ **435**, 274 (1994)

Part VII

Supernova Nucleosynthesis and Chemical Evolution

Supernova Nucleosynthesis
and Galactic Evolution

F.-K. Thielemann[1], D. Argast[1], F. Brachwitz[1], W.R. Hix[2], P. Höflich[3],
M. Liebendörfer[2], G. Martinez-Pinedo[1], A. Mezzacappa[2], K. Nomoto[3], and
I. Panov[1]

[1] Department of Physics & Astronomy, Univ. of Basel, CH-4056 Basel, Switzerland
[2] Physics Division, Oak Ridge National Laboratory, Oak Ridge, TN 37831-6371, USA
[3] Department of Astronomy, University of Texas, Austin, TX 78712, USA
[4] Department of Astronomy, University of Tokyo, Tokyo 113-0033, Japan

Abstract. The understanding of the abundance evolution in the interstellar medium,
and especially the enrichment of heavy elements, as a function of space and time reflects
the history of star formation and the lifetimes of the diverse contributing stellar objects.
Therefore, the understanding of the endpoints of stellar evolution is essential. These
are mainly planetary nebulae and type II/Ib/Ic supernovae as evolutionary endpoints
of single stars, but also events in binary systems can contribute, like e.g. supernovae of
type Ia, novae and possibly X-ray bursts and neutron star or neutron star - black hole
mergers. Despite many efforts, a full and self-consistent understanding of supernovae
(the main contributors to nucleosynthesis in galaxies) is not existing, yet. However,
observed spectra, light curves, radioactivities/decay gamma-rays and galactic evolution
witness the composition of their ejecta and constrain model uncertainties. We focus
on (i) neutrino-induced explosions for type II supernovae and the innermost ejected
layers, (ii) electron captures in type Ia supernovae and neutron-rich Fe-group nuclei
and finally (iii) galactic chemical evolution and possible r-process sites.

1 Introduction

The chemical composition of the interstellar medium in galaxies acts as a witness
of its stellar sources. The evolution of the elements in the range O through Ni
is dominated by two alternative explosive stellar sources, i.e. by the combined
action of type II and type Ia supernovae (SNe II and Ia). SNe II, originating from
massive stars $(M > 8 M_\odot)$, leave a central neutron star and expell large amounts
of the alpha elements O, Ne, Mg, Si, S, Ar, Ca, Ti. Oxygen yields can be of
the order $1 M_\odot$ or more while smaller amounts of Fe (and Fe-group elements
like Cr, Mn, Fe, Co, Ni, Cu, Zn, Ga, Ge), typically about $0.1 M_\odot$, are produced
in the innermost ejected mass zones. SNe Ia originate from those intermediate
to low mass stars which end as mass accreting white dwarfs in binary stellar
systems with high mass accretion rates $(>10^{-8}\ M_\odot\ yr^{-1})$, leading to a growing
C/O white dwarf mass which approaches the Chandrasekhar mass. The complete
thermonuclear disruption of the white dwarf after carbon is ignited in the center
produces predominantly Fe-group elements, on average of the order $0.6\text{-}0.8\ M_\odot$,
and smaller amounts of Si, S, Ar, and Ca.

The site for the production of the heaviest elements up to U and Th (the r-process based on rapid neutron captures in environments with very high neutron densities) is still debated and most probably related to type II supernovae and/or neutron star ejecta (from binary mergers or jets).

Nuclear reactions and nuclear properties enter in different ways. Low and intermediate temperatures in stellar evolution require individual cross sections to follow the reaction flows. High temperatures in explosions lead to chemical equilibria but are affected by weak interactions, i.e. beta-decays, electron captures and neutrino-nucleus interactions and can experience freeze-out effects. The proper knowledge of nuclear masses and fission properties far from stability has mainly an influence on the r-process. The nuclear equation of state enters directly into the formation of shock waves in supernovae, the formation of hydrodynamical instabilities and convection and has also a strong impact on the amount of ejecta from neutron star mergers.

In the following sections we want to present recent advances, especially in supernova nucleosynthesis, and discuss constraints for the ejecta compositions from the chemical evolution of our Galaxy.

2   Type II Supernovae

2.1   The Explosion Mechanism

The late phases of stellar evolution (O- and Si-burning) are less prone to cross section uncertainties from strong interactions due to chemical equilibria. However, the high densities result in partially or fully degenerate electrons with increasing Fermi energies [37]. This allows for electron capture on an increasing number of Si-group and later Fe-group (pf-shell) nuclei. The recent progress in pf-shell rates [28] led to drastic changes in the late phases of Si-burning [16], thus also setting new conditions for the subsequent Fe-core collapse after Si-burning, the size of the Fe-core and its electron fraction $Y_e = <Z/A>$ [32].

The electron capture on pf-shell nuclei (and later on free protons!) in the central Fe-core trigger a pressure decrease and the collapse of the Fe-core. The size of the homologous core, where infall velocities are proportional to the radius, which turns into nuclear matter during bounce, is dependent on the amount of prior electron captures. The total energy released, $2\text{-}3 \times 10^{53}$erg, equals the gravitational binding energy of a neutron star and depends on the nuclear equation of state (EOS). The apparently most promising mechanism for supernova explosions after Fe-core collapse of a massive star is based on neutrino heating beyond the hot proto-neutron star via the dominant processes $\nu_e + n \rightarrow p + e^-$ and $\bar{\nu}_e + p \rightarrow n + e^+$ with about a 1% efficiency in energy deposition [2]. The neutrino heating efficiency depends on the neutrino luminosity, which in turn is affected by neutrino opacities. Aspects of the explosion mechanism are still uncertain and related to Fe-core sizes from stellar evolution, electron capture rates of pf-shell nuclei, the supranuclear equation of state, as well as the details of neutrino transport and Newtonian vs. general relativistic calculations [45,33,29].

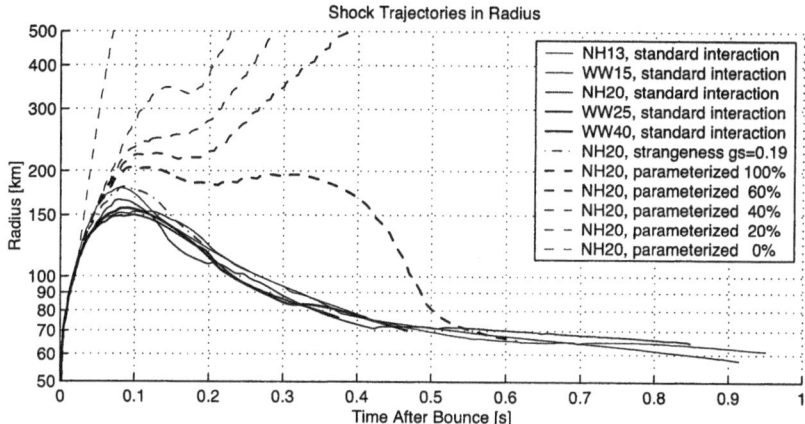

Fig. 1. Collapse calculations for different progenitor masses, showing in each case the position of the shock front after bounce. All 1D calculations result in non-explosions with receding shock fronts. A reduction in neutrino-nucleon elastic scattering to the given percentage, leading to higher luminosities, can cause explosions [30]

The present situation is that 1D spherically symmetric calculations without convective turnover (neither in the neutron star nor in the region of the neutrino sphere) do not show explosions. In order to evaluate properties of exploding models, we changed the neutrino luminosity by reducing the neutrino-nucleon elastic scattering cross sections [30] (see Fig. 1).

2.2 Composition of Ejecta

As long as uncertainties are still existing in self-consistent models, light curve as well as explosive nucleosynthesis calculations have been performed by introducing a shock of appropriate energy in the pre-collapse stellar model [59,54,19,6,56]. Such induced calculations, making use of strong and weak interaction nuclear rates [22,46,28], lack self-consistency and cannot predict the ejected ^{56}Ni-masses from the innermost explosive Si-burning layers (powering supernova light curves by the decay chain ^{56}Ni-^{56}Co-^{56}Fe), due to the missing knowledge of the mass cut between the neutron star and the supernova ejecta as well as the neutron-richness (or $Y_e = <Z/A>$) in the ejected matter.

Earlier calculations [54] showed the composition after explosive processing with an artificially induced shock of the order 10^{51}erg (see Fig. 3). The outer ejected layers ($M(r) > 2M_\odot$) are unprocessed by the explosion and contain results of prior H-, He-, C-, and Ne-burning in stellar evolution. The interior parts of SNe II contain products of explosive Si, O, and Ne burning. In the inner ejecta, which experience explosive Si-burning, such (not self-consistent) calculations made use of a Y_e of the order 0.4989 to 0.494. This Y_e originated from beta-decays and electron captures in the pre-explosive hydrostatic fuel but ommitted possible alterations via neutrino reactions during the explosion in these

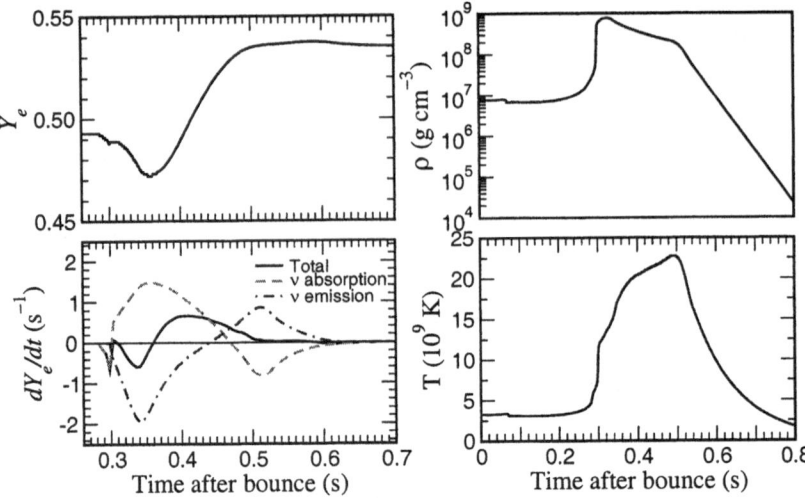

Fig. 2. Time evolution after core bounce for the innermost ejected layer of a core collapse supernova [30] (model NH20, 40% from Fig. 1). $Y_e=<Z/A>$ indicates the neutron-richness of ejected matter, dY_e/dt the time derivative due to $\nu_e + n \leftrightarrow p + e^-$ and $\bar{\nu}_e + p \leftrightarrow n + e^+$ in the directions of neutrino (and antineutrino) absorption or emission, $\rho(t)$ and $T(t)$ the density and temperature evolution. What can be noticed is that Y_e is strongly dependent on both neutrino absorption and emission reactions and that in exploding models the Y_e in the innermost zones is larger than 0.5, i.e. proton-rich

layers. This can cause huge changes in the Fe-group composition for the abundances of ^{58}Ni and ^{56}Ni, neutron-rich isotopes increase (^{57}Ni, ^{58}Ni, ^{59}Cu, ^{61}Zn, and ^{62}Zn) and N=Z isotopes like ^{40}Ca, ^{44}Ti, ^{48}Cr, and ^{52}Fe decrease. The latter are also strongly affected by the entropy obtained in those mass zones. The present calculations [30] show that the innermost zones in successfully exploding models avoid too neutron-rich ejecta when including neutrino-induced reactions (see Fig. 2). This permits an Fe-group composition consistent with supernova observations [54]. The slightly proton-rich environment leads to an alpha-rich and proton-rich freeze-out. The density is apparently not sufficient for an rp-process and unburned hydrogen survives deep in the ejected envelope.

Three types of uncertainties are inherent in the Fe-group ejecta, related to (i) the total amount of Fe(group) nuclei ejected and the mass cut between neutron star and ejecta, mostly measured by ^{56}Ni decaying to ^{56}Fe, (ii) the total explosion energy which influences the entropy of the ejecta and with it the amount of radioactive ^{44}Ti as well as ^{48}Cr, the latter decaying later to ^{48}Ti and being responsible for elemental Ti, and (iii) finally the neutron richness or $Y_e=< Z/A >$ of the ejecta, dependent on stellar structure, electron captures and neutrino interactions.

The intermediate mass elements Si-Ca are only dependent on the explosion energy and the stellar structure of the progenitor star, while abundances for

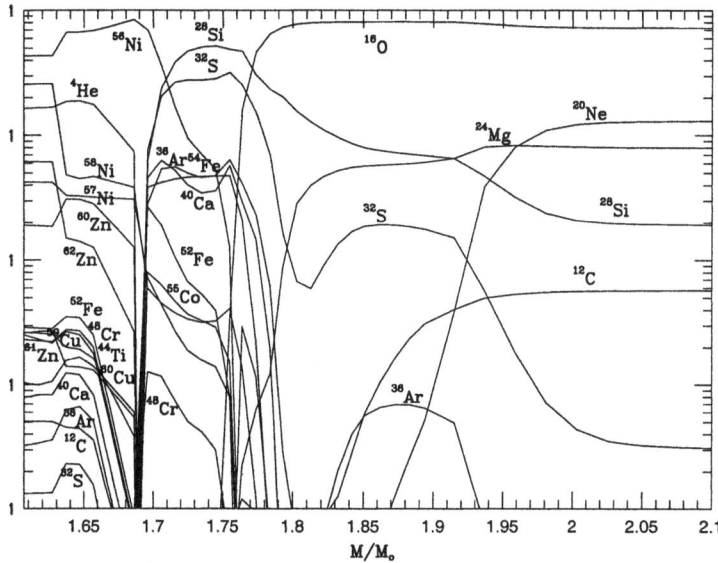

Fig. 3. Isotopic composition for the explosive C-, Ne-, O- and Si-burning layers of a core collapse supernova from a 20M$_\odot$ progenitor and an induced net explosion energy of 10^{51} erg [54]. The exact mass cut in $M(r)$ between neutron star and ejecta and the entropy and Y_e in the innermost ejected layers depend on the details of the (still open) explosion mechanism. The abundances of O, Ne, Mg, Si, S, Ar, and Ca dominate strongly over Fe (decay product of ^{56}Ni), if the mass cut is adjusted to 0.07M$_\odot$ of ^{56}Ni ejecta as observed in SN 1987A. The innermost ejecta show a strong dependence on entropy (indicated by the variations in ^{4}He and e.g. ^{44}Ti or ^{48}Cr) and Y_e (indicated by the variations in ^{56,57,58}Ni or ^{60,62}Zn

elements like O and Mg are essentially determined by the stellar progenitor evolution. Thus, when moving in from the outermost to the innermost ejecta of SN II explosions, we see an increase in the complexity of our understanding, depending (a) only on stellar evolution, (b) on stellar evolution and explosion energy, and (c) on stellar evolution and the complete explosion mechanism (see Figs. 3 and 2). The possible complexity of the explosion mechanism, including multi-D effects, may not affect this (spherically symmetric) discussion of explosive nucleosynthesis severely. 2D-calculations [25] show a spherically symmetric shock front after the explosion is initiated, leading to spherical symmetry in explosive nuclear burning when passing through the stellar layers. After freeze-out of nuclear reactions, the final nucleosynthesis products can be distributed in non-spherical geometries due to mixing by hydrodynamic instabilities. Thus, the total mass of nucleosynthesis yields shown in Fig. 3 is not changed, only its geometric distribution, with the possible expection of the innermost layers where mass cut, later fall back and convective turnover interfere with each other.

3   Type Ia Supernovae

There are strong observational and theoretical indications that SNe Ia are thermonuclear explosions of accreting white dwarfs in binary stellar systems [18,40,39]
with high rates of H-accretion causing quasi-stable H-burning and subsequent
He-burning in shells surrounding the white dwarf. The increasing mass of the
white dwarf, consisting of C and O, towards the maximum stable Chandrasekhar
mass leads finally to contraction.

3.1   Ignition and Burning Front Propagation

Contraction causes carbon ignition in the central region and a thermonuclear
runway with a complete explosive disruption of the white dwarf [38,61]. High
accretion rates cause a higher central temperature and pressure, favoring lower
ignition densities. A flame front then propagates at a subsonic speed as a deflagration wave due to heat transport across the front [36]. Here the most uncertain
quantity is the flame speed which depends on the development of instabilities
of various scales at the flame front. Multi-dimensional hydro simulations of the
flame propagation have suggested that a carbon deflagration wave might propagate at a speed v_{def} as slow as a few percent of the sound speed v_s in the central
region of the white dwarf. The nucleosynthesis consequences witness the actual
burning front velocities and can thus serve as a constraint. Electron capture affects the central electron fraction Y_e, which determines the composition of the
ejecta from such explosions. The amount of electron capture depends on (i) the
electron capture rates on nuclei, (ii) v_{def}, influencing the time duration of matter at high temperatures (and with it the availability of free protons for electron
captures), and (iii) the central density of the white dwarf ρ_{ign} (increasing the
electron chemical potential i.e. their Fermi energy) [21,3,28].

After an initial deflagration in the central layers, the deflagration can turn
into a detonation (supersonic burning front) at lower densities ρ_{tr} [24,35]. The
transition from a deflagration to a detonation (delayed detonation model) leads
to a change in the ratios of Si-burning subcategories with varying entropies.
This also leaves an imprint on the Fe-group composition. The observed bright-
ness/decline relation for SNe Ia can be related to variations in ρ_{ign}, v_{def} and ρ_{tr}
[9]. Whether an initial deflagration in the central layers can turn into a detonation (supersonic burning front) at lower densities [35] is still strongly debated.
Very high resolution (converged) 3D-calculations [14] seem not to be able to
obtain the correct amount of radioactive ^{56}Ni in order to reproduce typical type
Ia lightcurves, while different 3D-approaches of turbulent flame fronts seem to
achieve this goal with a pure deflagration model [47] (see also Niemeyer, this
conference).

3.2   Nucleosynthesis Details

Nucleosynthesis constraints can help to find the "average" SN Ia conditions
responsible for their contribution to galactic evolution, i.e. especially the Fe-
group composition. SNe Ia contribute essentially no elements lighter than Al in

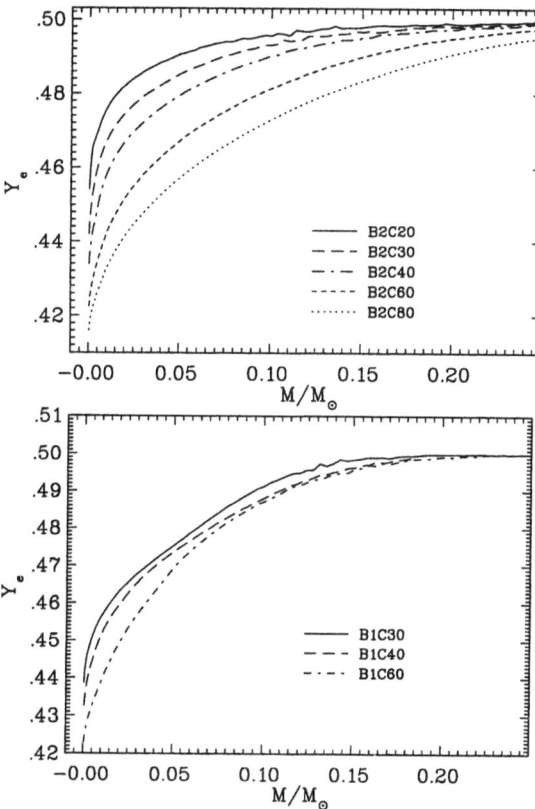

Fig. 4. Y_e after freeze-out of nuclear reactions measures the electron captures on free protons and nuclei. Small burning front velocities lead to steep Y_e-gradients which flatten with increasing velocities for the models B1 vs. B2 [4]. B1 and B2 stand for a factor of 0.1 and 0.2 in front of the square root in Eq.(1) of Dominguez et al. [9]. Lower central ignition densities shift the curves up (see the series of models 20, 30, 40, 60, which stand for ignition densities of 2, 3, 4, and 6×10^9 g cm^{-3})

important amounts, about 1/3 of the elements from Si to Ca, and the dominant amount of Fe group nuclei. In addition, the average Fe-group yields of SNe II and SNe Ia differ.

Explosive (complete) Si-burning with Y_e values of 0.47-0.485 leads to dominant abundances of ^{54}Fe and ^{58}Ni, values between 0.46 and 0.47 produce dominantly ^{56}Fe, values in the range of 0.45 and below are responsible for ^{58}Fe, ^{54}Cr, ^{50}Ti, ^{64}Ni, and values below 0.43-0.42 are responsible for ^{48}Ca. The intermediate Y_e-values 0.47-0.485 exist in all cases (see Fig. 4), but the masses encountered which experience these conditions depend on the Y_e-gradient and thus v_{def}. Whether the lower vales with $Y_e < 0.45$ are attained, depends on the central ignition density ρ_{ign}. Therefore, ^{54}Fe and ^{58}Ni are indicators of v_{def} while ^{58}Fe, ^{54}Cr, ^{50}Ti, ^{64}Ni, and ^{48}Ca are a measure of ρ_{ign}.

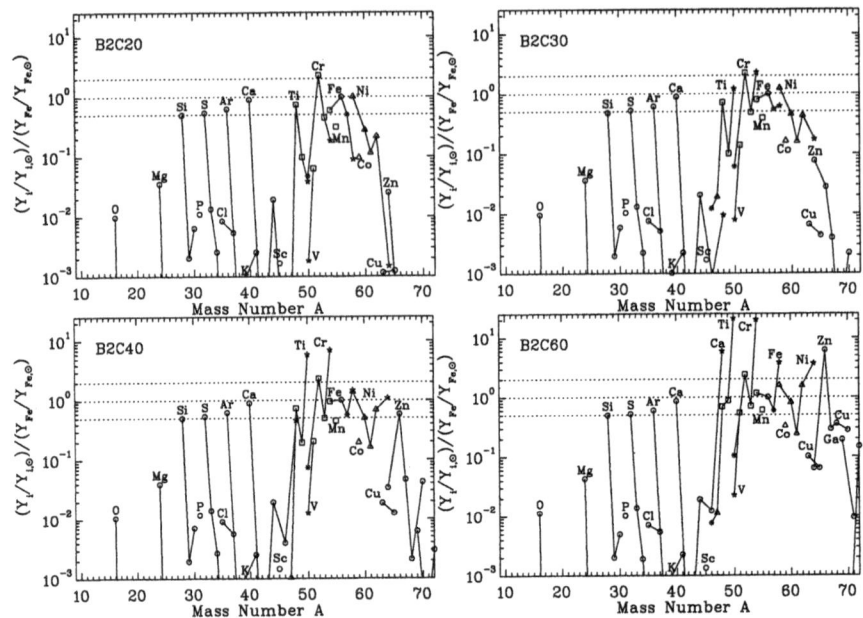

Fig. 5. Abundance ratios in comparison to solar for a series of central ignition densities of 2, 3, 4, and 6 ×10^9 g cm^{-3} (the models B2C20-60). While ignition densities in the range (2 - 3) ×10^9 g cm^{-3} lead to relative solar abundance ratios in the Fe-group, higher ignition densities can explain overabundances of ^{48}Ca, ^{50}Ti, ^{54}Cr, ^{58}Fe, ^{64}Ni, and ^{66}Zn, which should happen occasionally to account for these isotopes which are not produced in SNe II

Our nucleosynthesis efforts are still based on a 1D spherically symmetric approach with central ignition density of the white dwarf ρ_{ign} and a deflagration speed v_{def} [21,3]. Figure 5 shows the influence of central ignition densities ρ_{ign} of 2 (20), 3 (30), 4 (40) and 6×10^9 g cm^{-3} (60) at the onset of the thermonuclear runaway for the deflagration speed description B2 [4]. In none of the models abundances of ^{54}Fe or ^{58}Ni are problematic, supporting B2 as the correct choice. The Fe-group abundances are within a factor of 2 of solar ratios for ignition densities in the range (2 - 3) ×10^9 g cm^{-3}. This range should be typical for the dominant number of SNe Ia. However, it can be noticed that occasionally occuring SNe Ia at higher ignition densities (or lower accretion rates) can also account for nuclei like ^{48}Ca, ^{50}Ti, ^{54}Cr, ^{58}Fe, ^{64}Ni, and ^{66}Zn. For some of these nuclei no other production site is imaginable.

The calculations were performed with the new electron capture rates from large-scale shell models calculations for pf-shell nuclei [28,32]. The reduction in these rates [13] led to changes in comparison to [21], permitting higher ignition densities than 2×10^9 g cm^{-3} to obtain solar Fe-group ratios [3]. For this reason we also repeated the nucleosynthesis calculations for the original pure deflagration model W7 [38]. This model had apparently too large a deflagration speed,

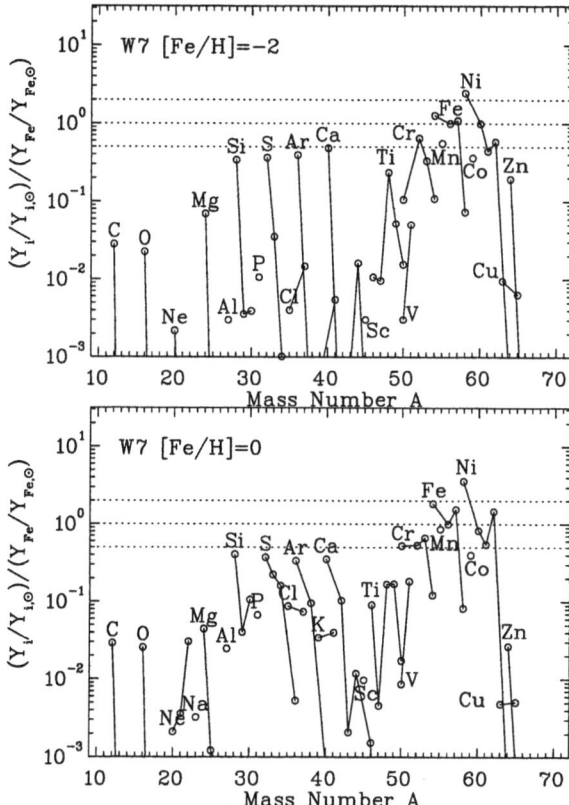

Fig. 6. Nucleosynthesis results for the pure deflagration model W7 [38] with new pf-shell electron capture rates as a function of metallicity. Notice the rise of ^{55}Mn, which actually reflects the behavior in galactic evolution for metallicities beyond [Fe/H]=-1

leading to too extended mass zones in the Y_e range 0.47-0.485 which caused an overproduction of ^{54}Fe and ^{58}Ni. This is also still visible in Fig. 6, where ^{54}Fe is close to the acceptable limit within a factor of 2 of solar, while ^{58}Ni exceeds this value. As discussed earlier, a reduction from solar metallicities [Fe/H]=0 down to one tenth of solar [Fe/H]=-1 can reduce this overproduction, as these relatively moderate Y_e-values can also be affected by the metallicities in the outer part of a SN Ia, where Y_e is not caused by electron captures, but rather the amount of ^{22}Ne resulting from ^{14}N$(\alpha, \gamma)^{18}$F$(e^+\nu)^{18}$O$(\alpha, \gamma)^{22}$Ne in Helium burning of the progenitor star, acting on the ^{14}N ashes from Hydrogen (CNO)-burning. Observations of low metallicity stars show a rise of [Mn/Fe] from -0.3 to 0 for [Fe/H]>-1. Such a behavior can be explained by the trend of SN Ia yields as a function of metallicity (see Fig. 6).

The future will lie in a consistent modeling in 3D of SNe Ia [47,14], rather than parametrizations as presented here. The nucleosynthesis constraints on e.g.

[54]Fe and [58]Ni will have to be met by these models as well and might also lead to insight into the exact working of Ia explosions (see Niemeyer et al., this volume).

4   Galactic Evolution and Sites of the r-Process

In sections 2 and 3 we described the present status of nucleosynthesis calculations in SNe II and SNe Ia, focussing on uncertainties (a) in the explosion mechanism of SNe II, the amount of Fe-group ejecta and the Fe-group composition and (b) the Fe-group composition in SNe Ia as a function of ignition density. Results from parametrized, induced SN II explosion calculations with assumed mass cut [54] and average delayed detonation models of SNe Ia [21] have been utilized before for galactic evolution calculations with instantaneous mixing [5]. More recent approaches consider the finite amount of mixing in supernova remnants [1] and can explain a scatter of abundance ratios at low metallicities due to their variation in supernova events as a function of progenitor mass and metallicity (see Fig. 7). The too extended scatter in [O/Fe] and [Mg/Fe] in comparison to observations is related to our remaining lack in the understanding of the SN II mechanism. However, the tendency of an increasing scatter towards low metallicities and its value is roughly correct.

The fact that r-process abundances behave quite differently (a scatter by almost three orders of magnitude in r/Fe [51,8] rather than about a factor of 10 as shown here for O or Si) gives rise to questions about the origin of the r-process. Site-independent classical analyses [26,44], based on neutron number density n_n, temperature T, and duration time τ, as well as entropy based calculations with the parameters entropy S, proton/nucleon-ratio Y_e, and expansion timescale τ [11] have shown that the solar r-process can be fit by a continuous superposition of components with neutron separation energies at freeze-out in the range 4-1 MeV [7]. Besides nuclear structure, related to masses far from stability, and beta-decay half-lifes, both discussed in many reviews [44], neutrino-induced reactions [27,17,42] and fission properties [43] have been investigated recently and should be investigated further [34,31].

A different question is related to the actual astrophysical realization of such conditions. The observations of stellar spectra of low metallicity stars, stemming from the very early phases of galactic evolution, are all consistent with a solar r-abundance pattern for elements heavier than Ba, and the relative abundances among heavy elements do apparently not show any time evolution [7,51,8]. However, there are also indications [50] that the heaviest elements (Th, U) might vary among different sites. Meteoritic abundances and low metallicity stars also support that at least two r-process sources have to contribute to the solar r-process abundances [58,51,8], leading to large variations in the nuclear mass range A≈80-120.

Suggested scenarios are the ejecta from neutron star mergers [12,49,48] with rates estimated to be of the order $10^{-6} - 10^{-4} \mathrm{yr}^{-1}$ per galaxy [10,23] and the ejection of neutron-rich elements of the order of $10^{-(2-3)} M_\odot$ in Newtonian and relativistic calculations [41]. If SNe II are also responsible for the solar r-process

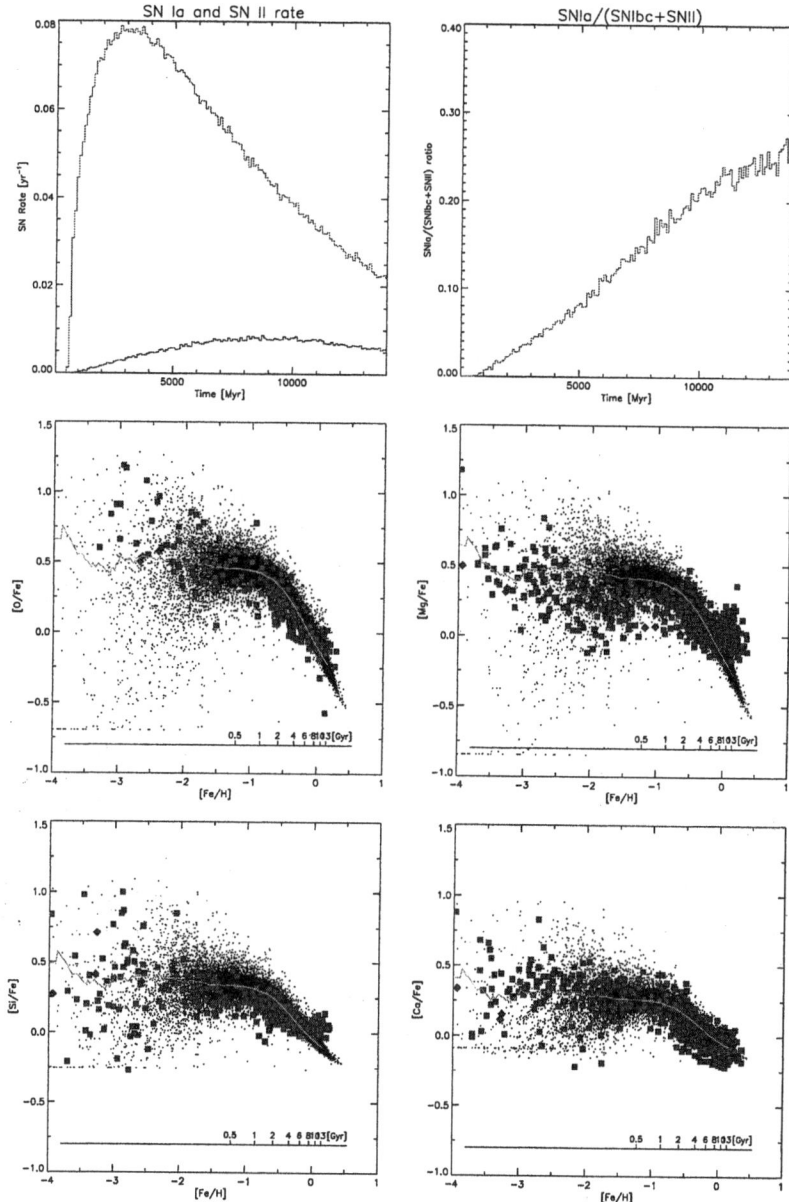

Fig. 7. Results from inhomogeneous galactic evolution calculations for the Milky Way [1]: (a) type II and type Ia supernova rates, (b) ratio of SNe Ia/SNe II, (c)-(f) [O/Fe], [Mg/Fe], [Si/Fe], and [Ca/Fe] as a function of metallicity [Fe/H]; filled squares denote observational values, dots display model stars from the inhomogeneous galactic evolution

abundances, given the galactic occurrence frequency, they would need to eject about 10^{-5} $M_\odot$ of r-process elements per event (if all SNe II contribute equally) in the so-called neutrino wind following a successful SNe II explosion [60,52]. However, the r-process by neutrino wind ejecta of SNe II faces two difficulties. (i) Whether the required high entropies for reproducing heavy r-process nuclei can really be attained in supernova explosions has still to be verified [55]. Presently it seems that only (unrealistically?) large or compact neutron stars with masses in excess of 2 $M_\odot$ can provide the high entropies required, unless ad hoc assumptions [53] can be verified. (ii) The mass region 80–110 experiences difficulties to be reproduced adequately [11,57].

The two possible sites discussed above (SNe II and neutron star mergers) have different occurrence frequencies ($\approx 10^{-2} \text{yr}^{-1}$ and $10^{-(4-5)} \text{yr}^{-1}$ per galaxy) and different amounts of r-process ejecta (10^{-5} $M_\odot$ and $10^{-(2-3)}$ $M_\odot$), if a successful r-process actually occurs. These properties will enter into the enrichment pattern of r-process element in galactic evolution. Observations in very metal-poor stars give clear indications of the [r/Fe] scatter as a function of metallicity. The increasing numbers of observations and the upcoming attempts to model inhomogeneous galactic evolution in the very early phases of the galaxy [20,1] will finally decide whether one of the above mentioned models or both in combination will meet the observational constraints.

References

1. Argast, D. et al. 2002, A&A 388, 842 and Ph.D. thesis Univ. Basel, unpublished
2. Bethe, H.A. 1990, Rev. Mod. Phys. 62, 801
3. Brachwitz, F. et al. 2000, ApJ, 536, 934
4. Brachwitz, F. et al. 2002, Ph.D. thesis, Univ. Basel and in preparation
5. Chiappini, C., Matteucci, F., Beers, T. C., Nomoto, K. 1999, ApJ 515, 226
6. Chieffi, A., Limongi, M., Straniero, O. 1998, ApJ 502, 737
7. Cowan, J.J. et al. 1999, ApJ 521, 194
8. Cowan, J.J. et al. 2002, ApJ 572, 861
9. Dominguez, I., Höflich, P., Straniero, O. 2001, ApJ 557, 279
10. Eichler, D. et al. 1989, Nature 340, 126
11. Freiburghaus, C. et al. 1999, ApJ 516, 381
12. Freiburghaus, C., Rosswog, S., Thielemann, F.-K. 1999, ApJ 525, L121
13. Fuller, G.M., Fowler, W.A., Newman, M. 1985, ApJ 293, 1
14. Gamezo, V. N., Khokhlov, A. M., Oran, E. S. 2002, AAS 200.1401G
15. Hashimoto, M., Iwamoto, K., Nomoto, K., 1993, ApJ 414, L105
16. Heger, A. et al. 2001, ApJ 560, 307
17. Hektor, A. et al. 2000, Phys. Rev. C 6105, 5803
18. Höflich, P., Khokhlov, A. 1996, ApJ, 457, 500
19. Hoffman, R.D. et al. 1999, ApJ 521, 735
20. Ishimaru, Y. et. al 2002, IAU SYMP 187, 117; MPA/P13, p.224
21. Iwamoto, K. et al. 1999, ApJ Suppl. 125, 439
22. Käppeler, F., Thielemann, F.-K., Wiescher, M. 1998, Ann. Rev. Nucl. Part. Sci. 48, 175
23. Kalogera, V. et al. 2001, ApJ 556, 340

24. Khokhlov, A.M. et al. 1999, ApJ 524, L107
25. Kifonidis, K., Plewa, T., Janka, H.-T., Müller, E. 2000, ApJ 531, L123
26. Kratz, K.-L. et al. 1993, ApJ 402, 216
27. Langanke, K., Kolbe, E. 2001, ADNDT 79, 293
28. Langanke, K., Martinez-Pinedo, G. 2002, Rev. Mod. Phys, in press
29. Liebendörfer, M. et al. 2001, Phys. Rev. D 63, 3004
30. Liebendörfer, M. et al. 2002, MPA/P13, p.126
31. Mamdouh, A. et al. 2001, Nucl. Phys. A 679, 337
32. Martinez-Pinedo, G. et al. 2000, ApJS, 126, 493
33. Mezzacappa, A. et al. 2001, Phys. Rev. Lett. 86, 1935
34. Myers, W.D., Swiatecki, W.J. 1999, Phys. Rev. C 60, 4606
35. Niemeyer, J.C. 1999, ApJ 523, L57
36. Niemeyer, J.C., Bushe, W. K., Ruetsch, G.R. 1999, ApJ 524, 290
37. Nomoto, K., Hashimoto, M. 1988, Phys. Rep. 163,
38. Nomoto, K., Thielemann, F.-K., Yokoi, K. 1984, ApJ 286, 644
39. Nomoto, K. et al. 2000, in *Type Ia Supernovae: Theory and Cosmology*, Cambridge Univ. Press, eds. J.C. Niemeyer & J.W. Truran, p.63
40. Nugent, P. et al. 1997, ApJ 485, 812
41. Oechslin, R. et al. 2002, Phys. Rev. D65, 3005
42. Pastor, S., Raffelt, G. 2002, Phys. Rev. Lett. 89, 191101
43. Panov I, Freiburghaus C, Thielemann F-K 2001, Nucl. Phys. A688, 587
44. Pfeiffer, B., Kratz, K.-L., Thielemann, F.-K., Walters, W.B. 2001, Nucl. Phys. A 693, 282
45. Rampp, M., Janka, H.T. 2000, ApJ 539, L33
46. Rauscher, T., Thielemann, F.-K. 2000, At. Data Nucl. Data Tables, 75, 1
47. Reinecke, M., Hillebrandt, W., Niemeyer, J. C. 2002, A&A 391, 1167
48. Rosswog, S.K., Davies, M. 2002, MNRAS 334, 481
49. Ruffert, M., Janka, H.-T. 2001, A&A 380, 544
50. Schatz, H. et al. 2002, ApJ 579, 626
51. Sneden, C. et al. 2000, ApJ 533, 139
52. Takahashi, K., Witti, J., Janka, H.-T. 1994, A&A, 286, 857
53. Terasawa, M. et al. 2002, ApJ 578, L137
54. Thielemann, F.-K., Nomoto, K., Hashimoto, M. 1996, ApJ 460, 408
55. Thompson, T.A. et al. 2001, ApJ 562, 887
56. Umeda, H., Nomoto, K. 2002, ApJ 565, 385
57. Wanajo, S. et al. 2001, ApJ 554, 578
58. Wasserburg, G., Busso, M., Gallino, R. 1996, ApJ 466, L109
59. Woosley, S.E., Weaver, T.A. 1995, ApJ Suppl. 101, 181
60. Woosley, S.E. et al. 1994, ApJ 433, 229
61. Woosley, S.E., Weaver, T.A. 1994, in *Les Houches, Session LIV, Supernovae*, eds. S.R. Bludman, R. Mochkovitch, J. Zinn-Justin, Elsevier, p. 63

Extremely Metal-Poor Stars: Tracing the Early Stages of Nucleosynthesis and Chemical Evolution

Francesca Primas

European Southern Observatory, Karl-Schwarzschild-Strasse 2,
Garching bei München, Germany

Abstract. Observational knowledge of the origin, history, and distribution of the chemical elements rests upon abundance analyses of interstellar gas and stars of various populations. When looking in detail at the abundance patterns of the most metal-deficient stars of the Milky Way, we are looking at the fossil record of nucleosynthesis processes that took place in the first generation(s) of stars. Each elemental abundance is a powerful diagnostic: from the very light elements that test homogeneous vs inhomogeneous BBN scenarios to the very heavy nuclei that test and constrain the signature of individual supernova (SN) events. Here, the remarkable (observational) progress recently achieved in the field of extremely metal-poor stars will be presented and critically reviewed.

1 A Brief Overview

The classical paradigm of the chemical evolution of the Galactic halo was first put forth by [34] and is based on one fundamental concept: the two main classes of supernova (type I and II) have differing timescales and eject different products. Type Ia SNe produce mainly Fe-group elements and have typical timescales on the order of 10^9 yr, whereas Type II SNe produce mainly lighter elements (e.g. the α-group) and are characterized by shorter timescales, on the order of 10^7 yr. SN Ia progenitor lifetimes are consequently too long to explain the iron content observed in halo stars, that must then arise in massive stars via explosive nucleosynthesis. By investigating the chemical composition of the low-mass stars still observable today we can then probe the ejecta of these earliest SNe and constrain how and when they were incorporated into subsequent stellar generations. Having formed at an epoch corresponding to redshifts $z \geq 4$-5, these old metal-poor halo stars represent our main Galactic fossils.

But what is a metal-poor star? It is a star with a metal content lower than the Sun, which is our main reference system (abundance ratios are usually given in the [A/B] notation, where [A/B] = log(A/B)$_*$ - log(A/B)$_\odot$). Although the magnitude of this "lack" of metals has not been univocally classified (one finds in the literature different definitions like *very*, *ultra* or *extremely* metal-poor, sometimes referring to the same metallicity regime), here I will use the term extremely metal-poor (EMP, in short) to indicate all those stars with a metallicity around or below 1/1000 solar (cf. Fig. 1), where metallicity is assumed to be equivalent to the stellar [Fe/H] value.

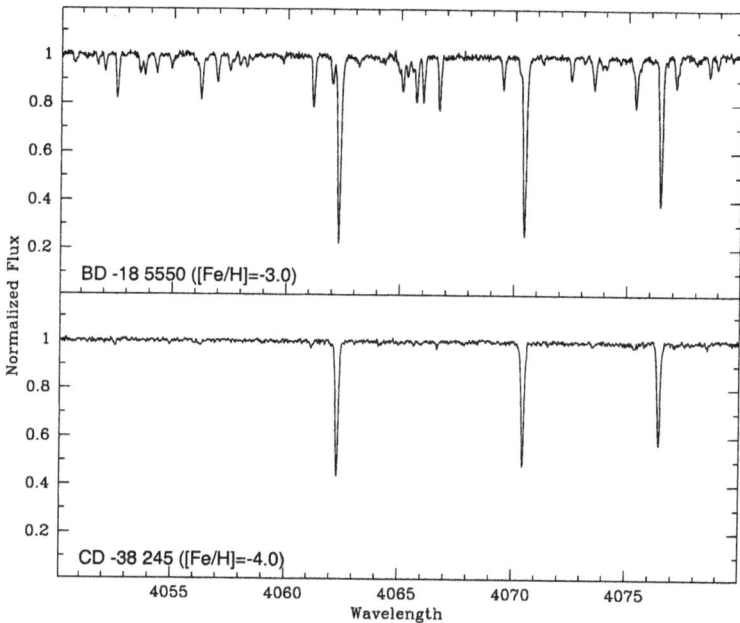

Fig. 1. Stellar spectra of two EMP stars, but of different metallicities: the top star has a metallicity 1/1000 solar (BD -18 5550), the star at the bottom has a metal content 1/10000 solar (CD -38 245). Observations were taken at the VLT, with UVES (courtesy of the UVES Large Program "The First Stars")

1.1   Chasing EMP Stars

Large efforts have been invested in the last decades to the search for metal-poor stars. Knowledge of the halo metallicity distribution function provides direct information about the initial stages of galaxy formation, and constrains the primordial rate of Type II SN, star formation rate (SFR), and the timescale for the re-distribution of elements in the early Galaxy. To do that, one needs large samples of stars.

Despite the variety of techniques used to discover metal-poor stars in the Galaxy, these searches can be divided in two main groups: proper-motion surveys (e.g. [31] and [8]) or objective-prism surveys (e.g. the HK survey of [4]). The HK survey, so far the most successful one, has produced $\simeq$ 10,000 candidate metal-poor stars, of which 5,000 have been looked at moderate resolution (1-2Å) and more than 3,000 have now broad-band photometry. Thanks to the HK and the other systematic surveys, the progress in searching for (and finding) metal-poor objects has been astonishing: from the (first) two metal-deficient stars (HD 19445 and HD 140283, $\approx$ 1/100 less than solar) studied by [11] we have now $\simeq$100 stars that have been confirmed to have [Fe/H] $\leq$ −3.0 (none was known only two decades ago!).

The more recent Hamburg-ESO Survey (HES in short, cf. [39]) promises to do even better: it covers a larger area of the southern high-latitude sky (8000 square degree vs the 4500 of the HK survey), and it goes $\simeq$ 2 mag deeper than the HK (thus reaching B$\simeq$ 17.5). Although it was initially intended for bright QSOs, exploitation of its stellar content (cf. [13]) has shown that the survey will be able to provide 20,000 metal-poor candidates.

A key feature of all surveys is the ability to fully explore all the information there contained in a timely manner. The newly developed classification schemes such as the Artificial Neural Networks (e.g. [33]) offer significant advantages to account for the non-linear dependences between temperature, gravity, and metallicity indicators.

2   The Chemical Signature of EMP Stars

Extremely metal-poor stars have atmospheres that consist of the most primitive matter in the Universe. Each elemental abundance helps us understanding the physics involved at the time of the SN explosion and before. Variations in any given abundance ratio help us constraining the number and types of SNe that have exploded at these early times. In other words, we are "stellar archaeologists" (cf. [14,9]) because we try to characterize previous stellar generation(s) (that no longer exists) by studying the oldest stellar population still observable today.

Elements lighter than Ca (i.e. C, O, Ne, Mg) are essentially produced in 10-30 $M_\odot$ during the hydrostatic phases of the Ne- and C-burning (and partly also in the corresponding explosive phases); they strongly depend on the progenitor mass. Si, S, Ar, and Ca originate from explosive O- and Si-burning and they are used to test the progenitor model and the explosion energy. Elements with higher atomic numbers like those belonging to the iron-group (e.g. V, Cr, Mn) are produced during the explosion, and their relative production rate depends directly on the explosion mechanism, which in turn is connected to the size of the collapsing Fe-core. Until the explosion mechanism is quantitatively constrained, the ejected amount of Fe-group material can vary very strongly. Heavy elements (i.e. $Z > 30$) are instead synthesized through neutron capture reactions by iron-peak nuclei which are exposed to *slow* or *rapid* neutron fluxes. Late stellar evolutionary stages are believed to be one of the main sources of the *s*-process elements, whereas SNe II have long been suspected to be the source of the *r*-process elements. Direct information about the production site for *r*-process elements should ideally come from measurements of the elemental abundances present in the ejecta of individual SN. Absorption lines due to Ba and Sr (10% of which are of *r*-process origin in the solar distribution) were indeed detected in the spectra of SN 1987A by [38] and [24] noted that the shape of the Ba line indicated the lack of this element at the very surface of the ejecta, thus implying that the observed Ba must had been synthesized inside the star and did not exist in the ISM from which the star formed.

3   New Results from New Observational Efforts

A major breakthrough in the field came in the mid 90s, when [25] looked at a sample of about 20 EMP stars and for the first time detected sudden variations in the run of some elemental abundance ratios (Cr, Mn, Co, and Cu) and large dispersion in some of the heavy elements (like Sr). These findings are still challenging our understanding of the early chemical evolution of the Galaxy.

Nowadays, the major observing campaigns for high resolution follow-ups are carried out at 8-10m class telescopes. Roger Cayrel (Observatoire de Paris) and collaborators have conducted a Large Program ("The First Stars") at the VLT with UVES [16] with the aim of covering the entire sample of the 100 most metal-poor stars (both dwarfs and giants) found by the HK survey. The first results have recently started to appear (cf. [10,22,18]). Judith Cohen (Caltech) and collaborators had similar goals for their Pilot Program at the Keck telescope with HIRES [35]: this is now completed [14,9] and it will likely extend into a long-term program, which will take advantage of the yet unexplored metal-poor stellar content of the HES database.

One of the most significant outcomes of these new studies is probably the unprecedented high quality of the data. The new results mostly confirm the earlier findings by [25]. For instance, in the case of the α-elements, [9] find Mg to follow the common halo overabundance (i.e. $[Mg/Fe]\simeq+0.4$, except for 4 objects which need to be discussed separately), some dispersion among the Si data-points (but Si abundances have been derived from one transition only), and they confirm the possible presence of scatter in the Ca abundance already noted by [25]. The UVES LP team (2002, in prep) also finds similar results but it also shows the possible presence of dispersion among the Mg abundances (see Fig. 2).

For the iron-group elements Cr and Mn, the important results presented in [25] are entirely confirmed by [9] and by the UVES LP team: as metallicity decreases, both $[Cr/Fe]$ and $[Mn/Fe]$ keep decreasing (Fig. 3). Different scenarios have been proposed to account for these trends: from the α-rich freeze-out to a possible dependence of the production of Cr and Mn on the position of the mass-cut, to a different neutron excess present in this metallicity range, or even hypernovae (the latter generically defined as massive objects having very large explosion energies). By moving the cut relative to the regions of complete and incomplete Si burning, [26] are indeed able to reproduce the observed trends of Cr, Mn, and Co. This is true also for their hypernova scenario, but in this case Ni is predicted to follow Co, which has not yet been confirmed observationally. The hypernova scenario has also been given lower priority by [9], based on the finding of a correlation between $[Cr/Fe]$ and $[Mn/Fe]$, not predicted by the models.

The abundances of heavy elements in the metal-poor halo population of the Galaxy were first surveyed by [20], who showed the operation of the r-process at low metallicity and revealed a significant star-to-star scatter. These findings have been recently confirmed also by [7] who have found that at very low metallicity the abundance ratio $[Ba/Eu]$ (i.e. the abundance ratio between a typical - for solar material - s- and r-process element) indicates that the heavy n-capture

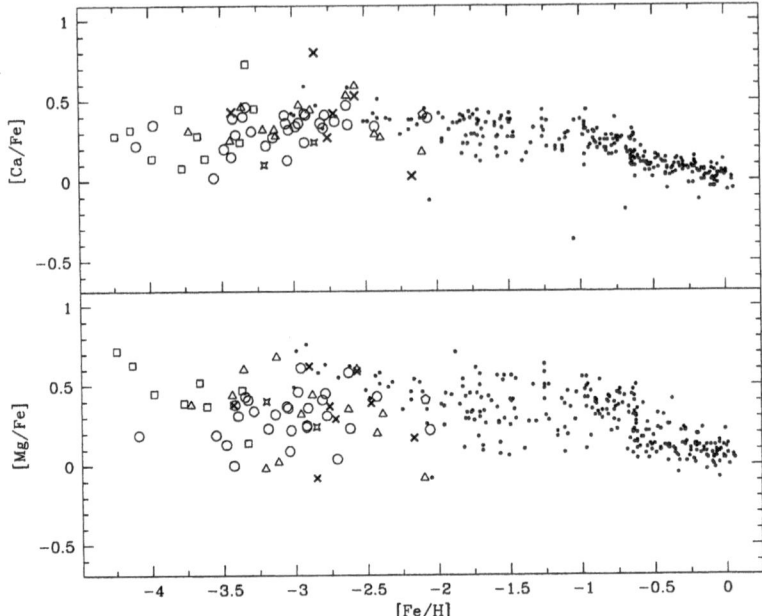

Fig. 2. The abundance ratios [Ca/Fe] and [Mg/Fe]. Data sources: [9] (triangles), the UVES LP team (2002 in prep., but also [22,18]) (circles), [15] (pentagons), [2] (crosses), [17] (four-sided stars), [29,28] (squares), [23,12,19] (dots)

elements were formed predominantly via the r-process at the earliest times in the Galactic history. The scatter in the abundances of all the n-capture elements from star to star is of astrophysical origin and the scatter increases as we go to lower metallicity.

Also [9] have confirmed the presence of dispersion, especially of a large scatter in [Sr/Fe] ratio (up to 3 dex), both among giant and dwarf stars. For Ba they find a decrease below [Fe/H]$\leq$ −2.0, much less scatter than Sr, although few objects (probably of different origin) are found at [Ba/Fe]$\geq$1.0, but they note that analyses of Ba abundances are plagued by the treatment of hyperfine splitting. They suggest that the different behaviour observed for [Sr/Fe] and [Ba/Fe] may point to the need for two distinct processes: the basic r-process, which is clearly necessary in stars with [Fe/H]$\leq$ −2.5 to account for the abundance pattern at large atomic number, and a second process that favours production of the elements with lower atomic number.

4   Some Highlights

There are some topics that, from my personal (and likely biased) point of view, I consider hotter than others. Here, I would like to spend few words on two of

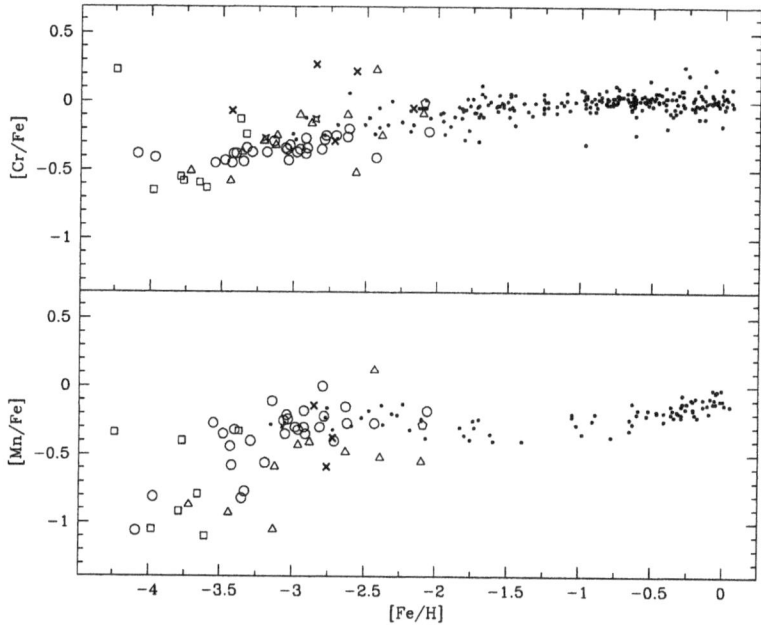

Fig. 3. The abundance ratios [Cr/Fe] and [Mn/Fe]. Data sources and symbols are the same as in Fig. 2

them, with no intention to diminish the value of all the other interesting projects currently under way in the field of EMP stars.

4.1 r-Process-Rich EMP Stars and Cosmo-Chronometry

The relative proportions of the n-capture elements in halo stars are in good agreement with production by the r-process, with little or no contribution from the s-process. This was noted already by [37] and strongly suggests their origin in a fast, explosive event. The already mentioned results by [20], i.e. the presence of dispersion among n-capture elements in metal-poor stars up to [Fe/H]= −2.0, provided direct evidence that local nucleosynthesis events affected the early galactic halo ISM that remained poorly mixed until the formation of stars at this much higher metallicity.

Although many of the key neutron-capture elements are observable from ground, such as Eu, Gd, Dy, Th, and quite recently, U as well [10], observations with HST are crucial to obtain abundances of n-capture elements that have no, or only extremely weak, lines in the visible domain. These elements include Pb, as well as Bi, Pt, Os and Ir – the heaviest r-process elements, which occupy the so-called third r-process peak. However, the biggest difficulty in observing heavy elements in EMP stars is the extreme weakness of their transitions.

This is why the serendipitous discovery of the n-capture rich star CS 22892-052 [32] is one of the most remarkable achievements of the last decade. It has allowed us not only to reveal the presence of 'never detected before' elements (like Au, Ag) in the atmospheres of such an old star (its metallicity is around 1/1000 solar), but also to further constrain possible sites for the r-process and the conditions under which it operates (from the agreement between the heavier n-capture elements and the scaled solar system r-process curve).

Only three EMP halo stars with large enhancements of the n-capture elements with respect to iron (by factors of about 50) are presently known: CS 22892-052, HD 115444 ([Fe/H]~ −3.0; [36]), and the newly discovered CS 31082-001 ([Fe/H]~ −2.9, [10,22]). Very interesting is also the finding of another n-capture rich star, BD +17 3248 [15], but at a much higher metallicity ([Fe/H] ~ −2.1).

A direct link to this distant universe is provided by dating methods that use cosmo-chronology (age estimates based on the radioactive decay of unstable heavy nuclei, like thorium, in EMP stars) to provide independent measurements of the ages of the oldest stars in the Galaxy. So far, cosmo-chronology has relied on the comparison of the abundances of the unstable Th to some other stable n-capture element, such as Eu (97% of which is of r-process origin in solar system material). One of the most serious limitations of this approach is that theoretical predictions of the production ratio of these elements suffer from the lack of a realistic astrophysical model of the r-process. One possible improvement may come from replacing Eu with some other unstable element, like uranium, since the [Th/U] ratio varies twice as fast with time as does [Th/Eu], and the production ratio of two nuclei with similar masses is much more stable (e.g. [21]).

CS 31082-001 currently represents our best opportunity to understand the nature of the r-process in the early Galaxy and to validate the use of radioactive cosmo-chronometers for age determinations of the oldest stars. Although very similar to CS 22892-052, it has the advantage of having a much lower content of C, hence much less blending of its atomic lines by CH and CN molecular bands. This has made possible the first detection of the uranium line U II 3859.59 Å (see Fig. 10 of [22]). An age of 12.5±3.3 Gyr has been derived from the [Th/U] ratio ([10], recently revised to 14.0±2.4 Gyr by [22] thanks to improved atomic physics [27]), and detailed knowledge of 41 elemental abundances [22] has now provided the first solid evidence that the r-process abundance patterns in the actinide region (Z>90) differ from star to star, making the conventional [Th/Eu] chronometer to fail.

Worldwide efforts are now underway for the detection of perhaps as many as 50 new r-process enhanced metal-poor stars, each of which could be used to obtain age limits, if the U/Th chronometer in CS 31082-001 can be shown to provide accurate information.

4.2   The Role of C-Rich EMP Stars in the Early Galaxy

A (somewhat unexpected) result of the HK survey is the high percentage (approx. 10-15%) of very metal-poor stars to exhibit anomalously strong CH bands ([C,N/Fe] ratios up to +2.0 above the solar value have been measured), which is in contrast with the commonly accepted picture of a [C/Fe] ratio close to solar.

Chemical peculiarities in cool stars (B-V>0.4) are often interpreted as a result of mixing nucleosynthesis products to the stellar surface. The nucleosynthesis may have taken place either in the star itself or in an evolved companion from which mass has been accreted. However, despite the small number of C-rich EMP stars that have been studied at high resolution, none of the so far proposed scenarios (i.e. transfer of C-enhanced material across a binary system containing an AGB star, self-pollution of internally processed material in individual stars, strong mixing of material enriched in C and CN-cycle burning due to He core flash in very low abundance stars) is able to account for the broad family we have discovered: there are objects which have been found to be enhanced in s-process elements (like the classical CH and Ba stars [3,29]), others which have the r-process elements enhanced (e.g. [Eu/Fe]=+1.7 in CS 22892-052) and stars characterized by the lack of any n-capture element enhancement [6].

Only one star, LP 625-44, has been confirmed to be a binary and strongly supports the mass-transfer scenario, though its [Pb/Ba] ratio differs from the predictions. Otherwise, a long-term radial velocity survey by [30] has shown that none of the three C-rich metal-poor stars they surveyed exhibited velocity variations (at least not exceeding 0.5 km/s over a period of 8 years). These authors have proposed an enhanced mixing episode occurring at the end of the giant branch evolution, that recycled these stars to the base of the subgiant branch (because of increased H mixing into their cores, as originally suggested by [5]).

Among the most recent observational overviews, the work by [2] (see their Figs. 6 and 8) attempted the following classification: s-process-rich C-enhanced metal-poor stars are likely to be the companion of a higher mass star from which n-capture-element-rich material has been accreted, whereas a C-rich metal-poor star with normal n-capture element abundances would be a low mass (0.8 $M_\odot$) star in which C and N have been self-enhanced during its AGB evolution without affecting the abundance of its n-capture elements. An alternative would be if these stars were the companions of slightly higher mass stars (0.8-1.0 $M_\odot$) from which C-rich material without excess of n-capture elements has been accreted.

5   Concluding Remarks

This is a very active field, that has just started benefiting enormously from the advent of large telescopes and much more efficient spectrographs. More than concluding remarks, I think that for the time being we are left with some certainties but also with some uncertainties.

It is well established that the vast majority of metal-poor Galactic halo stars show (observational) evidence for enrichment by solely massive stars. Simulations of the collapse of primordial molecular clouds also suggest that the first generation of stars contained many massive members from $100 M_\odot$ up to $1000 M_\odot$ [1]. Therefore, EMP stars represent the living witnesses of this first generation of high-mass, zero-metallicity objects that evolved and polluted the proto-galaxy. The metal enrichment of the halo then depends on how many SN II exploded and how efficiently the ejected gas was mixed with the surrounding ISM, which in turn depends on the volume in which these ejecta were distributed. It is important to note that halo stars in the metallicity range [Fe/H]=−2.5 and −1.0 appear instead to have fairly uniform abundance mixes indicating the existence of substantial amounts of nuclear processing and gas mixing before their formation.

Several elements (e.g. Al, Cr, Mn, Co, Sr, Ba) show a sudden change in the slope [X/Fe] vs. [Fe/H] near [Fe/H]=−2.4. Others, like Sr, show considerable scatter from star to star at very low [Fe/H], perhaps indicating that the yield from individual or at most a few SN events could be responsible for the metals observed in these stars. It is clear that a proper understanding of such intrinsic scatter (observed in the element-to-element ratios) will provide important information about the typical number of SNe polluting the first clouds that underwent some chemical evolution in the early Galaxy. This is the main challenge awaiting in front of us and that the new observing campaigns have decided to accept.

Acknowledgements

The author is indebted to the UVES LP Team ("The First Stars"), in particular to E. Depagne, M. Spite, and V. Hill (Observatoire de Paris-Meudon) for having kindly provided abundance results in advance of publication (some of which belonging to E. Depagne's PhD Thesis).

References

1. T. Abel, G.L. Bryan, M.L. Norman: ApJ **540**, 39 (2000)
2. W. Aoki, J.E. Norris, S.G. Ryan, T.C. Beers, H. Ando: ApJ **567**, 1166 (2002)
3. B. Barbuy, R. Cayrel, M. Spite, T.C. Beers, F. Spite, B. Nordstroem, P.E. Nissen: A&A **317**, L63 (1997)
4. T.C. Beers, G.W. Preston, S.A. Schectman: AJ **90**, 2089 (1985)
5. H.E. Bond: ApJ **194**, 95 (1974)
6. P. Bonifacio, P. Molaro, T.C. Beers, G. Vladilo: A&A **332**, 672 (1998)
7. D.L. Burris, C.A. Pilachowski, T.E. Armandroff, C. Sneden, J.J. Cowan, H. Roe: ApJ **544**, 302 (2000)
8. B.W. Carney, D.W. Latham, J.B. Laird, L.A. Aguilar: AJ **107**, 2240 (1994)
9. E. Carretta, R. Gratton, J.G. Cohen, T.C. Beers, N. Christlieb: AJ **124**, 481 (2002)
10. R. Cayrel et al: Nature **409**, 691 (2001)
11. J.W. Chamberlain, L.H. Aller: ApJ **114**, 52 (1951)
12. Y.Q. Chen, P.E. Nissen, G. Zhao, H.W. Zhang, T. Benoni: A&AS **141**, 491 (2000)
13. N. Christlieb, D. Reimers, L. Wisotzki, J. Reetz, T. Gehren, T.C. Beers: 'Finding the First Stars: The Hamburg/ESO Objective Prism Survey', in: *The First Stars, Proceedings of the MPA/ESO Workshop at Garching, Germany, August 4–6, 1999*, ed. by A. Weiss, T.G. Abel, V. Hill (Springer, Heidelberg 2000), pp. 49–50
14. J.G. Cohen, N. Christlieb, T.C. Beers, R. Gratton, E. Carretta: AJ **124**, 470 (2002)
15. J.J. Cowan, et al: ApJ **572**, 861 (2002)

16. H. Dekker, S. D'Odorico, A. Kaufer, B. Delabre, H. Kotzlowski: SPIE **4008**, 534 (2000)
17. E. Depagne, V. Hill, N. Christlieb, F. Primas: A&A **364**, L6 (2000)
18. E. Depagne et al: A&A **390**, 187 (2002)
19. J.P. Fulbright: AJ **120**, 1841 (2000)
20. K.K. Gilroy, C. Sneden, C.A. Pilachowski, J.J. Cowan: ApJ **327**, 298 (1988)
21. S. Goriely, B. Clairbaux: A&A **346**, 798 (1999)
22. V. Hill et al: A&A **387**, 560 (2002)
23. J.A. Johnson: ApJS **139**, 219 (2002)
24. P.A. Mazzali, L.B. Lucy, K. Butler: A&A **258**, 399 (1992)
25. A. McWilliam, G.W. Preston, C. Sneden, L. Searle: AJ **109**, 2757 (1995)
26. T. Nakamura, H. Umeda, K. Iwamoto, K. Nomoto, M. Hashimoto, W.R. Hix, F.K. Thielemann: ApJ **555**, 880 (2002)
27. H. Nilsson, S. Ivarsson, S. Johansson, H. Lundberg: A&A **381**, 1090 (2002)
28. J.E. Norris, R.C. Peterson, T.C. Beers: ApJ **415**, 797 (1993)
29. J.E. Norris, S.G. Ryan, T.C. Beers: ApJ **489**, L169 (1997)
30. G.W. Preston, C. Sneden: AJ **122**, 1697 (2001)
31. S.G. Ryan, J.E. Norris: AJ **101**, 1865 (1991)
32. C. Sneden, A. McWilliam, G.W. Preston, J.J. Cowan, D.L. Burris, B.J. Armosky: ApJ **467**, 819 (1996)
33. S. Snider, C. Allende Prieto, T. von Hippel, T.C. Beers, C. Sneden, Y. Qu, S. Rossi: ApJ **562**, 528 (2001)
34. B.M. Tinsley: ApJ **229**, 1046 (1979)
35. S. Vogt et al: SPIE **2198**, 362 (1994)
36. J. Westin, C. Sneden, B. Gustafsson, J.J. Cowan: ApJ **530**, 783 (2000)
37. J.C. Wheeler, C. Sneden, J.W. Truran: ARA&A **27**, 186 (1989)
38. R.E. Williams: ApJ **320**, L117 (1987)
39. L. Wisotzki, T. Koehler, D. Groote, D. Reimers: A&AS **115**, 227 (1996)

Supernova Signatures in Meteorites

Ulrich Ott and Günter W. Lugmair

Max-Planck-Institut für Chemie (Otto-Hahn-Institut),
Becherweg 27, D-55128 Mainz, Germany

Abstract. This review focuses on signatures of nucleosynthesis in supernovae that can be traced via isotope abundance anomalies in meteorites. Primary examples of such anomalies are found

- in high-temperature Ca-Al-rich inclusions (CAIs), especially those of the FUN type. Most prominent here are excesses and deficits of the n-rich isotopes of Fe peak elements.
- in grains of circumstellar origin (e.g. diamond, silicon carbide, graphite, aluminum oxide) that have survived in meteorites. Effects here are much larger and are observed in many major (C, N, Si, O) as well as minor and trace elements.
- in bulk meteorites as well as specific meteoritic phases as anomalies resulting from *in-situ* decay of extinct radionuclides that were present in the early solar system.

1 Introduction

Signatures of nucleosynthesis processes as they occur in supernovae can be recognized in meteorites in

- the overall solar abundance distribution that is largely based on analyses of primitive carbonaceous chondrites of type CI; and
- isotopic abundance anomalies where nuclides are overabundant that are produced in supernovae.

This review will focus on the latter type of contributions. Individual effects as discussed below have been recognized in Ca-Al-rich inclusions (CAIs), especially those of the FUN type and in grains of circumstellar origin that have survived in meteorites. Further discussed below are anomalies found in bulk meteorites as well as specific meteoritic phases that result from *in-situ* decay of extinct radionuclides present in the early solar system. In addition, there is evidence for small variations in non-radiogenic isotopic compositions on the bulk meteorite level. Some of these may be caused by an inhomogeneous distribution of presolar grains.

2 Ca-Al-Rich Inclusions (CAIs)

Primitive carbonaceous meteorites (excepting CI meteorites which consist of matrix only) consist of three major types of materials: chondrules, Ca-Al-rich inclusions (CAIs; of high-temperature origin), and the matrix (of low-temperature

origin), which fills the space in between (Fig. 1). CAIs are typically of mm-
to cm-size and consist of refractory Ca-Al-rich oxides and silicates. They are
especially abundant in type 3 carbonaceous chondrites like the Allende meteorite shown in Fig. 1, where they also occur in large size. In all likelihood they
represent the oldest dated material that formed *within* the Solar System [1,2].

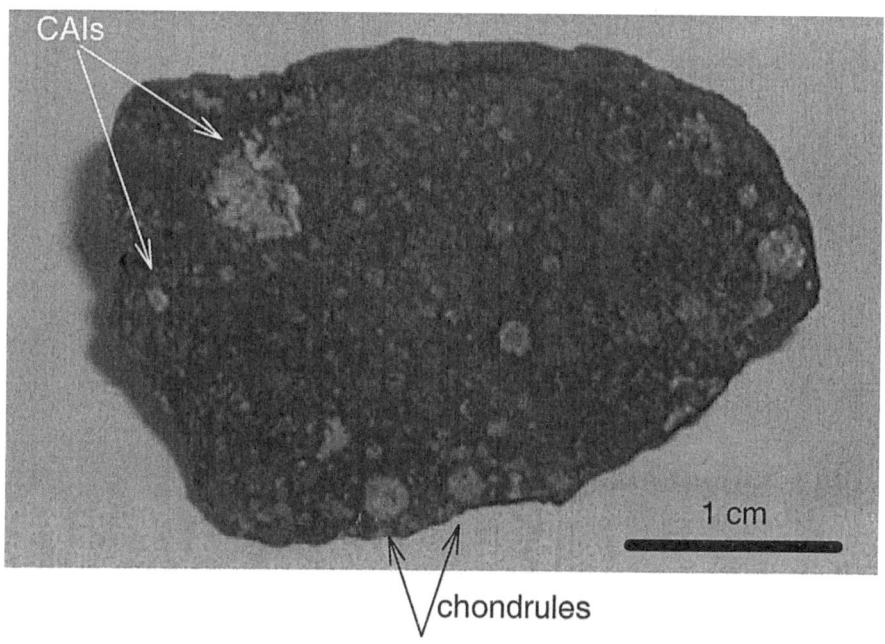

Fig. 1. A slice of the Allende meteorite. Indicated are CAIs (Ca-Al-rich inclusions)
and chondrules, which are surrounded by matrix. Photograph taken by S.P. Schwenzer

Not counting effects in noble gases (known at the time, but largely considered
curiosities by the larger community), the first isotope abundance anomalies were
detected in CAIs in the 1970's. The first was the oxygen isotopic anomaly [3,4]:
deviations in $^{17}O/^{16}O$ and $^{18}O/^{16}O$ deviating from the "mass-dependent" trend
defined by terrestrial rocks that are best explained by variable admixture of a
component of pure ^{16}O. The effect was taken as evidence for admixture of supernova debris without complete homogenization into the solar nebula. It thus
paved the way for the general acceptance of the existence of isotopic inhomogeneities established in the early Solar System due to contributions from stellar
nucleosynthesis. It is somewhat ironic that according to our best current knowledge exactly this isotopic effect in oxygen seems to have a non-nuclear, probably
physico-chemical origin [5–7].

Isotope abundance anomalies detected in the CAIs subsequent to the oxygen
anomaly clearly have a nuclear origin: among others, the ^{26}Mg overabundances

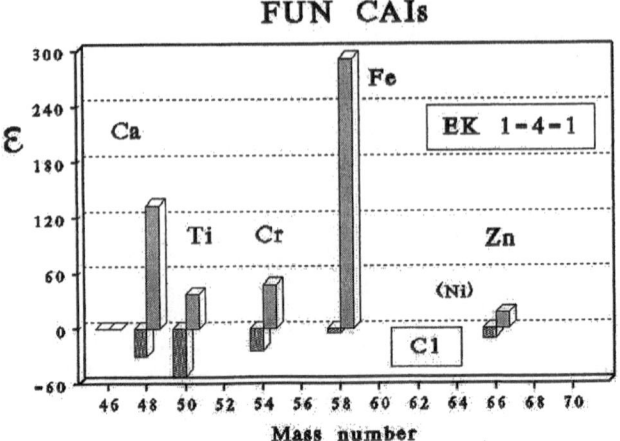

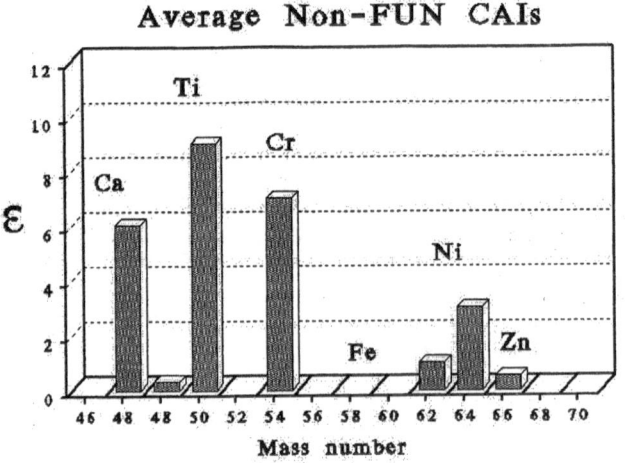

Fig. 2. Isotope abundance anomalies in neutron-rich isotopes of the Fe peak for FUN (top) and average non-FUN CAIs [9,10]. Shown are deviations in ε-units (parts in 10^4)

([8]; see also Sec. 5 below) from the decay of ^{26}Al, and correlated overabundances of the neutron-rich isotopes of the Fe peak elements (Fig. 2; cf. [9]). The latter are especially large in a subgroup of CAIs called "FUN inclusions" (the "<u>F</u>" in the acronym FUN stands for the large mass dependent mass fractionation of oxygen and magnesium observed in these inclusions; "UN" stands for the fact that at the time of the discovery the anomalies had an <u>U</u>nknown <u>N</u>uclear origin). The effects in the Fe peak elements for the two prime examples of FUN inclusions (named C-1 and EK1-4-1) as summarized in [10] are also shown in Fig. 2. Note that the effects not only are much larger than in "normal inclusions", but also that in the case of C-1 (as well as the isotopically virtually identical CAI 1623-5 from the Vigarano meteorite [11]) the neutron-rich isotopes are deficient rather than overabundant. Results for non-FUN CAIs seem rather well explained by the MZM model of Hartmann et al. ([12]; cf. also [9]). But other processes need to be considered as well, e.g. a type of weak r-process is being considered for FUN inclusions like EK1-4-1 ([13]; cf. also [14]). Even larger overabundances in the neutron-rich isotopes ^{48}Ca and ^{50}Ti (percents rather than the ε-units used in Fig. 2) are found in individual grains of hibonite (CaAl$_{12}$O$_{19}$) within CAIs [15].

Results are rare for isotopic abundances of the heavy elements for which the nuclear origin lies in p-, s- and r-process. In the case of FUN inclusions C-1 and EK1-4-1 results have been obtained for Sr, Ba, Nd and Sm (cf. [16]). Because of the small size of effects (up to several per mil only) interpretations are sensitive to the correction for mass-dependent fractionation processes (natural as well as instrumental), but overall the observations are consistent with excesses and/or deficits of r- and/or p-process matter of "normal" p- or r-process composition (e.g., Nd shown in Fig. 3). The situation is, however, somewhat puzzling in that the effects in different elements are not always coherent. In Sm from EK1-4-1, e.g., there is evidence for both p- and r-process enhancements, while Ba shows only evidence for r-excesses and the effect seen in Sr is a deficit rather than an enhancement of p-products [16].

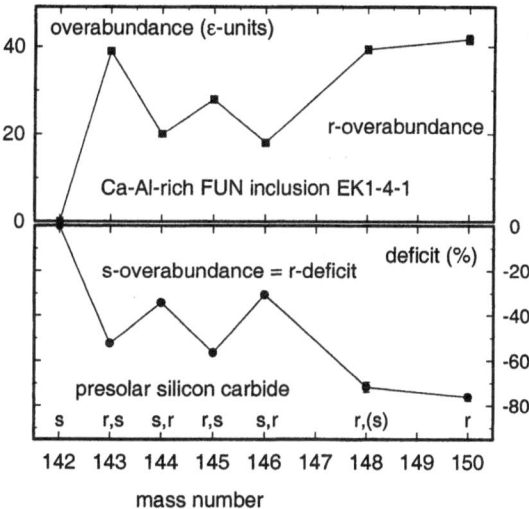

Fig. 3. Comparison of size of effects in FUN-CAI EK1-4-1 and typical presolar SiC grains from AGB stars. The inclusion shows an overabundance of r-process products in neodymium of $\sim$ 4 per mil, while the overabundance of s-products in SiC grains amounts to a factor $\sim$ 4

3   Presolar Grains

The detection of grains of presolar origin in primitive meteorites has been one of the highlights of laboratory cosmochemistry (cf. [17–19]). Diamonds were the first to be detected [20], followed shortly thereafter by silicon carbide [21] and graphite [22] grains. Abundances of SiC and graphite (of typically μm-size) in primitive meteorites are on the order of 10 parts per million, with the abundance of nanodiamonds (average size 2.6 nm) about a factor hundred higher. However, only a fraction of the nanodiamonds may be truly presolar (cf., e.g., [23,24]). Discovery of these grains was the result of the search for the host phases of isotopically anomalous noble gases; additional types of grains (rarer so far) have been found because they were present in the same mineral separates (obtained primarily by acid dissolution of more common phases) as the noble gas carriers. All these presolar grains are in fact not simply of "interstellar origin" but rather "circumstellar condensates" that formed directly from material ejected by stars. In their isotopic compositions – which differ from our "normal compositions" not just by fractions of per mil as in the case of CAIs, but often by factors of several (Fig. 3) – they carry information on nucleosynthesis processes in individual stars.

Table 1 gives on overview of presolar grains as far as they appear to be related to supernovae (assignments mostly taken from [25,26]). Commonly, these minerals, which are all-acid-resistant, are isolated for isotopic analysis from the bulk of the meteorites using physical and chemical separation techniques that include dissolution of the majority ($>99\%$) of the meteoritic material by acids [17–22]. *In situ* searches indicate that they are hosted by the low-temperature part of the meteorites (the matrix; Fig. 1) [27,23]. Analysis of single grains is possible for major and some minor elements by secondary ion mass spectrometry (SIMS; e.g. [28]) and for some trace elements also by resonance ionization mass spectrometry (RIMS; e.g. [29]). In addition, He and Ne have been analyzed by noble gas mass spectrometry in a suite of large graphite and SiC grains [30]. The diamonds are only 2.6 nm in size on average (about 1000 atoms of carbon) and single grain analysis is not possible. Accordingly, it is also not possible to make an assignment of what fraction of the diamonds is likely to derive from supernovae (or to be pre-solar at all, for that matter).

Because of size, abundance and content of trace elements, silicon carbide grains have been most thoroughly investigated (for a summary of results and assignment to stellar sources see [25,26]). The majority clearly is tied to carbon stars as indicated by, among others, the isotopic composition of carbon, nitrogen and neon as well as the occurrence of heavy trace elements with s-process signature [25,26,28–31]. About one percent of the SiC grains, however, clearly exhibits isotopic properties expected for supernova condensate material ("SiC-X grains"). This is most clearly indicated by the isotopic composition of Si which is highly enriched in ^{28}Si (Fig. 4). The overabundances of ^{44}Ca and ^{49}Ti detected in many SiC-X grains are in all likelihood due to the decay within the grains of ^{44}Ti ($T_{1/2} = 60$ a) and ^{49}V ($T_{1/2} = 330$ d), i.e. presence of nuclides thought to be produced in supernovae only [32,33]. High ^{12}C/^{13}C ratios, enhanced ^{15}N and high levels of ^{26}Al ($T_{1/2} = 0.7$ Ma), up to an inferred ^{26}Al/^{27}Al of ~ 0.6 at

Table 1. Major types of presolar dust grains identified in meteorites and some of their properties as they relate to supernovae

mineral	"typical" (total) abundance [ppm][a]	"typical" size [μm][b]	isotopic supernova signature[c]	contribution of supernova source[c]
diamond	~1400	~0.0026	Xe-HL, Te-H	?
silicon carbide	~14	~0.3–20	enhanced ^{12}C, ^{15}N, ^{28}Si; extinct ^{26}Al, ^{44}Ti	~1%
graphite	~10	~0.8–20	enhanced ^{12}C, ^{15}N, ^{28}Si; extinct ^{26}Al, ^{41}Ca, ^{44}Ti	>80% [d]
corundum	~0.01	~0.5–5	enhanced ^{16}O	~1%
silicon nitride	≥0.002	~1	enhanced ^{12}C, ^{15}N, ^{28}Si; extinct ^{26}Al	~100%

[a] Listed abundances are from the compilation of [66] and generally refer to the CI chondrite Orgueil (cf. [24]). Additional types of grains less well-characterized and/or abundant include grains of spinel ($MgAl_2O_4$; cf. [67]), hibonite ($CaAl_{12}O_{19}$), possibly a grain of rutile (TiO_2) and metal and carbides within graphite and SiC grains [24]. Silicates have been found in interplanetary dust [39] but not in meteorites so far.

[b] Size ranges are from [25,26].

[c] Isotopic signatures indicating a supernova origin as well as abundance of grains with probable supernova origin relative to total abundance of each type of presolar mineral phase as inferred and listed in [25,26]. Enhancements and depletions are expressed relative to solar system composition.

[d] Including possible contributions from Wolf-Rayet stars.

the time of grain formation, are consistent with this interpretation [25,26]. However, the fact that ^{44}Ti and ^{28}Si rich silicon occur connected with carbon points to formerly unexpected mixing processes in supernova ejecta, because ^{28}Si and ^{44}Ti are produced in the innermost zones while ^{12}C is located farther out [25,26]. Similar features as in SiC-X are observed in many of the graphite grains and in all of the presolar silicon nitride grains investigated in detail so far (Table 1).

Some puzzles are posed by the isotopic composition of the heavy trace elements in grains that appear to be tied to supernovae. This is different from the SiC grains from AGB stars (>90% of presolar SiC). These show s-process signatures typical of what also is observed in solar system matter. It is also different from FUN-CAIs, where overabundances in r-process nuclides indicate additions of the "average solar-system" r-process. Instead the overabundances in r-isotopes of Mo in SiC-X grains ([34]; Fig. 5), of Te in diamonds [35] and of p- and r-isotopes of xenon in diamonds ([36]; Fig. 6) are not consistent with such

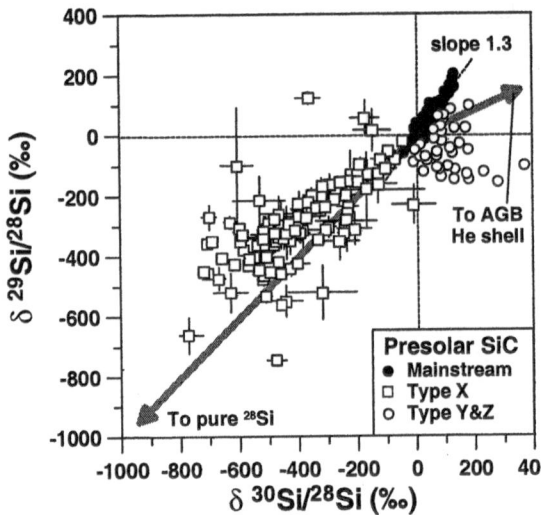

Fig. 4. Si isotopic composition of presolar SiC grains. Shown are δ-values, i.e. deviations of the $^{29}Si/^{28}Si$ and $^{30}Si/^{28}Si$ ratios from the terrestrial standard in per mil. Type X grains, thought to be of supernova origin, are characterized by extremely negative values corresponding to extreme enrichment of ^{28}Si

a simple scenario. Rather they may record contributions from a neutron burst – intermediate between s- and r-process – caused by the passage of the shock wave through the He/C shell of an exploding type II supernova (e.g., [37,38]) or some early separation process between r-process nuclides and precursors yet to decay into stable end products [36].

The search is intensive and additional types of grains have been discovered and are being discovered in addition to the major well-characterized types in Table 1 (see footnote to Table 1), but these still await full isotopic characterization. The same is true for presolar silicates which have been discovered in interplanetary dust particles [39] but not (yet) in meteorites.

4   Bulk Anomalies (Non-Radiogenic)

Isotopic abundance anomalies have been seen not only during analyses of specific meteoritic phases such as CAIs or presolar grains, but also in the analysis of bulk meteorites. This includes the elements titanium [40,41] and chromium [42,43], where the variations in the abundances of the most neutron-rich isotopes hint at the involvement of nucleosynthesis products from supernovae – analogous to the case of CAIs. Another recent case is that of molybdenum [44,45], where variations have been found in the ratio of p- and r-process products relative to s-process products. At least some of the differences in isotopic compositions observed in these studies may be due to differences between the individual meteorites in the abundances of CAIs and of presolar grains of the various types discussed above.

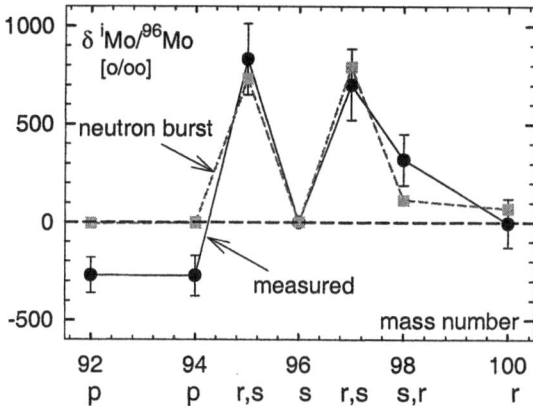

Fig. 5. Mo isotopic pattern of an X grain as measured by RIMS ([34]; with slight revisions on data based on A. Davis, pers. comm., cf. [65]). Shown are deviations from normal Mo composition in per mil. Also indicated is the standard assignment of the individual isotopes to p-, s-, and r-processes. The observed pattern can be reproduced reasonably well by a mixture of ∼ 17% neutron burst material [38] and 83% normal Mo [65]

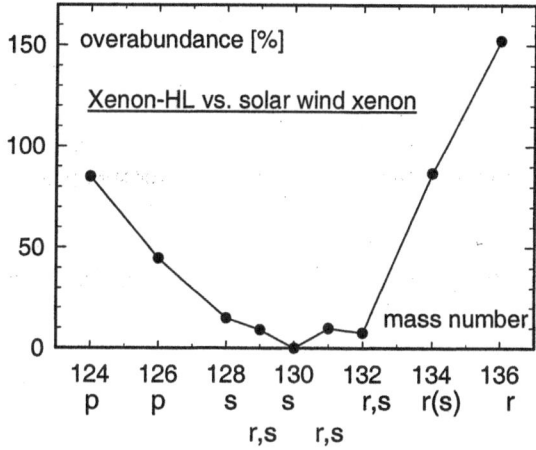

Fig. 6. Isotopic pattern of Xe-HL as found in presolar diamonds. Also indicated is the standard assignment of the individual isotopes to p-, s-, and r-processes. Shown are deviations from the solar wind Xe composition in percent. Because the enhancement of the two p-only isotopes is not identical, the pattern on the low-mass side cannot be caused by addition of "normal" p-process matter. The same situation applies to the high-mass r-only isotopes [36]

5 Extinct Radionuclides

The formation of the Solar System was a remarkably swift process. This is shown by the inferred presence in the Early Solar System of now extinct radionuclides with half-lives in the range $10^5 - 10^8$ a, such as ^{129}I [46] and ^{26}Al [8]. These effects express themselves as overabundances of the daughter nuclei (here: ^{129}Xe and ^{26}Mg) in matter with enhanced parent/daughter element ratios. Table 2 summarizes the current status of extinct radionuclides in the Early Solar System (for other recent summaries see, e.g., [47–51]).

Table 2. Extinct radionuclides in the Early Solar System. Mostly as compiled in [51,68]. Be-10 from [55,69]. In addition there is evidence for the presence in CAIs of ^{7}Be ($T_{1/2} = 53$ d; [57])

nuclide	$T_{1/2}$	occurrence	reference nucl.[b]	$\sim$ ratio	pref. process
^{10}Be	1.5 Ma	CAIs	^{9}Be	10^{-3}	spallation
^{26}Al	0.7 Ma	mostly CAI	^{27}Al	5×10^{-5}	H burn.
^{36}Cl(?)[c]	0.3 Ma	carb. chondr.	^{35}Cl	1.4×10^{-6}	s (?)
^{41}Ca	0.1 Ma	CAI	^{40}Ca	1.5×10^{-8}	s (?)
^{53}Mn	3.7 Ma	widespread	^{55}Mn	$1\text{-}2\times10^{-5}$	NSE, Si burn., spall.
^{60}Fe	1.5 Ma	HED met.[a]	^{56}Fe	4×10^{-9}	s, n-rich NSE
^{92}Nb	36 Ma	widespread	^{93}Nb	2×10^{-5}	p
^{107}Pd	6.5 Ma	iron met.	^{108}Pd	2×10^{-5}	s,r
^{129}I	16 Ma	widespread	^{127}I	1×10^{-4}	r
^{146}Sm	103 Ma	widespread	^{144}Sm	7×10^{-3}	p
^{182}Hf	9 Ma	widespread	^{180}Hf	1×10^{-4}	r
^{244}Pu	81 Ma	widespread	^{238}U	7×10^{-3}	r

[a] Basaltic (not primitive) meteorites, presumably from the Asteroid Vesta.

[b] Reference nuclide to which the abundance of the radionuclide (column 5) is normalized.

[c] Reported evidence probably due to experimental artifact [70].

Clearly, several of these radionuclides are being manufactured by supernovae, which has led to the suggestion that formation of the Solar System was triggered by a nearby exploding supernova [52,53]. But other sources may have played a role as well. AGB stars as a possible major source for a late injection and as trigger for solar system formation have been discussed in detail by [48]. Another important source are spallation reactions in the Early Solar System [54], at least for ^{10}Be [55], which is unlikely to be made in stellar nucleosynthesis (for an alternative view see [56]) and the recently detected ^{7}Be [57] with a halflife of only 53 d (not listed in Table 1). Surely not all contributions can come from a single source, and, even if Solar System formation was actually triggered by a supernova explosion [52,53] or an AGB star [48,58], for many of the nuclides in Table 2 alone the background found in the interstellar medium from "continuous nucleosynthesis" is important. This has been studied in detail by [51] who conclude that most likely only some of the shortest-lived nuclides in Table 2 (^{26}Al, ^{36}Cl, ^{41}Ca, ^{60}Fe, ^{182}Hf) require an extra late stellar input.

An interesting case is that of the r-process nuclides for which supernovae are commonly regarded as the most likely source. As was pointed out by [59] the observed abundance of ^{182}Hf in the Early Solar System seems consistent with quasi-continuous r-process nuclesosynthesis up to ~ 10 Ma before formation of the Solar System (~ 20 Ma with the newly revised ^{182}Hf abundance [60,61]), while the observed abundance of ^{129}I requires a much longer free-decay interval (on the order of 108 a). This has led to the suggestion of two types of r-processes, which occur with different frequencies: a more frequent one that produces the bulk of the r-process nuclides above $A \sim 130$, and a rarer type responsible for the lighter r-nuclides [62]. While similar conclusions can be drawn from recent observations of r-process elemental abundances in extremely metal-poor Halo stars (e.g. [63]), the extremely clear separation as suggested by [62] appears problematic on nuclear physics grounds [64]. Maybe a different "non-standard" type of nucleosynthesis is required to explain the high abundance of ^{182}Hf in the Early Solar System [51], possibly similar to what is recorded in presolar SiC grains of type X and presolar diamonds [65].

6 Summary

An overview has been presented on various types of isotope abundance anomalies in meteorites that may be tied to supernovae. Prominent effects occur

- in mm- to cm-sized high-temperature Ca-Al-rich inclusions (CAIs), especially those of the FUN type. Most prominent here are the overabundances of the n-rich isotopes of the Fe peak elements from Ca to Zn. FUN inclusions also exhibit variations in the relative contributions from the p-, s- and r-processes.

- in grains of circumstellar origin that have survived in meteorites (e.g. diamonds, silicon carbide of type X, graphite, silicon nitride). Anomalies here are much larger and comprise major elements such as C and Si, but also trace elements such as Te and Xe. In addition, anomalies are observed due to the decay of short-lived nuclides (^{26}Al, ^{41}Ca, ^{44}Ti) within the grains. The results bear on mixing processes in supernovae and may require a new type of nucleosynthesis for heavy elements.

- in bulk meteorites as well as in specific meteoritic phases resulting from *in-situ* decay of extinct radionuclides that were present in the early solar system. While for numerous radionuclides the former presence in the Solar System has been established, there are only a few that are unambiguously tied to supernovae only (^{60}Fe, r-process nuclides ^{129}I, ^{182}Hf, ^{244}Pu). The relative abundances of the r-process radionuclides suggest complexity within the r-process.

Acknowledgements

Thanks go to a large number of colleagues in the fields of cosmochemistry and astrophysics for many discussions on topics relating to isotope abundance anomalies in meteorites. Peter Hoppe contributed Fig. 3, Fig. 1 is from S.P. Schwenzer.

References

1. C.J. Allègre, G. Manhès, C. Göpel: Geochim. Cosmochim. Acta 59, 1445 (1995)
2. Y. Amelin, L. Grossman, A.N. Krot, T. Pestaj, S.B. Simon, A.A. Ulyanov: Lunar Planet. Sci. XXXIII, #1151 CD-ROM (2002)
3. R.N. Clayton, L. Grossmann, T.K. Mayeda: Science 182, 485 (1973)
4. R.N. Clayton: Annu. Rev. Earth Planet. Sci. 21, 115 (1993)
5. M.H. Thiemens, J.E. Heidenreich III: Science 219, 1073 (1983)
6. M.H. Thiemens: Science 283, 341 (1999)
7. R.N. Clayton: Nature 415, 860 (2002)
8. T. Lee, D.A. Papanastassiou, G.J. Wasserburg: Ap.J. 211, L107 (1977)
9. R.D. Loss, G.W. Lugmair: Ap.J. 360, L59 (1990)
10. J. Völkening, D.A. Papanastassiou: Ap.J. 358, L29 (1990)
11. R.D. Loss, G.W. Lugmair, A.M. Davis, G.J. MacPherson: Ap.J. 436, L193 (1994)
12. D. Hartmann, S.E. Woosley, M.F. El Eid: Ap.J. 297, 837 (1985)
13. K.-L. Kratz, W. Böhmer, C. Freiburghaus, P. Möller, B. Pfeiffer, T. Rauscher, F.-K. Thielemann: Mem.S.A.It. 72, 453 (2001)
14. B.S. Meyer, T.D. Krishnan, D.D. Clayton: Ap.J. 462, 825 (1996)
15. T. Ireland: Geochim. Cosmochim. Acta 54, 3219 (1990)
16. T. Lee: 'Implications of Isotopic Anomalies for Nucleosynthesis'. In: Meteorites and the Early Solar System, ed. by J.F. Kerridge, M.S. Matthews (University of Arizona Press, Tucson 1988) pp. 1063-1089
17. E. Anders, E. Zinner: Meteoritics 28, 490 (1993)
18. U. Ott: Nature 364, 25 (1993)
19. T.J. Bernatowicz, E. Zinner (editors): Astrophysical Implications of the Laboratory Study of Presolar Materials (AIP Conf. Proceedings 402, American Institute of Physics, Woodbury 1997)
20. R.S. Lewis, M. Tang, J.F. Wacker, E. Anders, E. Steel: Nature 326, 160 (1987)
21. T. Bernatowicz, G. Fraundorf, M. Tang, E. Anders, B. Wopenka, E. Zinner, P. Fraundorf: Nature 330, 728 (1987)
22. S. Amari, E. Anders, A. Virag, E. Zinner: Nature 345, 238 (1990)
23. Z.R. Dai., J.P. Bradley, D.J. Joswiak, D.E. Brownlee, H.G.M. Hill, M.J. Genge: Science 418, 157-159 (2002)
24. U. Ott: 'The Most Primitive Material in Meteorites'. In: Astromineralogy, ed. by Th. Henning (Springer, Berlin, Heidelberg, New York 2002), in press
25. E. Zinner: Annu. Rev. Earth Planet. Sci. 26, 147 (1998)
26. P. Hoppe, E. Zinner: J. Geophys. Res. 105, 10371 (2000)
27. F. Banhart, Y. Lyutovich, A.Braatz, C. Jäger, T. Henning, J. Dorschner, U. Ott: Meteorit. Planet. Sci. 33, A12 (1998)
28. P. Hoppe, S. Amari, E. Zinner, T. Ireland, R.S. Lewis: Ap.J. 430, 870 (1994)
29. G.K. Nicolussi, A.M. Davis, M.J. Pellin, R.S. Lewis, R.N. Clayton, S. Amari: Science 277, 1281 (1997)

30. R.H. Nichols Jr., K. Kehm, R. Brazzle, S. Amari, C.M. Hohenberg, R.S. Lewis: Meteoritics 29, 510 (1994)

31. P. Hoppe, U. Ott: 'Mainstream Silicon Carbide Grains from Meteorites'. In: Astrophysical Implications of the Laboratory Study of Presolar Materials, ed. by T.J. Bernatowicz, E. Zinner (AIP Conf. Proceedings 402, American Institute of Physics, Woodbury 1997) pp. 27-58

32. L.R. Nittler, S. Amari, E. Zinner, S.E. Woosley, R.S. Lewis: Ap.J. 462, L31 (1996)

33. P. Hoppe, A. Besmehn: Ap.J. 576, L69 (2002)

34. M.J. Pellin, A.M. Davis, R.S. Lewis, S. Amari, R.N. Clayton: Lunar Planet. Sci. XXX, #1969 CD-ROM (1999)

35. S. Richter, U. Ott, F. Begemann: Nature 391, 261 (1998)

36. U. Ott: Astrophys. Journ. 463, 344 (1996)

37. W.M. Howard, B.S. Meyer, D.D. Clayton: Meteoritics 27, 404 (1992)

38. B.S. Meyer, D.D. Clayton, L.-S. The: Ap.J. 540, L49 (2000)

39. S. Messenger, L.P. Keller, R.M. Walker: Lunar Planet. Sci. XXXIII, #1887 CD-ROM (2002)

40. F.R. Niederer, D.A. Papanastassiou, G.J. Wasserburg: Geochim. Cosmochim. Acta 49, 851 (1985)

41. S. Niemeyer: Geochim. Cosmochim. Acta 52, 2941 (1988)

42. M. Rotaru, J.-L. Birck, C.J. Allègre: Nature 358, 465 (1992)

43. F.A. Podosek, U. Ott, J.C. Brannon, C.R. Neal, T.J. Bernatowicz., P. Swan, S.E. Mahan: Meteorit. Planet. Sci. 32, 617 (1997)

44. Q. Yin, S.B. Jacobsen, K. Yamashita: Nature 415, 881 (2002)

45. N. Dauphas, B. Marty, L. Reisberg: Ap.J. 565, 640 (2002)

46. J.H. Reynolds: Phys. Rev. Lett. 4, 8 (1960)

47. A.G.W. Cameron, F.-K. Thielemann, J.J. Cowan: Phys. Rep. 227, 283 (1993)

48. G.J. Wasserburg, M. Busso, R. Gallino, C.M. Raiteri: Ap.J. 424, 412 (1994)

49. F.A. Podosek and R.H. Nichols Jr.: 'Short-lived Radionuclides in the Solar Nebula'. In: Astrophysical Implications of the Laboratory Study of Presolar Materials, ed. by T.J. Bernatowicz, E. Zinner (AIP Conf. Proceedings 402, American Institute of Physics, Woodbury 1997) pp. 617-647

50. J.N. Goswami: 'Chronology of Early Solar System Events: Dating with Short-Lived Nuclides'. In: Origins of the Elements in the Solar System: Implications of Post 1957 Observations, ed. by O. Manuel (Kluwer Academic / Plenum Publishers, New York 2000) pp. 407-430

51. B.S. Meyer, D.D. Clayton: Space Sci. Rev. 92, 133 (2000)

52. A.G.W. Cameron, J.W. Truran: Icarus 30, 447 (1977)

53. A.G.W. Cameron, P. Höflich, P.C. Myers, D.D. Clayton: Ap.J. 447, L53 (1995)

54. M. Gounelle, F.H. Shu, H. Shang, A.E. Glassgold, K.E. Rehm, T. Lee: Ap.J. 548, 1051 (2001)

55. K.D. McKeegan, M. Chaussidon, F. Robert: Science 289, 1334 (2000)

56. A.G.W. Cameron: Ap.J. 562, 456 (2001)

57. M. Chaussidon, F. Robert, K.D. McKeegan: Metorit. Planet Sci. 37, A61 (2002)

58. P.N. Foster, A.P. Boss: Ap.J. 468, 784 (1996)

59. G.J. Wasserburg, M. Busso, R. Gallino: Ap.J. 466, L109 (1996)

60. Q. Yin, S.B. Jacobsen, K. Yamashita, J. Blichert-Toft, P. Télouk, F. Albarède: Nature 418, 949 (2002)

61. T. Kleine, C. Münker, K. Mezger, H. Palme: Nature 418, 952 (2002)

62. Y.-Z. Qian, P. Vogel, G.J. Wasserburg: Ap.J. 494, 285 (1998)

63. C. Sneden, J.J. Cowan, I.I. Ivans, G.M. Fuller, S. Burles, T.C. Beers, J.E. Lawler: Ap.J. 533, L139 (2000)

64. B. Pfeiffer, U. Ott, K.-L. Kratz: Nucl. Phys. A688, 575c (2001)
65. U. Ott: New Astron. Rev. 46, 513 (2002)
66. G.R. Huss: 'The Survival of Presolar Grains in Solar System Bodies'. In: Astrophysical Implications of the Laboratory Study of Presolar Materials, ed. by T.J. Bernatowicz, E. Zinner (AIP Conf. Proceedings 402, American Institute of Physics, Woodbury 1997), pp. 721-748
67. E. Zinner, S. Amari, R. Guinness, A. Nguyen, R.S. Lewis: Meteorit. Planet. Sci. 37, A154 (2002)
68. M. Busso, R. Gallino, G.J. Wasserburg: Annu. Rev. Astron. Astrophys. 37, 239 (1999)
69. N. Sugiura, Y. Shuzou, A.A. Ulyanov: Meteorit. Planet. Sci. 36, 1397 (2001)
70. V.K. Rai, S.V.S. Murty, U. Ott: subm. to Geochim. Cosmochim. Acta (2002)

The Chemical Yields Produced by a Zero Metal Generation of Core Collapse Supernovae and Their Comparison with the Chemical Composition of Extremely Metal Poor Stars

Marco Limongi[1] and Alessandro Chieffi[2]

[1] INAF - Osservatorio Astronomico di Roma,
  Via Frascati 33, I-00040, Monteporzio Catone, Rome, Italy
[2] Istituto di Astrofisica Spaziale e Fisica Cosmica (CNR),
  Via Fosso del Cavaliere 100, I-00133, Roma, Italy

Abstract. We present a detailed comparison between an extended set of elemental abundances observed in some of the most metal poor stars presently known and the ejecta produced by a generation of primordial core collapse supernovae.

1 Introduction

Extremely metal poor stars probably formed in an environment enriched by just the first generation of stars hence they give us the unique opportunity of observing directly the ejecta of a single stellar generation and not the complex superimposition of many generations of stars of different metallicity. Observations of element abundance ratios show that, for several elements, real star-to-star abundance differences appear common at lowest metallicities [7,10]. This scatter has been interpreted [2] as an indication of the highly inhomogeneity of the primordial interstellar medium (ISM) [2]. In this scenario each primordial cloud was polluted by just one or, at least, a mixture of two to three SNII, hence the newly forming stars may retain information of the explosion of a single zero metal supernova event. Under this assumption, element abundance ratios observed in extremely metal poor stars can be directly compared with theoretical yields of metal free core collapse supernovae of different mass to determine which have contributed to the initial Galactic chemical enrichment and when. They can be used also to constrain theoretical models and nucleosynthesis when unexpected elemental ratios are found. Here we present a detailed comparison of an extended set of elemental abundances observed in some of the most metal poor stars presently known and the ejecta produced by a generation of primordial core collapse supernovae. This approach in which we will try to fit the (available) chemical composition of specific stars is complementary to most of the analysis performed up to know in which mainly the global trends with the metallicity are discussed [11,8].

2   The Sample

One of the most recent databases of surface abundances of extremely metal poor stars available is the one given by [9] and consists of five stars of metallicity lower than [Fe/H]=−3.3, i.e., Three out of these five stars, i.e. CD − 38°245 ([Fe/H]=−3.98), CS 22172-002 ([Fe/H]=−3.61) and CS 22885-096 ([Fe/H]= −3.66) show a remarkably similar pattern with the exception of [C/Fe] which shows significant differences between CS 22172-002 and CS 22885-096 (no C abundance is available for CD − 38°245). Given this similarity we define an "average" (hereinafter AVG) star which represents all three of them. The fourth star of the sample, i.e. the CD − 24°17504 ([Fe/H]=−3.37), shows a chemical pattern which closely matches that of the AVG star with the exception of Cr and Mn that are significantly more abundant (by a factor three to four) in CD − 24°17504 than in the AVG star. No C abundance determination is available for this star. A comparison between the last star of the sample, i.e. CS 22949-037 ([Fe/H]=−3.79), and the AVG star shows a good agreement for the elements from Ca to Ni. On the contrary the lighter elements are systematically overabundant (by a factor of five in the mean) in the CS 22949-037 compared to the AVG star, although the relative pattern resembles that of the AVG star. Leaving apart this last star, which certainly shows significant differences with respect to the other ones, we feel confident to say that the other four stars are similar enough to be well represented by the AVG star. Therefore in the following we will consider the abundance pattern of this "template" star as the one to be compared with the theoretical yields.

3   Theoretical Yields

The theoretical yields are based on a new set of zero metallicity massive stars (Y=0.23, Z=0) ranging in mass between 15 and 80 $M_\odot$. These models have been computed by means of the last version of the FRANEC evolutionary code [5,6]. A nuclear network of 179 isotopes (from neutrons to ^{70}Zn) fully coupled to the physical evolution of the star has been adopted. The adopted rate of the $^{12}C(\alpha, \gamma)^{16}O$ reaction is the one provided by [3], as suggested by [4]. The explosive nucleosynthesis has been computed in the frame of the radiation dominated shock approximation [13,1] using the same nuclear network adopted in the hydrostatic evolutions. The element yields coming from this set of models have been published in [4] for different choices of the amount of ^{56}Ni ejected, after all the unstable isotopes have been fully decayed into their stable form.

4   Fit to the AVG Star

Here we will discuss the comparison between the theoretical yields produced by a generation of zero metallicity core collapse supernovae and the element abundance ratios of the AVG star and under the hypothesis that they are the result of a single SNII event. The lack of a realistic description of the explosion of a core collapse supernova allows some freedom about the choice of the mass cut. Under the hypothesis that each cloud of pristine material is polluted by just a single supernova event, we choose the mass cut by simply imposing the fit to [Mg/Fe]. The reason of choosing the [Mg/Fe] log ratio is that Fe (mostly ^{56}Ni), that is produced in the innermost zones, significantly depends on the location of the mass cut, while Mg, on the contrary, is not affected at all by this quantity. It goes without saying that, since the yield of Mg varies with the initial mass, also the mass cut will vary with the mass of the star. The fit to the AVG star which is obtained with this choice for the mass cut is shown in Fig. 1.

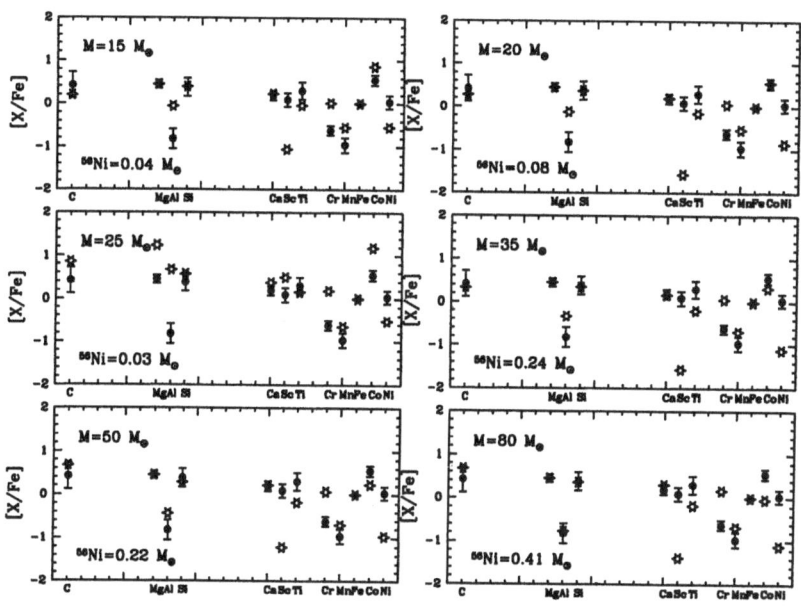

Fig. 1. Comparison between the AVG star (*filled dots*) and the ejecta of the six masses in our grid (*open stars*). The ^{56}Ni shown in each panel is the one needed to fit the observed [Mg/Fe]

This figure shows that the fit to the two light elements from C to Ca (Mg is imposed) is remarkably good for all the six masses except for [Al/Fe] which is reproduced only by the most massive star, i.e. the $80 M_\odot$ model. On the contrary the heavy elements (Sc, Ti, Cr, Mn, Fe, Co and Ni) are not fitted by any model. The only exception is [Co/Fe] which shows a trend with the initial mass (in particular it decreases as the initial mass increases) and a good fit is obtained for a mass of the order of 20 $M_\odot$. By the way, note that the 25 $M_\odot$ does not produce enough ^{56}Ni to fit the observed [Mg/Fe]; therefore in this case we simply fixed the mass cut at the border of the Fe core, a choice which corresponds to the largest possible amount of ^{56}Ni. All the elements between Sc and Ni are produced by either the incomplete and/or the complete explosive Si burning in the deepest layers of the star and hence they are those that are most affected by the adopted mass cut. As a consequence the lack of a good fit to these elements could be simply due to a wrong choice of the mass cut. On the other hand a change of the location of the mass cut alters the amount of Fe without modifying the amount of Mg ejected hence the observed [Mg/Fe] log ratio could be not fitted anymore. The best overall fits to the block of available elements is shown in Fig. 2. The closest match to the AVG star is obtained with the ejecta of the 35 and 50 $M_\odot$; in these cases 8 out of the 11 observed element abundances ratios are reproduced by the models. The systematic lack of a good fit to Cr, Ti and Ni clearly indicates that something crucial has been missed in the computation of the yields.

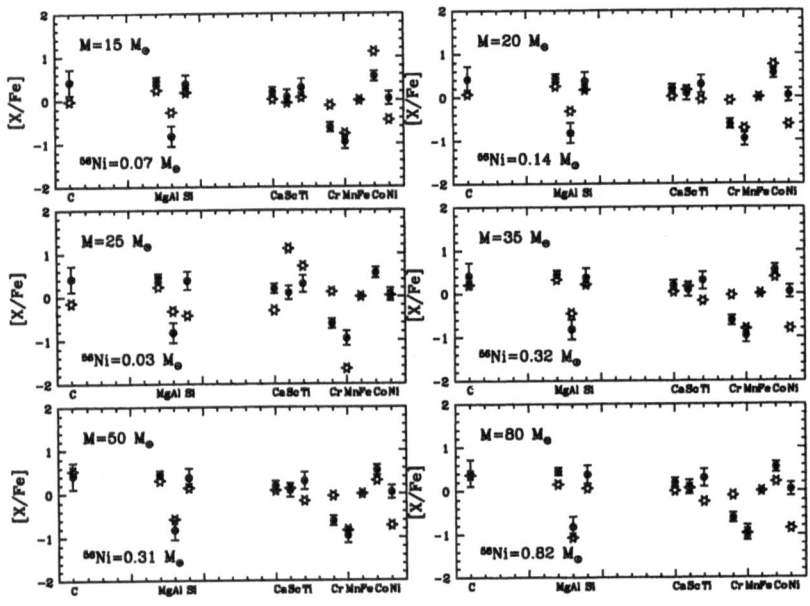

Fig. 2. Same as figure 1 but with a choice of the mass cut which minimizes the overall differences between the AVG star and the theoretical values

Alternative yields for zero metal massive stars have been published by [14], hereinafter WW95, and [12], hereinafter UN02. None of the two papers provides yields as a function of the ^{56}Ni ejected and hence there is no way to impose the fit to any of the [X/Fe].

Figure 3 shows, for the four masses 15, 20, 25 and 30 $M_\odot$, the fits obtained by adopting the UN02 yields. The first thing worth noting is that none of the four masses provides an overall fit to the data. In particular, the light elements from C to Ca are fit only by the 20 and the 25 $M_\odot$ models while the heavy elements are not fit by any model, except [Mn/Fe] which is fit only by the two most massive stars. It is interesting to note that the discrepancies are qualitatively very similar to the one we obtain with our yields: in particular, also these yields

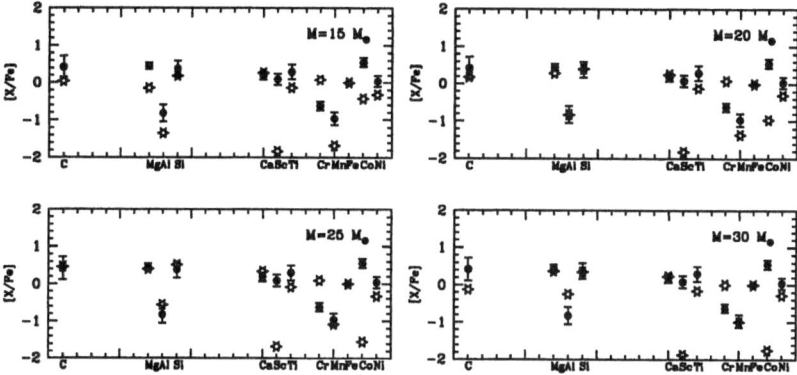

Fig. 3. Comparison between the AVG star (*filled dots*) and the yields provided by [12] (*open stars*)

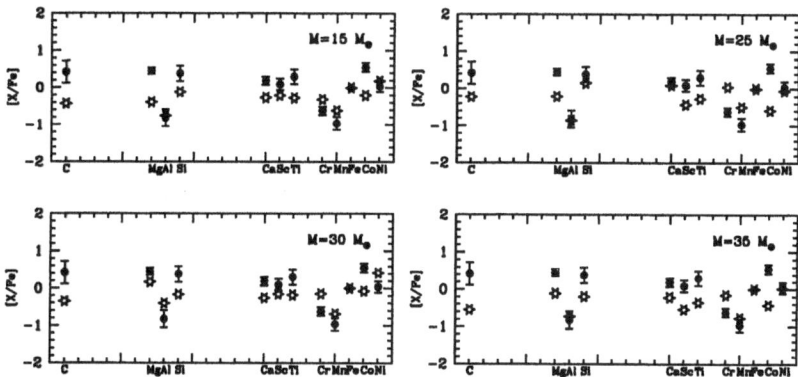

Fig. 4. Comparison between the AVG star (*filled dots*) and the yields provided by [14] (*open stars*)

show that [Ti/Fe] and [Ni/Fe] are systematically underproduced with respect to the observed value for all the masses, while [Cr/Fe] is always overproduced by the models. On the contrary Sc and Co are heavily underproduced by the models in this case.

Figure 4 shows the fit to the AVG star with the yields provided by WW95. The grid is now formed by a 15, 25, 30 and 35 $M_\odot$. Also in this case none of the models reproduces the data. The element C to Ca are systematically underproduced with respect to the observed values and in relative proportions that do not resemble the observations either. [Al/Fe] is vice versa well fitted by three out of the four stellar models, i.e., the 15, 25 and 35 $M_\odot$. The fit to the heavy elements shows significant similarities with those shown above. Once again Sc, Ti and Co are always underproduced with respect to the observed values while Cr and Mn are always overproduced. There is however a noticeable difference in these models: [Ni/Fe] is now well fit or even overproduced while neither our models nor those of UN02 are able to produce Ni in sufficient amount.

5   Conclusion

We have compared the ejecta of a primordial generation of core collapse supernovae with the observed element abundances in extremely metal poor stars under the hypothesis that these stars formed in clouds enriched by just the first generation of core collapse supernovae. Our main findings are the following. (1) Our theoretical yields allow a reasonable fit to 8 of the 11 observed element abundance ratios with the ejecta of the 35 and 50 $M_\odot$ models. (2) Within the present grid of models it is not possible to find a good match of the remaining three elements, Ti, Cr and Ni (even for an arbitrary choice of the mass cut). Ti and Ni are always underestimated while Cr is always overestimated by our models. (3) The adoption of other theoretical yields available in literature (WW95, UN02) does not improve the fit. Though the three sets of yields (ours, WW95 and UN02) differ significantly from each other, all of them show embarrassing systematic similarities in missing the fit of the AVG star. For example, all three sets tend systematically to underproduce Ti and Ni and to overproduce Cr. (4) Since no mass in our grid provides a satisfactory fit to these three elements, even an arbitrary choice of the initial mass function would not improve the fit. (5) At variance with current beliefs, our models easily and naturally allow a very good fit to [Co/Fe].

References

1. D. Arnett: 'Supernovae and Nucleosynthesis', Princeton University Press, p. 252 (1996)
2. J. Audouze, J. Silk: Astrophysical Journal, 451, L49 (1995)
3. G.R. Caughlan, W.A. Fowler: A.D.N.D.T., 40, 283 (1988)
4. A. Chieffi, M. Limongi: Astrophysical Journal, 577, 281 (2002)
5. A. Chieffi, M. Limongi, O. Straniero: Astrophysical Journal, 502, 737 (1998)
6. M. Limongi, O. Straniero, A. Chieffi: Astrophysical Journal Suppl. Ser., 219, 625 (2000)
7. A. McWilliam, G.W. Preston, C. Sneden, L. Searle: Astronomical Journal, 109, 2757 (1995)
8. T. Nakamura, H. Umeda, K. Nomoto, F.K. Thielemann, A. Burrows: Astrophysical Journal, 517, 193 (1999)
9. J.E. Norris, S.G. Ryan, S.G., T.C. Beers: Astrophysical Journal, 561, 1034 (2001)
10. S.G. Ryan, J.E. Norris, T.C. Beers: Astrophysical Journal, 471, 254 (1996)
11. T. Shigeyama, T. Tsujimoto: Astrophysical Journal, 507, L135 (1998)
12. H. Umeda, K. Nomoto: Astrophysical Journal, 460, 408 (2002)
13. T.A. Weaver, S.E. Woosley: Ann. NY Acad. Sci., 336, 335 (1980)
14. S.E. Woosley, T.A. Weaver: Astrophysical Journal Suppl. Ser., 101, 181 (1995)

Bipolar Jets in Hypernova Explosions and Abundance Patterns in Metal-Poor Stars

Keiichi Maeda and Ken'ichi Nomoto

Department of Astronomy, School of Science, University of Tokyo,
Hongo 7-3-1, Bunkyo-ku, Tokyo 113-0033, Japan

Abstract. Bipolar explosion models for hypernovae (very energetic supernovae) are presented. These models provide a favorable situation to explain some peculiar features in observations of hypernovae, e.g. high velocity Fe-rich matter and low velocity O-rich matter. The overall abundance of these models gives a good fit, at least qualitatively, to abundances in extremely metal-poor stars. We suggest hypernovae driven by bipolar jets have significantly contributed to the early Galactic chemical evolution.

1 Introduction

Discovery of a new class of supernovae, called 'hypernovae', which have distinctly large kinetic energies (E; $E_{51} \gtrsim 10$, where $E_{51} = E/10^{51}$ergs) compared with previously known supernovae ($E_{51} \sim 1$) is one of the most exciting topics in recent studies on supernovae. For a review of hypernovae see [11].

Modeling the spectra and light curves of these objects has uncovered interesting features on the relation between progenitor masses (M_{ZAMS}) and outcomes (E, $M(^{56}Ni)$; see also [9]). (1) There seems to be a transition in those properties around $20 - 22$ $M_\odot$. Below it are normal supernovae. (2) Hypernovae are placed on $M_{ZAMS} \gtrsim 20 - 22$ $M_\odot$. The modeling of hypernovae suggests that E and $M(^{56}Ni)$ be larger for larger M_{ZAMS} in this 'hypernova branch'. (3) On the other hand, SNe with very low E and $M(^{56}Ni)$ ('faint SN branch') also seem to exist for $M_{ZAMS} \gtrsim 20 - 22$ $M_\odot$ (e.g. SNeII 1997D, 1999br) [2,17,18].

The transition implies that the explosion mechanism of stars with $M_{ZAMS} \gtrsim 20 - 22$ $M_\odot$ might be different from those with $M_{ZAMS} \lesssim 20$ $M_\odot$, especially for hypernovae given their large energies. (For faint SNe the small E may simply be attributed to a large gravitational binding energy).

Further studies on hypernovae have indeed suggested that their ejecta deviate from spherical symmetry to a large extent. (1) Late time spectra of SN1998bw showed that the OI] 6300 lines were narrower than the FeII] 5200 blend, indicating Fe was ejected with larger velocity than O along the line of sight [6,8]. (2) The light curve modeling of SN1998bw and the spectroscopic analysis of SN1997ef reached the same conclusion, i.e. the existence of high density material near the center below the mass cut (cutoff dividing a compact remnant and ejecta) assumed in conventional 1D models [7,10]. (3) The detection of optical polarization in SN1998bw and SN2002ap [4,12].

To explain these properties, we have constructed bipolar explosion models for hypernovae. In this paper, we describe the hydrodynamical and nucleosynthetic results and their contribution to the early Galactic chemical evolution.

2   Hydrodynamics

The main ingredient of our models is a pair of jets propagating through a stellar mantle. At the beginning of each calculation the central part ($M_r \leq 1.5 \ M_\odot$) of a progenitor star is displaced by a compact remnant. The jets are injected at the interface with the opening half-angle being 15°. The energy injected by the jets is assumed to be proportional to the accretion rate (i.e. $\dot{E}_{jet} = \alpha \dot{M}_{accretion} c^2$). The efficiency coefficient α is set to be 0.01.

The outcome of the explosion depends on the interaction between the jets and the stellar mantle, and on the accretion rate which is affected by the interaction itself. We follow the hydrodynamical evolution and nucleosynthesis in two dimensions. At each time step, the properties of the jets and the mass of the central remnant are updated according to the accretion rate determined by hydrodynamics.

Figure 1 shows snapshots of the density and velocity distribution of the explosion of the star with $M_{ZAMS} = 40 \ M_\odot$. The jets propagate through the stellar mantle, depositing their energies into ambient material at the working surface. The bow shock expands laterally to push the stellar mantle sideways, which reduces the accretion rate. The strong outflow occurs along the z-axis (the jet direction), while matter accretes from the side. As the accretion rate decreases, the jets are turned off. Then the inflow along the r-axis turns to the weak outflow.

The outcome is a highly aspherical explosion. This is consistent with the detection of polarization in SN 1998bw [12] and SN 2002ap [4]. The accretion forms a central dense core. Densities near the center are much higher than those in spherical models. This feature is indeed what is suggested from the spectroscopic [7] and light curve [10] modeling of hypernovae.

Other properties of these models are as follows: (1) A more massive star makes a more energetic explosion. The reason is that a more massive star has a stronger gravity to make the accretion rate larger. This is consistent with the relation seen in the hypernova branch (see Fig. 3 of [9]). (2) A more massive star forms a more massive compact remnant. The remnant of the explosion of the 40 $M_\odot$ star is a black hole with $\sim 6 \ M_\odot$ in the model presented here. The bipolar models provide the way of explosions with a black hole formation in a consistent manner. Given the discovery of the evidence of a hypernova explosion accompanied by a formation of a black hole of $\sim 5 \ M_\odot$ (X-ray Nova Sco [3,13]), it offers an interesting possibility.

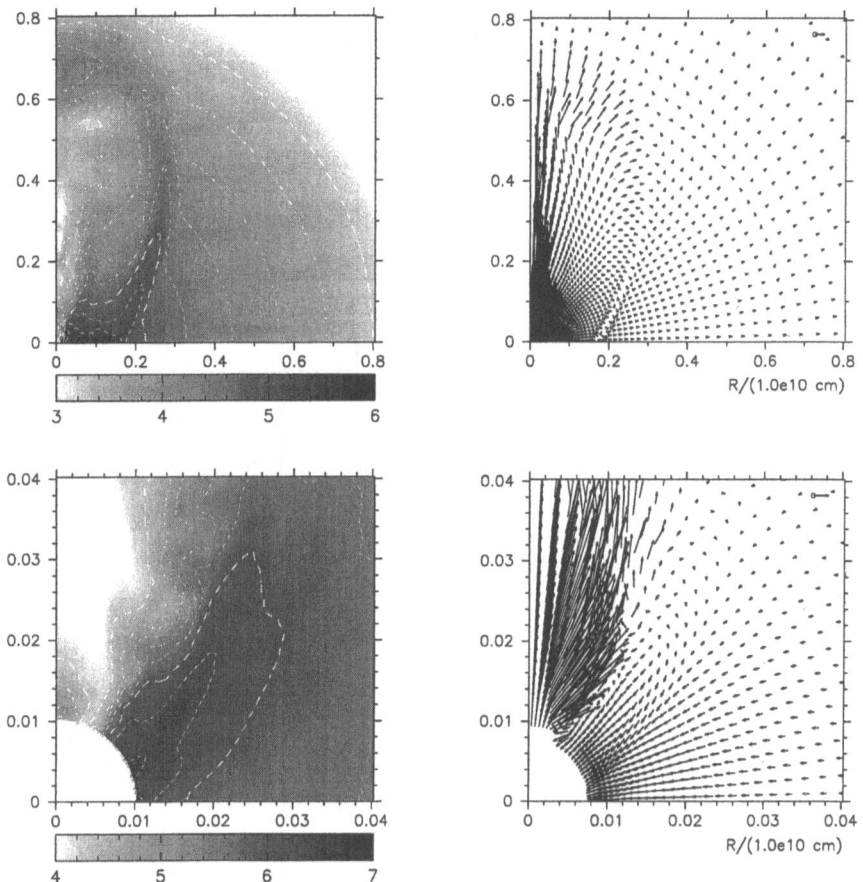

Fig. 1. Density (left columns; plotted logarithmically) and velocity (right columns) distribution of the bipolar explosion of a star with $M_{\mathrm{ZAMS}} = 40\ M_\odot$ at 1.5 second after the initiation of the jets. The lower two columns show the distribution in the central region on an expanded scale

3   Nucleosynthesis and Early Galactic Chemical Evolution

The high temperatures along the z-axis and the low temperatures along the r-axis observed in the hydrodynamical models have large effects on nucleosynthesis.

First, it results in highly aspherical distribution of nucleosynthetic products as shown in Fig. 2. The distribution of ^{56}Ni (which decays into ^{56}Fe) is elongated along the z-axis, where matter experiences higher temperatures than in the r-axis. Such a configuration, i.e. high velocity Fe and low velocity O, has been suggested to be responsible for the feature in the late phase spectra of SN1998bw, where the OI] 6300 was narrower than the FeII] 5200 blend [6,8].

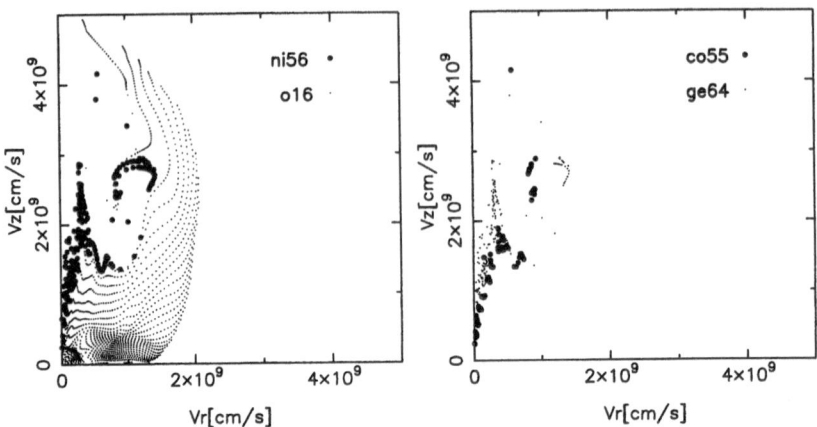

Fig. 2. Left: Distribution of ^{56}Ni (decays into ^{56}Fe) and ^{16}O. Right: ^{55}Co (decays into ^{55}Mn) and ^{64}Ge (into ^{64}Zn). The mass elements in which the mass fraction exceed 0.1 for ^{56}Ni and ^{16}O and 0.001 for ^{55}Co and ^{64}Ge are plotted

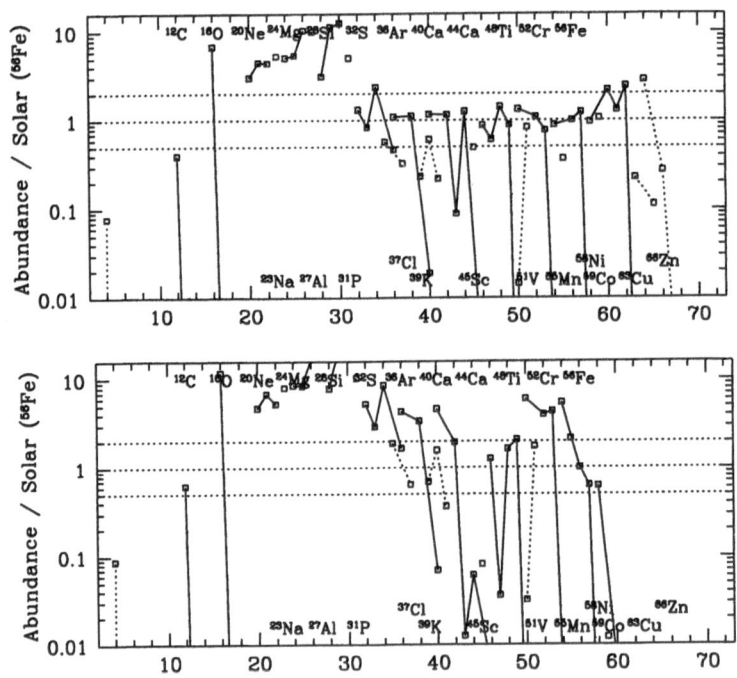

Fig. 3. Isotropic yields of the bipolar model (upper) and the spherical model (lower) for $E_{51} = 10$, $M(^{56}$Ni$) = 0.1\ M_\odot$, and $M_{\mathrm{ZAMS}} = 40\ M_\odot$

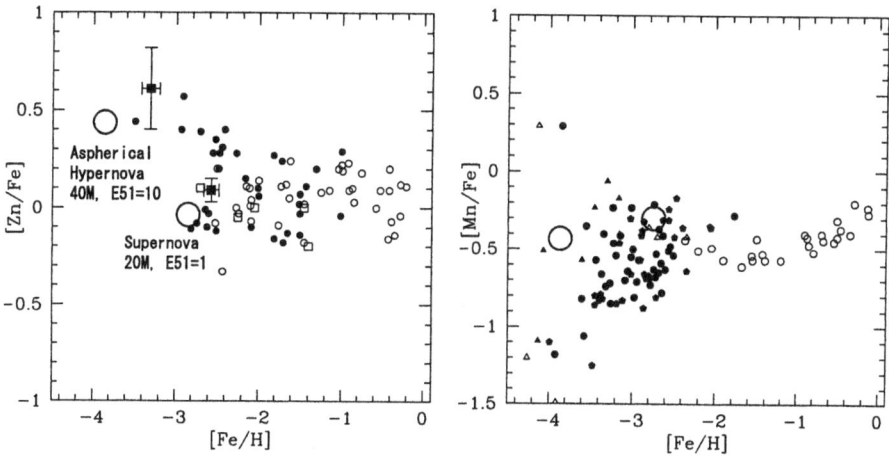

Fig. 4. Observed abundance rations of [Zn/Fe] and [Mn/Fe] and the theoretical abundance patterns for a normal supernova $(20M_\odot, E_{51} = 1)$ and a bipolar hypernova $(40M_\odot, E_{51} = 10)$ models

Second, the overall abundance patterns differ from those in spherical models as shown in Fig. 3. Matter which experiences higher entropy is preferentially ejected along the z-axis, while matter of lower entropy accretes. Therefore, [Zn/Fe] and [Co/Fe] $([X/Fe] \equiv \log 10(X/Fe) - \log 10(X/Fe)_\odot)$ are enhanced because they are produced through strong α-rich freezeout, while [Mn/Fe] and [Cr/Fe] produced through incomplete Si-burning are suppressed. This is evident in Fig. 2, which shows that the velocity of ^{64}Ge (Zn) is higher on average than that of ^{55}Co (Mn), the latter more easily accreting onto the central remnant. In other word, they can blow up high entropy material containing abundant Zn, Co without ejecting a large amount of lighter elements, e.g. Fe and Mn.

The unique yields of our models may have important implications on the early Galactic chemical evolution. Abundances in extremely metal-poor stars store the chemical information of an individual supernova, because those stars are likely to have been contaminated by only a single supernova at the early phase of the Galaxy before it was chemically well mixed.

Studies on metal-poor halo stars have uncovered that there exist interesting trends in the abundances of iron peak elements for [Fe/H] $\lesssim -2.5$ (Fig. 4, [1,5,14,15]). Both [Cr/Fe] and [Mn/Fe] decrease toward smaller [Fe/H], while [Co/Fe] and [Zn/Fe] increase to reach $\sim 0.3 - 0.5$ at [Fe/H] ~ -3. Also [O, Mg/Fe] are large (typically ~ 0.5) for [Fe/H] $\lesssim -2.5$, suggesting that the amount of Fe ejected in supernovae which are responsible for those stars is not so large.

We point out that the bipolar models presented here can naturally account for these features. Large [Zn, Co/Fe], small [Mn, Cr/Fe] and large [O, Mg/Fe] are simultaneously realized in the bipolar models (see Fig. 3). Figure 4 demonstrates how the trends in [Zn/Fe] and [Mn/Fe] are reproduced with our models. Given that the formation of metal-poor stars was driven by a supernova shock, [Fe/H] is determined by the ratio of the ejected mass of Fe to the explosion energy [15,16]. This places more energetic (bipolar) supernovae on lower [Fe/H]. The trends, therefore, are consistently reproduced.

4 Conclusions

We have presented hydrodynamical and nucleosynthetic features of the bipolar models. We have discussed advantages of our models in explaining the observed features of hypernovae, which conventional spherical models fail to reproduce. The overall abundances are consistent with those observed in extremely metal-poor stars. Therefore, we suggest they could make a significant contribution in the early Galactic chemical evolution.

References

1. L.A.J. Blake, S.G. Ryan, J.E. Norris, T.C. Beers: 2001, Nucl. Phys. A., 688, 502
2. M. Hamuy: 2002, ApJ, submitted
3. G. Israelian, R. Rebolo, G. Basri, et al.: 1999, Nature, 401, 142
4. K. Kawabata, et al.: ApJ, 580, L39
5. A. McWilliam, G.W. Preston, C. Sneden, L. Searle: 1995, AJ, 109, 2757
6. K. Maeda, T. Nakamura, K. Nomoto, P. Mazzali, F. Patat, I. Hachisu: 2002, ApJ, 565, 405
7. P. Mazzali, K. Iwamoto, K. Nomoto: 2000, ApJ, 545, 407
8. P. Mazzali, K. Nomoto, F. Patat, K. Maeda: 2001, ApJ, 559, 1047
9. P. Mazzali, K. Nomoto, J. Deng, K. Maeda, Y. Qiu: 2002, in this volume
10. T. Nakamura, P. Mazzali, K. Nomoto, K. Iwamoto: 2001, ApJ, 550, 991
11. K. Nomoto, K. Maeda, H. Umeda, T. Ohkubo, J. Deng, P. Mazzali: 2002, in 'A massive Star Odyssey, from Main Sequence to Supernova,' eds. V.D. Hucht, A. Herrero, & C. Esteban, (San Francisco; ASP), in press (astro-ph/0209064)
12. F. Patat, et al.: 2001, ApJ, 555, 900
13. Ph. Podsiadlowski, K. Nomoto, K. Maeda, T. Nakamura, P. Mazzali, B. Schmidt: 2002, ApJ, 567, 491
14. F. Primas, E. Brugamyer, C. Sneden, et al.: 2000, in 'The First Stars,' eds. A. Weiss, et al. (Springer), 51
15. S.G. Ryan, J.E. Norris, T.C. Beers: 1996, ApJ, 471, 254
16. T. Shigeyama, & T. Thujimoto: 1998, ApJ, 507, L135
17. M. Turatto, P. Mazzali, T. Young, K. Nomoto, et al.: 1998, ApJ, 498, L129
18. L. Zampieri, A. Pastorello, M. Turatto, et al.: 2002, MNRAS, submitted

Type Ia Supernova Progenitors, Lifetime, and Cosmic Supernova Rate

Chiaki Kobayashi[1] and Ken'ichi Nomoto[2]

[1] Max-Planck-Institute for Astrophysics, Karl-Schwarzschild-Strasse 1, D-85741 Garching, Germany
[2] Department of Astronomy, School of Science, University of Tokyo, Hongo 7-3-1, Bunkyo-ku, Tokyo 113-0033, Japan

Abstract. We make a comparison of the lifetime distribution derived from several models of Type Ia Supernovae (SNe Ia) and evaluate the SN Ia models from the observational constraints. Our SN Ia model, which is based on the single degenerate scenario and includes the metallicity effect, gives better reproduction of the [O/Fe]-[Fe/H] relation in the solar neighborhood if compared to, e.g., Matteucci & Recchi (2001). We also provide predictions of the cosmic supernova rate history.

1 Introduction

Since the explosion mechanism of Type Ia Supernovae (SNe Ia) is still debated, the progenitor model of SNe Ia is the most crucial uncertainty to understand many astrophysical problems, such as the determination of the cosmological constant, the chemical enrichment of galaxies and the thermal histories of the interstellar medium and intracluster medium. The progenitors of the majority of SNe Ia are most likely the Chandrasekhar mass C+O white dwarfs (WDs). For the evolution of accreting WDs toward the Chandrasekhar mass, two scenarios have been proposed: one is a double-degenerate (DD) scenario, i.e. merging of two WDs with a combined mass surpassing the Chandrasekhar mass limit, and the other is a single-degenerate (SD) scenario, i.e. accretion of hydrogen-rich matter via mass transfer from a binary companion. The issue of DD versus SD is still debated, but theoretical modeling has indicated that the merging of WDs does not make typical SNe Ia.

2 Type Ia Supernovae Models

In galactic chemical evolution models several SN Ia models have been proposed.

1) The formulation of an SN Ia model has been firstly proposed by [1] for the DD scenario, and is introduced as "most popular" model by [2] (hereafter M01; M-model).

$$\mathcal{R}_{\mathrm{Ia}} = A \int_{M_{\mathrm{B,inf}}}^{M_{\mathrm{B,sup}}} \phi(M_B) \int_{\mu_{\min}}^{\mu_{\max}} f(\mu)\, \psi(t - \tau_{M_2})\, d\mu\, dM_{\mathrm{B}}. \qquad (1)$$

The first integral is for the total mass of the binary $M_B = M_1 + M_2$, and the second integral is for the mass fraction of the secondary star $\mu = M_2/M_B$. $\phi(M_B)$ is the initial mass function (IMF), and $f(\mu)$ is the distribution function of the mass fraction. A is the binary fraction, which is the fraction of binary stars. M01 assumed that the initial mass range of secondary stars is $\leq 8M_\odot$. These systems must produce both DD and SD systems. If the companion mass exceeds $3M_\odot$, the mass accretion rate, which is determined by the thermal timescale of the star, exceeds $10^{-5} M_\odot/\text{yr}$. Such a rapid accretion forms a common envelope and the secondary star loses its envelope to become a WD. If the companion mass is smaller than $3M_\odot$, the primary WD is able to explode by the accretion from the companion star. In this case, however, it is necessary to blow optically thick winds from the primary WD to avoid the common envelope phase. To blow the winds the metallicity effect should be taken into account.

2) The other formulation was proposed by [3] and [4] (hereafter K98 and K00; K-model):

$$\mathcal{R}_{\text{Ia}} = b \int_{\max[m_{p,\ell},\, m_t]}^{m_{p,u}} \frac{1}{m} \phi(m)\, dm \int_{\max[m_{d,\ell},\, m_t]}^{m_{d,u}} \frac{1}{m} \psi(t - \tau_m)\, \phi_d(m)\, dm. \tag{2}$$

The first and second integrals are respectively for the primary and secondary stars (m_p and m_d). ϕ is the IMF, and ϕ_d is the distribution function of the companion mass. b is the binary parameter, which is the total fraction of primary stars that eventually produce SNe Ia. The mass ranges of the companion stars are given by the simulation of binary evolution and are $0.9 - 1.5M_\odot$ for the RG+WD system and $1.8 - 2.6M_\odot$ for the MS+WD system. This is based on the SD scenario and the metallicity effect is taken into account. The lowest metallicity to produce SNe Ia was [Fe/H]$= -1.1$ both in K98 and K00.

3) The DD scenario provided the distribution function of SN Ia lifetimes from the detailed calculation of the binary population synthesis [5]. We adopt this function into our galactic evolution model.

The lifetime of SNe Ia is determined from the mass of companion stars, because the time interval from the start of mass accretion to the explosion is very short ($\lesssim 10^7$ yr). Figure 1 shows the distribution functions of the SN Ia lifetime for the three SN Ia models. The solid and short-dashed lines show K-model for the metallicity $Z = 0.002$ and 0.02, respectively. In the case of $Z = 0.02$, the SN Ia lifetime has wide ranges ($0.5 - 20$ Gyr). The long-dashed line shows the distribution function derived by the DD scenario. The typical lifetime is ~ 0.1 Gyr, the fraction with $\gtrsim 1$ Gyr is very small and the fraction with $\gtrsim 10$ Gyr is almost zero. The dot-dashed line shows M-model. The typical lifetime is ~ 0.3 Gyr, and the fraction with $\gtrsim 10$ Gyr is small but not zero. The fraction with $\gtrsim 1$ Gyr is larger than the DD model.

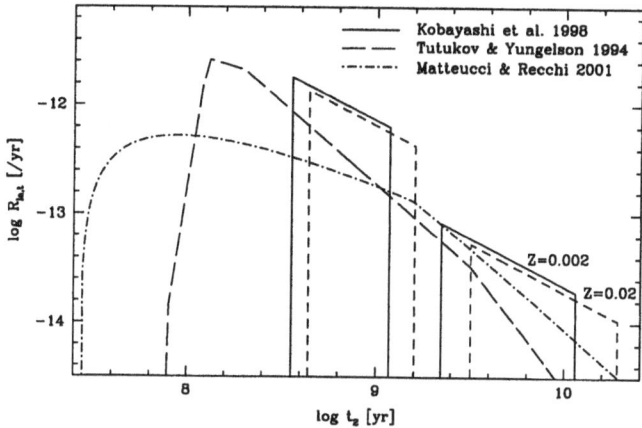

Fig. 1. The distribution functions of SN Ia lifetime for K-model with $Z = 0.002$ (solid) and $Z = 0.02$ (short-dash), the DD model (long-dash) and M-model (dot-dash)

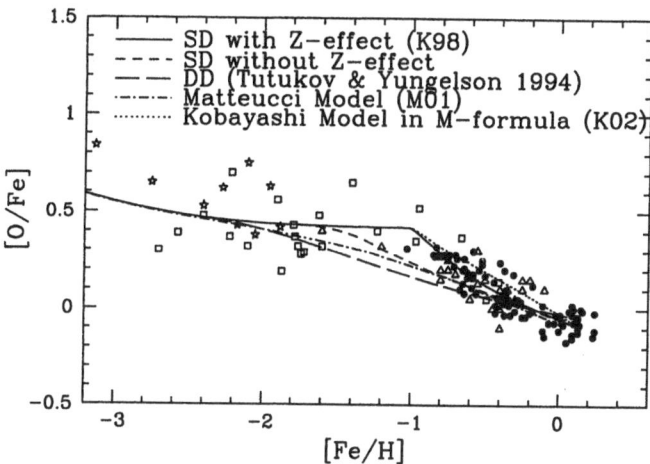

Fig. 2. [O/Fe] vs. [Fe/H] relation in the solar neighborhood for K-model with (solid) and without (short-dash) the metallicity effect, the DD model (long-dash) and M-model (dot-dash). The dotted line is for K-model calculated with Eq.[1]

3   Chemical Evolution in the Solar Neighborhood

The SN Ia models should be tested by comparison with observation. The element abundances of stars in the solar neighborhood gives the most stringent constraints. Figure 2 shows the [O/Fe] vs. [Fe/H] relation in the solar neighborhood. The dots are the observational data derived from oxygen forbidden lines, which show a constant [O/Fe] of ~ 0.45 at [Fe/H] $\lesssim -1$ and a decrease in [O/Fe] with increasing metallicity. Although the observations of OH lines show a linear increase towards lower metallicity, the other α elements (Mg, Si, S, Ca, and Ti) show very similar relations as the O relation for forbidden lines. The K-model gives the best reproduction of the [O/Fe] vs. [Fe/H] relation. Since the lowest metallicity to produce SNe Ia is [Fe/H]$= -1.1$, the decrease is caused at [Fe/H]~ -1 by the companion stars with $2.6 M_\odot$ after ~ 0.5 Gyr. If we do not include the metallicity effect, the companion stars with $M \sim 2.6 M_\odot$ evolve off the main-sequence to give rise to SNe Ia at the age of ~ 0.5 Gyr. The resultant decrease in [O/Fe] starts too early to be compatible with the observations. In the DD model the lifetime of the majority of SNe Ia is shorter than 0.3 Gyr. Then the decrease in [O/Fe] starts at [Fe/H] ~ -2, which is too early compared with the observation. We can reproduce M-model (Mod.3 in M01) with our code. The decrease in [O/Fe] starts at [Fe/H] ~ -2 as for the DD model because the starting time is determined from the largest companion mass of $8 M_\odot$. The slope of [O/Fe] against [Fe/H] is a little shallower than the DD model because the fraction of lifetime with $\gtrsim 1$ Gyr is larger.

Matteucci & Recchi [2] claimed that they failed to reproduce the K98 results (Mod.1 in M01). Here we explain their misunderstanding. The definition of b is quite different from that of A. Thus, A should be determined freely from the observational constraints. M01 adopted $A = 0.02$ and 0.05 for the RG+WD and MS+WD systems, respectively, which are the same values of b in K00. If we adopt $A = 0.5$, we can almost reproduce our results by using Eq.[1] with $\gamma = -1.35$, which is shown by the dotted line in Fig. 2. We note that they adopted $\gamma = -0.35$ to try to reproduce our results, but -1.35 is correct, because ϕ_d is the mass function and the number function should be multiplied by $1/m$. There is an important difference in the treatment of the IMF; in the M-model, the IMF is applied to the total mass of the binary, while it is applied to the mass of the primary star in K-model. If we adopt not $\phi(m_B)$ but $\phi(m_1)$, the normalization is changed and thus $A = 0.2$ gives the best fit. There is another important uncertainty in the distribution function of the mass fraction $f(\mu)$. The function in M01 is weighted for larger mass of secondary stars (i.e., the positive value of γ). However, the observed distribution suggests that the function should be weighted for smaller mass of secondary stars.

4   Cosmic Supernovae Rate

Here we predict the cosmic supernova rate history with our SN Ia model. Galaxies have different timescales for the heavy-element enrichment and the occurrence of supernovae depends on the metallicity therein. Therefore we calculate the cosmic supernova rate by summing up the supernova rates in spirals and ellipticals with the ratio of the relative mass contribution [4]. Here we adopt $H_0 = 65$ km s^{-1} Mpc^{-1}, $\Omega_0 = 0.3$, $\lambda_0 = 0.7$, and the formation epoch of $z_f = 5$.

Figure 3 shows the cosmic supernova rates in cluster galaxies. The SN Ia rate in spirals drops at $z \sim 1.9$ because of the low-metallicity inhibition of SNe Ia. We can precisely test the metallicity effect by finding this drop of the SN Ia in spirals. In ellipticals, the chemical enrichment takes place so early that the metallicity is large enough to produce SNe Ia at $z \gtrsim 2$. The two peaks of

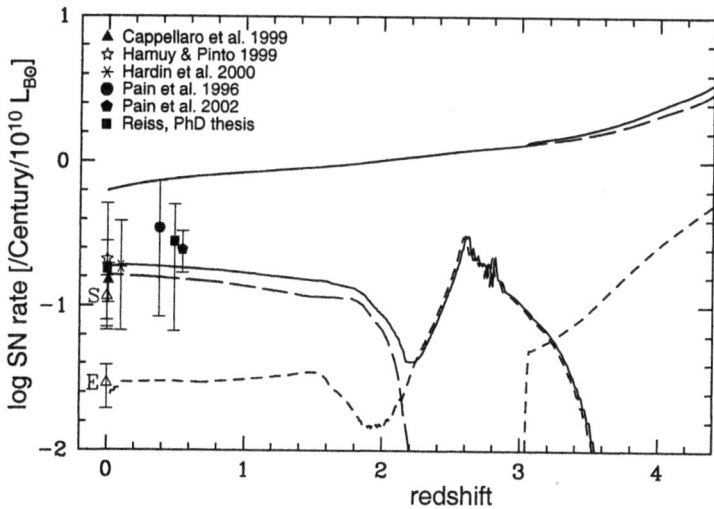

Fig. 3. The cosmic supernova rates (solid line) as the composite of ellipticals (short-dashed line) and spirals (long-dashed line). The upper three lines show SN II rates, the lower three lines show SN Ia rates

SN Ia rates at $z \sim 2.6$ and $z \sim 1.6$ come from the MS+WD and the RG+WD systems, respectively. The SN Ia rate in ellipticals decreases at $z \sim 2.6$, which is determined from the shortest lifetime of SNe Ia of ~ 0.5 Gyr. Thus, the total SN Ia rate decrease at the same redshift as ellipticals, i.e., $z \sim 2.6$.

References

1. Greggio, L. & Renzini, A. 1983, *Astron. Astrophys.*, 118, 217
2. Matteucci, F. & Recchi, S. 2001, *Astrophys. J.*, 558, 351
3. Kobayashi, C., et al. 1998, *Astrophys. J.*, 503, L155
4. Kobayashi, C., Tsujimoto, T., & Nomoto, K. 2000, *Astrophys. J.*, 539, 26
5. Tutukov, A. V., & Yungelson, L. R. 1994, *Mon. Not. R. Astr. Soc.*, 268, 871

Galaxy Formation: Feedback from Supernovae

Dieter H. Hartmann, Jeannette M. Myers, and Lih-Sin The

Department of Physics and Astronomy, Clemson University,
Clemson, SC 29634-0978, USA

Abstract. We simulate the collapse of a spherically symmetric proto-galaxy with a SPH/N-body code for a non-rotating, inhomogeneous initial density profile. We include dark matter, (cooling) gas, star formation and energy feedback from supernovae. An energy of 10^{51} ergs per supernova is deposited in the gas as thermal energy, but this does not significantly alter the dynamics of the collapse, as radiative losses are efficient.

1 Introduction

Understanding the formation and evolution of galaxies is a major challenge for modern observational cosmology and one should perhaps begin with an investigation of the Milky Way. For the Galaxy detailed abundance and kinematic data exist for disk, bulge and halo stars, so that we may be able to develop a self-consistent chemo-dynamical model of the origin, evolution and current structure of our Galaxy. Here we focus on the early history of the Milky Way, the first few Gyrs, during which the halo was assembled. The metal-poor halo globular clusters are assumed to be representative tracer particles of this stage, as their ages indicate formation during an era lasting only a few Gyrs that started about 12-14 Gyrs ago [11]. In the pioneering Galaxy formation scenario of Eggen, Lynden-Bell, and Sandage [3] (hereafter ELS), rapid collapse of a proto-galactic cloud (dynamic timescale of order 10^{8-9} yrs) is expected to create age and metallicity gradients. Truran & Burkert [17] pointed out that relative ages of halo clusters imply an age spread at least twice as large as the dynamic time scale of the ELS scenario. Furthermore, a recent study of halo clusters [11] suggests a spread of 2.5 Gyrs and shows that there is no significant age gradient in the outer halo ($R \geq 8$ kpc). HST studies of NGC 2419 and M92 also lend support to this result [4] and contribute to the presently favored scenario of Searle & Zinn [12], who argued for a longer duration assembly phase via successive mergers of relatively small and independently evolved gas clouds. The recent discovery of the Sagittarius Dwarf Galaxy by Ibata et al. [6] provides strong evidence for such mergers taking place throughout the full life of a galaxy.

In addition to globular clusters the field star population can be used to address the same type of questions, although it is much harder to establish ages for these stars. Chiba & Joshi [2] analyzed Hipparcos stars and find no correlation between metallicity, [Fe/H], and orbital eccentricities, in conflict with the original motivation for the ELS scenario. The mean metallicity of halo field stars is lower than that of the halo clusters, but like the clusters the field stars show no evidence for a significant metallicity gradient [1]. In contrast, M31 appears to exhibit a small gradient in its outer stellar halo [10].

In an attempt to explain the large age spread amongst halo globulars, their abundance distribution function (ADF), and the lack of an abundance gradient, Malinie et al. [8] simulated the dissipative collapse of a spherically symmetric proto-galaxy with a non-uniform density profile (thus the label "inhomogeneous halo collapse"). These authors included efficient heating feedback from exploding supernovae, which delayed the collapse and allowed time scales nearly compatible with the observed age spread of halo globulars. They were also able to match the ADF, given some freedom of choice in the initial density profile. Hartmann et al. [5] followed the orbital evolution of newly formed globulars in the evolving potential, and showed that the present-day metallicity gradient is indeed likely to be small. These simulations showed that the ELS scenario can in fact explain most of the observational features of the halo, although star streams are not to be expected in this pure collapse scenario. However, the model of Malinie et al. was limited to spherical symmetry and did not include a dark matter component. Here we address these shortcomings with a 3D SPH simulation.

2   Halo Collapse Simulations

The simulations were performed with the N-body/SPH code GADGET [14]. The gas and dark matter particles are initially distributed with an inhomogeneous (n=3)-density profile according to the parametrization of Malinie et al. [8]

$$\rho_i \propto \exp^{-(r_i/100\text{kpc})^3} \quad .$$

Initial velocities were set to zero causing the system to collapse with a purely radial flow. The cooling function employed in our simulation is based on the Sutherland and Dopita [15] calculations for a primordial composition with the adopted formalism of Katz et al. [7]. Star formation is determined by a smooth function of the form $\dot{\rho}_\star = c_\star \rho_{\text{gas}} \tau_{\text{SF}}^{-1}$, where $c_\star = 0.06$, and the star formation timescale is assumed to be equal to the dynamic timescale: $\tau_{\text{SF}} = (4\pi G\rho)^{-1/2}$. Initial conditions for the gas/DM particle distribution are: Radius = 100 kpc, Mass = 1×10^{12} M$_\odot$, v$_{\text{init}}$ = 0. The mass ratio of dark matter to gas is 9. The number of SPH particles is fixed to 90,000 with 50,000 being gas particles.

With no angular momentum in the system, all particles fall toward the center on a free-fall time ($\sim$ 0.26 Gyrs). Significant star formation begins after $\sim$ 0.35 Gyr and steadily increases through the collapse. During the evolution of the system, star formation occurs in the gas particles, but star particles are only created when half the mass of a gas particle has undergone star formation. Due to this prescription, the simulation is populated by particles that are part star part gas, but are treated as dissipative gas particles by the hydro code. The release of dissipationless star particles is thus artificially delayed, which introduces errors in the final chemo-dynamic properties of stars (here identified with globulars). When conversion does take place, the remaining gas mass is distributed to the gas particles nearby.

Because the hybrid gas-star particles are slowed by gas pressure, full star particles tend to be created with small velocities near the inner region of the spherical distribution. This does not produce an acceptable spatial distribution of halo clusters, but we are still able to address the issue of age (metallicity) gradients in a qualitative way. To measure ages of star forming particles we simply determine the duration of star formation prior to conversion of a particle from gas/star to full star (really an object containing a population of stars).

3   Results and Conclusions

The free-fall time of our simulated 'proto galaxy' is ~0.26 Gyrs, which is much shorter than the age spread of halo clusters. Feedback from supernova heating may pressurize the gas and thus delay the collapse. However, SN feedback energy is efficiently radiated away so that the collapse takes place essentially on the dynamic timescale. This is different from the results of Malinie et al., who found that SN feedback can dramatically affect the collapse even if only a small fraction of the generic 10^{51} ergs per supernova is fed back into the gas. The reason for this difference is that Malinie et al. stored the feedback energy in turbulent motion of the gas, which was dissipated slowly and thus provided pressure support for a longer time, while we treat the feedback as thermal energy, which is in fact rapidly lost through radiation. This causes an efficiently cooling infall without much slowdown due to feedback, and a rapidly increasing density. As a result of this "cooling problem" (and the assumption of purely radial collapse), the star

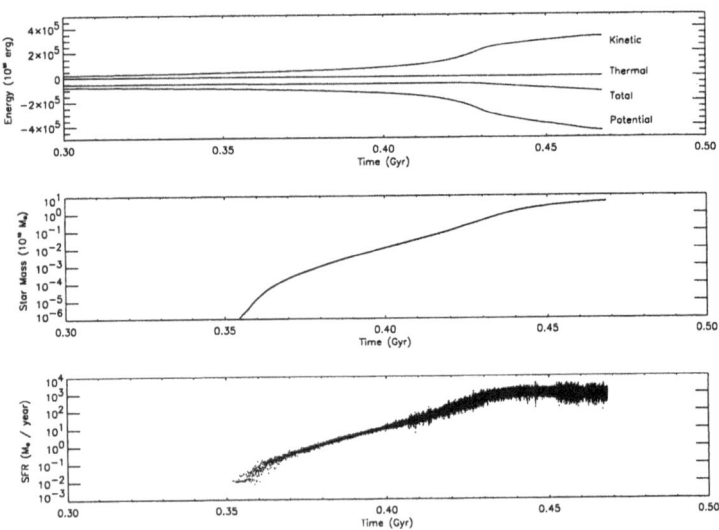

Fig. 1. Evolution of the collapsing proto-galaxy. Top to Bottom: various forms of energy, total stellar mass and global star formation rate ($M_\odot$/year)

formation rate reaches unacceptably large values ($\sim$1000 $M_\odot$/year) (see Fig. 1) and the age distribution of stars formed during the collapse is inconsistent with the observed age spread of globulars. This "cooling problem" has been noted previously by several groups (for a survey of methods for implementing feedback in simulations of galaxy formation see Thacker & Couchman[16]). Springel's work [13] on interacting galaxies showed that partial feedback of supernova energy into a reservoir of turbulence (an assumption also made by Malinie et al.) leads to realistic merger models.

A valuable diagnostic tool is the age-radius relationship. The age distribution of star forming particles develops a gradient, which becomes more shallow with time. We suspect that violent relaxation leaves a virialized halo with a small or zero age gradient [5], but our simulations have not yet been carried long enough to confirm this. Another note of caution: It is not clear how to relate star particles in the SPH simulation to actual globular clusters; here we made the assumption that a Schmidt-like star formation law applies globally and produces halo field stars and globular cluster stars in similar fashion. The calculations discussed here represent our initial effort at building a realistic model of early Galactic chemo-dynamics. Much work remains to be done, e.g. inclusion of chemical evolution, a meta galactic radiation field, and most importantly: a careful investigation of the treatment of energy feedback by supernova explosions and massive star winds (see [9] for a recent discussion). Supernovae are expected to play a significant role in the collapse dynamics of young galaxies, but the details of the interaction have yet to be clarified.

References

1. Carney, B., et al. 1990, AJ 99, 201
2. Chiba, M., & Yoshii, Y. 1997, ApJ 490, L73
3. Eggen, O., Lynden-Bell, D. & Sandage, A. 1962, ApJ 136, 748
4. Harris, W.E., et al. 1997, AJ 114, 1030
5. Hartmann, D.H., Miller, N., Matthews, G., and Malinie, G. 1993, in *The Globular Cluster Galaxy Connection*, eds. G.H. Smith and J.P. Brodie, PASP 48, p. 819
6. Ibata, R.A., Gilmore, G., Irwin, M.J. 1995, MNRAS 277, 781
7. Katz, N., Weinberg, D., & Hernquist, L. 1996, ApJSS 105, 19,
8. Malinie, G., Hartmann, D.H. & Mathews, G. 1991, ApJ 376, 520,
9. Marri, S. & White, S.D.M. 2002, MNRAS, submitted
10. Reitzel, D.B. & Guhathakurta, P. 2002, AJ 124, 234
11. Salaris, M. & Weiss, A. 2002, A&A 388, 492
12. Searle, L. & Zinn, R. 1978, ApJ 225, 357
13. Springel, V. 2000, MNRAS 312, 859
14. Springel, V., Yoshida, N. & White, S.D.M. 2001, New Ast. 6, 79
15. Sutherland, R. & Dopita, M. 1993, ApJS 88, 253
16. Thacker, R.J., & Couchman, H.M.P. 2000, ApJ 545, 728
17. Truran, J.W. & Burkert, A. 1993, Phys. Rep. 227, 235

Part VIII

Supernova Searches

The Search for Nuclear Supernovae

Seppo Mattila[1], Peter Meikle[1], and Robert Greimel[2]

[1] Imperial College of Science, Technology and Medicine
   Prince Consort Road, London SW7 2BW, UK
[2] Isaac Newton Group of Telescopes
   Apartado de correos 321, E-38700 Santa Cruz de la Palma,
   Tenerife, Spain

1 Why Search for Supernovae in Starburst Galaxies?

In the nuclear (central kiloparsec) regions of starburst galaxies, the presence of young supernova remnants (SNRs) suggests that there should be a correspondingly high rate of core-collapse supernovae (CCSNe). CCSN rates of 0.1-0.3 per year have been inferred for nearby starburst galaxies. Moreover, in luminous and ultraluminous IR galaxies, the number of bright radio supernovae detected indicates CCSN rates an order of magnitude higher than in the less luminous nearby starbursts. Very little is currently known about the behaviour of SNe in a starburst environment. The enhanced metallicity in starburst regions is expected to result in large mass-loss rates for the SN progenitor stars. In addition, many of these events are expected to occur within molecular clouds [1], adding further to the circum-supernova densities. It is also likely that such events will be heavily obscured by dust and so only discoverable at IR and longer wavelengths [2]. In general, a dense (but non-starburst) circumstellar environment has been observed to produce bright CCSNe with slow decline rates in the near-IR (e.g. SN 1998S, [3]). However, the behaviour of SNe in nuclear starburst regions may be much more extreme [4]. Discovering such supernovae is important since, as well as providing an opportunity to study the behaviour of SNe in the dusty and dense starburst environment, it will enable us to derive a more complete estimate of the core-collapse rate in galaxies in the local universe. In addition, supernovae provide bright point sources, and thus a novel way for probing the host galaxy extinctions.

2 The William Herschel Telescope Search

Since August 2001 we have been carrying out a near-IR search programme for CCSNe in the nuclear regions of nearby starburst galaxies. We use the INGRID near-IR camera at the William Herschel Telescope (WHT) on La Palma. We repeatedly image a sample of the $\sim$40 most-luminous nearby (d < 45 Mpc) starburst galaxies in the Ks-band where the extinction is greatly reduced and the sensitivity of ground-based observations is still good. These galaxies have been selected such that their far-IR luminosities are comparable to or greater than those of the two prototypical nearby starburst galaxies, M 82 and NGC 253.

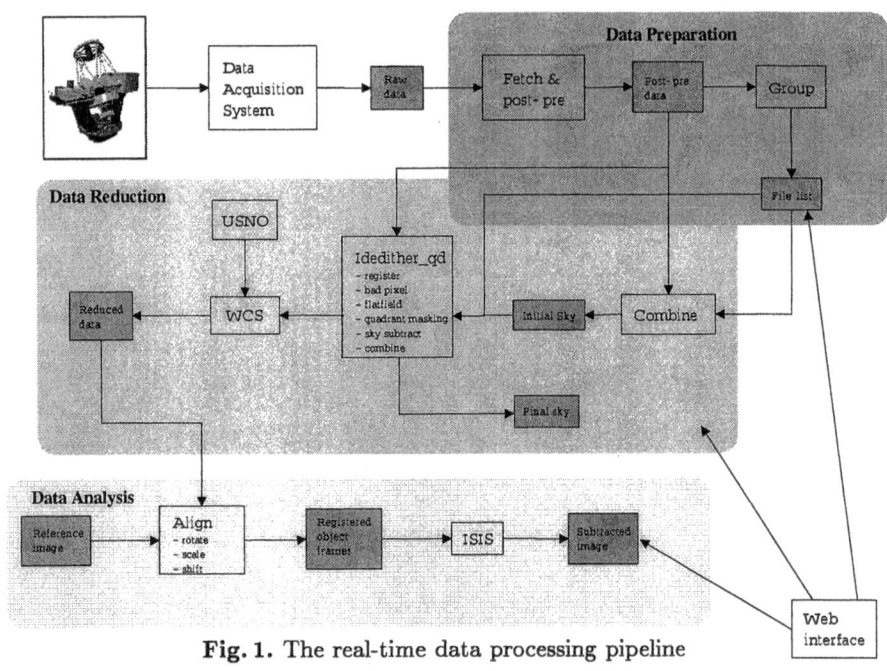

Fig. 1. The real-time data processing pipeline

Galaxies whose far-IR luminosity is expected to be powered by a population of old stars or an AGN have been excluded from the sample. One night of observations is carried out every 3 months. 20-30 galaxy images per clear night are obtained. More details of the search can be found in the 'nuclear SN search' web pages at http://astro.ic.ac.uk/nSN.html.

3   Processing the Supernova Search Data

We are currently developing a pipeline [5] to allow quick and effective reduction and analysis of the SN search data during the night of observation. This is essential to allow rapid follow-up of discovered SNe while they are still bright enough for near-IR photometry and spectroscopy. A flow diagram of the pipeline is shown in Fig. 1. The IRAF-based pipeline performs the standard reduction steps and applies a World Coordinate System (WCS) solution to the reduced image. The new image is then compared with a reference image using image subtraction. After aligning the images carefully, the image with the better seeing is convolved to match the one with poorer seeing. For this we use the Optimal Image Subtraction software (called ISIS) [6]. A web-based interface will allow easy steering of the pipeline as well as immediate access to the data by off-site collaborators.

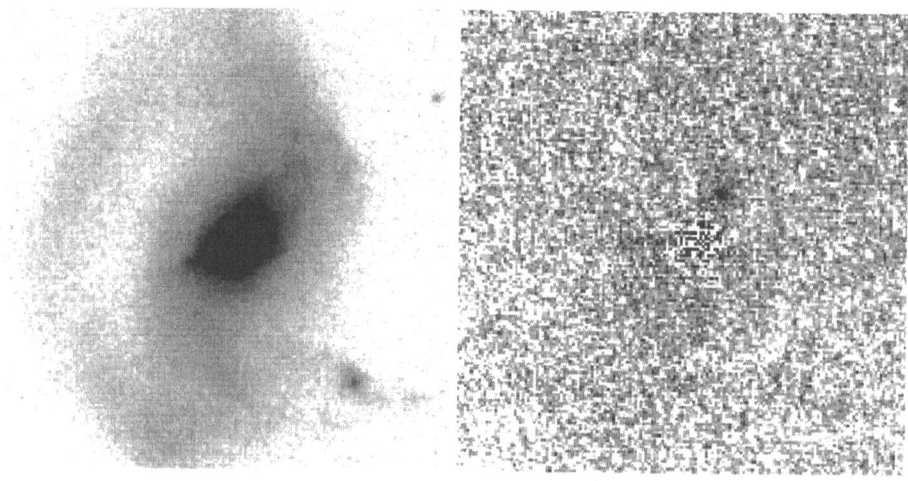

Fig. 2. The possible SN in NGC 7714. **Left:** A UKIRT K-band image (8.7kpc × 8.7kpc) of the starburst galaxy NGC7714. **Right:** The result of image subtraction using a WHT K_S image as the reference

4   Preliminary Results from the Search

4.1   Candidate Supernova in NGC 7714

An example of image subtraction between a WHT K_S image (Aug. 2001) and a UKIRT K image (5 Sept. 1998) of the starburst galaxy NGC 7714 is shown in Fig. 2. This reveals a supernova candidate [7], $m_K = 17.3$, in the UKIRT image, lying 4.9″ North and 1.8″ West (1 kpc projected distance) from the centre of the galaxy nucleus. A similar result was obtained when a K_S-band image observed with NTT (A. Lancon, private communication) on Oct. 2000 was subtracted from the UKIRT image. Moreover, subtraction of the NTT image from the WHT image yielded no point source at the supernova position. Separating the UKIRT data into two images taken 3 mins apart showed no detectable movement for the candidate SN. In addition, the detection was not due to any known minor planet. Image-subtraction procedures (with the ref. image from NTT) reveal no object at the SN location in a UKIRT H image obtained on the same night as the K image. From the non-detection in the H-band we estimate a lower limit for the H-K colour of 1.2. For a linearly declining "ordinary" CCSN or a slow-decliner [2] at an early epoch this limit implies an extinction towards the SN of $A_V > 6$. However, such a red H-K colour could also be produced by a slowly-declining SN (*e.g.* SN 1998S) at a later epoch with a much lower A_V. No optical detection of this possible event was reported in the KAIT images taken on 22 Oct. 1998 [8]. This is in agreement with a high optical extinction towards the SN. As the SN was more than 3 years old at the time of the discovery no useful near-IR follow-up was possible. However, in radio wavelengths CCSNe are often still luminous several years from the explosion. Furthermore, the most luminous radio SNe have been observed to reach their peak luminosities up to ~4 years from the explosion [9]. Therefore, we are currently working on a radio follow-up of this possible SN. We also encourage the search of archives for other detections of this event.

4.2 Expected Number of Supernova Discoveries

The Nuclear SN search with the WHT commenced quite recently and so only two (clear) nights of useful near-IR imaging data have been obtained so far. Nevertheless, including the data acquired from the archives we have now collected useful reference images for 90% of our targets, and at least one repeat image for ~50% of the galaxies. The expected number of SN discoveries from the data collected so far is ~1.3 [10]. This is consistent with the discovery of one probable event from the data. In the future we anticipate a discovery rate of around 0.5 SNe in each clear night's observations [10]. Therefore, a large amount of near-IR imaging remains to be done before we can measure, with statistical significance, the SN rate in nearby starburst galaxies. A number of infrared SN search campaigns [e.g. 11,12] have begun in the past couple of years. Therefore, the number of infrared SN discoveries is expected to increase substantially in the near future. This will eventually allow *complete* SN rates to be estimated for galaxies in the local Universe.

References

1. R. A. Chevalier, C. Fransson: ApJL **558**, L27 (2001)
2. S. Mattila, W. P. S Meikle: MNRAS **324**, 325 (2001)
3. A. Fassia, W. P. S. Meikle, W. D. Vacca, et al.: MNRAS **318**, 1093 (2000)
4. R. Terlevich, G. Tenorio-Tagle, J. Franco, J. Melnick: MNRAS **255**, 713 (1992)
5. S. Mattila, R. Greimel R., W.P.S Meikle, N. Walton, S. Ryder, R.D. Joseph: ING Newsl. **5**, in press (2002)
6. C. Alard, R. Lupton: ApJ **503**, 325 (1988)
7. S. Mattila, P. Meikle, N. Walton, R. Greimel, S. Ryder, C. Alard, A. Lancon: IAUC **7865** (2002)
8. W. Li: IAUC **7865** (2002)
9. K.W. Weiler, S.D., Van Dyk, M.J., Montes, N. Panagia, R.A. Sramek: ApJ **500**, 51 (1998)
10. S. Mattila: Supernovae as probes of their host galaxies and circumstellar environments: search strategies in nuclear starbursts and spectroscopy of SN 1987A CSM. PhD Thesis, Imperial College, London (2002)
11. F. Vitali, A. Arkharov, F. D'Alessio, A. di Paola, Y. Kuzmin, D. Lorenzetti, F. Pedichini, R. Speziali: Memorie della Societa Astronomica Italiana **71**, 565 (2000)
12. R. Maiolino, L. Vanzi, F. Mannucci, G. Cresci, F. Ghinassi, M. Della Valle: A&A **389**, 84 (2002)

An Intermediate Redshift Supernova Search at ESO: Preliminary Results*

Giuseppe Altavilla[1,2], Marco Riello[1,2], Enrico Cappellaro[3], Stefano Benetti[1], Andrea Pastorello[1,2], Ferdinando Patat[4], Marco Prevedello[1], Massimo Turatto[1], and Luca Zampieri[1]

[1] INAF - Astronomical Observatory of Padova, Vicolo dell'Osservatorio 5, I-35122 Padova, Italy
[2] Department of Astronomy, University of Padova, Vicolo dell'Osservatorio 2, I-35122 Padova, Italy
[3] INAF - Astronomical Observatory of Capodimonte, Via Moiariello 16, I-80131 Napoli, Italy
[4] European Southern Observatory, Karl-Schwarzschild-Str. 2, D-85748 Garching bei München, Germany

Abstract. We present the preliminary results of our Supernova Search. We observed periodically several fields, surveying about two square degrees in each run at the MPG/ESO2.2 m, and we checked our candidates by spectroscopic observations at VLT. From the beginning of the search we discovered 21 Supernovae (SNe), the redshift ranging up to 0.6, with a peak distribution at about $z = 0.2$. We found both type Ia and type II, almost in equal proportions. Our preliminary estimate of the SN rate at redshift ~ 0.2 is: $R_{Ia} = 0.20^{+0.20}_{-0.18}$, $R_{II} = 0.27^{+0.28}_{-0.25}$. Selection effects have not been considered yet.

1 Introduction

The supernova rate (SNR) is strongly linked to the star formation rate (SFR): because of the short lifetime of massive progenitors ($M > 8M_\odot$), Core Collapse SNe (CC-SNe: type II-Ib/c) are correlated with the instantaneous SFR while Thermonuclear SNe (type Ia), originating in evolved binary system, can give information about the long term star formation history. The SNR is also one of the fundamental ingredients in the theories of galactic chemical evolution. While the local SNR for different SN classes and morphological galaxy types is reasonably well known [1], the rate at higher redshift is still rather uncertain. Two pioneering works [3,4] were based on three and four SNe respectively. In a more recent study, Pain et al. [5] have presented the results based on a larger sample, including 38 SNe, of type Ia only.

We present the preliminary results of our Supernova Search at intermediate redshift ($z \sim 0.2$), especially designed to determine the SNR of both CC-SNe and type Ia SNe.

---

* Based on observations collected at the European Southern Observatory, Chile

2   Observational Data and Preliminary Results

The search was performed using the MPG/ESO 2.2 m telescope equipped with
the Wide Field Imager; the spectroscopic classification and follow up was performed with ESO VLT + FORS1/2.

We surveyed 21 fields and the effective searched area is $\sim$ 5.1 square degrees;
we confirmed spectroscopically 21 supernovae (9 type Ia, 10 type II, 1 type IIn
and 1 type Ic), with redshift up to $z = 0.6$ (Table 1) and V magnitude at
discovery ranging in the interval 20–24 (Fig. 1a). We found CC-SNe and SNe
Ia in almost equal proportion but, as expected, due to the brighter absolute
magnitude the latter are on average at higher redshift than the former ones
(Fig. 1b).

Table 1. SNe spectroscopically confirmed

SN	type	z	IAUC	SN	type	z	IAUC	SN	type	z	IAUC
1999ey	IIn	0.09	7310	2001be	Ia	0.24	7615	2001io	Ia	0.19	7780
1999gt	Ia	0.27	7346	2001ge	Ia	0.22	7762	2001ip	Ia	0.54	7780
1999gu	II	0.15	7346	2001gf	Ia	0.10	7762	2002cl	Ic	0.07	7885
2000fc	Ia	0.42	7537	2001gg	II	0.61	7762	2002cm	II	0.09	7885
2000fp	II	0.30	7549	2001gh	II	0.16	7762	2002cn	Ia	0.30	7885
2001bc	II	0.19	7615	2001gi	Ia	0.20	7762	2002co	II	0.32	7885
2001bd	II	0.10	7615	2001gj	II	0.27	7762	2002du	II	0.21	7929

2.1   SN Rate

To estimate the SN rate in SNu ($1SNu = 1SN(100yr)^{-1}(10^{10}L_{B\odot})^{-1}$) we need
to count the number of SNe exploded in a given time interval (control time)
in a distance limited volume of the Universe containing a known integrated
luminosity $L_{B\odot}$. The control time is the interval of time during which a SN
apparent magnitude is brighter than the magnitude limit of the search. The
magnitude limit depends on the seeing, on the transparency of the sky, even on
the position in the same frame and it is in general different for each observation
(cf. the contribution by Riello et al. in this volume). On the other side the SN
apparent magnitude depends on the absolute magnitude, on the distance and on
the luminosity evolution for the particular SN type.

Let's consider type Ia SNe first. To estimate the control time we calculated the
expected SN light curves at different redshifts, taking into account K correction
and time dilatation.

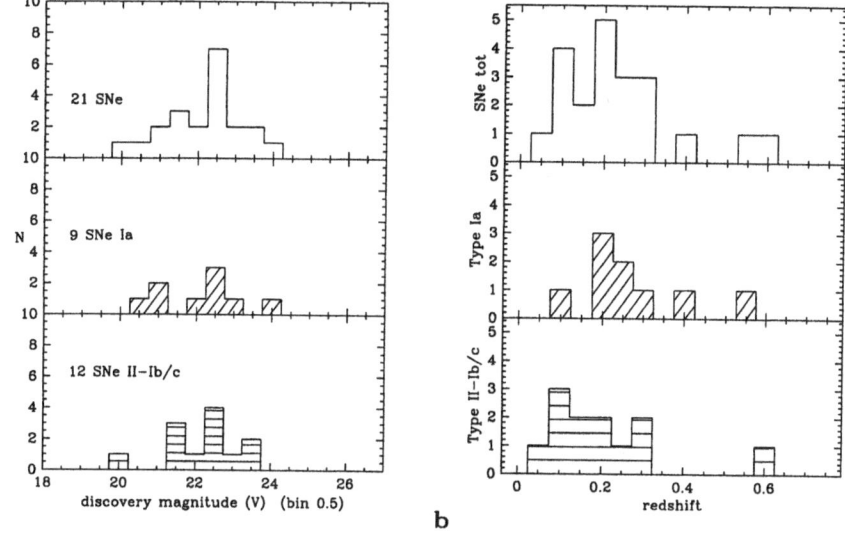

Fig. 1. (a) V magnitude distribution and (b) redshift distribution of the 21 SNe spectroscopically confirmed. Upper panels: all SNe; middle panels: SNe Ia; lower panels: CC-SNe

In this preliminary analysis we considered a subsample of the search, including 13 SNe found from April 2001 to April 2002. Before this date we did not have VLT time available and spectroscopic confirmation was possible only for very few bright candidates. We divided the observed volume in redshift bins and computed the integrated galaxy luminosity in each bin after the counts of Lilly et al. [6], Maddox et al. [8] and Metcalfe et al. [7]. We adopted $H_0 = 75 kms^{-1} Mpc^{-1}$ and an Einstein-De Sitter cosmology.

In the period considered we detected about 100 SN candidates but only a fraction could be observed spectroscopically. As a rule we selected the brightest candidates, thus introducing a severe bias in the magnitude distribution. After observation $\sim 70\%$ of the candidates turned out to be SNe, while the other $\sim 30\%$ turned out to be AGNs-QSOs with redshifts in the range $0.3 < z < 2.3$. In order to take into account the bias in the magnitude of the spectroscopically confirmed candidates, we compared their magnitude distribution with that of all the candidates detected (Fig. 2). The comparison of the two distributions gives us the correction factors to apply to the confirmed candidates counts.

To compute the SNR we need to derive the redshift distribution of the SN candidates from the magnitude distribution. Actually, the magnitude distribution results from the spread in redshift and the phase of the candidates. If the epoch of explosions were known, it would be possible to derive the redshift from the observed apparent magnitude. However the epochs are not known, so we derived several distributions assuming different phases. The SNR R_{Ia} could then be computed as the average of the rates derived from the different red-

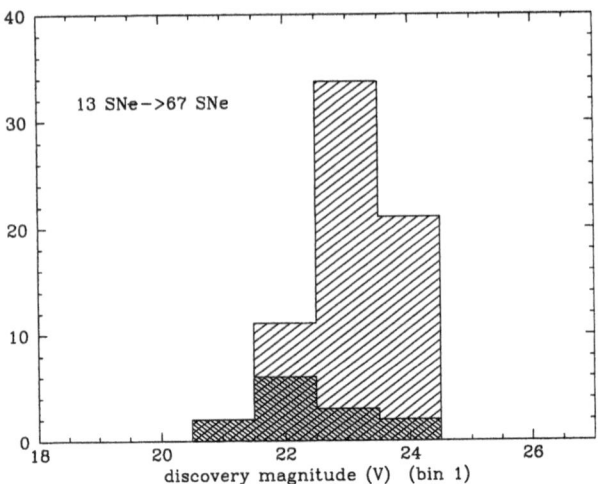

Fig. 2. Magnitude distribution of 13 SNe of the subsample of classified SNe (double shaded histogram) and magnitude distribution of all the candidates (single dashed histogram)

Table 2. SNe Rate (SNu, scaling as $(H_0/75)^2$)

redshift	Total # of SNe	Rate			reference
		Ia	Ib	II	
< 0.01	137	$0.20^{+0.06}_{-0.06}$	$0.08^{+0.04}_{-0.04}$	$0.40^{+0.19}_{-0.19}$	Cappellaro et al. [1]
∼ 0.10	4	$0.25^{+0.27}_{-0.16}$			Hardin et al. [4]
∼ 0.20	5[a]	$0.20^{+0.20}_{-0.18}$		$0.27^{+0.28}_{-0.25}$	present work
∼ 0.40	3	$0.46^{+0.51}_{-0.35}$			Pain et al. [3]
∼ 0.55	38	$0.51^{+0.15}_{-0.13}$[c]			Pain et al. [5]

[a] we considered 3 type Ia and 2 type II of our subsample (13 SNe) with redshift $0.15 < z < 0.25$
[c] model E (Einstein De Sitter Universe)

shift distributions. In this contribution we restricted the computation to the range $0.15 < z < 0.25$, since SNe at these distances the distribution corrections mentioned above are negligible (< 5%).

Considering the ratio between the number of SNe of type Ia and the CC-SNe and the ratio between the control time of the two classes, we can calculate the CC-SNe rate R_{II} from the value obtained for the type Ia SNe. The derived rates agree, within the error bars, with the values found in literature [1,4,3,5] (Table 2).

We find a ratio $R_{II}/R_{Ia} = 1.35$ somewhat smaller than the value derived for the local universe $R_{II}/R_{Ia} = 2$ [1] but similar to the value $R_{II}/R_{Ia} = 1.12$ uncorrected for absorption [2]. Work is in progress to reduce the uncertainties and to extend the estimate of the SNR at other redshifts.

References

1. E. Cappellaro, R. Evans, M. Turatto: A&A **351**, 459 (1999)
2. E. Cappellaro et al.: A&A **268**, 472 (1993)
3. R. Pain et al.: ApJ **473**, 356 (1996)
4. D. Hardin et al.: A&A **362**, 419 (2000)
5. R. Pain et al.: ApJ **577**, 120 (2002)
6. S.J. Lilly, L.L. Cowie, J.P. Gardner: ApJ **369**, 79 (1991)
7. N. Metcalfe et al.: MNRAS **273**, 257 (1995)
8. S.J. Maddox et al.: MNRAS **247**, 1 (1990)

An Intermediate Redshift Supernova Search at ESO: Reduction Tools and Efficiency Tests

Marco Riello[1,2], Giuseppe Altavilla[1,2], Enrico Cappellaro[3], Stefano Benetti[1], Andrea Pastorello[1,2], Ferdinando Patat[4], Marco Prevedello[1], Massimo Turatto[1], and Luca Zampieri[1]

[1] INAF - Astronomical Observatory of Padova, Vicolo dell'Osservatorio 5, I-35122 Padova, Italy
[2] Department of Astronomy, University of Padova, Vicolo dell'Osservatorio 2, I-35122 Padova, Italy
[3] INAF - Astronomical Observatory of Capodimonte, Via Moiariello 16, I-80131 Napoli, Italy
[4] European Southern Observatory, Karl-Schwarzschild-Str. 2, D-85748 Garching bei München, Germany

Abstract. We present the reduction and archiving tools developed for our search for supernovae at intermediate redshifts at ESO as well as the efficiency tests performed. The data reduction recipes developed for the SN candidates selection are described. All the variable sources detected are stored using a MySQL database which enables the identification of previously detected variable sources during past observing runs. Finally, experiments performed with artificial stars have shown that seeing plays a crucial role for the limiting magnitude of detection. Crucial is also the detection threshold used by Sextractor.

1 Introduction

In the last couples of years we have carried out a SN search with the main scientific aim of determining the SN rate at intermediate redshift ($0.2 \leq z \leq 0.5$) (for details see the contribution by Altavilla et al. in this volume). For the detection of SN candidates we use the Wide Field Imager at ESO/MPE 2.2m telescope (La Silla, Chile). Our strategy is to obtain deep images in the V and R bands of 21 different fields, covering a total useful area of ~ 5.1 square degrees. For the spectroscopic confirmation of the SN candidates we use FORS1/2 at ESO VLT. Due to the restriction of time allocation with the VLT only a subsample (the brightest candidates) could be spectroscopically confirmed.

2 Data Reduction

Pre-Reduction. For each search field we take three exposures of 900s with a dithering of few arcsec. This allows to remove cosmic rays, detector cosmetic defects and moving objects. The pre-reduction, bias and flat fielding, are performed using IRAF and MSCRED[1]. Whenever it was possible, standard sky flats

---

[1] MSCRED is an IRAF package designed to deal with mosaic images in multi-extension fits (MEF) format. See the manual by F. Valdes for further details.

are improved by constructing a super flat using the science frames. To combine the dithered frames, these are mapped to a common geometric grid, properly scaled in intensity and finally stacked. An accurate astrometric calibration (rms ~ 0.3 arcsec) is attached to the stacked image using the USNO2 catalogue. The astrometry is used to identify each object in the fields.

Image Subtraction. Our SN search technique is based on the subtraction between the image of a given field and a template image obtained at a previous epoch. The frame to be searched and the template are first geometrically matched and trimmed to the overlapping area. To perform the image subtraction we make use of the ISIS 2.1 package [1]. First, ISIS computes the convolution kernel to match the image with the best seeing to the other, then the frames are photometrically scaled and subtracted. Our experience is that, in any case, the best results are obtained if the two images have similar seeing.

SN Candidates Selection. To search the difference image for residuals due to variable sources we use Sextractor [2]. This produces a list which, in general, is heavily contaminated by spurious detections due to residuals of bright/saturated stars, poorly removed cosmic rays and detector cosmetic defects. In a typical field (eight frames) over a thousand detections are usually found. To clean as much as possible the spurious detections we have developed an automatic procedure which attributes a score to each detection based on several parameters (see the next section). After some experience with the WFI frames, the scores were calibrated through artificial star experiments. The aim is to drastically reduce the number of candidates which are finally verified by direct inspection. Eventually, after visual inspection we attach a provisional classification (supernovae, active galactic nuclei, variable stars and moving objects) to each reliable variable source after considering all the available information like the stellarity index or the distance from the host galaxy nucleus. Figure 1 shows our search software output for a confirmed SN (SN 2001io) and an AGN.

Scoring Algorithm. The scoring procedure takes into account several parameters measured by Sextractor in the new, template and subtracted images. Saturated objects produce many spurious residuals in the subtracted image so the first step is to purge the list from all detections which are located near bright sources. A rank list of the remaining detections is then created according to: the stellarity index measured by Sextractor, the FWHM of the object with respect to the mean, the distance of the residual with respect to the center of the associated object, if any (the host galaxy in the case of SN), and the difference between the object magnitudes measured with different prescriptions. After some tuning we were able to obtain that the scoring algorithm reduces the detection number by a factor 10 without significant loss of efficiency in the detection of good SN candidates.

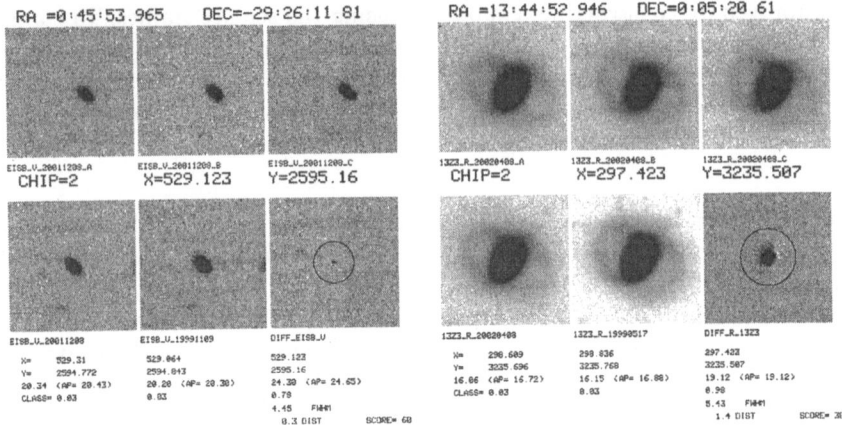

Fig. 1. Output of the search software. Each image is made of six panels: the upper three are the dithering set; at the bottom, from left: the stacked image, the reference image and the difference image. The panels on the left show SN 2001io, one of the 21 spectroscopically confirmed SNe. Those on the right show a bright AGN

3   Variable Sources and SN Candidates Database

Our search technique detects SN candidates but also other variable sources like AGN, QSO and variable stars. To reduce such contamination as much as possible we exploit the variability history of the fields: the fact that each of our monitored fields has been observed several times enables us to remove from the follow–up list the long term variable sources which are not SNe.

To record all the detected variable sources and check their variability history, we developed a MySQL database with a web user interface. The database contains all information for each field, the observed epochs and the complete list of the variable sources detected. During an observing run, with a simple database query from the web interface, it is possible to check if a given object has already been detected in the past. We found that this tool is very effective for the removal of AGNs, which are the only source of contamination we found. This translates into high efficiency in terms of telescope time spent on targets of interest.

4   Artificial Star Experiments

In order to obtain an accurate estimate of the search efficiency control time, which is needed to derive SN rates (cf. by Altavilla et al. in this volume), we need to construct the completeness curve as a function of magnitude for any given observation. To build this curve we performed a number of artificial star experiments. Artificial stars with the proper PSF and with different magnitudes were added to a given image, which is then searched using the same recipe, as

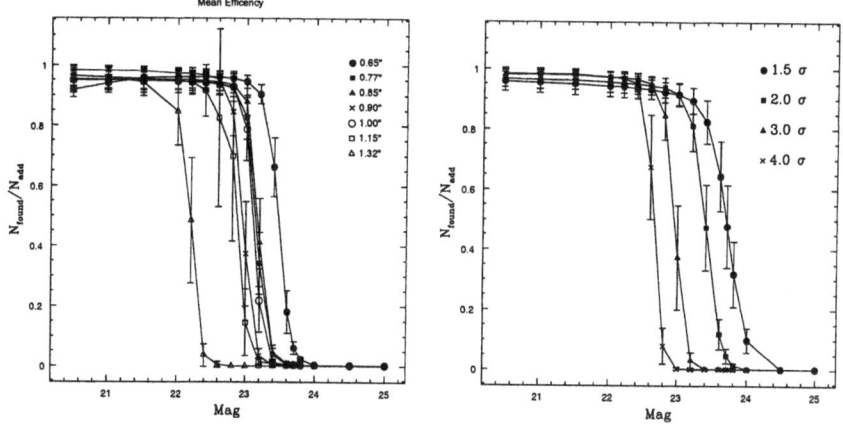

Fig. 2. left: Dependence of the completeness curves from the image seeing. **right:** Dependence of the completeness curves from Sextractor threshold. The error bars show the dispersion of the efficiency between the eight frames of the mosaic

for the real search. These stars were added to each galaxy of the frame. The distances from the host galaxy nucleus are distributed randomly assuming a gaussian distribution with σ equal to the FWHM/2.36 of the galaxy profile. To test the dependence of the SN detection efficiency from the seeing we ran several artificial star experiments on images of the same field taken at seven different epochs and with seeing in the range 0.65–1.32 arcsec (left panel of Fig. 2). For a given seeing, the efficiency drops abruptly from 95% to 5% in less than ~ 0.5 mag: this helps defining a limiting magnitude for each observation because a variation of a few tenths of the limiting magnitude has a negligible effect on the control time. We define as limiting magnitude the point where the detection efficiency is 90%. It turns out that an improvement of the seeing of a factor 2 arcsec increases by about 1.5 mag the magnitude corresponding to a given efficiency. Also, in order to check the dependence of the detection efficiency on the threshold adopted during the search we ran four experiments on the epoch with 0.90 arcsec seeing using different thresholds (in units of the background rms). The different detection efficiencies obtained are shown in the right panel of Fig. 2. As expected, the limiting magnitude is fainter for a lower threshold, ~ 1 mag for 1.5σ compared with 4.0σ. However, using a lower threshold has the important drawback of increasing significantly the number of spurious detections. Usually a 2.0–3.0σ threshold is a reasonable compromise. Further experiments will be performed to check for variations in efficiency between different fields.

References

1. C. Alard, R.H. Lupton: ApJ **503**, 325 (1998)
2. E. Bertin, S. Arnouts: A&AS **117**, 393 (1996)

The Nearby Supernova Factory

Emmanuel Pécontal[1], Greg Aldering[2], Gilles Adam[1], Pierre Antilogus[3], Pierre Astier[3], Yannick Copin[4], François Hénault[1], Jean-Pierre Lemonnier[1], Peter Nugent[2], Reynald Pain[3], Arlette Pécontal[1], Saul Perlmutter[2], Richard Quimby[2], Gérard Smadja[4], and Michael Wood-Vasey[2]

[1] Centre de Recherches Astronomiques de Lyon,
    9, Avenue Charles André, F-69230 Saint-Genis-Laval, France
[2] Lawrence Berkeley National Laboratory, Berkeley, CA, USA
[3] Laboratoire de Physique Nucléaire et des Hautes Energies, Paris, France
[4] Institut de Physique Nucléaire, Université Lyon1, Lyon, France

Abstract. The Nearby Supernova Factory is an international project dedicated to the study of the nearby thermonuclear (type Ia) supernovæ. Based upon the NEAT search for the target discovery and the dedicated integral field spectrograph SNIFS for the follow-up, the goal is to study, on a continuous period of 4 years, the spectro-photometric evolution of $\sim$ 300 SNe Ia at $z < 0.08$ from -15 to $+50$ days on the extended optical range ($320 - 1000$ nm). This will allow to probe in detail the local Hubble diagram, the SNe Ia physics, the SNe-host galaxy correlations, and will serve as an unprecedented nearby benchmark for the high-z cosmological studies to come.

1 Introduction

The type Ia supernovae (SNe Ia) are well known for having been used to measure the accelerating expansion of the universe [2,3]. However, their intrinsic properties are poorly known, and their use as standard candles still debated. The Nearby Supernova Factory (SNFactory) is a fully integrated project, initiated by scientists at LBNL (Berkeley, USA), LPNHE (Paris), IPNL (Lyon) and CRAL (Lyon), to discover and obtain detailed light curve and spectro-photometric observations for $\sim$ 300 nearby SNe Ia during the period 2003-2007.

2 Science Goals

The primary goal of the SNFactory will be to anchor the low-z portion of the SNe Ia Hubble diagram, and to reinforce the usage of SNe Ia as cosmological standard candles, e.g., by refining the standard light curve stretch *vs.* brightness relation [2] and exploring correlations between intrinsic luminosity and spectral features. Moreover, it will get a better understanding of the SN Ia explosion mechanisms, and correlate SN properties with host galaxy characteristics. As a consequence, the SNFactory will lay the foundation for the next generation of high-z cosmological SNe Ia studies to measure the expansion history of the Universe and the precise nature of the dark energy.

"Integral Field Unit" Spectrograph

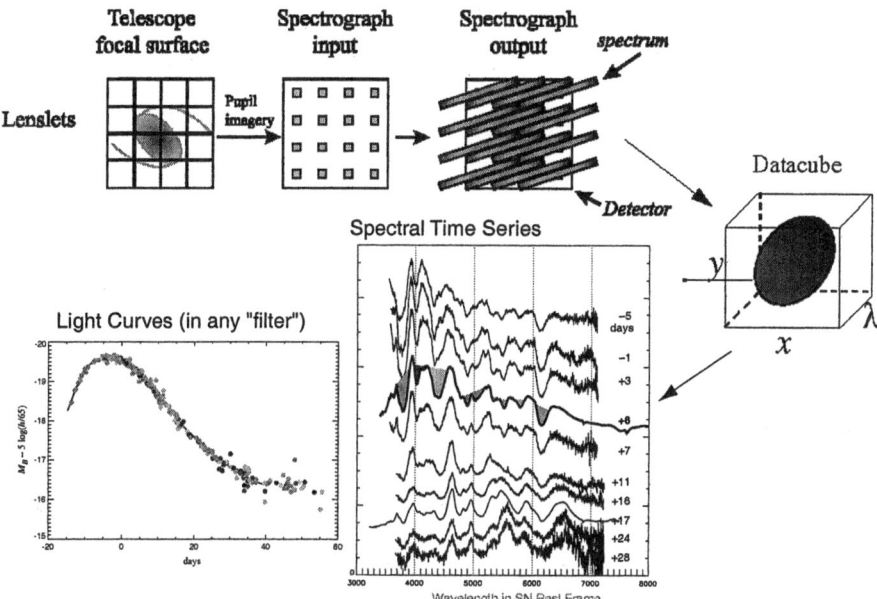

Fig. 1. The IFS is well known for its high spectro-photometric qualities and its ease of use for observations of point-like sources. The SNFactory will use the 3D-spectrograph SNIFS to obtain calibrated spectra of nearby SNe Ia and of their host galaxy

The SNFactory project will also be used to address the potential use of the SNe II as standard candles, as well as the $z \sim 0$ galaxy properties (e.g. luminosity function).

3   The SNFactory Project

SNe discovery with NEAT. The nearby SNe are searched for in unbiased wide-field images from the JPL's Near Earth Asteroid Tracking (NEAT) program, obtained on three 1.2 m telescopes (Haleakala & Palomar). The properties of the imagers used by NEAT are listed in Table 1. In 15 months of operation, the first Palomar imager has already surveyed over 20,000 square degrees of sky. Each patch of sky is revisited frequently (about every 6 days, since this is the "refresh rate" for NEA's); this enables early discovery. In recent test runs, 22 certain SNe have been discovered. An important goal of the SNfactory is to make the selection of supernova candidates quantitative and traceable. Without this, supernova rates and peculiarity fractions are extremely difficult to calculate accurately.

Site:	Haleakala	Palomar I	Palomar II
Aperture	1.2m	1.2m	1.2m
Imager format	4k × 4k	3 × 4k × 3k	112 × 2.4k × 0.6k
Imager scale	1.33″/pixel	1.01″/pixel	0.87″/pixel
Field of view	1.5° × 1.5°	1.1° × 3.4°	2.3° × 4.0°
Filters	open	open	open
Exposures	3 × 20 sec	3 × 60 sec	TBD
Readout	20 sec	20 sec	30 sec
Nightly coverage	300 sq deg	500 sq deg	(1000 sq deg)
Period	Mar 2000 -	Apr. 2001 - Aug. 2002	~Sept. 2002 -
Data(compressed)	12 Gbytes/night	40 Gbyte/night	(80 Gbyte/night)

Table 1. NEAT search facilities

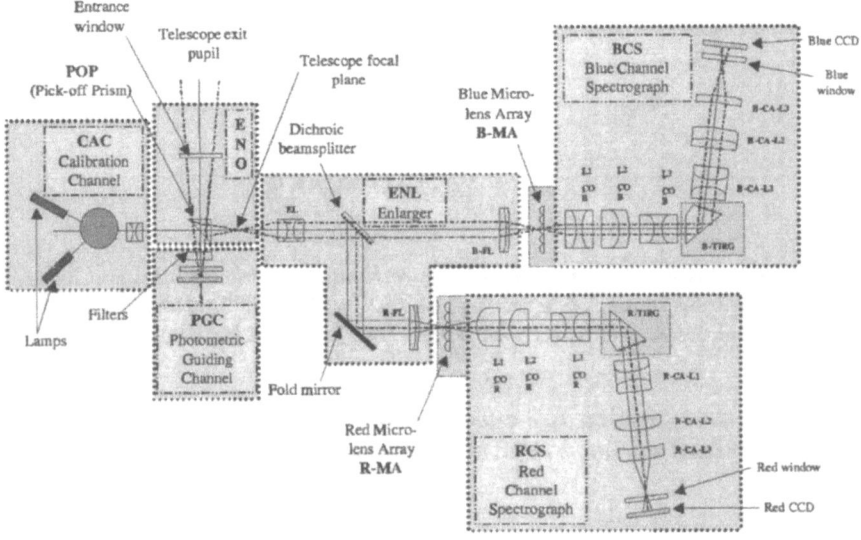

Fig. 2. Break-down of SNIFS component modules: The on-axis f/10 beam from the telescope impinges on a total internal reflection pick-off prism which directs light to an enlarger. The light is then split by dichroic, reimaged onto blue and red microlens arrays, sent through custom collimators, grism, camera, and CCD for each of the two channels (**BCS** and **RCS**). Light from the surrounding area passes through a filter wheel and then illuminates the photometry camera (**PGC**); a special multiple-band filter allows the monitoring of relative atmospheric extinction during the spectroscopy observations. Off-axis light also illuminates the guider camera, providing fast guiding during spectroscopy, and allowing focusing or offsetting between exposures. On the left, a calibration unit supplies a continuum lamp and arc lamps for calibration of the spectrograph

3D observations with SNIFS. The dedicated SuperNova Integral Field Spectrograph (SNIFS) mounted on the UH 2.2 m telescope (Mauna-Kea) is a fully integrated instrument comprising:

- a Tiger-like IFS [1], covering a 6″ × 6″ FoV (spatial sampling 0.4″) on the extended optical range (320 − 1000 nm, spectral sampling 2-3 Å)
- a multi-filter photometric channel, surrounding the IFS FoV. This channel will perform the atmospheric transmission monitoring necessary to ensure a high photometric accuracy during the whole life of the supernovae.
- a guiding & focusing channel because no guiding is provided at the bent cassegrain focus of the UH 2.2m telescope where the spectrograph will be mounted.

A schematic design of the instrument modules is displayed in Fig. 2.

Complementary observations. Near-IR (*JHK*) photometry will be obtained from the Yalo 1 m telescope, and HST will acquire UV spectrophotometry. Furthermore, very late observations could be acquired from major telescopes.

References

1. Bacon et al.: Astron. & Astrophys. Suppl. **113**, 347 (1995)
2. Perlmutter et al.: ApJ **517**, 565 (1999)
3. Riess et al.: Astron. J. **116**, 1009 (1998)

The Supernova Program of the Canada-France-Hawaii-Telescope Legacy Survey

Reynald Pain[1] for the SNLS Collaboration[2]

[1] LPNHE, Universités de Paris VI & VII, 4, Place Jussieu,
   Tour 33 RdC, F-75252 Paris cedex, France
[2] See http://snls.in2p3.fr

Abstract. The CFHT Legacy Survey (CFHTLS) is a large-scale multi-component program intended to operate for 5 years beginning February 2003. The CFHTLS will consist of 3 surveys: a very wide shallow survey, a wide synoptic survey and a deep synoptic survey. Covering 4 square degrees in four MegaCam fields through the whole filter set (u*g'r'i'z') this deep synoptic survey will be sequenced at an average rate of 5 nights a dark run for 5 runs in a row each year on each field. It will make it possible to observe each 1 square degree field every 2-3 nights in 4 colors for 5 months and aims primarily at detecting and monitoring as many as 2000 SNe up to redshift $z \sim 1$. This unprecedented supernova data set will permit to obtain a precise measurement of the cosmic star formation rate as well as of the cosmological parameters including a first measurement of the equation of state of Dark Energy.

1 Measuring the Equation of State of Dark Energy

Evidence for acceleration of the expansion of the universe was found almost four years ago by two independent teams using the Hubble diagram of type Ia supernovae [1,2]. This approach remains the most direct way of measuring the cosmic acceleration, though indirect measurements using CMB anisotropies, gravitational weak shear, and studies of galaxy clusters confirm the overall picture. The "Standard Model" that emerges from this is that the present universe is flat and the energy component that dilutes with the expansion, i.e. matter, amounts to one third of the total energy content today. The remainder, often called "Dark Energy", does not dilute with expansion: its density decreases slowly or not at all as time goes on. The central aim of the Supernova program of the CFHT Legacy Survey is to place constraints on the slow time variation of the Dark Energy density.

Among available experimental approaches, the Hubble diagram of SNe remains the most sensitive to the acceleration, because the observed flux of a standard candle at a given redshift mainly traces the time it took the light to reach us: the acceleration of the expansion causes the supernovae to be observed dimmer than in a decelerated universe. Rather than the acceleration itself, we aim to measure the Dark Energy equation of state, i.e. the pressure-energy density ratio $w \equiv p_X/\rho_X$, which also defines the expansion rate of this component: $\rho_X \sim a^{-3(1+w)}$, where a is the scale factor. Current constraints on w [3,2] are consistent with a very wide range of Dark Energy models. Among them, the cosmological constant ($w = -1$) remains 10^{120} to 10^{60} smaller than plausible vacuum energies of fundamental particle theories. It also cannot explain why matter and Dark Energy have comparable densities today. Many "dynamical Λ" models have been proposed (quintessence, k-essence) to solve these problems. They are based on speculative field models, and predict w values above -0.8, significantly different from -1. Dynamical Dark Energy models can therefore be separated from a cosmological constant by measuring the average value of w over moderate redshifts with a precision better than 0.1.

2   A Second Generation Supernova Survey

Improving very significantly over present results requires an ambitious effort of the SN Ia community to build up the necessary SN dataset: first a ten-fold larger sample (i.e. more than 500) at $0.2 < z < 1$. where w is best measured. This sample has to be of excellent quality in order to allow precise comparisons from redshift to redshift and with nearby samples; second a significant improvement of systematic uncertainties by a detailed study of samples of nearby and distant SNe Ia. The nearby objects are also mandatory to anchor the low z part of the Hubble diagram. No compelling estimation of cosmological parameters can be obtained without them.

2.1   A "Rolling Search Mode"

The traditional method to measure distances to SNe Ia involves different types of observations at about 10 different epochs spread over about 3 months: discovery via image subtraction, spectroscopic identification, and photometric follow-up. A lot of objects are lost along this perilous path because of bad weather. The SN program of the CFHTLS was designed to improve significantly over this traditional strategy by: (1) carrying out the discovery and photometric follow-up with a wide field imager used in the "Rolling Search" mode, where a given field is observed every second or third night as long as it remains visible; (2) resorting to service observing for both spectroscopy and imaging, in order to reduce the impact of bad weather. This "Rolling Search" strategy also impacts on the efficiency of time allocated to spectroscopy:

- it will allow us to monitor the rise of the object and trigger spectroscopy at maximum light;
- the detection of objects uses a light curve rather than a single flux difference;
- the flux of the target is well known since it was measured one or two days before; this allows one to choose the exposure time without accounting for a safety margin due the extrapolation of its flux;
- repeating observations of the same fields over and over again allows one to identify long term varying objects such as AGN's and variable stars (an important background to supernovae detection) once and for all.

We wish to perform most of the spectroscopic observations with ESO telescopes.

2.2   The Supernova Program of the CFHT Legacy Survey

The French and Canadian astronomical communities have committed themselves to a large imaging survey at the Canada-France-Hawaii Telescope to be done with the new one square degree imager MegaCam. This "CFHT Legacy Survey" will represent a total amount of 474 nights over 5 years and will start in February 2003 (see [6])

As described previously, one of the leading scientific objectives of this survey is the detection and monitoring of more than 2000 supernovae from a repetitive observation of each observable field every 2 or 3 nights during grey/dark period for 5-6 months per year during 5 years, for a total amount of 1320 hours ($\sim$ 202 nights) *allocated in service mode*. Four one square degree fields evenly distributed in RA will be monitored that way in four photometric bands (g', r', i', z') with excellent S/N at each visit. We have formed a collaboration, the "SuperNova Legacy Survey (SNLS)" [4] which aims primarily at: (1) built a high statistics high quality Hubble diagram of Type Ia supernovae up to $z \sim 1$; (2) perform a precise measurement of SN rates *versus* redshift.

3   Expected Precision on Cosmological Parameters

We have conducted detailed Monte-Carlo simulations of the survey to optimize it and assess its scientific capabilities, accounting for full moon gaps and using observed cloud coverage extinction and seeing sequences. We have used the recently measured rate of SN Ia [5] which indicates a total explosion rate of about 100 SN Ia supernovae per square degree and per year up to $z = 1$. For type II SNe, current observations show that about 1-2 SN II or SN Ib/c are detected for 10 SN Ia. This is mostly due to the fact that although core collapse SNe are about 5 times more numerous, they are much fainter than SNe Ia and usually embedded in their host galaxy with a lower contrast. However, with the performance of our image subtraction software (systematic subtraction residuals of the order of 1%), and the Rolling Search scheme (which provides repeatedly new images with different psf's and hence different subtraction residuals) we will be able to greatly enrich our sample of candidates in core collapse supernovae which are excellent tracers of the the star formation rate. We estimated that $\sim$ 50% of the candidates detected at any given time will be of Type II or Ib/c. Therefore, including the detailed observational setup and procedure, we have computed the expected number of supernovae that will be detected during the 5 years of this program (for the details of the simulation see [7]):

Redshift	0.3	0.4	0.5	0.6	0.7	0.8	0.9	1.0	1.1	1.2	Total
Number of SN Ia	68	80	108	132	144	148	152	156	160	160	1304
Number of SN Ib/c+SN II	120	160	200	160	120	-	-	-	-	-	760

Applying strict quality cuts, about 700 distant SNe Ia up to redshifts $z \sim 1$ will be put on a Hubble diagram after the 5 years of the survey. The accessible constraints in the space of cosmological parameters are displayed in Fig. 1: w can

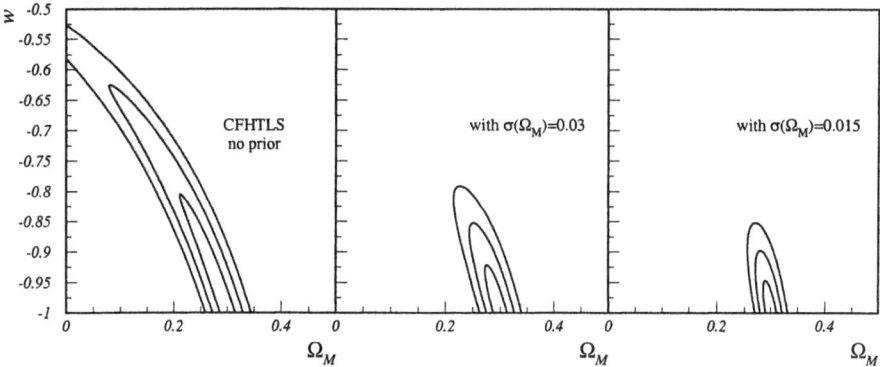

Fig. 1. One, two and three σ (Ω_M, w) joint contours (39, 86 and 99% CL) for: *Left* The full 5 years CFHT Legacy Survey statistics + 200 nearby objects with no prior on Ω_M, *center* with a prior on Ω_M of 0.03; *right* with a prior on Ω_M of 0.015. The values assumed for the cosmological parameters are $\Omega_M = 0.3$, $w = -1$ in a flat Universe. The contours account for photometric errors, a conservative 0.15 mag intrinsic dispersion and were marginalized over the SN Ia intrinsic brightness. For the middle case, $\sigma(w)$ = 0.07, which means that $w = -1$ and $w = -0.8$ are 3 σ away from each other

be measured to ± 0.07, with Ω_M known to 0.03 (i.e. 10%). A better determination of Ω_M will improve the resolution on w.

3.1 Improving on Systematics

High statistics also benefit to systematics: we will gather on average 100 objects per $\delta z = 0.1$ bin. This allows us to study samples in redshift bins, by populating the brightness, decline-rate, and color 3-dimensional parameter space, in order to detect possible drifts in "SNe Ia demographics" and avoid biases of the average distance in each bin. The size of the sample even allows us to correlate photometric and spectroscopic features, and may lead to an improvement of the distance estimator.

References

1. Riess, A. et al. 1998, AJ. 116, 1009
2. Perlmutter, S. et al. 1999, ApJ 517, 565
3. Garnavich, P. et al. 1998, ApJ. 509, 74
4. The "SuperNova Legacy Survey". See http://snls.in2p3.fr/
5. Pain, R. et al. 2002, ApJ 577, 120
6. See http://www.cfht.hawaii.edu/Science/CFHLS/
7. See http://snls.in2p3.fr/notes/general/proposal_140901.ps

Author Index

ESO ASTROPHYSICS SYMPOSIA
European Southern Observatory

Series Editor: Bruno Leibundgut

Series homepage – http://www.springer.de/phys/books/eso/

ESO ASTROPHYSICS SYMPOSIA
European Southern Observatory

Series Editor: Bruno Leibundgut

W. Hillebrandt, B. Leibundgut (Eds.),
From Twilight to Highlight:
The Physics of Supernovae
Proceedings, 2002. XVII, 414 pages. 2003.

Series homepage – http://www.springer.de/phys/books/eso/